Pseudo-Differential Operators, Singularities, Applications

Yuri V. Egorov
Bert-Wolfgang Schulze

Springer Basel AG

Authors

Yuri V. Egorov
Université Paul Sabatier
UFR MIG, MIP
118 route de Narbonne
31062 Toulouse
France

Bert-Wolfgang Schulze
Universität Potsdam
Institut für Mathematik
Am Neuen Palais 10
14469 Potsdam
Germany

1991 Mathematics Subject Classification 35S30, 47G30, 58G15

A CIP catalogue record for this book is available from the Library of Congress, Washington D.C., USA

Deutsche Bibliothek Cataloging-in-Publication Data

Egorov, Jurij V.:
Pseudo-differential operators, singularities, applications / Yuri
V. Egorov ; Bert-Wolfgang Schulze. - Basel ; Boston ; Berlin :
Birkhäuser, 1997
  (Operator theory ; Vol. 93)
    ISBN 978-3-0348-9820-1   ISBN 978-3-0348-8900-1 (eBook)
    DOI 10.1007/978-3-0348-8900-1
NE: Schulze, Bert-Wolfgang:; GT

© 1997 Springer Basel AG
Originally published by Birkhäuser Verlag in 1997
Softcover reprint of the hardcover 1st edition 1997
Printed on acid-free paper produced from chlorine-free pulp. TCF ∞
Cover design: Heinz Hiltbrunner, Basel

ISBN 978-3-0348-9820-1

9 8 7 6 5 4 3 2 1

Contents

5 Kondratiev's theory

6 Non-elliptic operators; propagation of singularities

Preface

Pseudo-differential operators belong to the most powerful tools in the analysis of partial differential equations. Basic achievements in the early sixties have initiated a completely new understanding of many old and important problems in analysis and mathematical physics. The standard calculus of pseudo-differential and Fourier integral operators may today be considered as classical. The development has been continuous since the early days of the first essential applications to ellipticity, index theory, parametrices and propagation of singularities for non-elliptic operators, boundary-value problems, and spectral theory. The basic ideas of the calculus go back to Giraud, Calderón, Zygmund, Mikhlin, Agranovich, Dynin, Vishik, Eskin, and Maslov. Subsequent progress was greatly stimulated by the classical works of Kohn, Nirenberg and Hörmander.

In recent years there developed a new vital interest in the ideas of microlocal analysis in connection with analogous fields of applications over spaces with singularities, e.g. conical points, edges, corners, and higher singularities. The index theory for manifolds with singularities became an enormous challenge for analysists to invent an adequate concept of ellipticity, based on corresponding symbolic structures. Note that index theory was another source of ideas for the later development of the theory of pseudo-differential operators. Let us mention, in particular, the fundamental contributions by Gelfand, Atiyah, Singer, and Bott.

This book was written in connection with the DMV-Seminar 'Pseudo-Differential Operators, Singularities, Applications', in Reisensburg-Günzburg, 12–19 July 1992. The choice of material and the division into two parts of specific character was motivated by the desire to give first a thorough introduction to the recent development. The theory stated in the first part may also be the subject of special university courses in partial differential equations.

The second part presents pseudo-differential operators on manifolds with conical and edge singularities. This theory contains as special cases pseudo-differential operators on $\mathbb{R}_+$ (with the origin interpreted as a conical singularity) and pseudo-differential boundary value problems (with the boundary being the corresponding edge).

There are many textbooks on pseudo-differential operators in the literature, devoted either to a comprehensive treatment such as Hörmander [H6], Taylor [Ta], Treves [Tr], and Kumano-go [Ku] or to more specific subjects such as Palais [P] (index theory), Shubin [Sh] (spectral theory), Egorov [Eg2] (subelliptic operators), Rempel-Schulze [RSc] (boundary value problems and index theory).

The intention of our book is to match the spirit of the DMV Seminar, namely, on the one hand to introduce young mathematicians to an interesting field and to focus attention on typical structures and applications in an accessible form. On the other hand it presents the new ideas of current research and invites to participation.

Among this book's specific features is a new way of generating the pseudo-differential calculus on a manifold with edges. First, a 'conification' of the standard pseudo-differential analysis on a closed compact C^∞ manifold yields the operator algebra on an associated infinite cone that is taken as the model cone of a wedge. Then, an 'edgification' of the cone algebra obtained gives rise to the edge calculus, where the operators are locally interpreted as pseudo-differential operators along the edge with cone operator-valued symbols that are parametrized by the points of the cotangent bundle of the edge. This is to some extent analogous to a possible way for introducing Boutet de Monvel's algebra of boundary value problems for operators with the transmission property, cf. Schrohe, Schulze [Sr]. Also Vishik and Eskin's theory [VE1], [VE2], [VE3] of general pseudo-differential boundary value problems may be embedded into such a concept, where the boundary corresponds to the edge and the inner normal to the model cone of the wedge, cf. Schulze [Sc5]. This, in particular, gives an idea why also in the general edge calculus there appear additional edge conditions of trace and potential type, analogously to boundary conditions, with an analogue of the Shapiro-Lopatinskij condition as part of the ellipticity.

The approach fits into a general program for establishing pseudo-differential operator algebras on manifolds with piece-wise smooth geometry ("manifolds with singularities"), in terms of repeated conifications and edgifications of the operator algebras already achieved also containing the additional trace and potential operators along the lower-dimensional skeletons of the configuration. This, however, has not yet been finished for general singularities, and future progress will depend on adequate axiomatic ideas being deduced from existing theories for lower singularity orders. The essential steps of the scheme may be found in some generality in Schulze [Sc1].

Let us emphasize that there are other independent concepts for dealing with pseudo-differential algebras on singular manifolds, in particular, those of Plamenevskij [Pl1], [Pl2] (Mellin calculus, C^*-algebra methods) and Melrose [M1], [M2] (b-calculus), with applications to index theory, geometry, mathematical physics and concrete problems in partial differential equations connected with degenerate symbols. Such applications also belong to the motivations for the conification-edgification concept for higher singularity orders, though the methods involved and the whole strategy are rather different from those of the authors mentioned. Of course, there are common starting points, such as the nature of degenerate symbols (Fuchs type, edge- and corner-degenerate, ...) or the role of the Mellin transform in $\mathbb{R}_+$, the distance variable to the singular set. Note also that our Mellin operator conventions are completely different from those in Plamenievskij's work.

The operators in the calculus here are given in terms of a hierarchy of symbols (interior as well as edge symbols in the edge case) which describe also the additional trace and potential conditions on the lower-dimensional skeletons. The symbolic (complete and principal) structures of the operators (as well as the smoothing operators) in the algebras also reflect the nature of asymptotics of solutions to elliptic problems obtained in terms of parametrices within the algebras. As mentioned, this is done here for cone and edge singularities. In contrast to the earlier book [Sc1], here we allow edge-degenerate symbols in full generality, and we apply new Mellin operator conventions as well as many of the technical achievements of the past years for making the iterative approach more transparent and manageable. More efficient descriptions of the edge algebras are necessary for starting the next higher algebra in the hierarchy of singularities and also for interpreting the algebras themselves as ranges of corresponding operator-valued symbols in this process. Because of the complexity of the phenomena, an independent exposition was devoted in [Sc5] to the special case of pseudo-differential boundary value problems.

The cone theory of Chapter 8 may also be regarded as a special case of the edge analysis, namely when the edge is of dimension zero. At the same time it plays the role of the subordinate edge symbolic structure for the edge case of higher dimensions. Cone theory, on the other hand, contains many substructures, so we found it essential to prepare both the standard pseudo-differential analysis together with the functional analytic background in a self-contained way in Chapters 1–3, and Kondrat'ev's theory in its classical understanding in Chapter 5, employing the boundary value problems of Chapter 4. In Chapter 6 we deal with non-elliptic theory and elements of micro-local analysis. The edge pseudo-differential algebra in general form is presented for the first time in detail here in Chapter 9. We have emphasized the formal relations to boundary value problems, although the methods in [Sc5] cannot simply be applied to the new situation of a non-trivial model cone. Nevertheless, various obvious similarities to boundary value problems allowed us to organize parts of the assertions as excersises. We hope that this strategy will help the reader to concentrate on the axiomatic ideas which are the basis for new applications. In particular, for performing an analogue of the Atiyah-Singer index theorem in K theoretic terms, cf. Atiyah, Singer [AtS] and Boutet de Monvel [BdM2], it becomes clear immediately that our symbolic hierarchies are adequate objects for the stable homotopy classification of ellipticity in the corresponding algebras, though an extension of the index theorem in Boutet de Monvel's sense itself is still an open problem for manifolds with edges.

Acknowledgement: The authors thank Dr. F. Mantlik (University of Dortmund) and J.B. Gil, J. Seiler (Max-Planck-Arbeitsgruppe "Partielle Differentialgleichungen und Komplexe Analysis", University of Potsdam) for valuable remarks about the manuscript.

Yu.V. Egorov
(Toulouse)

B.-W. Schulze
(Potsdam)

Chapter 1

Sobolev spaces

1.1 Fourier transform

1.1.1 Definition

The theory of pseudo-differential operators systematically uses Fourier transformation. Recall that for a function u from $L_1(\mathbb{R}^n)$ its *Fourier transform* is defined as

$$v(\xi) = \mathcal{F}u(\xi) = \tilde{u}(\xi) = \int_{\mathbb{R}^n} u(x)e^{-ix\xi}dx. \tag{1}$$

It is obvious that the function $v(\xi)$ is continuous. Moreover, if $v(\xi) \in L_1(\mathbb{R}^n)$, then the following formula for the inverse Fourier transform,

$$u(x) = \mathcal{F}^{-1}v(x) = (2\pi)^{-n} \int_{\mathbb{R}^n} v(\xi)e^{ix\xi}d\xi, \tag{2}$$

is valid.

If $u \in L_2(\mathbb{R}^n)$, then its Fourier transform is defined as follows: one can prove that there exists a function $v(\xi)$ in the space $L_2(\mathbb{R}^n)$ such that

$$\lim_{R\to\infty} \|v(\xi) - \int_{|x|<R} u(x)e^{-ix\xi}dx\|_{L_2(\mathbb{R}^n)} = 0.$$

Furthermore,

$$\|v\|_{L_2(\mathbb{R}^n)} = (2\pi)^{n/2}\|u\|_{L_2(\mathbb{R}^n)}$$

and

$$\lim_{R\to\infty} \|u(x) - (2\pi)^{-n} \int_{|\xi|<R} v(\xi)e^{ix\xi}d\xi\|_{L_2(\mathbb{R}^n)} = 0.$$

In what follows we will use formulas (1) and (2) for the Fourier transform of functions from $L_2(\mathbb{R}^n)$, understanding them in the way stated above. It will be convenient for us to use the operator $D^\alpha = D_1^{\alpha_1}\ldots D_n^{\alpha_n}$, putting $\alpha = (\alpha_1,\ldots,\alpha_n) \in \mathbb{Z}_+^n$ and as $D_j = \partial/i\partial x_j$. We will also use the notation $\partial^\alpha = \partial_{x_1}^{\alpha_1}\ldots\partial_{x_n}^{\alpha_n}$.

1.1.2 The Fourier transform in the Schwartz spaces

Definition 1. The space $\mathcal{S}(\mathbb{R}^n)$, called the *Schwartz space* in $\mathbb{R}^n$, is the set of all $u \in C^\infty(\mathbb{R}^n)$ such that

$$p_{\alpha,\beta}(u) := \sup_{x \in \mathbb{R}^n} |x^\alpha D^\beta u(x)| < \infty$$

for all multi-indices $\alpha, \beta \in \mathbb{Z}_+^n$.

These expressions form a countable system of semi-norms in the space $\mathcal{S}(\mathbb{R}^n)$. A linear map $T : \mathcal{S}(\mathbb{R}^n) \to \mathcal{S}(\mathbb{R}^n)$ is called continuous if for every $\gamma, \delta \in \mathbb{Z}_+^n$ there are $\alpha, \beta \in \mathbb{Z}_+^n$ such that

$$p_{\gamma,\delta}(Tu) \le c\, p_{\alpha,\beta}(u)$$

for all $u \in \mathcal{S}(\mathbb{R}^n)$, with constants $c = c(\alpha, \beta, \gamma, \delta) > 0$.

Exercise 2. Show that $\mathcal{S}(\mathbb{R}^n)$ equals the subspace of all $u(x) \in L_2(\mathbb{R}^n)$ such that

$$q_{\alpha,\beta}(u) := \|x^\alpha D^\beta u(x)\|_{L_2(\mathbb{R}^n)} < \infty$$

for all multi-indices $\alpha, \beta \in \mathbb{Z}_+^n$.

Theorem 3. *The Fourier transform is a linear continuous operator in the Schwartz space $\mathcal{S}(\mathbb{R}^n)$.*

Proof. Let φ be a function from $\mathcal{S}(\mathbb{R}^n)$ and $\tilde{\varphi}$ its Fourier transform. It is clear that $\tilde{\varphi} \in C^\infty(\mathbb{R}^n)$ and

$$D_\xi^\alpha \tilde{\varphi}(\xi) = \int_{\mathbb{R}^n} (-x)^\alpha \varphi(x) e^{-ix\xi}\,dx.$$

Moreover,

$$|\xi|^{2j} D_\xi^\alpha \tilde{\varphi}(\xi) = \int_{\mathbb{R}^n} (-x)^\alpha \varphi(x)(-\Delta_x)^j e^{-ix\xi}\,dx$$

$$= \int_{\mathbb{R}^n} (-\Delta_x)^j \big((-x)^\alpha \varphi(x)\big) e^{-ix\xi}\,dx.$$

Since $\varphi \in \mathcal{S}$, there exists a constant $A_{j,\alpha}$ such that

$$|(-\Delta_x)^j \big((-x)^\alpha \varphi(x)\big)| \le A_{j,\alpha}(1 + |x|)^{-n-1},$$

and hence

$$|\xi|^{2j} |D_\xi^\alpha \tilde{\varphi}(\xi)| \le A_{j,\alpha} \int_{\mathbb{R}^n} (1 + |x|)^{-n-1}\,dx,$$

which means that $\tilde{\varphi} \in \mathcal{S}(\mathbb{R}^n)$ and the operator $\mathcal{F} : \mathcal{S}(\mathbb{R}^n) \to \mathcal{S}(\mathbb{R}^n)$ is continuous. $\qquad\square$

Definition 4. The dual space $\mathcal{S}'(\mathbb{R}^n)$ of $\mathcal{S}(\mathbb{R}^n)$ (consisting by definition of all linear continuous functionals on $\mathcal{S}(\mathbb{R}^n)$) is called the Schwartz space of *temperate distributions* on $\mathbb{R}^n$.

We will also use distribution theory notations common such as $\mathcal{D}'(\Omega)$ for the dual of $C_0^\infty(\Omega)$ for any open subset $\Omega \subseteq \mathbb{R}^n$ and $\mathcal{E}'(\Omega)$ for the subspace of all elements of $\mathcal{D}'(\Omega)$ having compact supports. For $\Omega = \mathbb{R}^n$ we shall also write either $\mathcal{S}', \mathcal{D}'$ or $\mathcal{E}'$ as appropriate. The sign of integral without indicating the domain of integration means in what follows the integral over $\mathbb{R}^n$.

Theorem 3 enables the Fourier transform of a functional from $\mathcal{S}'$ to be defined as the element $f \in \mathcal{S}'$ for which

$$\tilde{f}(\varphi) = f(\tilde{\varphi}), \quad \varphi \in \mathcal{S}. \tag{3}$$

Example 5. Let $f(x) = \delta(x)$. Then

$$\tilde{f}(\varphi) = \delta(\tilde{\varphi}) = \tilde{\varphi}(0) = \int \varphi(x)dx.$$

Hence $\tilde{\delta} = 1$.

Example 6. Let $f(x) = x^\alpha, x \in \mathbb{R}^n, \alpha \in \mathbb{Z}_+^n$. Then for $\varphi \in \mathcal{S}$,

$$\tilde{f}(\varphi) = x^\alpha(\tilde{\varphi}) = \int x^\alpha \tilde{\varphi}(x)dx = \int x^\alpha (\int \varphi(\eta)e^{-i\eta x}d\eta)dx$$

$$= \int (\int \varphi(\eta)(-D_\eta)^\alpha e^{-i\eta x}d\eta)dx$$

$$= \int (\int e^{-i\eta x} D_\eta^\alpha \varphi(\eta)d\eta)dx = (2\pi)^n D_\xi^\alpha \varphi(0).$$

Therefore, $\tilde{f}(\xi) = (-1)^{|\alpha|}(2\pi)^n D_\xi^\alpha \delta(\xi)$.

Theorem 7. *If $u \in \mathcal{E}'$, then $\tilde{u}(\xi) = u_x(e^{-ix\xi})$.*

Proof. Consider the *averaging* $u_h(x)$ of the distribution u with the help of a smooth function $\omega(x) \in C_0^\infty(\mathbb{R}^n)$, such that $\omega \geq 0$, $\omega(x) = \omega(-x)$, $\int \omega(x)dx = 1$:

$$u_h(x) = u_y(\omega_h(x - y)), \quad \omega_h(x) = h^{-n}\omega(\frac{x}{h}), \ 0 < h < 1. \tag{4}$$

The function u_h is infinitely differentiable, vanishes outside the h-neighbourhood of the compact $K = \operatorname{supp} u$, and $u_h \to u$ in $\mathcal{D}'(\mathbb{R}^n)$. Therefore, $u_h \to u$ in $\mathcal{E}'(\mathbb{R}^n)$ and hence in $\mathcal{S}'(\mathbb{R}^n)$. By Theorem 3, $\tilde{u}_h \to \tilde{u}$ in $\mathcal{S}'(\mathbb{R}^n)$. However,

$$\tilde{u}_h(\xi) = \int u_h(x)e^{-ix\xi}dx = (u_h)_x(e^{-ix\xi}b(x)),$$

where $b \in C_0^\infty(\mathbb{R}^n), b(x) = 1$ in the 1-neighbourhood of K. Therefore,

$$\tilde{u}(\xi) = u_x(e^{-ix\xi}b(x)) = u_x(e^{-ix\xi}). \qquad \square$$

Corollary 8. If $v \in \mathcal{S}', u \in \mathcal{E}'$, then $\tilde{u}\tilde{v} \in \mathcal{S}'$.

Proof. By Theorem 3 we have $\tilde{v} \in \mathcal{S}'$. From Theorem 7 we see that $\tilde{u}(\xi) = u_x(h(x)e^{-ix\xi})$, where $h(x) = 1$ in a neighbourhood of $\operatorname{supp} u$. Hence $\tilde{u}(\xi) \in C^\infty(\mathbb{R}^n)$, and for any $\alpha \in \mathbb{Z}^n_+$ the inequalities $|D^\alpha \tilde{u}(\xi)| \leq C_\alpha (1 + |\xi|)^m$ hold with some constants m, C_α, independent of ξ. Therefore, from $\varphi \in \mathcal{S}$ it follows that $\tilde{u}\varphi \in \mathcal{S}$ and hence the product $\tilde{u}\tilde{v}$ is a functional from $\mathcal{S}'$, the value of which on φ is defined as $\tilde{u}\tilde{v}(\varphi) = \tilde{v}(\tilde{u}\varphi)$. $\qquad\square$

1.2 The first definition of the Sobolev space

1.2.1 The classical definition

The Sobolev spaces $H^s(\Omega)$ serve as a very useful tool in the linear theory of partial differential equations.

The basic space $H^0(\Omega)$ is the Hilbert space $L_2(\Omega)$, consisting of all measurable functions f for which $\int_\Omega |f(x)|^2 dx < \infty$. The norm $\|f\|_0$ in this space is defined as

$$\|f\|_0 = \left(\int_\Omega |f(x)|^2 dx \right)^{1/2},$$

and the inner product of two elements f and g is

$$(f,g)_0 = \int_\Omega f(x)\overline{g(x)}dx.$$

Recall that by the Parseval theorem

$$(f,g)_0 = (2\pi)^{-n} \int \tilde{f}(\xi)\overline{\tilde{g}(\xi)}d\xi,$$

where $\tilde{f}$ and $\tilde{g}$ are the Fourier transforms of f and g, respectively, extended by zero values in $\mathbb{R}^n \backslash \Omega$.

If $s \in \mathbb{N}$, we set

$$H^s(\Omega) = \{f \in L_2(\Omega), D^\alpha f \in L_2(\Omega), |\alpha| \leq s\}.$$

Here $D^\alpha f$ are the distributions obtained by differentiation of the distribution f. In this space the inner product is defined as

$$(f,g)_s = \int_\Omega \sum_{|\alpha| \leq s} D^\alpha f(x)\overline{D^\alpha g(x)}dx.$$

1.2.2 The completeness of the classical Sobolev space

Theorem 9. *Let $u_k \in H^s(\Omega), k = 1, 2, \ldots$, be a sequence, weakly converging to u in $L_2(\Omega)$, and $\|u_k\|_s \leq M$, where the constant M is independent of k. Then $u \in H^s(\Omega)$ and $\|u\|_s \leq M$.*

Proof. Let $v_{k,\alpha}$ denote the derivative $D^\alpha u_k$, which is, by our condition, a function of $L_2(\Omega)$ for $|\alpha| \le s$, $k = 1, 2, \ldots$. By the definition

$$\int v_{k,\alpha}\varphi dx = (-1)^{|\alpha|} \int u_k D^\alpha \varphi dx, \ \varphi \in C_0^\infty(\Omega).$$

Hence

$$\int v_{k,\alpha}\varphi dx \to v_\alpha(\varphi)$$

and

$$v_\alpha(\varphi) = (-1)^{|\alpha|} \int u D^\alpha \varphi dx, \ \varphi \in C_0^\infty(\Omega).$$

This means that $v_\alpha = D^\alpha u$ in the sense of the theory of distributions. The set $C_0^\infty(\Omega)$ is dense in $L_2(\Omega)$. Hence the convergence of the sequence $\int v_{k,\alpha}\varphi dx$ with $\varphi \in L_2(\Omega)$ follows, i.e. the weak convergence of $v_{k,\alpha}$ in $L_2(\Omega)$ to some $w_\alpha(x)$. From the equality $v_\alpha(\varphi) = \int w_\alpha \varphi dx$ it follows that the distribution v_α coincides with the function $w_\alpha \in L_2(\Omega)$. Thus $u \in H^s(\Omega)$. Moreover, $\|v_\alpha\|_0 \le \underline{\lim}_{k\to\infty}\|v_{k,\alpha}\|_0$ (see, for instance, [KA, p. 287]), and hence

$$\|u\|_s^2 = \sum_{|\alpha|\le s} \|v_\alpha\|_0^2 \le \overline{\lim_{k\to\infty}} \sum_{|\alpha|\le s} \|v_{k,\alpha}\|_0^2 \le M^2. \qquad \square$$

1.3 General definition of Sobolev spaces in $\mathbb{R}^n$

1.3.1 General definition

This definition has been proposed by Slobodetzky, as a generalization of the one given by Sobolev.

Definition 10. A distribution $u \in \mathcal{S}'$ belongs to the space $H^s = H^s(\mathbb{R}^n), s \in \mathbb{R}$, if $\tilde{u}(\xi)$ is a function and

$$\int |\tilde{u}(\xi)|^2(1 + |\xi|^2)^s d\xi < \infty.$$

The inner product of two elements u and v from H^s is defined by the equality

$$(u, v)_s = (2\pi)^{-n} \int \tilde{u}(\xi)\overline{\tilde{v}(\xi)}(1 + |\xi|^2)^s d\xi,$$

and the norm by the equality

$$\|u\|_s = (u, u)_s^{1/2}.$$

It is easy to see that for $s \in \mathbb{N}$ this norm is equivalent to the norm introduced earlier.

Theorem 11. *The space H^s coincides with the completion of the space $C_0^\infty(\mathbb{R}^n)$ with respect to the norm $\| \cdot \|_s$.*

Proof. If $u \in H^s$, then

$$v(\xi) = \tilde{u}(\xi)(1 + |\xi|^2)^{s/2} \in L_2(\mathbb{R}^n).$$

Therefore, there exist functions $v_k(\xi) \in C_0^\infty(\mathbb{R}^n)$, such that $v_k(\xi)(1 + |\xi|^2)^{s/2} \to v(\xi)$ in $L^2(\mathbb{R}^n)$. It is evident that $v_k \in \mathcal{S}$, v_k are the Fourier transforms of functions u_k from $\mathcal{S}$ and $\|u_k - u\|_s \to 0$ as $k \to \infty$. It remains to verify that the space C_0^∞ is dense in $\mathcal{S}$ with respect to the norm $\| \cdot \|_s$. Let $l \in \mathbb{N}, l \geq s$. It is sufficient to verify that C_0^∞ is dense in $\mathcal{S}$ with respect to the norm $\| \cdot \|_l$. If $v \in \mathcal{S}$, we put $v_\varepsilon(x) = h(\varepsilon x)v(x)$, where $h \in C_0^\infty, h(x) = 1$ for $|x| \leq 1$. It is evident that $v_\varepsilon(x) \in C_0^\infty(\mathbb{R}^n)$ and

$$\|v_\varepsilon - v\|_l^2 = \|v(x)[1 - h(\varepsilon x)]\|_l^2 \to 0$$

as $\varepsilon \to 0$, since $v_\varepsilon \neq v$ only if $|x| > \varepsilon^{-1}$. Now let $u_k \in C_0^\infty(\mathbb{R}^n)$ and $\|u_k - u_m\|_s \to 0$ as $k, m \to \infty$. The function sequence $(1 + |\xi|^2)^{s/2}\tilde{u}_k(\xi)$ is a Cauchy sequence in L_2 and therefore converges to a function $v(\xi) \in L_2$. The function $w(\xi) = v(\xi)(1 + |\xi|^2)^{-s/2}$ is the Fourier transform of an element u of H^s. Thus each element of the completion of the space $C_0^\infty(\mathbb{R}^n)$ in the norm $\| \cdot \|_s$ defines an element of the space H^s. $\qquad\square$

1.3.2   Some properties of Sobolev spaces in $\mathbb{R}^n$

Theorem 12. *If $s > t$, then for all $\varepsilon > 0$ and all functions $u \in C_0^\infty(\mathbb{R}^n)$ the inequality*

$$\|u\|_t \leq \varepsilon\|u\|_s + C_\varepsilon\|u\|_{t-1},$$

where $C_\varepsilon = C_{s,t}\varepsilon^{1/(t-s)}$, holds.

Proof. The statement follows from the inequality

$$\lambda^t \leq \varepsilon\lambda^s + C_\varepsilon\lambda^{t-1},$$

which is true for all $\lambda > 0$. $\qquad\square$

Lemma 13. *If $f \in L_1(\mathbb{R}^n), g \in L_p(\mathbb{R}^n)$, where $p \geq 1$, then $f * g \in L_p(\mathbb{R}^n)$ and $\|f * g\|_{L_p} \leq \|f\|_{L_1} \cdot \|g\|_{L_p}$.*

Proof. If $p = 1$, then

$$\|f * g\|_{L_1} = \int |\int f(y)g(x - y)dy|dx \leq \int |f(y)| \int |g(x - y)|dxdy$$

$$\leq \int |f(y)| \int |g(x)|dxdy = \|f\|_{L_1} \cdot \|g\|_{L_1}.$$

If $p > 1$, then by the Hölder inequality

$$\|f * g\|_{L_p}^p = \int |\int f(y)g(x - y)dy|^p dx$$

$$\leq \int (\int |f(y)|dy)^{p/q}(\int |f(y)||g(x - y)|^p dy)dx,$$

where q is a number such that $1/p + 1/q = 1$. Therefore,

$$\|f * g\|_{L_p}^p \le \|f\|_{L_1}^{p/q} \int |f(y)| \left(\int |g(x-y)|^p dx \right) dy$$

$$= \|f\|_{L_1}^{p/q} \int |f(y)| \left(\int |g(x)|^p dx \right) dy = \|f\|_{L_1}^{p/q+1} \|g\|_{L_p}^p.$$

Hence $\|f * g\|_{L_p} \le \|f\|_{L_1} \cdot \|g\|_{L_p}$. $\square$

Theorem 14. *If $u \in H^s, \varphi \in \mathcal{S}$, then $\varphi u \in H^s$ and $\|\varphi u\|_s \le A\|u\|_s$, where the constant A is independent of u.*

Proof. We have

$$\mathcal{F}(\varphi u)(\xi) = (2\pi)^{-n} \int \tilde{u}(\eta) \tilde{\varphi}(\xi - \eta) d\eta.$$

Therefore

$$\|\varphi u\|_s^2 = (2\pi)^{-2n} \int (1 + |\xi|^2)^s \left| \int \tilde{u}(\eta) \tilde{\varphi}(\xi - \eta) d\eta \right|^2 d\xi.$$

Note that

$$1 + |\xi|^2 \le 2(1 + |\eta|^2)(1 + |\xi - \eta|^2),$$
$$1 + |\eta|^2 \le 2(1 + |\xi|^2)(1 + |\xi - \eta|^2),$$

so that

$$(1 + |\xi|^2)^s \le 2^{|s|}(1 + |\eta|^2)^s(1 + |\xi - \eta|^2)^{|s|}.$$

Therefore, we have

$$\|\varphi u\|_s^2 \le (2\pi)^{-2n} 2^{|s|} \int \left| \int (1 + |\eta|^2)^{s/2} \int (1 + |\xi - \eta|^2)^{|s/2|} |\tilde{u}(\eta)| \right.$$

$$\left. \times |\tilde{\varphi}(\xi - \eta)| d\eta \right|^2 d\xi \le 2^{|s|} (2\pi)^{-2n} \|u\|_s^2 \left(\int (1 + |\zeta|^2)^{|s|/2} |\tilde{\varphi}(\zeta)| d\zeta \right)^2,$$

where the latter inequality is valid by Lemma 13 with $p = 2$. Thus

$$A = (2\pi)^{-n} 2^{|s|/2} \int (1 + |\eta|^2)^{|s|/2} |\tilde{\varphi}(\eta)| d\eta. \qquad \square$$

1.4   Representation of a linear functional over H^s

Theorem 15. *The space $(H^s)^*$ is topologically isomorphic to H^{-s}.*

This theorem is a corollary of the following three theorems.

Theorem 16. *If $u \in H^s$, $v \in H^{-s}$, then there exists*

$$\lim_{k \to \infty} \int u_k(x) \overline{v_k(x)} dx,$$

where

$$u_k, v_k \in C_0^\infty(\mathbb{R}^n) \text{ and } \|u_k - u\|_s \to 0, \ \|v_k - v\|_{-s} \to 0$$

as $k \to \infty$. This limit is denoted by $\int u(x)\overline{v(x)}dx$. Moreover, the inequality

$$\left| \int u(x)\overline{v(x)}dx \right| \leq \|u\|_s \|v\|_{-s}$$

is valid.

Proof. Since

$$\int u_k(x)\overline{v_k(x)}dx = (2\pi)^{-n} \int \tilde{u}_k(\xi)\overline{\tilde{v}_k(\xi)}d\xi$$

$$= \int \tilde{u}_k(\xi)(1 + |\xi|^2)^{s/2}\overline{\tilde{v}_k(\xi)}(1 + |\xi|^2)^{-s/2}d\xi,$$

and the sequences

$$\tilde{u}_k(\xi)(1 + |\xi|^2)^{s/2}, \quad \tilde{v}_k(\xi)(1 + |\xi|^2)^{-s/2}$$

converge as $k \to \infty$ in $L_2(\mathbb{R}^n)$ to the functions

$$\tilde{u}(\xi)(1 + |\xi|^2)^{s/2}, \quad \tilde{v}(\xi)(1 + |\xi|^2)^{-s/2},$$

respectively, we have

$$\int u(x)\overline{v(x)}dx = (2\pi)^{-n} \int \tilde{u}(\xi)\overline{\tilde{v}(\xi)}d\xi,$$

whence our statement follows. $\square$

Theorem 17. *If $u \in H^s$, then*

$$\|u\|_s = \sup_{v \in C_0^\infty(\mathbb{R}^n)} \frac{\int u(x)\overline{v(x)}dx}{\|v\|_{-s}}.$$

Proof. By the definition

$$\int u(x)\overline{v(x)}dx = (2\pi)^{-n} \int \tilde{u}(\xi)\overline{\tilde{v}(\xi)}d\xi.$$

Let $u_1(\xi) = \tilde{u}(\xi)(1 + |\xi|^2)^{s/2}, v_1(\xi) = \tilde{v}(\xi)(1 + |\xi|^2)^{-s/2}$, so that $u_1, v_1 \in L_2$. Since

$$\|u\|_s = (2\pi)^{-n/2}\|u_1\|_0, \quad \|v\|_{-s} = (2\pi)^{-n/2}\|v_1\|_0,$$

our statement follows from the well-known equality

$$\|u_1\|_0 = \sup_{v_1 \in C_0^\infty(\mathbb{R}^n)} \frac{\int u_1(\xi)\overline{v_1(\xi)}d\xi}{\|v_1\|_0}. \qquad \square$$

Theorem 18. *Each linear continuous functional l over H^s has the form*

$$l(u) = (2\pi)^n \int u(x)\overline{v(x)}dx,$$

where $v \in H^{-s}$ and $\|l\| = \|v\|_{-s}$.

Proof. Let $A_s u = (2\pi)^{-n/2}\tilde{u}(\xi)(1 + |\xi|^2)^{s/2}$.

If $u \in H^s$, then $A_s u \in L_2$ and $\|A_s u\|_0 = \|u\|_s$, i.e., A_s gives an isometric map of H^s on L_2. Hence the functional l corresponds to a linear continuous functional over L_2, i.e., the function $w \in L_2$, such that $\|w\|_0 = \|l\|$ and $l(u) = \int \overline{w} \cdot A_s u d\xi$. It is clear that

$$l(u) = \int \overline{w(\xi)}\tilde{u}(\xi)(1 + |\xi|^2)^{s/2}d\xi = (2\pi)^n \int u(x)\overline{v(x)}dx,$$

where v is a function of H^{-s}, whose Fourier transform is equal to $(2\pi)^{-n/2}w(\xi)(1+|\xi|^2)^{3/2}$. Moreover, $\|l\| = \|w\|_0 = \|v\|_{-s}$. $\qquad\square$

1.5 Embedding theorems

1.5.1 Sobolev's theorem

In the proof of Theorem 8 we defined the averaging of a distribution, which we now use to show that the space of $C_0^\infty(\mathbb{R}^n)$ functions is dense in H^s.

Theorem 19. *The space $C_0^\infty(\mathbb{R}^n)$ is dense in H^s.*

Proof. Let $u \in H^s$ and u_h be the averaging of u. Since $u_h = u * \omega_h$, we have $\tilde{u}_h(\xi) = \tilde{u}(\xi) \cdot \tilde{\omega}_h(\xi)$. Moreover, $\tilde{\omega}_h(\xi) = \tilde{\omega}(h\xi)$, the functions $\tilde{\omega}(h\xi)$ are bounded uniformly with respect to h, $\tilde{\omega}_h(\xi) \to \tilde{\omega}_h(0) = 1$, and the convergence is uniform on each compact set of ξ. Thus

$$\|\tilde{u}(\xi)[\tilde{\omega}(h\xi) - 1](1 + |\xi|^2)^{s/2}\|_0 \to 0,$$

i.e., $\|u_h - u\|_s \to 0$. Therefore, the space $\mathcal{S}$ is dense in H^s.

Let us show that the space $C_0^\infty(\mathbb{R}^n)$ is dense in $\mathcal{S}$ in the norm of H^s. Let $l \in \mathbb{N}$ and $l \geq s$. Given $v \in \mathcal{S}$, put $v_\varepsilon = h(\varepsilon x)v(x)$, where $h \in C_0^\infty(\mathbb{R}^n)$, $h(x) = 1$ for $|x| \leq 1$. Then $v_\varepsilon \in C_0^\infty(\mathbb{R}^n)$ and

$$\|v_\varepsilon - v\|_s \leq \|v_\varepsilon - v\|_l = \|v(x)(1 - h(\varepsilon x))\|_l \to 0$$

as $\varepsilon \to 0$, since $v_\varepsilon(x) \neq 0$ only for $|x| \geq \varepsilon^{-1}$. $\qquad\square$

Theorem 20. (Sobolev's embedding theorem.) *If $u \in H^s$ and $s > k + n/2$, where $k \in \mathbb{Z}_+$, then u coincides almost everywhere with a function u_1 from C^k, and $\|u_1\|_{C^k} \leq A\|u\|_s$, where A is independent of u.*

Proof. Assume first that $u \in C_0^\infty(\mathbb{R}^n)$. Then we have

$$D^\alpha u(x) = (2\pi)^{-n} \int \xi^\alpha \tilde{u}(\xi) e^{ix\xi} d\xi,$$

and therefore, for $|\alpha| \leq k$,

$$(2\pi)^n |D^\alpha u(x)| \leq \int |\xi^\alpha|(1 + |\xi|^2)^{-s/2} |\tilde{u}(\xi)|(1 + |\xi|^2)^{s/2} d\xi$$

$$\leq \left(\int |\xi^\alpha|^2 (1 + |\xi|^2)^{-s} d\xi \right)^{1/2} \left(\int |\tilde{u}(\xi)|^2 (1 + |\xi|^2)^s d\xi \right)^{1/2} = C\|u\|_s.$$

Thus $\|u\|_{C^k} \leq A\|u\|_s$.

Now let $u \in H^s$. By Theorem 19 there exists a sequence $\{u_m\}$ of elements of $C_0^\infty(\mathbb{R}^n)$, for which $\|u_m - u\|_s \to 0$ as $m \to \infty$. As we have proved, it follows that $\|u_m - u_{m'}\|_{C^k} \leq A\|u_m - u_{m'}\|_s$ and therefore, $u_m \to v$ in C^k. But

$$u(\varphi) = \lim_{m \to \infty} \int u_m(x)\varphi(x)dx = \int v(x)\varphi(x)dx, \quad \varphi \in C_0^\infty(\mathbb{R}^n).$$

Therefore, $u(x) = v(x)$ almost everywhere, and $\|v\|_{C^k} \leq A\|u\|_s$. $\qquad\square$

1.5.2 Distributions with compact supports

Theorem 21. *Every distribution with compact support is an element of a space H^s with some real s, i.e. $\mathcal{E}'(\mathbb{R}^n) \subset \bigcup H^s$.*

Proof. Let $u \in \mathcal{E}'(\mathbb{R}^n)$, so that $K = \operatorname{supp} u$ is a compact set. The distribution u has in K a finite order m, i.e.

$$|u(\varphi)| \leq C \sup_{x \in K} \sum_{|\alpha| \leq m} |D^\alpha \varphi(x)|, \quad \varphi \in C_0^\infty(\mathbb{R}^n).$$

From Theorem 20 we have the inequality

$$|u(\varphi)| \leq CA\|\varphi\|_s, \quad s = m + [n/2] + 1.$$

Therefore, $u \in (H^s)^*$, and by Theorem 15, there exists an element $v \in H^{-s}$ such that

$$u(\varphi) = \int v(x)\varphi(x)dx, \quad \varphi \in C_0^\infty(\mathbb{R}^n).$$

It is evident that u can be identified with v. $\qquad\square$

Remark 22. If $u \in \mathcal{D}'(\mathbb{R}^n)$, then for each function $\varphi \in C_0^\infty(\mathbb{R}^n)$ there is an s such that $\varphi u \in H^s(\mathbb{R}^n)$. However, the distribution $u(\varphi) = \sum_{k=1}^\infty D^k \varphi(k)$ on the line $\mathbb{R}$ does not belong to any $H^s(\mathbb{R})$.

1.5.3 Traces on the boundary

Theorem 23. *If $u \in H^s$ and $s > 1/2$, then one can define the trace of u on the plane $x_n = 0$, and the trace is an element of the space $H_{s-1/2}(\mathbb{R}^{n-1})$. There exists a constant $A > 0$ such that*

$$\|u(x', 0)\|_{s-1/2} \leq A\|u\|_s.$$

Proof. Let $u \in C_0^\infty(\mathbb{R}^n)$ and $v(x') = u(x', 0), x' = (x_1, \ldots, x_{n-1})$. Then $\tilde{v}(\xi') = (2\pi)^{-1} \int \tilde{u}(\xi) d\xi_n$, and therefore

$$|\tilde{v}(\xi')|^2 \leq \int_{\mathbb{R}} (1 + |\xi|^2)^s |\tilde{u}(\xi)|^2 d\xi_n \int_{\mathbb{R}} (1 + |\xi|^2)^{-s} d\xi_n$$

$$= C_0 (1 + |\xi'|^2)^{1/2 - s} \int_{\mathbb{R}} (1 + |\xi|^2)^s |\tilde{u}(\xi)|^2 d\xi_n.$$

Multiplying both sides of this inequality by $(1 + |\xi'|^2)^{s-1/2}$ and integrating with respect to ξ', we obtain the inequality

$$\|v\|_{s-1/2}^2 \leq C_0 \|u\|_s^2.$$

By virtue of Theorem 19, the proof is complete. $\qquad\square$

Corollary 24. *If $u \in H^s(\mathbb{R}^n), s > k + 1/2, k \in \mathbb{N}$, then one can define the trace on the plane $x_n = 0$ of the derivatives $D_n^j u$ for $j \leq k$, and the trace of $D_n^j u$ is an element of the space $H^{s-j-1/2}(\mathbb{R}^{n-1})$. There exists a constant $A > 0$ such that*

$$\|D_n^j u(x', 0)\|_{s-j-1/2} \leq A\|u\|_s, \quad j = 0, 1, \ldots, k.$$

Theorem 25. *Let us assume that functions $\varphi_0, \varphi_1, \ldots, \varphi_k$, are defined in $\mathbb{R}^{n-1}$ and $\varphi_j \in H^{s-j-1/2}(\mathbb{R}^{n-1})$ for $s > k + 1/2$. Then there exists a function $u \in H^s(\mathbb{R}^n)$ such that $D_n^j u(x', 0) = \varphi_j(x'), j = 0, 1, \ldots, k$. Moreover, there exists a constant $A > 0$ such that*

$$\|u\|_s \leq A \sum_{j=0}^k \|\varphi_j\|_{s-j-1/2}.$$

Proof. Let $h \in C_0^\infty(\mathbb{R}), h(t) = 1$ for $|t| \leq 1$. Put

$$V(\xi', x_n) = \sum_{j=0}^k \frac{1}{j!} (ix_n)^j h(x_n (1 + |\xi'|^2)^{1/2}) \tilde{\varphi}_j(\xi').$$

It is evident that $V(\xi', 0) = \varphi_0(\xi')$, $D_n^j V(\xi', 0) = \varphi_j(\xi')$ for $j = 1, \ldots, k$. Let us show that V is the Fourier transform with respect to x' of a function $u \in H^s$. Indeed,

$$\|u\|_s^2 \leq \sum_{j=0}^{k} \int |\tilde{\varphi}_j(\xi')|^2 (1 + |\xi'|^2)^{-j-1} |\tilde{h}^{(j)}(\xi_n(1 + |\xi'|^2)^{-1/2})|^2 (1 + |\xi|^2)^s d\xi,$$

since the Fourier transform of the function $(-ix_n)^j g(x_n)$ is equal to $\tilde{g}^{(j)}(\xi_n)$, and that of the function $g(\rho x_n)$ is equal to $\rho^{-1} \tilde{g}(\xi_n \rho^{-1})$. Substitute $\tau(1 + |\xi'|^2)^{1/2}$ for ξ_n. We obtain

$$\|u\|_s^2 \leq \sum_{j=0}^{k} \int |\tilde{\varphi}_j(\xi')|^2 (1 + |\xi'|^2)^{-j-1+s+1/2} |\tilde{h}^{(j)}(\tau)|^2 (1 + \tau^2)^s d\tau d\xi'$$

$$\leq C \sum_{j=0}^{k} \|\varphi_j\|_{s-j-1/2}^2. \qquad \Box$$

1.6 Sobolev spaces in a domain

1.6.1 Definition

If Ω is a domain in $\mathbb{R}^n$, then the space $H^s(\Omega)$ is the set of elements $u \in \mathcal{D}'(\Omega)$, which can be extended to an element $u \in H^s(\mathbb{R}^n)$. The norm on this space is defined as

$$\|u\|_s = \inf_{v \in H^s, v = u \, in \, \Omega} \|v\|_s.$$

The subspace $H_0^s(\Omega)$ is the set of elements of $H^s(\Omega)$ belonging to $H^s(\mathbb{R}^n)$ if they are extended on $\mathbb{R}^n \setminus \Omega$ by the zero values.

Theorem 26. *The completion with respect to the norm of H^s of the space $C_0^\infty(\mathbb{R}^n)$ is contained in $H_0^s(\Omega)$.*

Proof. Let $u_k \in C_0^\infty(\Omega)$ and $\|u_k - u_l\|_s \to 0$ as $k, l \to \infty$. By Theorem 11 this sequence defines an element $u \in H^s$. Moreover, if $\varphi \in C_0^\infty(\mathbb{R}^n \setminus \Omega)$, then $u(\varphi) = \lim_{k \to \infty} \int u_k(x) \varphi(x) dx = 0$, and hence $\operatorname{supp} u \subset \overline{\Omega}$. $\qquad \Box$

Let us show that the definition of the space $H^s(\Omega)$ for positive integers s for a domain Ω with smooth boundary Γ is compatible with the definition given in 1.2.1.

Theorem 27. *Let $s \in \mathbb{N}$. Each function $u(x)$ with a finite value of integral $\int_\Omega \sum_{|\alpha| \leq s} |D^\alpha u(x)|^2 dx$ can be extended to a function $l_s u$ from $H^s(\mathbb{R}^n)$.*

Proof. First let $u \in H^s(\mathbb{R}_+^n)$, where

$$\mathbb{R}_+^n = \{x \in \mathbb{R}^n, x_n > 0\}.$$

Set

$$l_s u = \begin{cases} u(x) & \text{for } x_n > 0, \\ \sum_{k=1}^{s} \lambda_k u(x', -kx_n) & \text{for } x_n < 0, \end{cases}$$

where $x' = (x_1, \ldots, x_{n-1})$, and the numbers λ_k are found as the solution of the system of equations

$$\sum_{k=1}^{s} \lambda_k (-k)^j = 1 \quad \text{for} \quad 0 \le j \le s-1.$$

If $u \in C^s(\overline{\mathbb{R}^n_+})$, then, because of the choice of the constants λ_k, we have that $l_s u \in C^{s-1}(\mathbb{R}^n)$ and $\int_{\mathbb{R}^n} |D^\alpha l_s u(x)|^2 dx \le C \int_{\mathbb{R}^n_+} |D^\alpha u(x)|^2 dx \le C$ if $|\alpha| \le s$, and the constant C is independent of u.

Let $\sum_{j=0}^{N} h_j(x) = 1$ in $\overline{\Omega}$ and $h_0 \in C_0^\infty(\Omega)$, $h_j \in C_0^\infty(\overline{\Omega})$ for $j = 1, \ldots, N$. It is evident that one can put $h_0 u = 0$ outside Ω. Therefore it is sufficient to construct a continuation for a function $h_j u$. Let $\operatorname{supp} h_j \subset \overline{\Omega}_j$ and the domain Ω_j be so small that there exists a smooth map $f_j : \Omega_j \to \mathbb{R}^n$, by which $\Gamma \cap \overline{\Omega}_j$ is mapped to a part of the plane $x_n = 0$, and the image of Ω_j is lying in $\mathbb{R}^n_+$. For $u \in H^s(\Omega)$ put

$$l_{s,\Omega} u = h_0 u + \sum_{j=1}^{N} (f_j)_* l_s (f_j)^* (h_j u),$$

where $f_j^* v(x) = v(f_j^{-1}(x))$, $f_{j*} w(x) = w(f_j(x))$. By the construction we have $l_{s,\Omega} u \in H^s$ and $\|u\|_s \le C\|u\|_{H^s(\Omega)}$. $\qquad \square$

Theorem 28. *For $0 < s < 1$ the norm $\|u\|_s$ is equivalent to the norm*

$$\|u\|_s' = \left(\int |u(x)|^2 dx + \int \int \frac{|u(x) - u(y)|^2}{|x-y|^{n+2s}} dx dy \right)^{1/2}.$$

Proof. The Fourier transform of the function $u(x) - u(x+z)$ is equal to $\tilde{u}(\xi)[1 - e^{iz\xi}]$. Thus

$$\int \int \frac{|u(x) - u(y)|^2}{|x-y|^{n+2s}} dx dy = \int \int \frac{|u(x) - u(x+z)|^2}{|z|^{n+2s}} dx dz$$

$$= (2\pi)^{-n} \int \int \frac{|1 - e^{iz\xi}|^2}{|z|^{n+2s}} |\tilde{u}(\xi)|^2 d\xi dz.$$

The function $\int |1 - e^{iz\xi}|^2 |z|^{-n-2s} dz$ is homogeneous in ξ of degree $2s$ and depends on $|\xi|$ only. So this function is equal to $A|\xi|^{2s}$ with a positive constant A. Therefore,

$$\|u\|_s'^2 = (2\pi)^{-n} \int |\tilde{u}(\xi)|^2 (1 + A|\xi|^{2s}) d\xi.$$

Since

$$C_1 (1 + |\xi|^2)^s \le 1 + A|\xi|^{2s} \le C_2 (1 + |\xi|^2)^s$$

with certain constants C_1, C_2, the proof is complete. $\qquad \square$

1.6.2 The invariance under diffeomorphisms

Let us prove the invariance of the space $H_0^s(\Omega)$ with respect to a diffeomorphism of Ω, extendible on $\overline{\Omega}$.

Let $f : \Omega \to \Omega'$ be a diffeomorphic map. Then each function $\varphi \in C_0^\infty(\Omega')$ has corresponding pull-back $f^*\varphi(x) = \varphi(f(x))$ which is a function of $C_0^\infty(\Omega)$. If v is a distribution of $\mathcal{D}'(\Omega)$, then $u = f^*v$ is defined by the formula

$$u(\varphi) = v(\varphi(f^{-1}(y))|f_x'|^{-1}), \quad \varphi \in C_0^\infty(\Omega).$$

Here $|f_x'|$ is the Jacobian of the map f. It is easy to see that $u \in \mathcal{D}'(\Omega)$. This formula generalizes in a natural way the equality which is obtained after the change of variable in the classical case, when $v(\psi) = \int v(x)\psi(x)dx$ and $v \in L_1(\Omega')$.

Theorem 29. *Let K' be an open subset of the domain Ω', $\overline{K}' \subset\subset \Omega'$, and $v \in H_0^s(K')$. Then $u = f^*v \in H_0^s(K)$, where K is a preimage of K' and $\|u\|_s \le A\|v\|_s$, where the constant A is independent of v.*

Proof. 1°. If $s = 0$, then the statement is well-known in classical calculus, which also considers the case of natural s.

2°. If $0 < s < 1$, then it is convenient to apply Theorem 28. Let $y = f(x)$, $y' = f(x')$. Then

$$\int\int \frac{|u(x) - u(x')|^2}{|x - x'|^{n+2s}}dxdx' = \int\int \frac{|v(y) - v(y')|^2}{|f^{-1}(y) - f^{-1}(y')|^{n+2s}} \frac{dy}{|f'(x)|} \frac{dy'}{|f'(x')|}.$$

Since x and x' belong to the compact $\overline{K}$, we have

$$|f'(x)||f'(x')| \ge C_1, \qquad |x - x'| \ge C_2|y - y'|.$$

Therefore,

$$\int\int \frac{|u(x) - u(x')|^2}{|x - x'|^{n+2s}}dxdx' \le C_3 \int\int \frac{|v(y) - v(y')|^2}{|y - y'|^{n+2s}}dydy',$$

whence our assertion follows.

3°. If $s > 0$, then the theorem can be proved by combining the cases 1° and 2°.

4°. Now let $s < 0$. Then by Theorem 17 we have

$$\|u\|_s = \sup |\int u(x)\overline{\varphi}(x)dx|, \; \varphi \in C_0^\infty(K), \|\varphi\|_{-s} = 1.$$

Let $y = f(x)$, $\varphi(f^{-1}(y)) = \psi(y), u(x) = v(y), |(f^{-1})'(x)| = D(y)$. Then

$$|\int u(x)\overline{\varphi}(x)dx| = |\int v(y)\overline{\psi}(y)\frac{dy}{|f'(x)|}| \le \|v\|_s\|\psi(y)D(y)\|_{-s}.$$

Since $-s > 0$, one can apply the estimate obtained above and write

$$\|\psi(f(x))|f'(x)|^{-1}\|_{-s} \le A\|\varphi\|_{-s}.$$

Therefore,
$$\|u\|_s \le \|v\|_s \cdot A\|\varphi\|_{-s} = A\|v\|_s, \qquad \text{since } \|\varphi\|_{-s} = 1. \quad \square$$

Definition 30. The space $H^s_{\text{loc}}(\Omega)$ is the set of the functions u, for which $\varphi u \in H^s(\mathbb{R}^n)$ for all functions $\varphi \in C^\infty_0(\Omega)$.

It is evident that $C^k(\Omega) \subset H^k_{\text{loc}}(\Omega)$.

Theorem 31. *If $u \in \mathcal{D}'(\Omega)$, and for each point $x_0 \in \Omega$ there exists a function $\varphi \in C^\infty_0(\Omega)$ such that $\varphi(x_0) \neq 0$ and $\varphi u \in H^s(\mathbb{R}^n)$, then $u \in H^s_{\text{loc}}(\Omega)$.*

Proof. Let $\varphi \in C^\infty_0((\Omega)$. Since the support of φ is a compact set in Ω, for each point x_0 from this compact set there exists a function φ_{x_0} with the indicated property. By virtue of the Heine-Borel theorem, one can find a finite set of functions $\varphi_1, \ldots, \varphi_N$ from $C^\infty_0(\Omega)$ such that $\varphi_j u \in H^s(\Omega)$ and $\Phi = \sum_1^N |\varphi_j|^2 > 0$ on the set $\operatorname{supp}\varphi$. It is obvious that $\psi = \varphi \Phi^{-1} \in C^\infty_0(\Omega)$ and

$$\varphi u = \sum_{j=1}^N \psi |\varphi_j|^2 u = \sum_{j=1}^k (\psi \varphi_j)(\overline{\varphi}_j u) \in H^s(\Omega),$$

since $\varphi_j u \in H^s$ (see Theorem 14). Therefore, $u \in H^s_{\text{loc}}(\Omega)$. $\qquad\square$

1.6.3 The compactness of embeddings

Theorem 32. *Let Ω be a bounded domain in $\mathbb{R}^n$ and $s < t$. Then the embedding operator $H^t_0(\Omega) \to H^s_0(\Omega)$ is compact.*

Proof. Let $u_k \in H^t_0(\Omega)$ and $\|u_k\|_t \leq 1$. Let $h \in C^\infty_0(\mathbb{R}^n), h(x) = 1$ in a neighbourhood of the set Ω. Continue u_k by zero on $\mathbb{R}^n \setminus \Omega$. It is evident that

$$hu_k = u_k, \quad \tilde{u}_k(\xi) = (2\pi)^{-n} \int \tilde{u}_k(\eta)\tilde{h}(\xi - \eta)d\eta,$$

$$D_{\xi_j}\tilde{u}_k(\xi) = (2\pi)^{-n} \int \tilde{u}_k(\eta)D_{\xi_j}\tilde{h}(\xi - \eta)d\eta.$$

Since

$$(1 + |\xi|^2)^s \leq 2^{|s|}(1 + |\eta|^2)^s(1 + |\xi - \eta|^2)^{|s|}$$

(see proof of Theorem 14), we have

$$(1 + |\xi|^2)^s|\tilde{u}_k(\xi)|^2$$

$$\leq 2^{|s|}(2\pi)^{-2n} \int (1 + |\eta|^2)^s|\tilde{u}_k(\eta)|^2(1 + |\xi - \eta|^2)^{|s|}|\tilde{h}(\xi - \eta)|^2 d\eta;$$

$$(1 + |\xi|^2)^s|D_{\xi_j}\tilde{u}_k(\xi)|^2$$

$$\leq 2^{|s|}(2\pi)^{-2n} \int (1 + |\eta|^2)^s|\tilde{u}_k(\eta)|^2(1 + |\xi - \eta|^2)^{|s|}|D_{\xi_j}\tilde{h}(\xi - \eta)|^2 d\eta.$$

Since $\|u_k\|_s \leq \|u_k\|_t \leq 1$, these inequalities imply that

$$(1 + |\xi|^2)^s\left(|\tilde{u}_k(\xi)|^2 + \sum_{j=1}^n |D_{\xi_j}\tilde{u}_k(\xi)|^2\right) \leq A,$$

where A is independent of k. Hence it follows that the sequence $\tilde{u}_k(\xi)$ is uniformly bounded and equicontinuous as $|\xi| \leq R$ for any $R > 0$. Let $\varepsilon > 0$ be any number, and R be so large that $(1 + R^2)^{(s-t)/2} < \varepsilon$. By virtue of the Arzela theorem one can choose a subsequence, convergent uniformly for $|\xi| \leq R$, which we will denote again as $\{\tilde{u}_k(\xi)\}$. Let N be so large, that for $k \geq N, m \geq N$ the inequality

$$\left(\int_{|\xi| \leq R} (1 + |\xi|^2)^s d\xi \right)^{1/2} \cdot \max_{|\xi| \leq R} |\tilde{u}_k(\xi) - \tilde{u}_m(\xi)| < \varepsilon$$

holds. Then

$$\|u_k - u_m\|_s \leq \varepsilon \|u_k - u_m\|_t + \left(\int_{|\xi| \leq R} (1 + |\xi|^2)^s |\tilde{u}_k(\xi) - \tilde{u}_m(\xi)|^2 d\xi \right)^{1/2} < 3\varepsilon,$$

i.e. the sequence $\{u_k\}$ is a Cauchy sequence in H^s. $\square$

Theorem 33. *Let $s > t, s \geq -n/2$, and Ω be a bounded domain in $\mathbb{R}^n$ with its diameter not larger than δ. Then for any $\varepsilon > 0$ there exists δ_0 such that for $\delta < \delta_0$ we have*

$$\|u\|_t \leq \varepsilon \|u\|_s \quad \text{for } u \in C_0^\infty(\Omega).$$

Proof. Assume that this is not true and that for any k there exists a function $u_k \in C_0^\infty(\omega_k)$, where ω_k is a neighbourhood of the origin and $diam(\omega_k) < 1/k, \|u_k\|_s = 1, \|u_k\|_t \geq c_0 > 0$. Without loss of generality we can suppose that $\omega_{k+1} \subset \omega_k$ for $k = 1, 2, \ldots$. By Theorem 32 the set $\{u_k\}$ is compact in H^t. Therefore, there exists a function $u \in H^s$ such that

$$\|u_{k_i} - u\|_t \to 0, \ \|u\|_s \leq 1, \ \|u\|_t \geq c_0 > 0,$$

and $\operatorname{supp} u = \{0\}$. But then $\tilde{u}(\xi) = P(\xi)$ is a polynomial in ξ of degree $m \geq 0$ and $P(\xi)(1 + |\xi|^2)^{s/2} \in L_2$, which is possible only if $m + s < -n/2$. Since $s \geq -n/2$, we get a contradiction. $\square$

Remark 34. The condition $s \geq -n/2$ is essential in the theorem, since $\delta(x) \in H^s$ for $s < -n/2$.

It is often useful to consider the space $H^{s,\delta}$, where $\delta \in \mathbb{R}$ with the norm

$$\|u\|_{s,\delta} = \left((2\pi)^{-n} \int (1 + |\xi|^2)^s (1 + |\delta\xi|^2)^{-1} |\tilde{u}(\xi)|^2 d\xi \right)^{1/2}.$$

The proof of the following theorem is rather obvious.

Theorem 35. *If $u \in H^s(\mathbb{R}^n)$ then*

$$\|u\|_{s-1} \leq \|u\|_{s,\delta} \leq C(s,\delta) \|u\|_{s-1},$$

and as $\delta \to 0$, then

$$\|u\|_{s,\delta} \to \|u\|_s.$$

If $u \in H^{s-1}(\mathbb{R}^n)$ and $\|u\|_{s,\delta} \leq C$ with a constant C independent of δ, then $u \in H^s(\mathbb{R}^n)$ and $\|u\|_s \leq C$.

Chapter 2

Pseudo-differential operators

In this chapter we briefly state the classical theory of pseudo-differential operators. This theory has its roots in the works of Giraud, Calderon-Zygmund, Mikhlin, Agranovich-Dynin, and Vishik-Eskin, but officially it started in 1965 with the remarkable work of Kohn-Nirenberg [KN], where it was stated in a very simple and attractive manner, showing its applications to the theory of boundary-value problems.

Originally this theory was constructed in order to provide an algebra of operators containing all smooth linear differential operators and the inverse operators to elliptic differential operators. Later it became clear that the scope of possible applications are much wider and the theory is very useful in many other problems of mathematical physics and geometry. In particular, one can apply it, after some modification, to solving boundary value problems in domains with singular points on the boundary. To show this is one of the goals of this book.

2.1 The algebra of differential operators

2.1.1 Differential operators in $\mathbb{R}^n$

A linear differential operator is an expression of the form

$$P(x, D) = \sum_{|\alpha| \leq m} a_\alpha(x) D^\alpha.$$

Here as usual

$$D^\alpha = D_1^{\alpha_1} \dots D_n^{\alpha_n}, \quad D_j = \frac{\partial}{i \partial x_j}, \quad j = 1, \dots, n.$$

The polynomial

$$P(x, \xi) = \sum_{|\alpha| \leq m} a_\alpha(x) \xi^\alpha$$

17

is called the *symbol* of the differential operator $P(x, D)$, and the polynomial

$$P_0(x, \xi) = \sum_{|\alpha|=m} a_\alpha(x)\xi^\alpha$$

its *principal symbol* or its *characteristic form*. The number m is *the order* of the differential operator $P(x, D)$.

The connection between the differential operator $P(x, D)$ and its symbol is clear from the formulas

$$P(x, D)e^{ix \cdot \xi} = e^{ix \cdot \xi} P(x, \xi),$$

$$P(x, D)f(x) = (2\pi)^{-n} \int P(x, \xi)\tilde{f}(\xi)e^{ix \cdot \xi} d\xi.$$

It is obvious that the composition of two differential operators $P(x, D)$ and $Q(x, D)$ is a differential operator $R(x, D)$ with a symbol $R(x, \xi)$. The function $R(x, D)f(x)$ can be represented as

$$(2\pi)^{-n} \int P(x, \xi)e^{ix \cdot \xi} d\xi (2\pi)^{-n} \int Q(y, \eta)\tilde{f}(\eta)e^{iy \cdot \eta} d\eta e^{-iy \cdot \xi} dy$$

$$= (2\pi)^{-n} \int R(x, \eta)\tilde{f}(\eta)e^{ix \cdot \eta} d\eta,$$

where

$$R(x, \eta) = (2\pi)^{-n} \int \int P(x, \xi)Q(y, \eta)e^{i(x-y) \cdot (\xi-\eta)} dy d\xi.$$

Put

$$z = y - x, \quad \zeta = \xi - \eta.$$

We will systematically use the following notation:

$$f^{(\alpha)}(x, \xi) = \frac{\partial^\alpha f(x, \xi)}{\partial \xi^\alpha}; \quad f_{(\alpha)}(x, \xi) = \frac{\partial^\alpha f(x, \xi)}{\partial x^\alpha}.$$

Then we have

$$R(x, \eta) = (2\pi)^{-n} \int \int P(x, \zeta + \eta)Q(z + x, \eta)e^{iz \cdot \zeta} dz d\zeta$$

$$= (2\pi)^{-n} \int \int \sum_{|\alpha| \leq m} \frac{1}{\alpha!} P^{(\alpha)}(x, \eta)\zeta^\alpha Q(z + x, \eta)e^{iz \cdot \zeta} dz d\zeta$$

$$= (2\pi)^{-n} \sum_{|\alpha| \leq m} \frac{1}{\alpha! i^{|\alpha|}} P^{(\alpha)}(x, \eta)Q_{(\alpha)}(x, \eta).$$

The formula

$$R(x, \eta) = (2\pi)^{-n} \sum_{|\alpha| \leq m} \frac{1}{\alpha! i^{|\alpha|}} P^{(\alpha)}(x, \eta)Q_{(\alpha)}(x, \eta)$$

is a natural generalization of the classical Leibniz formula.

The differential operator $P^*(x, D)$ is *the formal adjoint* to the operator $P(x, D)$ if for all functions $f, g \in C_0^\infty$ the equality

$$\int Pf(x) \cdot \overline{g(x)} dx = \int f(x) \cdot \overline{P^* g(x)} dx$$

is valid. Since

$$\int D^\alpha f(x) \cdot \overline{g(x)} dx = \int f(x) \cdot \overline{D^\alpha g(x)} dx,$$

it is clear that

$$\int a_\alpha(x) D^\alpha f(x) \overline{g(x)} dx = \int f(x) \overline{D^\alpha [\overline{a_\alpha(x)} g(x)]} dx,$$

and thus

$$P^*(x, D) g(x) = \sum_{|\alpha| \le m} D^\alpha [(\overline{a_\alpha(x)} g(x)].$$

Therefore, the symbol of the operator $P^*(x, D)$ is

$$P^*(x, \xi) = \sum_{|\alpha| \le m} \sum_{\beta \le \alpha} \frac{\alpha!}{\beta!(\alpha - \beta)! i^{|\beta|}} \overline{a}_{(\alpha)}^{(\beta)}(x) \xi^{\alpha - \beta} = \sum_{|\beta| \le m} \frac{1}{\beta! i^{|\beta|}} \overline{P}_{(\beta)}^{(\beta)}(x, \xi).$$

2.1.2 Differential operators on a manifold

We recall that an *n-dimensional manifold* is a Hausdorff topological space, every point of which has a neighbourhood, homeomorphic to an open n-dimensional open set.

A manifold M is *smooth* if *a smooth structure* is defined on it, i.e.,

1. Each point $x_0 \in M$ has a neighbourhood V_{x_0} such that a homeomorphic map f_{x_0} of this neighbourhood on a neighbourhood ω_{x_0} of the origin in $\mathbb{R}^n$ is defined;

2. If the intersection of two such neighbourhoods V_{x_0} and V_{x_1} is non-empty, then the map

$$f_{x_1} f_{x_0}^{-1} : f_{x_0}(V_{x_0} \cap V_{x_1}) \to f_{x_1}(V_{x_0} \cap V_{x_1})$$

is smooth.

Then V_{x_0} is called a *coordinate neighbourhood*, f_{x_0} a *coordinate map*, and ω_{x_0} *a local chart*. If $x_1, x_2, \ldots$ is such a sequence of points on M that $\cup V_{x_j}$ covers M, then the set $\{V_{x_k}, f_{x_k}, \omega_{x_k}\}$ is *an atlas* of M.

We will consider only *paracompact* manifolds, i.e. representable as a union of a countable set of compact subsets. A manifold M is *closed* if it is compact. For example, the smooth boundary of a bounded domain in $\mathbb{R}^n$ is closed.

A function φ defined on M is of the class $C^\infty(M)$ (and is often called smooth) if each function of the form $\varphi(f_{x_0}^{-1}(x))$ belongs to the class $C^\infty(\omega_{x_0})$.

The set of smooth functions on M vanishing outside a compact subset (depending on the function) is a linear space $C_0^\infty(M)$. We have $\varphi_k \to 0$ in $C_0^\infty(M)$, if $\varphi_k = 0$ outside a compact subset common to all k, and $D^\alpha \varphi_k(f_{x_0}^{-1}(x))$ tends to zero as $k \to \infty$ uniformly on each compact subset of ω_{x_0} for all α and x_0. The space of linear continuous functionals over $C_0^\infty(M)$ is denoted by $\mathcal{D}'(M)$.

A linear operator $P : C^\infty(M) \to C^\infty(M)$ is a *differential operator* of order m if each operator

$$(f_{x_0})_* P f_{x_0}^* : C_0^\infty(\omega_{x_0}) \to C^\infty(\omega_{x_0})$$

is a differential operator of order m. Here

$$(f_{x_0})^*(\varphi(x)) = \varphi(f_{x_0}^{-1}(x)), \quad (f_{x_0})_* g(x) = g(f_{x_0}(x)).$$

A linear operator $P : C^\infty(M) \to C^\infty(M)$ is *local* if $\operatorname{supp} Pu \subset \operatorname{supp} u$ for all $u \in C^\infty(M)$. For example, every differential operator is local. It turns out that the inverse statement is also true. In other words, the following theorem of L. Schwartz is valid.

Theorem 1. *Let M be a smooth manifold and $P : C_0^\infty(M) \to C_0^\infty(M)$ a linear continuous map which is local. Then for any compact subset $K \subset M$ there exists a non-negative integer m such that the restriction of P on the subspace of the functions from C_0^∞ with their support in K is a linear differential operator of order m.*

Proof. It is sufficient to prove this statement for the case when K is contained in a coordinate neighbourhood. Therefore, we can assume without loss of generality that K is a compact subset in $\mathbb{R}^n$. If a point x_0 is in K, then the functional $f_{x_0}(\varphi) = P\varphi(x_0)$ is a distribution. Moreover, if $\varphi \in C_0^\infty$ and $\varphi(x) = 0$ in a neighbourhood of the point x_0, then $f_{x_0}(\varphi) = 0$, by the definition of the locality. Hence the support of f_{x_0} is a point x_0. But then $f_{x_0} = \sum_{|\alpha| \leq m(x_0)} a_\alpha \delta^{(\alpha)}(x - x_0)$, with $\delta^{(\alpha)}(x - x_0) = \partial^\alpha \delta(x - x_0)$ and therefore

$$P\varphi(x_0) = \sum_{|\alpha| \leq m(x_0)} a_\alpha(x_0)(iD)^\alpha \varphi(x_0).$$

The function $m(x_0)$ is bounded on every compact subset, i.e. $m(x_0) \leq m$ for $x_0 \in K$. It remains to prove the smoothness of the coefficients $a_\alpha(x_0)$. Let $h \in C_0^\infty(K)$ and $h(x) = 1$ in K', where $K' \subset\subset K$. Let $\varphi_\alpha(x) = h(x) \cdot x^\alpha$. Then the functions $f_\alpha(x) = P\varphi_\alpha(x)$ are smooth in K'. But

$$f_\alpha(x) = a_\alpha(x) \cdot \alpha! + \sum_{|\beta| < |\alpha|} a_\beta(x)(iD)^\beta x^\alpha$$

for $|\alpha| \leq m$. Hence we have $a_\alpha \in C^\infty(K')$. Since K' is an arbitrary compact subset in K, it follows that $a_\alpha \in C^\infty(K)$. $\qquad\square$

2.1.3 The cotangent space and the characteristic form

Let x be a point of a smooth manifold M, and $T_x M$ be the space tangent to M at x. The space of linear continuous functionals on the space $T_x M$ is isomorphic to $\mathbb{R}^n$ and is denoted by $T_x^* M$. *The cotangent space* $T^* M$ is the set of the points of the union $\bigcup T_x^* M$ for all $x \in M$, whose topology is defined by the basic neighbourhoods of the form $V \times \mathbb{R}^n$ for all coordinate neighbourhoods V on M. Moreover, in $T^* M$ one can induce the smooth structure, defining charts for $T^* M$ as the products $\omega \times \mathbb{R}^n$ and coordinate maps as

$$f \times i : V \times \mathbb{R}^n \to \omega \times \mathbb{R}^n \subset \mathbb{R}^n \times \mathbb{R}^n,$$

where f is a coordinate map for M and i is the identity map in $\mathbb{R}^n$.

It is obvious that such products are compatible on the intersections of coordinate neighbourhoods.

A *section* of the cotangent bundle is a smooth map $M \to T^* M$, moving each point $x \in M$ to a vector from the cotangent space $T_x^* M$. In particular, the *null section* is the map moving each point $x \in M$ to the point $(x, 0) \in T^* M$.

After the change of coordinates $y = F(x)$ the image of a point $(x, \xi) \in T^* M$ is the point $(F(x), {}^t F'(x)^{-1} \xi)$, where ${}^t F'(x)$ is the transposed Jacobi matrix. This rule is the same as the rule of transformation of a covector $a = (a_1(x), \ldots, a_n(x))$ of the coefficients of a differential form $\sum a_j(x) dx_j$, so that the cotangent bundle can be identified with the space of such forms.

If $P(x, D) = \sum_{|\alpha| \le m} a_\alpha(x) D^\alpha$ is a differential operator in a domain $U \subset \mathbb{R}^n$, then its *characteristic form* is the polynomial

$$p_0(x, \xi) = \sum_{|\alpha| = m} a_\alpha(x) \xi^\alpha$$

of variables $\xi \in \mathbb{R}^n$. By the change of variables $y = F(x)$ the operator P passes through a differential operator $Q(y, D)$ of order m and $q_0(y, \eta) = p_0(F(x)^{-1}, {}^t F'(x) \eta)$, i.e.,

$$p_0(x, \xi) = q_0(F(x), {}^t F'^{-1}(x) \xi).$$

This formula means that the characteristic form is defined invariantly on the cotangent space $T^* M$ for each differential operator $P(x, D)$ defined on M.

A *vector bundle* $F(M)$ of dimension k over a manifold M is a manifold of dimension $n + k$, whose coordinate neighbourhoods can be chosen in the form $V \times \mathbb{R}^k$, where V is a coordinate neighbourhood of the manifold M, and the coordinate maps in the form

$$f = g \times h : V \times \mathbb{R}^k \to \omega \times \mathbb{R}^k,$$

where ω is a domain in $\mathbb{R}^n$, and the maps $g : V \to \omega, h : \mathbb{R}^k \to \mathbb{R}^k$ are diffeomorphisms.

Let π be the natural projection of $F(M)$ on M. Then the *fibre* over a point $x \in M$ is the set $\pi^{-1}(x)$. A *section* of $F(M)$ is the map $s : M \to F(M)$ such that $\pi \circ s = id$.

The set of all vector bundles over M is denoted by $\mathrm{Vect}(M)$. It is obvious that the tangent and cotangent bundles are n-dimensional vector bundles. If $F(M)$ and $G(M_1)$ are vector bundles over manifolds M and M_1 of dimensions l and m, respectively, then a differential operator

$$P(D) : F(M) \to G(M_1)$$

is a linear operator having in local coordinates the form

$$[p(D)u(x)]_j = \sum_{l=1}^{k} p_{jl}(x, D)u_l(x), \quad j = 1, \ldots, m,$$

where $p_{jl}(x, D)$ are linear differential operators with smooth coefficients, defined on M with their values in the space of smooth functions defined on M_1.

2.1.4 Fundamental solutions of differential operators with constant coefficients

Let us consider a differential operator

$$P(D) = \sum_{|\alpha| \leq m} a_\alpha D^\alpha,$$

where $a_\alpha \in \mathbf{C}$ and $\sum_{|\alpha|=m} |a_\alpha| \neq 0$. If $u \in \mathcal{S}'$ and $P(D)u = f$, then

$$P(\xi)\tilde{u}(\xi) = \tilde{f}(\xi),$$

and if $\tilde{f}(\xi)P(\xi)^{-1} \in \mathcal{S}'$, then the Fourier transform of the distribution u is equal to $\tilde{f}(\xi)P(\xi)^{-1}$, which allows u to be found. This method works if $|P(\xi)| \geq c_0 = \mathrm{const} > 0$. In the general case one can use the exit in the complex domain in the variable ξ, as will be shown below.

Definition 2. A distribution $E \in \mathcal{D}'$ is a *fundamental solution* of the operator $P(D)$, if $P(D)E(x) = \delta(x)$.

Note that for $f \in \mathcal{E}'(\mathbb{R}^n)$ the distribution $u = E * f$ satisfies the equation $P(D)u = f$, since $P(D)u = P(D)E * f = \delta * f = f$. Thus the solvability of the equation $P(D)u = f$ for any $f \in \mathcal{E}'$ follows from the existence of a fundamental solution.

Theorem 3. *Every differential operator $P(D)$ has a fundamental solution.*

Proof. First let $n = 1$. Without loss of generality we can assume that the leading coefficient of the polynomial $P(\xi)$ is equal to 1. The equation $P(\zeta) = 0$ has m complex roots, $\lambda_1, \ldots, \lambda_m$. Thus there exists a real number τ such that $|\tau| \leq m+1$ and $|\xi + i\tau - \lambda_j| > 1$ for all $\xi \in \mathbb{R}$ and $j = 1, \ldots, m$. Therefore,

$$|P(\xi + i\tau)| = \prod |\xi + i\tau - \lambda_j| > 1$$

for all $\xi \in \mathbb{R}$. Now let

$$E(\varphi) = \frac{1}{2\pi} \int_{-\infty}^{+\infty} \frac{\tilde{\varphi}(-\xi - i\tau)}{P(\xi + i\tau)} d\xi, \qquad \varphi \in C_0^\infty(\mathbb{R}^n).$$

The integral converges, since

$$|\tilde{\varphi}(-\xi - i\tau)| \le C_N(1 + |\xi|)^{-N}$$

for all N. It is obvious that $E \in \mathcal{D}'(\mathbb{R})$. On the other hand, since $\tilde{\varphi}$ is an entire analytic function,

$$P(D)E(\varphi) = E(P(-D)\varphi) = \frac{1}{2\pi} \int_{-\infty}^{+\infty} \frac{P(\xi + i\tau)\tilde{\varphi}(-\xi - i\tau)}{P(\xi + i\tau)} d\xi$$

$$= \frac{1}{2\pi} \int_{-\infty}^{+\infty} \tilde{\varphi}(-\xi - i\tau)d\xi = \frac{1}{2\pi} \int_{-\infty}^{+\infty} \tilde{\varphi}(-\xi)d\xi = \varphi(0),$$

i.e. $P(D)E(x) = \delta(x)$. If $n > 1$, the construction of a fundamental solution is similar. Using rotation in the space of variables ξ one can get the coefficient of ξ_1^m to be non-zero. Indeed, if $\xi_j = \sum a_{jk}\eta_k$, then the coefficient at η_1^m is equal to $P_m(a_{11},\ldots,a_{n1})$, where P_m is the homogeneous principal part of the polynomial $P(\xi)$. Since $P_m \ne 0$, there exists a unit vector $a_{11},\ldots,a_{n1}$, for which $P_m(a_{11},\ldots,a_{n1}) \ne 0$. Therefore there exists an orthogonal matrix $||a_{ij}||$ with this property. Without loss of generality we can assume that the coefficient of η_1^m is equal to 1.

Having fixed a vector $\xi' = (\xi_2,\ldots,\xi_n)$, one can, as above, find a real τ, for which $|P(\xi_1 + i\tau, \xi')| > 1$ for all $\xi_1 \in \mathbb{R}$. By continuity, $|P(\xi_1 + i\tau, \xi')| > 1$ for all $\xi_1 \in \mathbb{R}$ and all ξ' from some neighbourhood $\omega_{\xi_0'}$ of a fixed vector ξ_0'. The set of such neighbourhoods covers the space $\mathbb{R}^{n-1}$. One can choose from this covering a subcovering by a countable set of neighbourhoods $\omega_1, \omega_2, \ldots.$ Shrinking the sets ω_j, we can assume that $\cup\overline{\omega}_j = \mathbb{R}^{n-1}$ and ω_j are mutually disjoint. By construction, a number τ_j corresponds to each j, for which $|\tau_j| \le m + 1$ and $|P(\xi_1 + i\tau_j, \xi')| > 1$ for all $\xi \in \mathbb{R} \times \omega_j$. Now put

$$E(\varphi) = (2\pi)^{-n} \sum_j \int_{\omega_j} d\xi' \int_{-\infty}^{+\infty} \frac{\tilde{\varphi}(-\xi_1 - i\tau_j, -\xi')}{P(\xi_1 + i\tau_j, \xi')} d\xi_1.$$

It is easy to see that $E \in \mathcal{D}'(\mathbb{R}^n)$. On the other hand,

$$P(D)E(\varphi) = E(P(-D)\varphi)$$

$$= (2\pi)^{-n} \sum_j \int_{\omega_j} d\xi' \int_{-\infty}^{+\infty} \frac{P(\xi_1 + i\tau_j, \xi')\tilde{\varphi}(-\xi_1 - i\tau_j, -\xi')}{P(\xi_1 + i\tau_j, \xi')} d\xi_1$$

$$= (2\pi)^{-n} \sum_j \int_{\omega_j} d\xi' \int_{-\infty}^{+\infty} \tilde{\varphi}(-\xi_1 - i\tau_j, -\xi')d\xi_1$$

$$= \int \tilde{\varphi}(-\xi_1, -\xi')d\xi = \varphi(0),$$

i.e., $P(D)E = \delta.$ $\qquad\square$

The method of construction of a fundamental solution stated is not the only possible one. It is often more convenient to apply the Fourier transform with respect to some variables only and to solve the differential equation obtained with respect to other variables. For instance, if one applies the Fourier transform with respect to $n-1$ variables, then the following lemma is very useful.

Lemma 4. *Let $n = 1$ and*

$$P(x, D) = \sum_{j=0}^{m} a_j(x) D^j, \quad a_m(x) = 1, \quad a_j(x) \in C^\infty(\mathbb{R}).$$

Let a function $u(x)$ be such that

$$P(x, D)u(x) = 0, \ x \in \mathbb{R}, \ u(0) = 0, \ldots, u^{(m-2)}(0) = 0, \ u^{(m-1)}(0) = 1.$$

Then the function $E(x) = \theta(x)u(x)$, where $\theta(x) = 1$ for $x > 0, \theta(x) = 0$ for $x < 0$, satisfies the equation $P(x, D)E(x) = \delta(x)$.

Proof. We have
$$E'(x) = \theta'(x)u(x) + \theta(x)u'(x).$$

But $\theta'(x) = \delta(x)$, and $u(0) = 0$. Hence $E'(x) = \theta(x)u'(x)$. Similarly,

$$E^{(j)}(x) = \theta(x)u^{(j)}(x) \quad for \quad j = 2, \ldots, m-1.$$

Since $u^{(m-1)}(0) = 1$, we have

$$E^{(m)}(x) = \theta(x)u^{(m)}(x) + \delta(x).$$

Therefore,
$$P(x, D)E(x) = \theta(x)P(x, D)u(x) + \delta(x) = \delta(x). \qquad \square$$

2.1.5　Examples of fundamental solutions

1. **The wave operator.** Let us find a fundamental solution $E(t, x)$ of the wave operator

$$P(D) = \partial^2/\partial t^2 - a^2 \Delta_3$$

(where Δ_3 is the Laplace operator in $\mathbb{R}^3$), which vanishes for $t < 0$. Let $v(t, \xi)$ be its Fourier transform with respect to the variables x. Then

$$\frac{\partial^2 v}{\partial t^2} + a^2 |\xi|^2 v = \delta(t). \quad \text{By Lemma 4,} \quad v(t, \xi) = \theta(t) \cdot \frac{\sin at|\xi|}{a|\xi|}.$$

Consider the Fourier transform of the distribution

$$\delta_R(\varphi) = \int_{|x|=R} \varphi(x) dS_x.$$

Since $\delta_R \in \mathcal{E}'$, we have

$$\tilde{\delta}_R(\xi) = \int_{|x|=R} e^{-ix\xi} dS_x = R^2 \int_0^{2\pi} d\varphi \int_0^\pi \sin\theta \cdot e^{-iR|\xi|\cos\theta} d\theta,$$

where θ is the angle between the vectors x and ξ. Thus,

$$\tilde{\delta}_R(\xi) = 2\pi R^2 \int_0^\pi \sin\theta \cdot e^{-iR|\xi|\cos\theta} d\theta = \frac{2\pi R^2}{iR|\xi|}(e^{iR|\xi|} - e^{-iR|\xi|}) = \frac{4\pi R}{|\xi|}\sin(R|\xi|).$$

Therefore,

$$E(t,x) = \theta(t)\mathcal{F}_{\xi\to x}^{-1}\frac{\sin(at|\xi|)}{a|\xi|} = \theta(t)\frac{\delta_{at}(x)}{4\pi a^2 t}.$$

2. **The heat operator.** Let

$$P(D) = \frac{\partial}{\partial t} - a^2\Delta,$$

where Δ is the Laplace operator in $\mathbb{R}^n$. Put

$$v(t,\xi) = \mathcal{F}_{x\to\xi}E(t,x),$$

where $E(t,x)$ is a fundamental solution of the operator P, vanishing for $t < 0$. Then

$$\partial v/\partial t + a^2|\xi|^2 v = \delta(t),$$

and by Lemma 4 we find that

$$v(t,\xi) = \theta(t)e^{-a^2|\xi|^2 t}.$$

Since this function is fast decreasing for $t > 0$, we have

$$E(t,x) = (2\pi)^{-n}\theta(t)\int e^{-a^2|\xi|^2 t + ix\xi} d\xi.$$

The latter integral is the product of the one-dimensional integrals

$$I_j = \int_{-\infty}^{+\infty} e^{-a^2\xi_j^2 t + ix_j\xi_j} d\xi_j.$$

These integrals can be calculated by using the Cauchy theorem, so that

$$I_j = \frac{\sqrt{\pi}}{a\sqrt{t}}\exp(\frac{-x_j^2}{4a^2 t}).$$

Therefore,

$$E(t,x) = (2\pi)^{-n}\theta(t)(\frac{\sqrt{\pi}}{a\sqrt{t}})^n\exp(\frac{-|x|^2}{4a^2 t}) = \frac{\theta(t)}{2a(\sqrt{\pi t})^n}\exp(\frac{-|x|^2}{4a^2 t}).$$

3. **The Laplace operator.** A fundamental solution for the Laplace operator can be obtained from the fundamental solution of the heat operator by the method of descent, described in the following Lemma.

Lemma 5. *Let a function $u(t,x)$ satisfy the equation $P(D_t, D_x)u = f(t,x)$ in $\mathbb{R}^{n+1}$, where*

$$P(D_t, D_x) = \sum_{j+|\alpha|\leq m} a_{j,\alpha} D_t^j D_x^\alpha, \quad f \in \mathcal{E}'(\mathbb{R}^{n+1}).$$

Let the integral

$$v(x) = \int_{-\infty}^{+\infty} |u(t,x)|\,dt$$

converge for almost all x and $v \in L^1_{\mathrm{loc}}(\mathbb{R}^n)$, i.e.

$$\int_K |v(x)|\,dx < \infty$$

for all compacts $K \subset \mathbb{R}^n$. Then $P(0, D_x)v(x) = g(x)$, where $v(x) = \int u(t,x)dt$, $g(x) = \int f(t,x)dt \in \mathcal{D}'(\mathbb{R}^n)$.

Proof. Let $\mathcal{F}$ be the subspace in $C^\infty(\mathbb{R}^{n+1})$ of functions $\varphi(t,x)$ independent of t and belonging to $C_0^\infty(\mathbb{R}^n)$ for each $t \in \mathbb{R}$. It is evident that $\mathcal{F}$ is isomorphic to $C_0^\infty(\mathbb{R}^n)$. To every function $\varphi_0 \in \mathcal{D}(\mathbb{R}^n)$ there corresponds the function $\varphi \in \mathcal{F}$ with $\varphi(t,x) = \varphi_0(x)$. The space $\mathcal{F}'$ is isomorphic to $\mathcal{D}'(\mathbb{R}^n)$, and to each element $f \in \mathcal{F}'$ there corresponds the functional $f_0 = \int f(t,x)dt \in \mathcal{D}'(\mathbb{R}^n)$ such that $f_0(\varphi_0) = f(\varphi)$. We have for $\varphi_0 \in C_0^\infty(\mathbb{R}^n)$

$$P(0, D_x)v(\varphi_0) = v({}^t P(0, D_x)\varphi_0) = u({}^t P(D_t, D_x)\varphi)$$

$$= P(D_t, D_x)u(\varphi) = f(\varphi) = g(\varphi_0). \qquad \square$$

The conditions of Theorem 5 are fulfilled for the fundamental solution of the heat equation (with $f(t,x) = \delta(t,x)$) found in Section 2 if $n \geq 3$. Moreover,

$$\int_{-\infty}^{+\infty} \frac{\theta(t)}{(2a\sqrt{\pi t})^n} exp(\frac{-|x|^2}{4a^2 t})dt = \frac{|x|^{2-n}}{4a^2 \pi^{n/2}} \int_0^\infty e^{-y} y^{n/2-2}dy,$$

where we have put $y = |x|^2/4ta^2$. Since

$$\int_0^\infty e^{-y} y^{n/2-2}dy = \Gamma(\frac{n}{2} - 1),$$

where Γ is the Euler function, we can see that the function

$$E(x) = \frac{1}{\omega_n |x|^{n-2}(2-n)},$$

where $\omega_n = 2\pi^{-n/2}\Gamma(n/2)$ in $\mathbb{R}^n$, is a fundamental solution of the Laplace operator in $\mathbb{R}^n$.

Note that if $n = 2$, then the Laplace operator has a fundamental solution $E(x) = \ln|x|/(2\pi)$.

2.1.6 Hypoelliptic operators

A differential operator $P(D)$ is *hypoelliptic* if for any domain $\Omega \subset \mathbb{R}^n$ any solution u of the equation $P(D)u = 0$ from the class $\mathcal{D}'(\Omega)$ is a function from $C^\infty(\omega)$ for any open set $\omega \subset\subset \Omega$.

A complete algebraic description of all hypoelliptic differential operators has been obtained by Hörmander in [H1].

Theorem 6. *A differential operator $P(D)$ is hypoelliptic if and only if*

$$\lim_{|\xi|\to\infty} \frac{P(\xi + \eta)}{P(\xi)} = 1$$

for any $\eta \in \mathbb{R}^n$.

For example, the Laplace operator or the heat operator are hypoelliptic, but the wave operator is not.

2.2 Basic properties of pseudo-differential operators

The theory of pseudo-differential operators was originally created with the specific aim of obtaining an algebra which included all differential operators and the operators inverse to elliptic differential operators. Since these inverse operators may be represented as singular integral operators, the algebra of pseudo-differential operators can be regarded also as the algebra of singular integro-differential ones. However it turned out that the calculus of pseudo-differential operators becomes especially simple if their definition is based on the Fourier transform.

2.2.1 Definition and basic properties

An operator P is *pseudo-differential* if it has the form

$$Pf(x) = (2\pi)^{-n} \int a(x,\xi)\tilde{f}(\xi)e^{ix\xi}d\xi, \; f \in \mathcal{S}$$

where the function a called a *symbol* satisfies the following condition:

$$|D_\xi^\alpha D_x^\beta a(x,\xi)| \leq C_{\alpha,\beta}(1 + |\xi|)^{m-|\alpha|}$$

for all α, β from $\mathbb{Z}_+^n, x \in K, \xi \in \mathbb{R}^n$, where K is a compact subset in Ω. The constant C may depend on the compact K. The number m is called the *order* of the operator P.

The set of symbols of order m is denoted by S^m or $S^m(\Omega)$. We will denote by $Op(p)$ the operator corresponding to the symbol p.

It is easy to see that $P = I$ if $p(x,\xi) = 1$, and P is a differential operator of order m, if $p(x,\xi)$ is a polynomial in ξ of order m.

A pseudo-differential operator $P(x, D)$ of order m is called a *classical* pseudo-differential operator if the following two conditions are fulfilled:

1. For each $j = 0, 1, 2, \ldots$ there exists a function $p_j(x, \xi) \in C^\infty(\Omega \times \mathbb{R}^n)$, such that $p_j(x, t\xi) = t^{m-j} p_j(x, \xi)$ for all $t \geq 1, |\xi| \geq 1$ and $j = 0, 1, \ldots$;

2. For all $N = 0, 1, \ldots$ we have

$$p(x, \xi) - \sum_{j=0}^{N} p_j(x, \xi) \in S^{m-N}(\Omega).$$

The function $p_0(x, \xi)$ is called the *principal symbol.*

Pseudo-differential operators appear in many problems of the analysis in a very natural way. For example, when solving an elliptic differential equation

$$P(x, D)u(x) = f(x)$$

one uses the operator of the form

$$Af(x) = (2\pi)^{-n} \sum_j \int_{|\xi| \geq 1} \frac{(h_j \widetilde{f})(\xi)}{p(x_j, \xi)} e^{ix\xi} d\xi,$$

where $\sum_j h_j(x) = 1$ is a partition of the unity such that the diameters of the supports of the functions h_j are small, and x_j is a point of the support of h_j. One can prove (see Section 3.1) that

$$P(x, D)A = I + T_1 + T_2,$$

where T_1 is a compact operator, and the operator T_2 has a small norm. It seems to be natural enough to pass to the operator

$$Bf(x) = (2\pi)^{-n} \sum_j \int_{|\xi| \geq 1} \frac{(h_j \widetilde{f})(\xi)}{p(x, \xi)} e^{ix\xi} d\xi$$

$$= (2\pi)^{-n} \int_{|\xi| \geq 1} \frac{\widetilde{f}(\xi)}{p(x, \xi)} e^{ix\xi} d\xi,$$

which is a parametrix, since the operator $PB - I$ is compact in every space H^s (shown below).

Theorem 7. *A pseudo-differential operator is continuous as an operator from $C_0^\infty(\Omega)$ to $C^\infty(\Omega)$.*

Proof. If $u \in C_0^\infty(\Omega)$, then by the theorem of Paley-Wiener, for all N

$$|\widetilde{u}(\xi)| \leq C_N (1 + |\xi|)^{-N}.$$

Therefore, if $|\alpha| \leq k$, then

$$|D_x^\alpha [p(x, \xi) \widetilde{u}(\xi) e^{ix\xi}]| \leq C'_{k,N} (1 + |\xi|)^{m+k-N},$$

whence it follows that for $|\alpha| \leq N - m - n$ the function $D^\alpha Pu(x)$ is continuous. Since N is arbitrary, we see that $Pu(x) \in C^\infty(\Omega)$. $\qquad\square$

2.2.2 Pseudo-differential operators as integral operators

A pseudo-differential operator can be expressed also without using the Fourier transform. Namely,

$$Pu(x) = \int K(x, x - y)u(y)dy,$$

where

$$K(x, z) = (2\pi)^{-n} \int p(x, \xi)e^{i\xi z}d\xi.$$

The latter integral converges if $m < -n$. If $m \geq -n$, the integral can be interpreted as a distribution from $\mathcal{D}'(\Omega \times \mathbb{R}^n)$ in the following way. If $\varphi \in \mathcal{D}(\Omega \times \mathbb{R}^n)$, we put

$$K(\varphi) = (2\pi)^{-n} \int d\xi(\int \int p(x, \xi)\varphi(x, z)e^{i\xi z}dxdz).$$

Thus K is a distribution. If $p = 0$ for large $|\xi|$ or $m < -n$, then K is a function and this definition coincides with the one given above.

Theorem 8. *The distribution $K(x, z)$ is an infinitely differentiable function if $z \neq 0$.*

Proof. Let $\varphi(x, z) \in \mathcal{D}(\Omega \times \mathbb{R}^n)$ and $\varphi = 0$ for $|z| \leq \delta$. Note that

$$e^{i\xi z} = |z|^{-2}(-\Delta_\xi)e^{i\xi z} = |z|^{-2N}(-\Delta_\xi)^N e^{i\xi z}$$

for any natural N. If $h \in C_0^\infty(\mathbb{R}^n)$ and $h(\xi) = 1$ for $|\xi| \leq 1$, then

$$(2\pi)^n K(\varphi) = \lim_{\varepsilon \to 0} \int h(\varepsilon\xi)d\xi(\int \int p(x, \xi)\varphi(x, z)e^{i\xi z}dxdz)$$

$$= \lim_{\varepsilon \to 0} \int \int \int p(x, \xi)h(\varepsilon\xi)\varphi(x, z)|z|^{-2N}(-\Delta_\xi)^N e^{i\xi z}d\xi dxdz$$

$$= \lim_{\varepsilon \to 0} \int \int \int (-\Delta_\xi)^N[p(x, \xi)h(\varepsilon\xi)]e^{i\xi z}d\xi|z|^{-2N}\varphi(x, z)dxdz.$$

If N is so big that $2N > m + n$, then one can pass to the limit:

$$K(\varphi) = (2\pi)^{-n} \int \int \int [(-\Delta_\xi)^N p(x, \xi)]|z|^{-2N}\varphi(x, z)e^{i\xi z}dxdzd\xi,$$

and for $|z| > \delta$,

$$K(x, z) = (2\pi)^{-n}|z|^{-2N} \int [(-\Delta_\xi)^N p(x, \xi)]e^{i\xi z}d\xi.$$

Therefore, if $2N > m + n + t$, then $K \in C^t(\Omega \times (\mathbb{R}^n \setminus \omega_\delta))$, where ω_δ is the ball of radius δ with its center at the origin. Since N is arbitrarily large, and δ is arbitrarily small, we see that $K \in C^\infty(\Omega \times (\mathbb{R}^n \setminus 0))$. $\square$

The following theorem can help to explain why pseudo-differential operators are called also *integro-differential*.

Theorem 9. *A pseudo-differential operator P is a composition of a differential and of an integral operator $K : v \to \int K(x,y)v(y)dy$ with a continuous kernel $K(x,y)$ which is in C^∞ for $x \neq y$.*

Proof. Let $2N > m + n$ and f be a function of $C_0^\infty(\mathbb{R}^n)$. Then

$$Pf(x) = (2\pi)^{-n} \int \frac{p(x,\xi)}{(1+|\xi|^2)^N}(1+|\xi|^2)^N \tilde{f}(\xi)e^{ix\xi}d\xi$$

$$= K(1 - \Delta)^N f(x),$$

where

$$Kv(x) = (2\pi)^{-n} \int p(x,\xi)(1+|\xi|^2)^{-N} \int v(y)e^{i(x-y)\xi}dyd\xi$$

is the integral operator with the continuous kernel

$$K(x,y) = (2\pi)^{-n} \int p(x,\xi)(1+|\xi|^2)^{-N} e^{i(x-y)\xi}d\xi.$$

Repeating the same arguments as in the proof of Theorem 8, one can demonstrate the infinite smoothness of the kernel $K(x,y)$ for $x \neq y$. $\square$

2.2.3   Continuity in the Sobolev spaces

Lemma 10. *The norm of the operator*

$$A : L_2(\mathbb{R}^n) \to L_2(\mathbb{R}^n),$$

acting by the formula

$$Au(x) = \int K(x,y)u(y)dy,$$

is not larger than

$$N = \max\left(\sup_x \int |K(x,y)|dy, \ \sup_y \int |K(x,y)|dx\right).$$

Proof. It is sufficient to verify that

$$\left|\int\int K(x,y)u(x)v(y)dxdy\right| \leq N\|u\|\|v\|,$$

$$u,v \in C_0^\infty(\mathbb{R}^n), \|\cdot\| = \|\cdot\|_{L_2}.$$

By the Cauchy inequality we have

$$\left| \int \int K(x,y)u(x)v(y)dxdy \right|^2 \leq \left| \int \int |K(x,y)||u(x)|^2 dxdy \right|$$

$$\times \left| \int \int |K(x,y)||v(y)|^2 dxdy \right| \leq N\|u\|^2 N\|v\|^2. \qquad \square$$

Theorem 11. *If $p \in S^m(\Omega), h \in C_0^\infty(\mathbb{R}^n)$, then*

$$\|h(x)P(x,D)u\|_s \leq C\|u\|_{s+m}$$

for all $u \in H^{s+m}$ and all $s \in \mathbb{R}$ with a constant C independent of u. Here P is a pseudo-differential operator having symbol p.

Proof. Let $q(x,\xi) = h(x)p(x,\xi)$. It suffices to verify the inequality for $u \in C_0^\infty(\mathbb{R}^n)$. If $u \in C_0^\infty(\mathbb{R}^n)$, then

$$\mathcal{F}(hPu)(\eta) = (2\pi)^{-n} \int \tilde{q}(\eta - \xi, \xi)\tilde{u}(\xi)d\xi,$$

where

$$\tilde{q}(\eta,\xi) = \int q(x,\xi)e^{-ix\eta}dx.$$

It is obvious that for any N,

$$|\tilde{q}(\eta - \xi, \xi)| \leq C_N(1 + |\xi - \eta|^2)^{-N}(1 + |\xi|^2)^{m/2}.$$

Since

$$(1 + |\eta|^2)^{s/2}\mathcal{F}(hPu)(\eta) = (2\pi)^{-n} \int (1 + |\eta|^2)^{s/2}\tilde{q}(\eta - \xi, \xi)(1 + |\xi|^2)^{-(s+m)/2}$$

$$\times [(1 + |\xi|^2)^{(s+m)/2}\tilde{u}(\xi)]d\xi,$$

it is sufficient, by the Lemma, to verify that

$$\int (1 + |\eta|^2)^{s/2}|\tilde{q}(\eta - \xi, \xi)|(1 + |\xi|^2)^{-(s+m)/2}d\xi \leq C,$$

$$\int (1 + |\eta|^2)^{s/2}|\tilde{q}(\eta - \xi, \xi)|(1 + |\xi|^2)^{-(s+m)/2}d\eta \leq C.$$

From the inequality

$$1 + |\eta|^2 \leq 2(1 + |\xi|^2)(1 + |\eta - \xi|^2)$$

it follows that

$$(1 + |\eta|^2)^{s/2}(1 + |\xi|^2)^{-s/2} \leq 2^{|s|/2}(1 + |\eta - \xi|^2)^{|s|/2}.$$

Therefore, the required estimates follow from the inequality

$$(2\pi)^{-n} \int C_N 2^{|s|/2}(1 + |\eta - \xi|^2)^{-N+|s|/2}d\xi \leq C,$$

which is true if $2N > |s| + n$. $\qquad \square$

Therefore, the number m is the *order* of the operator P in the scale of the Sobolev spaces H^s. The set of symbols satisfying the inequalities (2) for any m is denoted by $S^{-\infty}(\Omega)$. Operators with such symbols are called *smoothing* operators, since the distributions

$$K(x, z) = (2\pi)^{-n} \int p(x,\xi) e^{i\xi z} d\xi$$

are infinitely differentiable functions for all x and z. These operators transform the distributions with compact supports in infinitely differentiable functions.

Theorem 11 allows us to define pseudo-differential operators of order m as operators acting from H^{s+m} to H^s for any real s. Since $\mathcal{E}'(\Omega) = \cup H^s(\Omega)$, this means that a pseudo-differential operator of order m is continuous as an operator from $\mathcal{E}'(\Omega)$ to $\mathcal{D}'(\Omega)$.

In Section 2.1.2 we saw that differential operators — and they alone — are *local*, i.e.,

$$\operatorname{supp} Pu \subset \operatorname{supp} u \text{ for } u \in \mathcal{D}'(\Omega).$$

The operators satisfying the relation

$$\operatorname{sing\,supp} Pu \subset \operatorname{sing\,supp} u$$

are called pseudo-local. This means that if $u \in \mathcal{E}'(\Omega)$ and $u \in C^\infty(\omega)$, where $\omega \subset\subset \Omega$, then $Pu \in C^\infty(\omega)$. This property is in some sense inverse to hypoellipticity, defined by the relation

$$\operatorname{sing\,supp} u \subset \operatorname{sing\,supp} Pu.$$

Theorem 12. *Pseudo-differential operators are pseudo-local.*

Proof. Let $u \in \mathcal{E}'(\Omega)$, $u \in C^\infty(\omega)$, where $\omega \subset\subset \Omega$. Let $h \in C_0^\infty(\omega)$ and $h(x) = 1$ for $x \in \omega'$, where ω' is a subset of ω. Then $hu \in C_0^\infty(\omega)$ and $P(hu) \in C^\infty(\omega)$ by Theorem 1. On the other hand, if $h_1 \in C^\infty(\omega')$, then

$$h_1(x)P((1 - h)u) = h_1(x)K((1 - \Delta)^N[(1 - h)u(x)]) \in C^\infty,$$

since the supports of the functions h_1 and $1 - h$ have an empty intersection, and the kernel $K(x, y)$ is smooth when $x \neq y$. $\square$

2.3 Calculus of pseudo-differential operators

2.3.1 A technical Lemma

In order to consider the composition of two pseudo-differential operators, we introduce the subclass $S_0^m(\Omega)$ of symbols from $S^m(\Omega)$, which vanish in a neighbourhood of the boundary of Ω. We will show that such symbols form an algebra, i.e. the symbol of the operator $P \cdot Q$ belongs to $S_0^{m+m'}(\Omega)$ if the symbol of P belongs to

$S_0^m(\Omega)$ and the symbol of Q belongs to $S_0^{m'}(\Omega)$. The calculus can be based on the following:

Lemma 13. *Let a function $A(x, y, \xi, \eta)$ be infinitely differentiable in $\Omega \times \Omega \times \mathbb{R}^n \times \mathbb{R}^n$ and vanish if $y \in \Omega \setminus K$, where $K \subset\subset \Omega$. Let*

$$|D_\xi^\alpha D_\eta^{\alpha'} D_x^\beta D_y^{\beta'} A(x, y, \xi, \eta)| \leq C_{\alpha, \alpha', \beta, \beta'} (1 + |\xi|)^{m - |\alpha|} (1 + |\eta|)^{m' - |\alpha'|}$$

for all $\alpha, \alpha', \beta, \beta'$. Then the function

$$a(x, \xi) = (2\pi)^{-n} \int \left(\int A(x, y, \xi, \eta) e^{i(x-y)(\eta - \xi)} \, dy \right) d\eta$$

is a symbol of the class $S^{m+m'}$. Furthermore, if

$$a_N(x, \xi) = \sum_{|\alpha| < N} \frac{1}{\alpha!} \partial_\eta^\alpha D_y^\alpha A(x, y, \xi, \eta)|_{y=x, \eta=\xi},$$

then $a - a_N \in S^{m+m'-N}(\Omega)$ for every N.

Proof. The fact that $a \in C^\infty$ follows from the Paley-Wiener theorem. Let

$$a_N'(x, \xi) = (2\pi)^{-n} \int \left(\int \sum_{|\alpha| < N} \frac{1}{\alpha!} (iD_\eta)^\alpha A(x, y, \xi, \xi)(\eta - \xi)^\alpha e^{i(x-y)(\eta - \xi)} \, dy \right) d\eta.$$

It is evident that

$$a_N'(x, \xi) = (2\pi)^{-n} \int \left(\int \sum_{|\alpha| < N} \frac{1}{\alpha!} (iD_\eta)^\alpha A(x, y, \xi, \xi)(-D_y)^\alpha e^{i(x-y)(\eta - \xi)} \, dy \right) d\eta$$

$$= (2\pi)^{-n} \int \left(\int \sum_{|\alpha| < N} \frac{1}{\alpha!} \partial_\eta^\alpha D_y^\alpha A(x, y, \xi, \xi) e^{i(x-y)(\eta - \xi)} \, dy \right) d\eta$$

$$= (2\pi)^{-n} \int \left(\int \sum_{|\alpha| < N} \frac{1}{\alpha!} (iD_\eta)^\alpha D_y^\alpha A(x, y, \xi, \xi) e^{i(y-x)\xi} e^{-iy\eta} \, dy \right) e^{ix\eta} d\eta.$$

Since the function $A(x, y, \xi, \xi)$ has a compact support for fixed x, ξ, the formulas for the inverse Fourier transform are valid, and therefore

$$a_N'(x, \xi) = \sum_{|\alpha| < N} \frac{1}{\alpha!} \partial_\eta^\alpha D_y^\alpha A(x, x, \xi, \xi) = a_N(x, \xi).$$

The conditions of the theorem imply that $a_N \in S^{m+m'}(\Omega)$. It remains to verify that the function

$$a(x, \xi) - a_N(x, \xi) = (2\pi)^{-n} \int \left(\int r_N(x, y, \xi, \eta) e^{i(x-y)(\eta - \xi)} \, dy \right) d\eta,$$

where

$$r_N(x, y, \xi, \eta) = \sum_{|\alpha|=N} \frac{N}{\alpha!} \int_0^1 (1-t)^{N-1} D_\eta^\alpha A(x, y, \xi, \xi + t(\eta - \xi)) dt \, (\eta - \xi)^\alpha$$

also belongs to $S^{m+m'-N}(\Omega)$. Since

$$(\eta - \xi)^\alpha e^{i(x-y)(\eta-\xi)} = (-D_y)^\alpha e^{i(x-y)(\eta-\xi)},$$

one can integrate by parts, obtaining

$$a(x, \xi) - a_N(x, \xi) = (2\pi)^{-n} \sum_{|\alpha|=N} \frac{N}{\alpha!} \int d\eta \int_0^1 (1-t)^{N-1} dt$$

$$\times \int \partial_y^\alpha D_\eta^\alpha A(x, y, \xi, \xi + t(\eta - \xi)) e^{i(x-y)(\eta-\xi)} dy.$$

Let k be a natural number. Then

$$e^{i(x-y)(\eta-\xi)} = (1 - \Delta_y)^k e^{i(x-y)(\eta-\xi)} (1 + |\eta - \xi|^2)^{-k},$$

and therefore,

$$a(x, \xi) - a_N(x, \xi) = (2\pi)^{-n} \sum_{|\alpha|=N} \frac{N}{\alpha!} \int (1 + |\eta - \xi|^2)^{-k} d\eta \int_0^1 (1-t)^{N-1} dt$$

$$\times \int B_\alpha(t, x, y, \xi, \eta) e^{i(x-y)(\eta-\xi)} dy,$$

where

$$B_\alpha(t, x, y, \xi, \eta) = (1 - \Delta_y)^k \partial_y^\alpha D_\eta^\alpha A(x, y, \xi, \xi + t(\eta - \xi)).$$

Since the modulus of the integrand is not larger than

$$C(1 + |\xi|)^m (1 + |\xi + t(\eta - \xi)|)^{m'-N} (1 + |\eta - \xi|^2)^{-k} h_1(y),$$

where $h_1 \in C_0^\infty(\mathbb{R}^n), h_1 \geq 0, h_1 = 1$ in a neighbourhood of the compact K, the following estimate

$$|a(x, \xi) - a_N(x, \xi)|$$

$$\leq C_1 \int_0^1 \int (1 + |\xi|)^m (1 + |\xi + t(\eta - \xi)|)^{m'-N} (1 + |\eta - \xi|^2)^{-k} d\eta dt$$

is true, if $2k > m' - N + n$. Regarding the latter integral as a double one, we will divide the domain of integration into two:

$$\Omega_1 = \{(t, \eta) : |t(\eta - \xi)| < |\xi/2|\},$$

$$\Omega_2 = \{(t, \eta) : |t(\eta - \xi)| > |\xi/2|\}.$$

In the domain Ω_1 the inequalities

$$|\xi| \leq 2|\xi + t(\eta - \xi)| \leq 3|\xi|$$

hold, and therefore the integral over Ω_1 does not exceed

$$C_2 \int_{\Omega_1} \int_0^1 (1 + |\xi|)^{m+m'-N}(1 + |\eta - \xi|^2)^{-k+|m'-N|/2} dt d\eta \leq C_3 (1 + |\xi|)^{m+m'-N},$$

if $2k > n + |m' - N|$.

In the domain Ω_2 the inequalities

$$|\xi| \leq 2|\eta - \xi|, 1 + |\xi + t(\eta - \xi)| \leq 1 + 3|\xi - \eta|$$

hold, so that the integral over Ω_2 does not exceed

$$C_4 \int_{\Omega_2} \int_0^1 (1 + |\xi|)^{m+m'-N}(1 + |\eta - \xi|^2)^{-k+|m'-N|/2} dt d\eta \leq C_5 (1 + |\xi|)^{m+m'-N},$$

if $2k > |m' - N| + n$. Indeed, if $m' \geq N$, then

$$(1 + |\xi + t(\eta - \xi)|)^{m'-N} \leq 3^{m'-N}(1 + |\eta - \xi|)^{m'-N}$$

$$\leq 3^{m'-N}(1 + |\xi|)^{m'-N}(1 + |\eta - \xi|)^{m'-N},$$

and if $m' < N$, then

$$(1 + |\xi + t(\eta - \xi)|)^{m'-N} \leq 1 \leq 2^{N-m'}(1 + |\xi|)^{m'-N}(1 + |\eta - \xi|)^{N-m'}.$$

Therefore,

$$|a(x, \xi) - a_N(x, \xi)| \leq C_6 (1 + |\xi|)^{m+m'-N}.$$

If one differentiates both parts of the equality defining r_N with respect to x and ξ several times and repeats the same argments, then the inequality

$$|D_\xi^\alpha D_x^\beta [a(x, \xi) - a_N(x, \xi)]| \leq C'_{\alpha, \beta}(1 + |\xi|)^{m+m'-N-|\alpha|}$$

will be obtained. $\qquad \square$

Remark 14. A similar result is valid for the integral

$$b(x, \xi) = (2\pi)^{-n} \int (\int A(x, y, \xi, \eta)e^{i(y-x)(\eta-\zeta)} dy) d\eta,$$

which differs from $a(x, \xi)$ by the sign of the exponent. Passing to the complex conjugate functions we will find at once that

$$b \in S^{m+m'}, \quad b - b_N \in S^{m+m'-N},$$

where $\qquad b_N(x, \xi) = \sum_{|\alpha| < N} \frac{1}{\alpha!} \partial_\eta^\alpha (-D_y)^\alpha A(x, y, \xi, \eta)|_{y=x, \eta=\xi}.$

Remark 15. The proof is also valid in the case when the condition on the function A is substituted by the following, weaker condition:

$$|D_\xi^\alpha D_\eta^{\alpha'} D_x^\beta D_y^{\beta'} A(x,y,\xi,\eta)|$$

$$\leq C_{\alpha,\alpha',\beta,\beta'}(1+|\xi|)^{m-|\alpha|}(1+|\eta|)^{m'-|\alpha'|}(1+|\xi-\eta|)^{t_{\alpha,\alpha',\beta}}$$

for all $\alpha,\alpha',\beta,\beta'$ with some numbers $t_{\alpha,\alpha',\beta}$. Indeed, the functions a_N are estimated as before, since they are defined by the values of a at $\xi=\eta$. To estimate the remainder it suffices to increase k on $t/2$, where $t=t_{\alpha,\beta}=\max t_{\alpha,\alpha',\beta}$ for $|\alpha'|\leq N$.

2.3.2 The composition of pseudo-differential operators

Theorem 16. *Let $a\in S_0^m(\Omega), b\in S_0^{m'}(\Omega)$. Then the operator $C=B\cdot A$ is pseudo-differential with its symbol $c\in S_0^{m+m'}(\Omega)$. Moreover, $c-c_N\in S_0^{m+m'-N}(\Omega)$ for each natural N, if*

$$c_N(x,\xi)=\sum_{|\alpha|<N}\frac{1}{\alpha!}\partial_\xi^\alpha b(x,\xi)D_x^\alpha a(x,\xi).$$

Proof. We can calculate formally that for $u\in C_0^\infty(\Omega)$,

$$Cu(x)=B\cdot(2\pi)^{-n}\int a(y,\xi)\tilde{u}(\xi)e^{ix\xi}d\xi$$

$$=(2\pi)^{-2n}\int b(x,\eta)e^{ix\eta}\left(\int e^{-iy\eta}\left(\int a(y,\xi)\tilde{u}(\xi)e^{iy\xi}d\xi\right)dy\right)d\eta$$

$$=(2\pi)^{-n}\int c(x,\xi)\tilde{u}(\xi)e^{ix\xi}d\xi,$$

where

$$c(x,\xi)=(2\pi)^{-n}\int\left(\int b(x,\eta)a(y,\xi)e^{i(y-x)(\xi-\eta)}dy\right)d\eta.$$

It is easy to see that Lemma 13 is applicable to the latter integral. $\square$

Theorem 16 can be regarded as a natural generalization of the Leibniz formula for differential operators P, see Section 2.2.2,

$$P(x,D)(a(x)u(x))=\sum_{|\alpha|\leq m}\frac{1}{\alpha!}(P^{(\alpha)}(x,D)a(x))D^\alpha u(x).$$

Corollary 17. The operator $BA-C_0$ has the order $m+m'-1$ if C_0 is the operator with the symbol $b(x,\xi)a(x,\xi)$.

Corollary 18. The operator $BA - AB$ has the order $m + m' - 1$, and the operator $BA - AB - C_1$ has the order $m + m' - 2$, if C_1 is the operator with the symbol

$$c_1(x, \xi) = \frac{1}{i}\{b(x, \xi), a(x, \xi)\},$$

where

$$\{b(x, \xi), a(x, \xi)\} = \sum_{k=1}^{n}\left(\frac{\partial b}{\partial \xi_k}\frac{\partial a}{\partial x_k} - \frac{\partial b}{\partial x_k}\frac{\partial a}{\partial \xi_k}\right)$$

is the *Poisson bracket* of the functions $b(x, \xi)$ and $a(x, \xi)$.

Corollary 19. If $b(x, \xi)a(x, \xi) \equiv 0$, then the operators BA and AB are smoothing. If A or B is smoothing, then BA and AB are smoothing too.

2.3.3 A more general definition

Theorem 20. *Let $A : C_0^\infty(\Omega) \to C^\infty(\Omega)$ be the operator, defined by the formula*

$$\langle Au(x)\varphi\rangle = (2\pi)^{-n}\int\int\int a(x, y, \eta)e^{i(x-y)\eta}u(y)\varphi(x)dxdyd\eta,$$

where $a \in C^\infty(\Omega \times \Omega \times \mathbb{R}^n)$, $a(x, y, \xi) = 0$ for $y \in \Omega \setminus K$, where $K \subset\subset \Omega$ and

$$|D_\eta^\alpha D_x^\beta D_y^\gamma a(x, y, \eta)| \le C_{\alpha,\beta,\gamma}(1 + |\eta|)^{m-|\alpha|} \text{ for all } \alpha, \beta, \gamma \in \mathbb{Z}_+^n.$$

Then the operator A is pseudo-differential with its symbol $b(x, \xi)$ from $S_0^m(\Omega)$ and $b - a_N \in S_0^{m-N}(\Omega)$ for all natural N, where

$$a_N(x, \xi) = \sum_{|\alpha|<N}\frac{1}{\alpha!}\partial_\xi^\alpha D_y^\alpha a(x, y, \xi)|_{y=x}.$$

Proof. Let $u \in C_0^\infty(\Omega)$. Then

$$Au(x) = (2\pi)^{-2n}\int\int a(x, y, \eta)\left(\int \tilde{u}(\xi)e^{iy\xi}d\xi\right)e^{i(x-y)\eta}dyd\eta$$

$$= (2\pi)^{-n}\int b(x, \xi)\tilde{u}(\xi)e^{ix\xi}d\xi,$$

where

$$b(x, \xi) = (2\pi)^{-n}\int\left(\int a(x, y, \eta)e^{i(x-y)(\eta-\xi)}dy\right)d\eta.$$

Now the statement of the theorem follows at once from the Lemma 13 of 2.3.1. $\qquad\square$

2.3.4   Formally adjoint operators

Theorem 21. *Let $a \in S_0^m(\Omega)$, and A be the operator with the symbol a and*

$$b_N(x,\xi) = \sum_{|\alpha| < N} \frac{1}{\alpha!} \partial_\xi^\alpha D_x^\alpha \overline{a}(x,\xi).$$

Then the operator A^*, defined by the formula

$$\int Au(x)\overline{v(x)}dx = \int u(x)\overline{A^*v(x)}dx; u, v \in C_0^\infty(\Omega),$$

is pseudo-differential with its symbol $b(x,\xi)$ from $S^m(\Omega)$ and $b - b_N \in S^{m-N}(\Omega)$ for all natural N.

Proof. Let $u, v \in C_0^\infty(\Omega)$. Then

$$\int Au(y)\overline{v}(y)dy = (2\pi)^{-n} \int \overline{v}(y)dy \int a(y,\eta)e^{iy\eta}d\eta \int u(x)e^{-ix\eta}dx,$$

so that

$$A^*v(x) = (2\pi)^{-n} \int \int \overline{a(y,\eta)}e^{i(x-y)\eta}v(y)dyd\eta.$$

The result now follows from Theorem 20. $\square$

2.4   Pseudo-differential operators on closed manifolds

2.4.1   Transformation of operators under a change of variables

Theorem 22. *Let $F : \Omega' \to \Omega$ be a diffeomorphism and A a pseudo-differential operator with its symbol $a(x,\xi) \in S^m(\Omega)$. Then the operator B, defined by the formula*

$$F^*(Au) = B(F^*u), \text{ where } F^*u(x) = u(F(x)), u \in C_0^\infty(\Omega),$$

is pseudo-differential with the symbol $b \in S^m(\Omega)$, and

$$b(x,\xi) - a(F(x),{}^t F'(x)^{-1}\xi) \in S^{m-1}(\Omega).$$

Proof. Let K be a compact subset in Ω, K' be its preimage in Ω' under the map F. Let $u \in C_0^\infty(K), h \in C_0^\infty(\Omega)$ and $h(x) = 1$ in a neighbourhood of the compact K. Then

$$Au(x) = (2\pi)^{-n} \int \int a(x,\xi)h(y)u(y)e^{i(x-y)\xi}dyd\xi.$$

Let

$$x = F(x'), \quad y = F(y'), \quad u(F(y')) = v(y'),$$

$$Au(F(x')) = w(x'), \quad h(F(y')) = h_1(y').$$

Then

$$w(x') = (2\pi)^{-n} \int\!\!\int a(F(x'),\xi)h_1(y')v(y')e^{i(F(x')-F(y'))\xi}|F'(y')|\,dy'\,d\xi,$$

where $|F'(y')|$ is the Jacobian. By the Hadamard lemma we have

$$F(x') - F(y') = \varphi(x',y')(x' - y'),$$

where

$$\varphi(x',y') = \int_0^1 \frac{\partial F(x' + t(y' - x'))}{\partial y'}\,dt.$$

Note that $\varphi(x',x') = \partial F(x')/\partial x'$, so that the matrix $\varphi(x',y')$ is invertible for $|x' - y'| < \varepsilon$, $y' \in K'$. Let $h_2 \in C_0^\infty(\mathbb{R}^n)$, $h_2(x) = 1$ for $|x| \le \varepsilon/2$, $h_2(x) = 0$ for $|x| \ge \varepsilon$. Let

$$w_1(x') = (2\pi)^{-n} \int\!\!\int a(F(x'),\xi)h_1(y')h_2(x' - y')v(y')$$

$$e^{i(x' - y')\varphi(x',y')^t\xi}|F'(y')|\,dy'\,d\xi.$$

Let $\varphi(x',y')^t\xi = \eta$. Then

$$w_1(x') = (2\pi)^{-n} \int\!\!\int a(F(x'),\varphi^t(x',y')^{-1}\eta)h_1(y')h_2(x' - y')$$

$$v(y')e^{i(x' - y')\eta}|F'(y')|/|\varphi(x',y')|\,dy'\,d\eta.$$

Applying Theorem 20 to this operator we see that $w_1 = Bv$, where B is a pseudo-differential operator with the symbol $b \in S_0^m(\Omega)$ and

$$b(x,\xi) - a(F(x),{}^t F'(x)^{-1}\xi) \in S^{m-1}(\Omega).$$

It remains to verify that the operator

$$B_1 v(x') = (2\pi)^{-n} \int\!\!\int a(F(x'),\varphi^t(x',y')^{-1}\eta)$$

$$h_1(y')[1 - h_2(x' - y')]v(y')e^{i(x' - y')\eta}|F'(y')|/|\varphi(x',y')|\,dy'\,d\eta$$

is smoothing. But this also follows immediately from Theorem 20 and Corollary 19. $\qquad\square$

2.4.2 Pseudo-differential operators on a manifold

Theorem 22 allows us to define pseudo-differential operators on a smooth manifold X. Let $\{U_i, f_i, \omega_i\}$ be an atlas on X. Let $\varphi_i \in C_0^\infty(\omega_i)$ and $\sum_i \varphi_i \equiv 1$. A linear operator P defined on $C_0^\infty(X)$ is a *pseudo-differential operator* of order m if the following two conditions fulfilled:

1. The operators

$$(f_i)_* P f_i^* \varphi_i : C_0^\infty(\omega_i) \to C^\infty(\omega_i)$$

are pseudo-differential operators of order m for all i.

2. If φ, ψ are functions of $C^\infty(X)$ with disjoint supports, then the operator $u \to \varphi P(\psi u)$ is smoothing.

The symbol of P is defined as the section of $T^*(x)$ and in a coordinate neighbourhood ω_i is equal to the symbol of the corresponding pseudo-differential operator $(f_i)_* P f_i^*$. The set of symbols of order m is denoted as $S^m(X)$. If one defines a *principal symbol* of the operator P as an element of the quotient space $S^m(X)/S^{m-1}(X)$, then from Theorem 22 it follows that the principal symbol of a pseudo-differential operator on a manifold X is defined invariantly on the space $T^*(X)$.

2.5 Gårding inequality

2.5.1 Gårding inequality for elliptic differential operators

Recall that a differential operator

$$P(x, D) = \sum_{|\alpha| \leq m} a_\alpha(x) D^\alpha$$

is elliptic in a domain $\Omega \subset \mathbb{R}^n$ if there is a positive constant c_0 such that $|p_0(x, \xi)| \geq c_0|\xi|^m$ for all $x \in \Omega$, $\xi \in \mathbb{R}^n$, where p_0 is the characteristic form of the operator P. It is evident that if Ω is bounded and the coefficients of P are continuous in $\overline{\Omega}$, then this condition is equivalent to each of the following: $(1)|p_0(x, \xi)| \neq 0$ for $x \in \overline{\Omega}$, $\xi \in \mathbf{R}^n \setminus \{0\}$; $(2)|p(x, \xi)| \geq c_0|\xi|^m - C_1$ for $x \in \Omega$, $\xi \in \mathbb{R}^n \setminus \{0\}$, $c_0 = \text{const} > 0$.

It is easy to see that for $n \geq 2$ the order m of an elliptic operator with real coefficients is always even. This follows from the fact that $p_0(x, -\xi) = (-1)^m p_0(x, \xi)$ and for $n \geq 2$ the unit sphere in $\mathbb{R}^n$ is connected. The order m of an elliptic operator with complex coefficients is even, too, if $n \geq 3$. If $n = 2$, there are elliptic operators of odd order; for example, the operators $D_1^k + iD_2^k$ are elliptic in $\mathbb{R}^2$ for any natural k. The operators

$$\Delta, \ \Delta^k, \ \sum_{i,j=1}^n a_{ij}(x) D_i D_j$$

are elliptic, if

$$|\sum_{i,j=1}^n a_{ij}(x)\xi_i\xi_j| \geq c_0|\xi|^2, c_0 = \text{const} > 0.$$

Theorem 23. *Let $P(x, D)$ be a differential operator of even order m with coefficients of the class $C^{m/2}(\overline{\Omega})$, and*

$$\operatorname{Re} p_0(x, \xi) \geq c_0 |\xi|^m$$

for $x \in \overline{\Omega}$, $\xi \in \mathbf{R}^n$, $c_0 = \text{const} > 0$, where Ω is a bounded domain in $\mathbb{R}^n$. Then for any $\epsilon > 0$ there is a constant C_ϵ such that

$$\operatorname{Re} \int P(x, D)u(x) \cdot \overline{u}(x)dx \geq (c_0 - \epsilon)\|u\|_m^2 - C_\epsilon\|u\|_0^2, \ u \in C_0^\infty(\Omega).$$

Proof. First let P be an operator with constant coefficients and $\varepsilon > 0$ be fixed. Then

$$\operatorname{Re} \int P(D)u(x) \cdot \overline{u}(x)dx = (2\pi)^{-n} \operatorname{Re} \int P(\xi)|\tilde{u}(\xi)|^2 d\xi$$

$$\geq \int [(c_0 - \varepsilon)(|\xi|^2 + 1)^{m/2} - C_\varepsilon]|\tilde{u}(\xi)|^2 d\xi = (c_0 - \varepsilon)\|u\|_{m/2}^2 - C_\varepsilon\|u\|_0^2,$$

and the statement has been proved.

Now let $\varepsilon_1 > 0$ and δ be so small that

$$|a_\alpha(x) - a_\alpha(y)| < \varepsilon_1,$$

if $|x - y| < \delta$, $x \in \overline{\Omega}$, $y \in \overline{\Omega}$, $|\alpha| = m$. Let the sequence of smooth functions $\{\varphi_j(x)\}, j = 1, \ldots, N$ be such that the diameters of their supports are not larger than δ,

$$\sum_j \varphi_j(x)^2 \equiv 1 \text{ for } x \in \overline{\Omega}, \quad \varphi_j(x)u(x) = u_j(x).$$

Note that

$$\operatorname{Re} \int P(x, D)u(x) \cdot \overline{u}(x)dx = \sum_{j=1}^N \operatorname{Re} \int \varphi_j(x)^2 P(x, D)u(x) \cdot \overline{u}(x)dx$$

$$= \sum_{j=1}^N \operatorname{Re} \int P_0(x_j, D)u_j(x) \cdot \overline{u}_j(x)dx + \sum_{j=1}^N \operatorname{Re} \int Q_j(x, D)u(x) \cdot \overline{u}_j(x)dx$$

$$+ \sum_{j=1}^N \operatorname{Re} \int \varphi_j(x) \sum_{|\alpha|=m} [a_\alpha(x) - a_\alpha(x_j)]D^\alpha u(x) \cdot \overline{u}_j(x)dx,$$

where Q_j are differential operators of order $m - 1$ and $x_j \in \operatorname{supp} \varphi_j$.

Integrating by parts one gets the relation

$$\sum_{j=1}^{N} \operatorname{Re} \int \varphi_j(x) \sum_{|\alpha|=m} [a_\alpha(x) - a_\alpha(x_j)] D^\alpha u(x) \cdot \overline{u}_j(x) dx$$

$$= \sum_{j=1}^{N} \operatorname{Re} \int \sum_{|\alpha|=|\beta|=m/2} [a_{\alpha,\beta}(x) - a_{\alpha,\beta}(x_j)] D^\alpha u_j(x) \overline{D^\beta u_j(x)} dx$$

$$+ \sum_{j=1}^{N} \operatorname{Re} \int \sum_{|\alpha|\leq(m-2)/2} \sum_{|\beta|\leq m/2} b^j_{\alpha,\beta}(x) D^\alpha u(x) \cdot \overline{D^\beta u_j(x)} dx,$$

and similarly,

$$\sum_{j=1}^{N} \operatorname{Re} \int Q_j(x,D)u(x) \cdot \overline{u}_j(x) dx$$

$$= \sum_{j=1}^{N} \operatorname{Re} \int \sum_{|\alpha|\leq(m-2)/2} \sum_{|\beta|\leq m/2} c^j_{\alpha,\beta}(x) D^\alpha u(x) \cdot \overline{D^\beta u_j(x)} dx.$$

As has been proved above, we have

$$\operatorname{Re} \int P(x_j,D)u_j(x) \cdot \overline{u}_j(x) dx \geq (c_0 - \frac{\varepsilon}{2})\|u_j\|^2_{m/2} - C'_0\|u_j\|^2_0.$$

Summing the estimates obtained, we get

$$\operatorname{Re} \int P(x,D)u(x) \cdot \overline{u(x)} dx \geq (c_0 - \frac{\varepsilon}{2}) \sum \|u_j\|^2_{m/2} - C_1\varepsilon_1 \sum \|u_j\|^2_{m/2}$$

$$- C_2 \sum \|u\|_{(m-2)/2}\|u_j\|_{m/2} - C_3\|u\|^2_0.$$

Note that

$$\sum_{j=1}^{N} \|u_j\|^2_{m/2} = \sum_{j=1}^{N} \|\varphi_j u\|^2_{m/2} = \sum_{j=1}^{N} \sum_{|\alpha|\leq m/2} \int |D^\alpha \varphi_j u(x)|^2 dx$$

$$\geq \sum_{j=1}^{N} \sum_{|\alpha|=m/2} \int \varphi_j^2(x)|D^\alpha u(x)|^2 dx$$

$$- C_4\|u\|^2_{(m-2)/2} = \|u\|^2_{m/2} - C_5\|u\|^2_{(m-2)/2}.$$

Furthermore, there is a constant C_6 such that

$$\sum_j \|u\|_{(m-2)/2}\|u_j\|_{m/2} \leq \varepsilon_1\|u\|^2_{m/2} + C_6\|u\|^2_{(m-2)/2}.$$

Finally, by virtue of Theorem 12 of 1.3.2, there is a constant C_7 such that

$$\|u\|^2_{(m-2)/2} \leq \varepsilon_1 \|u\|^2_{m/2} + C_7 \|u\|^2_0.$$

Therefore, we have obtained the estimate

$$\mathrm{Re} \int P(x,D)u(x) \cdot \overline{u(x)}dx$$

$$\geq (c_0 - \frac{\varepsilon}{2} - C_1\varepsilon_1 - C_2\varepsilon_1 - C_5\varepsilon_1 - C_2C_6\varepsilon_1)\|u\|^2_{m/2} - C_8\|u\|_0.$$

If ε_1 is so small that $(C_1 + C_2 + C_5 + C_2C_6)\varepsilon_1 = \frac{\varepsilon}{2}$, we get the result. $\qquad\square$

2.5.2 Sharp Gårding inequality for pseudo-differential operators

Theorem 24. *Let $P(x, D)$ be a pseudo-differential operator of order m such that*

$$\mathrm{Re}\, p(x, \xi) \geq c_0 |\xi|^m \ for \ |\xi| \geq C_1 = C_1(K),$$

where K is an arbitrary compact subset in Ω. Then the inequality

$$\mathrm{Re} \int P(x,D)u(x) \cdot \overline{u}(x)dx \geq c_0\|u\|^2_{m/2} - C_2(K)\|u\|^2_{(m-1)/2}$$

is valid for all functions u from $C_0^\infty(K)$.

Proof. a. Let $p = a + ib$, where a and b are real-valued. Let A and B be the operators with the symbols a and b, respectively. Then

$$\mathrm{Re}\, i \int B(x,D)u(x) \cdot \overline{u}(x)dx$$

$$= i/2 \int B(x,D)u(x) \cdot \overline{u}(x)dx - i/2 \int u(x) \cdot \overline{B}(x,D)u(x)dx$$

$$= i/2 \int [B(x,D) - B^*(x,D)]u(x) \cdot \overline{u}(x)dx.$$

The operator $B - B^*$ has the order $m - 1$. Thus, if Λ is the operator with the symbol $(1 + |\xi|^2)^{1/2}$, we have

$$\left| \int [B(x,D) - B^*(x,D)]u(x) \cdot \overline{u}(x)dx \right|$$

$$= \left| \int \Lambda^{(1-m)/2}[B(x,D) - B^*(x,D)]u(x) \cdot \overline{\Lambda^{(m-1)/2}u(x)}dx \right|$$

$$\leq \|\Lambda^{(1-m)/2}[B(x,D) - B^*(x,D)]u(x)\|_0 \cdot \|\Lambda^{(m-1)/2)}u(x)\|_0$$

$$\leq C\|u\|^2_{(m-1)/2}.$$

On the other hand,

$$\text{Re} \int \Lambda^m u(x) \cdot \overline{u}(x)dx = \int |\Lambda^{m/2}u(x)|^2 dx = \|u\|^2_{m/2}$$

and so it is sufficient to prove the theorem for the case $c_0 = 0$.

b. **Lemma 25.** *Let $f \in C^2(\mathbb{R}^n)$, $f(x) \geq 0$ and*

$$\sup_{|\alpha|=2,\, x\in\mathbb{R}^n} |D^\alpha f(x)| = M < \infty.$$

Then

$$\sum_{|\alpha|=1} |D^\alpha f(x)|^2 \leq 2M f(x)$$

for all $x \in \mathbb{R}^n$.

Proof. Taking into consideration at a point x the direct line passing through this point in the direction of the gradient of f, we will get the result from the one-dimensional inequality

$$f'(t)^2 \leq 2M f(t), \text{ where } M = \sup |f''(t)|.$$

Since for all $s \in \mathbb{R}$ we have

$$0 \leq f(t+s) = f(t) + sf'(t) + s^2/2f''(\theta)$$

$$\leq f(t) + sf'(t) + s^2/2M,$$

we can take $s = -f'(t)/M$ and will get

$$0 \leq f(t) - \frac{f'^2(t)}{2M}. \qquad \square$$

c. Now let $g_0, g_1, \ldots, g_k, \ldots$ be all the points in $\mathbb{R}^n$ with integer coordinates and $g_0 = 0$. Let

$$\varphi(x) \in C_0^\infty(\mathbb{R}^n),\ 0 \leq \varphi \leq 1,\ \varphi(x) = 1,\ \text{if } |x_j| \leq 1/2$$

for all $j = 1, \ldots, n$ and $\varphi(x) = 0$ if $|x_j| \geq 3/4$ for some j. Now put

$$\varphi_j(x) = \varphi(x - g_j)\Big(\sum_{k=0}^{\infty} \varphi(x - g_k)^2\Big)^{-1/2}.$$

Then $\varphi_j(x) \in C_0^\infty(\mathbb{R}^n)$ and

$$\sum_{k=0}^{\infty} \varphi_k^2(x) \equiv 1, \quad \text{but} \quad \sum_{k=0}^{\infty} |D^\alpha \varphi_k(x)|^2 \leq C_\alpha$$

for all α, and $|x - y| \leq C$ if $x \in \operatorname{supp} \psi_k$, $y \in \operatorname{supp} \psi_k$ for any k.

Similarly, put

$$\psi_j(\xi) = \varphi_j(\xi|\xi|^{-1/2}).$$

It is evident that $\psi_j(\xi) \in C_0^\infty(\mathbb{R}^n)$ and $\sum_{k=0}^\infty \psi_k^2(\xi) \equiv 1$, but

$$\sum_{k=0}^\infty |D^\alpha \psi(\xi)|^2 \le C_\alpha |\xi|^{-\alpha}$$

for all α and $|\xi - \eta| \le C|\xi|^{1/2}$ if $\xi \in \operatorname{supp}\psi_k$, $\eta \in \operatorname{supp}\psi_k$ for any $k \ne 0$. If $u \in C_0^\infty(K)$, then

$$\operatorname{Re}\int P(x,D)u(x) \cdot \overline{u(x)}dx$$

$$= \operatorname{Re}\int A(x,D)u(x) \cdot \bar{u}(x)dx + O(\|u\|_{(m-1)/2}^2)$$

$$= \operatorname{Re}\sum_k \int \psi_k(D)A(x,D)u(x) \cdot \overline{\psi_k(D)u(x)}dx + O(\|u\|_{(m-1)/2}^2).$$

Let us consider the integral

$$\int [\psi_k(D)A(x,D) - A(x,D)\psi_k(D)]u(x) \cdot \overline{\psi_k(D)u(x)}dx$$

$$= (2\pi)^{-n}\int [\psi_k(\xi)\widetilde{Au}(\xi) - \widetilde{A\psi_k u}(\xi)]\psi_k(\xi)\overline{\tilde{u}(\xi)}d\xi$$

$$= (2\pi)^{-2n}\int\int [\psi_k(\xi) - \psi_k(\eta)]\tilde{a}(\eta-\xi,\eta)\tilde{u}(\eta)\psi_k(\xi)\overline{\tilde{u}(\xi)}d\xi d\eta,$$

where

$$\tilde{a}(\xi,\eta) = \int a(x,\eta)e^{-ix\xi}dx.$$

We have the inequality

$$|\tilde{a}(\eta-\xi,\eta)| \le C_N(1+|\eta-\xi|)^{-N}(1+|\eta|)^m$$

$$\le C_N'(1+|\eta-\xi|)^{-N+|m-1|/2}(1+|\eta|)^{(m+1)/2}(1+|\xi|)^{(m-1)/2},$$

for any N, since

$$(1+|\eta|) \le (1+|\xi|)(1+|\eta-\xi|) \quad \text{and} \quad (1+|\xi|) \le (1+|\eta|)(1+|\eta-\xi|).$$

If $|\xi - \eta| > |\eta|/2$, then

$$\sum |\psi_k(\xi) - \psi_k(\eta)|^2$$

$$\le \sum(|\psi_k(\xi)|^2 + 2|\psi_k(\xi)\psi_k(\eta)| + |\psi_k(\eta)|^2) \le 4 \le 8|\xi-\eta||\eta|^{-1}.$$

If $|\xi - \eta| < |\eta|/2$, then

$$\sum_k |\psi_k(\xi) - \psi_k(\eta)|^2 = \sum_k |\sum_{i=1}^n D_i \psi_k(\theta_k)(\xi_i - \eta_i)|^2$$

$$\leq C_1(1 + |\eta|)^{-1}|\xi - \eta)|^2.$$

Therefore, for $|\eta| \geq 1$, the estimate

$$\sum |\psi_k(\xi) - \psi_k(\eta)|^2 \leq C_2(1 + |\eta|)^{-1}(1 + |\xi - \eta)|)^2$$

is valid. Note that

$$\sum_k [\psi_k(\xi) - \psi_k(\eta)]\psi_k(\xi) = 1 - \sum_k \psi_k(\xi) \cdot \psi_k(\eta)$$

$$= 1/2 \sum [\psi_k(\xi) - \psi_k(\eta)]^2.$$

Thus,

$$\sum_k |\int [\psi_k(D)A(x, D) - A(x, D)\psi_k(D)]u(x) \cdot \overline{\psi_k(D)u(x)}]dx|$$

$$\leq C_N \int \int (1 + |\eta - \xi|)^{-N+|m-1|/2+1}(1 + |\eta|)^{(m-1)/2}$$

$$(1 + |\xi|)^{(m-1)/2}|\tilde{u}(\xi)| \cdot |\tilde{u}(\eta)|d\xi d\eta \leq C_3\|u\|_{(m-1)/2}^2,$$

if $N > |m - 1|/2 + 1 + n$. The latter inequality is true by virtue of Lemma 10 of 2.2.3. Therefore,

$$\mathrm{Re} \int P(x, D)u(x) \cdot \overline{u}(x)dx$$

$$= \mathrm{Re} \sum_j \int A(x, D)\psi_j(D)u(x) \cdot \overline{\psi_j(D)u(x)}dx + O(\|u\|_{(m-1)/2}^2).$$

It is clear that the equality is true also after replacing the operator A by the operator A_0, whose symbol a_0 coincides with a for $|\xi| \geq C$ and is non-negative everywhere, since $a - a_0 \in S^{-\infty}$.

d. Let ξ^j be a point of supp ψ_j. Note that if

$$r_j(x, \xi) = a_0(x, \xi) - a_0(x, \xi^j) - \sum_{k=1}^n \frac{\partial a_0(x, \xi^j)}{\partial \xi_k}(\xi_k - \xi_k^j),$$

then

$$|D_x^\beta r_j(x, \xi)| \leq C_0(1 + |\xi|)^{m-1}$$

for all β and $\xi \in$ supp ψ_j.

Hence for all N and $\xi, \eta \in \operatorname{supp} \psi_j$ we have

$$|\tilde{r}_j(\eta - \xi, \xi)| = (2\pi)^{-n}\Big| \int r_j(x, \xi)e^{-ix(\eta-\xi)}dx\Big|$$

$$\leq C_N'(1 + |\eta - \xi|)^{-N}(1 + |\xi|)^{m-1}$$

$$\leq C_N''(1 + |\eta - \xi|)^{-N+|m-1|/2}(1 + |\eta|)^{(m-1)/2}(1 + |\xi|)^{(m-1)/2}.$$

Let R_j be the operator with the symbol r_j. Then

$$\sum_j \int R_j(x, D)\psi_j(D)u(x) \cdot \overline{\psi_j(D)u(x)}dx$$

$$\leq \sum_j \int\int |\tilde{r}_j(\eta - \xi, \xi)\psi_j(\xi)\tilde{u}(\xi) \cdot \overline{\psi_j(\eta)\tilde{u}(\eta)}|d\xi d\eta$$

$$\leq C_N \int\int (1 + |\eta - \xi|)^{-N+|m-1|/2}(1 + |\eta|)^{(m-1)/2}|\tilde{u}(\eta)|$$

$$(1 + |\xi|)^{(m-1)/2}|\tilde{u}(\xi)|d\xi d\eta \leq C_4\|u\|_{(m-1)/2}^2,$$

if N is chosen so large that $N > |m - 1|/2 + n$.

Therefore,

$$\operatorname{Re} \int P(x, D)u(x) \cdot \overline{u(x)}dx$$

$$= \operatorname{Re} \sum_j \int a_0(x, \xi^j)\psi_j(D)u(x) \cdot \overline{\psi_j(D)u(x)}dx$$

$$+ \operatorname{Re} \sum_j \sum_{i=1}^n \int \frac{\partial a_0(x, \xi^j)}{\partial \xi_i}(D_i - \xi_i^j)\psi_j(D)u(x) \cdot \overline{\psi_j(D)u(x)}dx$$

$$+ O(\|u\|_{(m-1)/2}^2).$$

By Lemma 25 we have for $i = 1, \ldots, n$

$$\Big|\frac{\partial a_0(x, \xi^j)}{\partial \xi_i}\Big|^2 \leq C_5 a_0(x, \xi^j)(1 + |\xi^j|)^{m-2}.$$

Thus,

$$\int \frac{\partial a_0(x, \xi^j)}{\partial \xi_i}(D_i - \xi_i^j)\psi_j(D)u(x) \cdot \overline{\psi_j(D)u(x)}dx$$

$$\leq C_5(1 + |\xi^j|)^{m/2-1}\Big(\int a_0(x, \xi^j)|\psi_j(D)u(x)|^2dx\Big)^{1/2}$$

$$\times \Big(\int (\xi_i - \xi_i^j)^2|\psi_j(\xi)\tilde{u}(\xi)|^2d\xi\Big)^{1/2}$$

$$\leq \frac{1}{2n} \int a_0(x, \xi^j)|\psi_j(D)u(x)|^2dx + C_6(1 + |\xi^j|)^{m-1}\int |\psi_j(\xi)\tilde{u}(\xi)|^2d\xi.$$

Since for $\xi \in \operatorname{supp}\psi_j$ for $j \geq 1$ the inequalities

$$1 + |\xi| \leq 2(1 + |\xi^j|) \leq 4(1 + |\xi|)$$

are true, we have

$$\sum_j (1 + |\xi^j|)^{m-1} \int |\psi_j(\xi)\tilde{u}(\xi)|^2 d\xi$$

$$\leq C_7 \sum_j \int (1 + |\xi|^2)^{(m-1)/2} |\psi_j(\xi)\tilde{u}(\xi)|^2 d\xi = C_7 \|u\|^2_{(m-1)/2}.$$

Thus,

$$\operatorname{Re} \int P(x, D)u(x) \cdot \overline{u}(x) dx$$

$$\geq \operatorname{Re} \sum_j \int a_0(x, \xi^j)|\psi_j(D)u(x)|^2 dx$$

$$- 1/2 \operatorname{Re} \sum_j \int a_0(x, \xi^j)|\psi_j(D)u(x)|^2 dx$$

$$\geq -C_8 \|u\|^2_{(m-1)/2}. \qquad \square$$

2.5.3 Some generalizations

A. Melin has established a stronger version of the Gårding inequality, using the positivity of the subprincipal symbol in characteristic points.

Theorem 26.[1] *Let the principal symbol $p_0(x, \xi)$ of a classical pseudo-differential operator $P(x, D)$ of order m have real non-negative values, and*

$$\operatorname{Re} p_1(x, \xi) + \sum_{j=1}^{n} [(\frac{\partial^2 p_0}{\partial x_j^2})_+ + (\frac{\partial^2 p_0}{\partial \xi_j^2})_+] \geq 0,$$

where $R_+ = \max(R, 0)$, at the characteristic set

$$\Sigma = \{(x, \xi);\ p_0(x, \xi) = 0\}.$$

If Σ is a C^∞ manifold, then for any compact subset $K \subset X$ one can find a constant C_K such that

$$\operatorname{Re}(Pu, u) \geq -C_K \|u\|^2_{m/2-1}, \quad u \in C_0^\infty(K).$$

Another generalization of the Gårding inequality has been proved by Ch. Fefferman and D.H. Phong.

---

[1]This is a simplified version of Melin's theorem.

Theorem 27. *Let $p \in S^2(X)$, $p(x,\xi) \geq 0$. Then for any compact subset $K \subset X$ one can find a constant C_K such that*

$$\mathrm{Re}(Pu, u) \geq -C_K \|u\|_0^2, \quad u \in C_0^\infty(K).$$

They also proposed the following conjecture, which has been proved for many specific cases.

Fefferman-Phong conjecture. *Let p be a real symbol from $S^2(X)$ and*

$$\mu = inf_{\chi \in \varphi} \int \int_{\chi(Q_0)} p(x,\xi) dx d\xi,$$

where Q_0 is the unit cube of $\mathbb{R}^n$, and φ is the set of canonical transformations of $T^*(x)$ (see 6.1.1. in below). Then

$$\mathrm{Re}(Pu, u) \geq \mu \|u\|_0^2, \quad u \in C_0^\infty(K).$$

Chapter 3

Elliptic pseudo-differential operators

3.1 Parametrices of the elliptic operators

3.1.1 Definitions and a technical lemma

A pseudo-differential operator P with a symbol $p \in S^m(\Omega)$ is *elliptic* in Ω, if for any compact subset $K \subset \Omega$ there exist positive constants c and C such that

$$|p(x,\xi)| \geq c|\xi|^m \text{ for } |\xi| \geq C.$$

Let $\Omega' \subset\subset \Omega, h \in C_0^\infty(\Omega)$ and $h(x) = 1$ in a neighbourhood of the domain Ω'. A pseudo-differential operator Q is a *right parametrix* for P in Ω', if

$$PhQu = u + Tu, u \in C_0^\infty(\Omega'),$$

and the operator T is smoothing (i.e. has an order $-\infty$). Similarly, a pseudo-differential operator Q' is a *left parametrix* for P in Ω', if

$$Q'hPu = u + T'u, u \in C_0^\infty(\Omega'),$$

and the operator T' is smoothing.

From Theorem 16 of 2.3.2 it follows that

$$p(x,\xi)q(x,\xi) - 1 \in S^{-1}(\Omega'),$$

if q is the symbol of the right parametrix. Therefore, $q \in S^{-m}(\Omega')$ and

$$|p(x,\xi)q(x,\xi)| \geq 1/2$$

for sufficiently large $|\xi|$. Thus, from the existence of a parametrix it follows that P is an elliptic operator. Moreover, the operator Q is also elliptic.

The following lemma is very useful. It allows an infinite series of symbols of decreasing orders to be summarized.

Lemma 1. *Let $a_j \in S^{m-j}(\Omega)$ for all non-negative integers j. Then there exists a symbol $a \in S^m(\Omega)$, such that for any $N \geq 1$,*

$$a - \sum_{j=0}^{N} a_j \in S^{m-N-1}(\Omega).$$

The symbol a is unique modulo smoothing operator.

Proof. Let $h \in C^\infty(\mathbb{R}^n)$ so that $h(\xi) = 0$ for $|\xi| \leq 1$ and $h(\xi) = 1$ for $|\xi| \geq 2$. Let Ω' be an open set containing the closure of Ω. We use h to cut away the support near $\xi = 0$. Let $\{t_j\}$ be a decreasing sequence of positive numbers such that $t_j \to 0$ and define

$$a(x,\xi) = \sum_{j=0}^{\infty} h(\xi t_j) a_j(x,\xi).$$

For any fixed ξ, $h(\xi t_j) = 0$ for all but a finite number of j, so this sum is well-defined and smooth at (x,ξ). For $j > 0$ we have

$$|a_j(x,\xi)| \leq C_j(1 + |\xi|)^{m-j} = C_j(1 + |\xi|)^m(1 + |\xi|)^{-j}.$$

If $|\xi|$ is large enough, $(1 + |\xi|)^{-j}$ is as small as we like and therefore, by passing to a subsequence of t_j, we can assume that

$$|h(\xi t_j) a_j(x,\xi)| \leq 2^{-j}(1 + |\xi|)^m \text{ for } j > 0.$$

This implies that

$$|a(x,\xi)| \leq (C_0 + 1)(1 + |\xi|)^m.$$

We can use a similar argument for the derivatives, and use a diagonalization argument on the resulting subsequences to conclude that $a \in S^m$. The supports of all a_j are contained compactly in Ω, so the support of a is contained in $\overline{\Omega}$ which is contained in Ω'.

Now we can apply exactly the same arguments to the sum $a_1 + \ldots$ to show that it belongs to S^{m-1}. We continue in this fashion and use a diagonalization argument on the resulting subsequences to conclude finally that

$$\sum_{j=N+1}^{\infty} h(\xi t_j) a_j(x,\xi) \in S^{m-N-1}.$$

Since $a_j - h(\xi t_j) a_j \in S^{-\infty}$, this implies that

$$a - \sum_{j=0}^{N} a_j \in S^{m-N-1},$$

and completes the proof. $\square$

Remark 2. Lemma 1 can be extended to the case of $a_j \in S^{m_j}(\Omega)$, $m_{j+1} < m_j, m_j \to -\infty$ for $j \to \infty, j \in \mathbb{Z}_+$. Then there is a symbol $a \in S^m(\Omega)$ with

$$a - \sum_{j=0}^{N} a_j \in S^{m_{N+1}}(\Omega)$$

for all N, and a is unique modulo a smoothing operator.

3.1.2 The construction of a parametrix

Theorem 3. *If a pseudo-differential operator P is elliptic in Ω, then in each domain Ω' whose closure is compact in Ω, there exists an operator Q which is a left and a right parametrix for P in Ω'.*

Proof. We first construct the right parametrix. From the definition of the ellipticity it follows that

$$|p(x,\xi)| \geq c(1 + |\xi|)^m \text{ for } |\xi| \geq C.$$

Let $\chi \in C_0^\infty(\mathbb{R}^n), \chi(\xi) = 0$ for $|\xi| \leq C$ and $\chi(\xi) = 1$ for $|\xi| \geq C+1$. Let $h \in C_0^\infty(\Omega)$ and $h(x) = 1$ in a neighbourhood of the compact Ω'.

The order of the operator Q_0 with the symbol

$$q_0(x,\xi) = \chi(\xi)p^{-1}(x,\xi)$$

is equal to $-m$, and the order of the operator

$$R_1 = Ph(x)Q_0 - h(x)I$$

with the symbol $r_1(x,\xi)$ is equal to -1. Let

$$q_1(x,\xi) = -r_1(x,\xi)\chi(\xi)p^{-1}(x,\xi).$$

The operator Q_1 with this symbol has an order $-m - 1$, and the order of the operator

$$R_2 = Ph(x)(Q_0 + Q_1) - h(x)I$$

is equal to -2, since

$$R_2 = h(x)I + R_1 + PhQ_1 - h(x)I = R_1 + PhQ_1.$$

Continuing this process, we get the sequence of the operators $Q_0, Q_1, \ldots$, in which the order of Q_j is equal to $-m - j$, and the order of the operator

$$R_j = Ph(x)(Q_0 + Q_1 + \ldots + Q_{j-1}) - h(x)I$$

is $-j$. Using Lemma 1 one can construct an operator Q of order $-m$, such that the order of the operator

$$Q - (Q_0 + Q_1 + \ldots + Q_{j-1})$$

is equal to $-m - j$ for any j. Moreover, the order of the operator

$$R = Ph(x)Q - h(x)I$$

is $-j$ for any j, i.e.this operator is smoothing.

The left parametrix is constructed in the same way. The order of the operator Q'_0 with the symbol $\chi(\xi)p^{-1}(x, \xi)$ is equal to $-m$, and the order of the operator

$$R'_1 = Q'_0 h(x)P - h(x)I$$

with a symbol $r'_1(x, \xi)$ is equal to -1. Let

$$q'_1(x, \xi) = -r'_1(x, \xi)\chi(\xi)p^{-1}(x, \xi).$$

The operator Q'_1 with this symbol has the order $-m - 1$, and the order of the operator

$$R'_2 = (Q'_0 + Q'_1)h(x)P - h(x)I$$

is equal to -2, since

$$R'_2 = h(x)I + R'_1 + Q'_1 h(x)P - h(x)I = R'_1 + Q'_1 h(x)P.$$

Continuing this process, we get the sequence of the operators $Q'_0, Q'_1, \ldots$, in which the order of Q'_j is equal to $-m - j$, and the order of the operator

$$R'_j = (Q'_0 + Q'_1 + \ldots + Q'_{j-1})h(x)P - h(x)I$$

is $-j$. Using the lemma one can construct an operator Q' of order $-m$ such that the order of the operator

$$Q' - (Q'_0 + Q'_1 + \ldots + Q'_{j-1})$$

is equal to $-m - j$ for any j. Moreover, the order of the operator

$$R' = Q'h(x)P - h(x)I$$

is $-j$ for any j, i.e., this operator is smoothing.

Let us show that the operator $(Q - Q')h$ is smoothing and we can pose $Q = Q'$. We have

$$Q'hPhQ = Q'h(hI + R) = Q'h^2 + T,$$

$$Q'hPhQ = (hI + R')hQ = h^2Q' + T',$$

where the operators T and T' are smoothing. Hence we see that the operator $Q'h^2 - h^2Q$ is smoothing, i.e. $q' - q \in S^{-\infty}(\Omega')$.　　　　$\square$

3.2 Elliptic operators on a manifold

3.2.1 Definitions

Let Ω be a smooth compact manifold without boundary. A pseudo-differential operator P is *elliptic* on Ω, if it is elliptic in each coordinate neighbourhood. This means that for any coordinate neighbourhood U and for a coordinate map $f : U \to \omega \subset \mathbb{R}^n$, the operator

$$f_* P f^* : C_0^\infty(\omega) \to C^\infty(\omega)$$

is elliptic.

3.2.2 The parametrix construction

Let $\sum h_j(x) = 1$ be a smooth partition of the unity on Ω and $\operatorname{supp} h_j$ lie in a coordinate neighbourhood for each j. Let g_j be smooth functions on Ω such that $\operatorname{supp} g_j$ lies in a coordinate neighbourhood and $g_j = 1$ in $\operatorname{supp} h_j$. The operator P can be represented in the form $\sum h_j P g_j + T$, where T is a smoothing operator, since the operators $h_j P(1 - g_j)$ are smoothing.

Theorem 3 allows us to construct an operator Q_k such that $Q_k g_k P = g_k I + T_k$, where T_k is a smoothing operator. Now set $Q = \sum h_k Q_k g_k$. Note that the order of the operator

$$g_k Q_k P - g_k I = (g_k Q_k - Q_k g_k)P + T_k$$

is equal to -1. The order of the operator

$$T_{kj} = Q_k(g_k h_j)P - g_k h_j I$$

is equal to -1, too, and therefore

$$QPu = \sum_{j,k} h_k Q_k g_k h_j P g_j u + QTu$$

$$= \sum_{j,k} h_k(g_k h_j I + T_{kj})g_j u + QTu = u + T_0 u,$$

where T_0 is an operator of order -1.

Analogously, if Q'_k is an operator such that $P h_k Q'_k = h_k I + T'_k$ and T'_k is a smoothing operator, then setting $Q' = \sum h_k Q'_k g_k$, we get

$$PQ'u = \sum_{j,k} h_j P g_j h_k Q'_k g_k u + TQ'u$$

$$= \sum_{j,k} h_j(g_j h_k I + T'_{kj})g_k u + TQ'u = u + T'_0 u,$$

where T'_0 is an operator of order -1.

As was shown in Theorem 3, we can pose $Q_k = Q'_k$ so that $Q = Q'$. Using the constructed parametrix we can easily investigate the principal properties of elliptic operators on a manifold.

3.2.3 A priori estimates and regularity of solutions

Let P be an elliptic pseudo-differential operator of order m on a smooth compact manifold X.

Theorem 4. *For any real number s there exists a constant $C = C(s)$ such that*

$$\|u\|_s \leq C(\|Pu\|_{s-m} + \|u\|_{s-1}), u \in C^\infty(X).$$

Proof. Let Q be a left parametrix for P, i.e.

$$QPu = u + Tu,$$

where the operator T has an order -1. Then we have

$$\|u\|_s \leq \|QPu\|_s + \|Tu\|_s \leq C(\|Pu\|_{s-m} + \|u\|_{s-1}),$$

since the order of Q is $-m$. $\square$

The following theorem shows that the inverse statement is also true, i.e. from the a priori estimate it follows the ellipticity.

Theorem 5. *For some real s let the estimate*

$$\|u\|_s \leq C(\|Pu\|_{s-m} + \|u\|_{s-1})$$

be true for all functions u from $C^\infty(X)$. Then P is an elliptic operator of order m.

Proof. Let us fix a point $x_0 \in X$ and a coordinate neighbourhood ω of this point. Let $h \in C_0^\infty(\omega)$. Substitute the function

$$u(x) = h(x)e^{i\lambda x \xi_0},$$

where λ is a positive real and $\xi_0 \in \mathbb{R}^n, |\xi_0| = 1$, in the a priori estimate. Let $\psi \in C_0^\infty(\omega)$ and $\psi(x) = 1$ in a neighbourhood of the support of h. Then

$$\tilde{u}(\eta) = \tilde{h}(\eta - \lambda \xi_0),$$

and as $\lambda \to \infty$ we have

$$(2\pi)^n \|u\|_s^2 = \int (1 + |\eta|^2)^s |\tilde{u}(\eta)|^2 d\eta$$

$$= \int (1 + |\eta + \lambda\xi_0|^2)^s |\tilde{h}(\eta)|^2 d\eta = \lambda^{2s}\|h\|_0^2 + O(\lambda^{2s-2}).$$

The operator $(1 - \psi(x))Ph$ is smoothing by virtue of Corollary 19 of 2.3.2, so its distributional kernel is infinitely differentiable. Therefore,

$$\psi(x)P(x,D)h(x)v = Q(x,D)v$$

for all $v \in C_0^\infty(\omega)$ and $q - p \in S^{m-1}(\omega)$, if q is the symbol of $Q(x,D)$. Note that

$$\psi(x)P(x,D)u = Q(x,D)[e^{i\lambda x \xi_0}] = q(x,\lambda\xi_0)e^{i\lambda x \xi_0}$$

and

$$\|\psi Pu\|_{s-m}^2 = (2\pi)^{-n} \int (1+|\eta|^2)^{s-m}|\tilde{q}(\eta - \lambda\xi_0, \lambda\xi_0)|^2 d\eta$$

$$= (2\pi)^{-n} \int (1+|\eta + \lambda\xi_0|^2)^{s-m}|\tilde{q}(\eta, \lambda\xi_0)|^2 d\eta$$

$$= \lambda^{2(s-m)} \int |q(x,\lambda\xi_0)|^2 dx + O(\lambda^{2(s-m)-2}).$$

Since

$$\|u\|_{s-1}^2 \leq \int (1+|\eta + \lambda\xi_0|^2)^{s-1}|\tilde{h}(\eta)|^2 d\eta = O(\lambda^{2s-2}),$$

we get the inequality

$$\lambda^{2s}\|h\|_0^2 \leq C\lambda^{2s-2m}\|p(x,\lambda\xi_0)h(x)\|_0^2 + C_1\lambda^{2s-2}.$$

The constant C_1 depends on h, but C does not. Thus we obtain that as $\lambda \to \infty$,

$$\varliminf_{\lambda\to\infty}|p(x_0,\lambda\xi_0)|\lambda^{-m} \geq 1/C,$$

and since it is true uniformly for all $\xi_0 \in S^{m-1}$, we get the inequality

$$|p(x_0,\xi)| \geq |\xi|^m/C \text{ for } |\xi| \geq C_0,$$

proving the ellipticity of P. $\qquad\square$

Theorem 6. *If $u \in \mathcal{D}'(X), Pu = f \in H^s(X)$ for some real s, then $u \in H^{s+m}(X)$.*

Proof. Let Q be a left parametrix of P, i.e.

$$QPu = u + Tu,$$

where T is an operator of order -1. Since X is compact, there exists a $t \in \mathbb{R}$ such that $u \in H^t(X)$ (see Theorem 21 of Chapter 1). Since $u = QPu - Tu$, we see that $u \in H^{t_1}(X)$, where $t_1 = \min(s+m, t+1)$. If $t+1 < s+m$, then from the same relation it follows that $u \in H^{t_2}(X)$, where $t_2 = \min(s+m, t_1+1)$ and so on. $\qquad\square$

Theorem 7. *Let $u \in \mathcal{D}'(X), Pu = f \in H^s(X^{(1)})$ for some real s, where $X^{(1)}$ is an open subset of X. Then $u \in H_{\mathrm{loc}}^{s+m}(X^{(1)})$.*

Proof. Let Q be a left parametrix of P, and $h \in C_0^\infty(X^{(1)})$. Then as above

$$QPhu = hu + Thu,$$

where T is an operator of order -1, and therefore

$$QhPu = hu + Thu + Q[h, P]u.$$

Let t be a real number such that $u \in H^t(X)$. Then $hu \in H^t(X)$ and $[h, P]u \in H^{t-m+1}(X)$. Therefore, $Q[h, P]u \in H^{t+1}(X)$ and $hu \in H^{t_1}(X)$, where $t_1 = \min(s + m, t + 1)$. We can assume that $h = 1$ in $X^{(2)}$, where $X^{(2)} \subset X^{(1)}$. Let $h_1 \in C_0^\infty(X^{(2)})$. Then

$$Qh_1Pu = h_1 + Th_1u + Q[h_1, P]u$$

and $h_1u \in H^{t_1}(X), [h_1, P]u = [h_1, P]hu \in H^{t_1-m+1}(X)$. Thus $h_1u \in H^{t_2}(X)$, where $t_2 = \min(s + m, t_2 + 1)$. Repeating the argument, we can prove that $gu \in H^{s+m}(X^{(N)})$ for all functions $g \in C_0^\infty(X^{(N)})$, where $N \leq s + m - t$. Since for any open subset $Y \subset X$ one can construct a sequence of domains $X^{(1)}, \ldots, X^{(N)}$ such that $X^{(i)} \subset X^{(i-1)}$ and $X^{(N)} = Y$, we see that $u \in H^{s+m}(Y)$, and therefore $u \in H_{\mathrm{loc}}^{s+m}(X)$. $\qquad\square$

3.2.4   The Fredholm property

Definition 8. Let H_1 and H_2 be Hilbert spaces, $A \in \mathcal{L}(H_1, H_2)$. The operator A is *Fredholm* if it has the following properties:
1. *Its range*
$$\mathrm{im}\, A \equiv \{u \in H_2 : u = Av \ for \ some \ v \in H_1\}$$

is closed in H_2 and has a finite codimension.
2. *Its kernel*
$$\ker A \equiv \{u \in H_1 : Au = 0\}$$

has a finite dimension.

Definition 9. *The index of a Fredholm operator A is an integer*

$$\mathrm{ind}\, A \equiv \dim \ker A - \mathrm{codim}\, \mathrm{im}\, A.$$

Example 10. Let A be the operator in the space l_2, moving a vector $x = (x_1, \ldots, x_n, \ldots)$ in the vector $(0, x_1, \ldots, x_n, \ldots)$. It is clear that $\ker A = \{0\}$ and $\mathrm{im}\, A$ is the set of vectors $(y_1, \ldots, y_n, \ldots)$ from l_2, having $y_1 = 0$. Thus $\mathrm{ind}\, A = -1$. It is obvious that for all $k \in \mathbf{N}$

$$\mathrm{ind}\, A^k = -k, \quad \mathrm{ind}\, A^{*k} = k,$$

if A^* is the adjoint operator, i.e.

$$A^*(y_1, \ldots, y_n, \ldots) = (y_2, \ldots, y_n, \ldots).$$

Let A be a Fredholm operator. If S is a closed complement to $\ker A$ in H_1, then the restriction of A on S gives an 1-1 correspondence between S and $\operatorname{im} A$, so that this restriction has an inverse operator B, which is Fredholm also, if we set $B = 0$ on the complement to $\operatorname{im} A$ in H_2.

The above enables the following properties of A to be proved easily:

1. $\operatorname{codim} \operatorname{im} A = \dim \ker A^*$;

2. *If $T \in \mathcal{L}(H_1, H_2)$ and $\dim \operatorname{im} T$ is finite, then the operator $I + T$ is Fredholm and $\operatorname{ind}(I + T) = 0$; more generally, if K is a compact operator, then $I + K$ is Fredholm and $\operatorname{ind}(I + K) = 0$;*

3. *If P_1 is the projection on $\ker A$ and $I - P_2$ is the projection on $\operatorname{im} A$, then there exists a Fredholm operator B such that*

$$BA = I - P_1, \quad AB = I - P_2;$$

4. *If operators A and B are Fredholm, then BA is Fredholm, too, and*

$$\operatorname{ind} AB = \operatorname{ind} BA = \operatorname{ind} A + \operatorname{ind} B;$$

5. *The set of Fredholm operators in $\mathcal{L}(H_1, H_2)$ is open and the function ind has a constant value on each connected component of this set;*

6. *If $A \in \mathcal{L}(H_1, H_2)$ and there exist operators B_1 and B_2 in $\mathcal{L}(H_2, H_1)$ such that*

$$B_1 A = I + R_1, \quad AB_2 = I + R_2,$$

and R_j is compact in $H_j, j = 1, 2$, then A is Fredholm;

7. *If A is Fredholm and K is compact in $\mathcal{L}(H_1, H_2)$, then $A + K$ is Fredholm and*

$$\operatorname{ind}(A + K) = \operatorname{ind} A.$$

The reader can prove the above properties 1–7 as exercises. The proofs can be found, for example, in the books of [H1], [P], [Sh].

Let $P(x, D)$ be an elliptic pseudo-differential operator of order m on a closed smooth manifold X.

Theorem 11. *The kernel of the operator $P : H^s(X) \to H^{s-m}(X)$ has a finite dimension and is independent of s.*

Proof. Theorem 6 implies that each solution of the homogeneous equation $Pu = 0$ from the space $H^s(X)$ with a real s is a function from the space $C^\infty(X)$. Thus the kernel of the operator P is independent of s. It is obvious that $\ker P$ is a closed subspace in $H^s(X)$, so the space $H^{s-m}(X) \cap \ker P$ is Hilbert. But the H^s-norms of the elements of the unit sphere in this space are uniformly bounded since

$$\|u\|_s \le C\|u\|_{s-1} \le C$$

by virtue of Theorem 4. Thus the unit sphere is compact, which is only possible in a finite-dimensional space. $\square$

Theorem 12. *The range of the operator $P : H^s(X) \to H^{s-m}(X)$ is the orthogonal complement to the kernel of the adjoint operator P^*, and it has a finite codimension which is independent of s.*

Proof. If v is orthogonal to $\operatorname{im} P$, i.e. $(Pu, v) = 0$ for all $u \in H^s(X)$, then $(u, P^*v) = 0$ and therefore $P^*v = 0$. The operator P^* is an elliptic operator of order m. Thus, as above, $v \in C^\infty(X)$, $\dim \ker P^*$ is finite and independent of s. It is obvious also that if $v \in \ker P^*$, then $(Pu, v) = 0$ for all $u \in H^s(X)$, i.e. $v \in (\operatorname{im} P)^\perp$. Therefore, $(\operatorname{im} P)^\perp = \ker P^*$.

Let us prove that for $u \in H^s(X) \cap (\ker P)^\perp$ the estimate

$$\|u\|_s \le C\|Pu\|_{s-m}$$

is valid. Indeed, if it is not true, then there exists a sequence $\{u_k\}$ of elements of $H^s(X) \cap (\ker P)^\perp$ such that

$$\|u_k\|_s = 1 \text{ and } \|Pu_k\|_{s-m} \le 1/k.$$

The sequence $\{u_k\}$ is compact in $H^{s-1}(X)$ and weakly compact in $H^s(X)$. So we can choose a subsequence $\{u'_k\}$, converging in $H^{s-1}(X)$ and weakly in $H^s(X)$. By Theorem 5 of 1.3.1 we have

$$\|u'_k - u'_l\|_s \le C(\|Pu'_k - Pu'_l\|_{s-m} + \|u'_k - u'_l\|_{s-1}).$$

Since $\|Pu'_k\|_{s-m} \to 0$, the sequence $\{u'_k\}$ is converging in $H^s(X)$ to an element $v \in H^s(X)$. However v must be orthogonal to $\ker P$, which contradicts the relations

$$\|v\|_s = 1, \quad Pv = 0.$$

Therefore, the above-mentioned a priori estimate is true.

Now let $Pu_k \to v$ in $H^s(X)$. Without loss of generality we can assume that $u_k \perp \ker P$ in $H^s(X)$. From the stated estimate it follows that

$$\|u_k - u_l\|_s \le C\|Pu_k - Pu_l\|_{s-m},$$

and thus the sequence $\{u_k\}$ is converging in $H^s(X)$ to an element w. But then $Pu_k \to Pw$ in $H^{s-m}(X)$, so that $Pw = v$, i.e. $v \in \operatorname{im} P$ and the range of P is closed. $\square$

Corollary 13. *The index of an elliptic pseudo-differential operator $P : H^s(X) \to H^{s-m}(X)$ on a closed smooth manifold X is independent of s.*

Theorem 14. *The elliptic pseudo-differential equation $Pu = f$ on a closed smooth manifold X with $f \in H^s(X)$ has a solution $u \in H_{s+m}(X)$ iff $\int f\overline{\varphi}d\mu = 0$ ($d\mu$ being the measure on X associated with a fixed Riemannian metric on X), for all functions φ from $\ker P^*$, where P^* is the operator adjoint to P with respect to the measure $d\mu$. If this condition is satisfied there exists a solution such that the following a priori estimate*

$$\|u\|_{s+m} \leq C\|Pu\|_s$$

holds with a constant C independent of f.

Proof. If $Pu = f, u \in H^{s+m}(X)$, then

$$(f, \varphi) = (Pu, \varphi) = (u, P^*\varphi), \quad \varphi \in H^{-s}(X),$$

and (f, φ) must vanish if $P^*\varphi = 0$.

On the other hand, let T be the linear subspace in $H^{-s}(X)$ of functions ψ, orthogonal to $\ker P^*$ with the norm $\|\psi\|_T = \|P^*\psi\|_{-s-m}$. As was shown in the proof of Theorem 12, this norm is equivalent to the norm $\|\psi\|_{-s}$. The linear functional $l(\psi) = (\psi, f)$ is continuous on T, if $f \in H^s(X)$, since

$$|l(\psi)| \leq \|f\|_s\|\psi\|_{-s} \leq C\|f\|_s\|\psi\|_T.$$

One can consider l also as a linear continuous functional on the subspace of all functions of the form $\{P^*\psi\}$ in $H^{-s-m}(X)$. By the Hahn-Banach theorem, this functional can be extended to a linear functional over $H^{-s-m}(X)$ with the same norm. Thus there exists an element $u \in H^{s+m}(X)$ such that $l(\psi) = (P^*\psi, u)$ for all $\psi \in T$. Since $(f, \psi) = 0$ for all $\psi \in \ker P^*$, we see that $(f, \varphi) = (u, P^*\varphi)$ for all $\varphi \in H^{-s}(X)$. Therefore, $Pu = f$ in $H^s(X)$ and

$$\|u\|_{s+m} \leq C\|f\|_s. \qquad \square$$

For completeness, let us also mention the following theorem.

Theorem 15. *Let P be a pseudo-differential operator of order m on X. If $P : H^s(X) \to H^{s-m}(X)$ is a Fredholm operator for a fixed $s \in \mathbb{R}$, then P is elliptic.*

3.2.5 Vanishing of the index

All the above theorems are true for general (non-scalar) elliptic operators, but the following one is specific.

Theorem 16. *If X is a simply-connected closed smooth manifold of dimension $n > 2$ and P is an elliptic pseudo-differential operator on X, then $\operatorname{ind} P = 0$.*

Proof. Let $|p(x, \xi)| \geq c_0|\xi|^m$ for $|\xi| \geq C_1$, where $c_0 > 0$. Let Q be a pseudo-differential operator of order $-m/2$ with the symbol $q(x, \xi) = p(x, \xi)^{-1/2}$ when

$|\xi| \geq C_1$. Since $n \geq 3$, the set $\{\xi : |\xi| \geq C_1\}$ is simply connected and the function q is well-defined for all x from any simply connected coordinate neighbourhood and for ξ with $|\xi| \geq C_1$. Using a partition of unity we can define the operator Q on the whole manifold X.

The order of the operator Q^*QP is equal to 0 and

$$\operatorname{ind} Q^*QP = \operatorname{ind} P + \operatorname{ind} Q^*Q.$$

But Q^*Q is a self-adjoint operator and $\operatorname{ind} Q^*Q = 0$. Hence $\operatorname{ind} P = \operatorname{ind} Q^*QP$. But

$$Q^*QP = I + T,$$

where T is an operator of order -1. Thus

$$\operatorname{ind} P = \operatorname{ind}(I + T) = 0$$

by Proposition 7 of 3.2.4.   □

Remark 17. Note that the condition on the dimension in Theorem 16 is essential. For instance, if $X = S^1$ is the unit circle, for every $k \in \mathbf{Z}$ there exists an elliptic pseudo-differential operator on S^1 of order zero with k as its index.

Remark 18. In the proof of Theorem 16, the fact that the operator in question is scalar was also employed. The assertion is not true in general for elliptic systems (that are $N \times N$-matrices, whose ellipticity means that the determinant of the principal symbols for $\xi \neq 0$ does not vanish).

Note that one can consider pseudo-differential operators acting between distributional sections of vector bundles on X, cf., for instance, [AtS], [P], [RSc]. In Palais' book one can find further concrete examples of the values of indices of elliptic operators on closed C^∞ manifolds.

Exercise 19. Show that for every $m \in \mathbb{R}$ there exists an elliptic pseudo-differential operator P on X of order m with $\operatorname{ind} P = 0$.

Exercise 20. Let P be an elliptic pseudo-differential operator on X of order m with $\operatorname{ind} P = 0$. Show that there exists a finite-dimensional operator T with kernel in $C^\infty(X \times X)$ such that

$$P + T : H^s(X) \to H^{s-m}(X)$$

is an isomorphism for every $s \in \mathbb{R}$.

Chapter 4

Elliptic boundary value problems

4.1 Model elliptic boundary value problems

4.1.1 Statement of the problem and the condition of ellipticity

Let $\mathbb{R}^n_+ = \{x \in \mathbb{R}^n, x_n > 0\}$ and $n \geq 3$. Let

$$P(D) = \sum_{|\alpha|=m} a_\alpha D^\alpha$$

be an elliptic operator of order m with constant complex coefficients, and the differential operator

$$B_j(D) = \sum_{|\alpha|=m_j} b_{j\alpha} D^\alpha$$

have an order $m_j < m$ for $j = 1, \ldots, k$. Consider the following boundary value problem:

$$P(D)u = 0 \text{ for } x_n > 0,$$

$$B_j(D)u = g_j(x) \text{ for } x_n = 0, j = 1, \ldots, k.$$

In order to solve it one can use the Fourier transform with respect to the variables $x' = (x_1, \ldots, x_{n-1})$. Let

$$v(\xi', x_n) = \int_{\mathbb{R}^{n-1}} u(x) e^{-ix'\xi'} \, dx'.$$

Then

$$P(\xi', D_n)v = 0 \text{ for } x_n > 0,$$

$$B_j(\xi', D_n)v = \tilde{g}_j(\xi') \text{ for } x_n = 0, \ j = 1, \ldots, k.$$

The equation $P(\xi', \lambda) = 0$ for λ cannot have real roots if $\xi' \neq 0$, since P is elliptic and

$$|P(\xi)| \geq c_0 |\xi|^m, \ \forall \, \xi \in \mathbb{R}^n, \ c_0 > 0.$$

Let $\lambda_1, \ldots, \lambda_r$ be the different roots of this equation, having the multiplicities $n_1, \ldots, n_r$ and $n_1 + \ldots + n_r = m$. The general solution of the ordinary differential equation $P(\xi', D_n)v = 0$ has the form

$$v(\xi', x_n) = \sum_{j=1}^{r} \sum_{s=0}^{n_j-1} C_{js}(\xi') x_n^s e^{i x_n \lambda_j(\xi')}.$$

It is clear that

$$\lambda_j(t\xi') = t\lambda_j(\xi')$$

for all j and all real $t \neq 0$. If $\mathrm{Im}\,\lambda_j(\xi') < 0$, then the function $e^{i x_n \lambda_j(\xi')}$ is such that

$$\left| e^{i x_n \lambda_j(\xi')} \right| \geq e^{b_0 x_n |\xi'|}, \ b_0 > 0.$$

Such a function cannot be a Fourier transform of a distribution. Therefore, one has to assume that $C_{js}(\xi') = 0$ if $\mathrm{Im}\,\lambda_j(\xi') < 0$. Since $\mathrm{Im}\,\lambda_j(\xi') \neq 0$ for $\xi' \neq 0$, it is sufficient to verify this condition for one value of $\xi' \neq 0$. Let $\mathrm{Im}\,\lambda_j(\xi') > 0$ for $j = 1, \ldots, r_0$ and $\mathrm{Im}\,\lambda_j(\xi') < 0$ for $j = r_0 + 1, \ldots, r$. Substituting the function

$$v(\xi', x_n) = \sum_{j=1}^{r_0} \sum_{s=0}^{n_j-1} C_{js}(\xi') x_n^s e^{i x_n \lambda_j(\xi')},$$

in the boundary conditions

$$B_j(\xi', D_n)v = \tilde{g}_j(\xi') \text{ for } x_n = 0, j = 1, \ldots, k,$$

we obtain a system of k linear equations for the unknown functions $C_{js}(\xi')$, whose number is

$$n_1 + \ldots + n_{r_0}.$$

This system is uniquely solvable if $n_1 + \ldots + n_{r_0} = k$ and the corresponding determinant of coefficients of this system does not vanish. We call the boundary value problem *elliptic* if this condition is fulfilled for all $\xi' \neq 0$. Note that the solutions $v(\xi', x_n)$ are exponentially decreasing as $x_n \to +\infty$, i.e.

$$|v(\xi', x_n)| \leq C e^{-a_0 |\xi'| x_n}$$

with some positive constants C and a_0.

4.1.2 Construction of a parametrix

Let $\mathcal{H}$ be the direct product

$$\mathcal{H} = L_2(\mathbb{R}_+^n) \times H^{m-m_1-1/2}(\mathbb{R}^{n-1}) \times \ldots \times H^{m-m_k-1/2}(\mathbb{R}^{n-1})$$

and $\mathcal{A}$ be the operator

$$\mathcal{A} : H^m(\mathbb{R}_+^n) \to \mathcal{H},$$

such that

$$\mathcal{A}u = (P(D)u, B_1(D)u, \ldots, B_k(D)u).$$

Definition 1. An operator

$$E : \mathcal{H} \to H^m(\mathbb{R}^n_+)$$

is a *right (left) parametrix*, if $\mathcal{A}E = I + T$ $(E\mathcal{A} = I + T_1)$, where

$$T \in \mathcal{L}(\mathcal{H}, \mathcal{H}'), \ (T_1 \in \mathcal{L}(H^m(\mathbf{R}^n_+), H^{m+1}(\mathbb{R}^n_+)));$$

$$\mathcal{H}' = H^1(\mathbb{R}^n_+) \times H^{m-m_1+1/2}(\mathbb{R}^{n-1}) \times \ldots \times H^{m-m_k+1/2}(\mathbb{R}^{n-1}).$$

Theorem 2. *If the boundary value problem is elliptic, then there exists an operator $E \in \mathcal{L}(\mathcal{H}, H^m(\mathbb{R}^n_+))$, which is a right and a left parametrix for $\mathcal{A}$.*

Proof. Consider first the equation

$$P(D)u = f(x) \quad \text{on} \quad \mathbb{R}^n_+.$$

Let $f(x) = 0$ for $x_n < 0$ and put

$$u_1(\xi', x_n) = \frac{1}{2\pi} \int_{\mathbb{R}} e^{ix_n\xi_n} h(\xi) \frac{\tilde{f}(\xi)}{P(\xi)} d\xi_n,$$

where $h \in C^\infty(\mathbb{R}^n)$, $0 \le h(\xi) \le 1$, $h(\xi) = 0$ for $|\xi| \le 1/2$, $h(\xi) = 1$ for $|\xi| > 1$. If $0 \le j \le m$, then

$$D_n^j u_1(\xi', x_n) = \frac{1}{2\pi} \int_{\mathbb{R}} h(\xi)\xi_n^j e^{ix_n\xi_n} \frac{\tilde{f}(\xi)}{P(\xi)} d\xi_n.$$

It is obvious that

$$\int_0^\infty |D_n^j u_1(\xi', x_n)|^2 dx_n \le C \int_{\mathbb{R}} |h(\xi)|^2 |\xi_n|^{2j} |\tilde{f}(\xi)|^2 |\xi|^{-2m} d\xi_n$$

$$\le C \int_{\mathbb{R}} |\xi'|^{2j-2m} |\tilde{f}(\xi)|^2 d\xi_n.$$

Therefore,

$$|\xi'|^{2m-2j} \int_0^\infty |D_n^j u_1(\xi', x_n)|^2 dx_n \le C_1 \int_0^\infty |\tilde{f}(\xi', x_n)|^2 dx_n.$$

Summing these inequalities for $j = 0, 1, \ldots, m$ and integrating them with respect to ξ', we obtain the inequality

$$\|u_1\|_m^2 \le C_2(\|f\|_0^2 + \|u_1\|_0^2).$$

Moreover, the equation

$$P(D)u_1(x) = h(D)f(x) = f(x) + Tf(x) \text{ for } x_n > 0$$

holds, where $T \in \mathcal{L}(L_2(\mathbb{R}^n_+), H^1(\mathbb{R}^n_+))$. Let u_2 be the solution of the equation

$$P(D)u_2 = 0 \text{ for } x_n > 0$$

satisfying the conditions

$$B_j(D)u_2 = g_j(x) - B_j(D)u_1(x) = h_j(x) \text{ for } x_n = 0, j = 1, \ldots, k.$$

It is evident that

$$h_j \in H^{m-m_j-1/2}(\mathbb{R}^{n-1}).$$

If

$$w(\xi', x_n) = \int_{\mathbb{R}^{n-1}} u_2(x)e^{-ix'\xi'} dx',$$

then

$$P(\xi', D_n)w = 0 \text{ for } x_n > 0,$$

$$B_j(\xi', D_n)w = \tilde{h}_j(\xi') \text{ for } x_n = 0, \ j = 1, \ldots, k.$$

By the definition of ellipticity this problem has a solution $v(\xi', x_n)$, decreasing exponentially as $x_n \to +\infty$. First let $|\xi'| = 1$. Then

$$\sum_{j=0}^{m} \int_0^\infty |D_n^j w(\xi', x_n)|^2 dx_n \leq C \sum_{j=1}^{k} |\tilde{h}_j(\xi')|^2$$

$$\leq C_1 \left(\int_0^\infty |\tilde{f}(\xi', x_n)|^2 dx_n + \sum_{j=1}^{k} |\tilde{g}_j(\xi')|^2 \right).$$

If $|\xi'| = \lambda$, we can substitute $\xi' = \lambda\eta'$ and $\lambda x_n = y$, and the function $w_1(\eta', y) = \tilde{u}_1(\xi', x_n) + w(\xi', x_n)$ will be a solution of the following problem:

$$P(\eta', D_y)w_1 = \lambda^{-m}\tilde{f}(\lambda\eta', y\lambda^{-1}) \text{ for } y > 0,$$

$$B_j(\eta', D_y)w_1 = \lambda^{-m}\tilde{g}_j(\lambda\eta') \text{ for } y = 0.$$

Therefore,

$$\sum_{j=0}^{m} \int_0^\infty |D_n^j w_1(\eta', y)|^2 dy$$

$$\leq C_1 \left(\lambda^{-2m} \int_0^\infty |\tilde{f}(\lambda\eta', \lambda^{-1}y)|^2 dy + \sum_{j=1}^{k} \lambda^{-2m_j} |\tilde{g}_j(\lambda\eta')|^2 \right).$$

With the original variables this gives

$$\sum_{j=0}^{m} |\xi'|^{2m-2j} \int_0^\infty |D_n^j \tilde{u}(\xi', x_n)|^2 dx_n$$

$$\leq C_1 \left(\int_0^\infty |\tilde{f}(\xi', x_n)|^2 dx_n + \sum_{j=1}^{k} |\xi'|^{2m-2m_j-1} |\tilde{g}_j(\xi')|^2 \right). \tag{1}$$

Adding the integral $\int_0^\infty |\tilde{u}(\xi', x_n)|^2 dx_n$ to both sides of the inequality and integrating with respect to ξ', we get the estimate

$$\|u\|_m^2 \leq C_2(\|Pu\|_0^2 + \sum_{j=1}^k \|B_j u\|_{m-m_j-1/2}^2 + \|u\|_0^2).$$

Let E be the operator, defined by $E(f, g_1, \ldots, g_k) = u$. The latter estimate shows that $E \in \mathcal{L}(\mathcal{H}, H^m(\mathbb{R}_+^n))$. By the construction,

$$P(D)u = f + Tf, \quad B_j(D)u = g_j(x), \quad j = 1, \ldots, k$$

and $T \in \mathcal{L}(L_2(\mathbb{R}_+^n), H^1(\mathbb{R}_+^n))$. Therefore, E is a right parametrix.

On the other hand, by virtue of our construction, the operator $T_2 = E\mathcal{A} - I$ belongs to $\mathcal{L}(H^m(\mathbb{R}_+^n), H^{m+1}(\mathbb{R}_+^n))$ so that E is also a left parametrix. $\qquad\square$

4.2 Elliptic boundary value problems in a domain

4.2.1 Ellipticity condition

Now let $P(x, D)$ be an elliptic differential operator with smooth (C^∞) coefficients in $\bar{\Omega}$, where Ω is a bounded domain in $\mathbb{R}^n$ with a smooth boundary Γ and $B_j(x, D)$ be a differential operator of order m_j with smooth coefficients, defined on Γ, for $j = 1, \ldots, k$.

Let x_0 be a point on Γ. There exists a smooth transformation of a neighbourhood ω of x_0, moving x_0 in the origin and $\Gamma \bigcap \omega$ in a part of the plane $y_n = 0$, such that all points of $\omega \bigcap \Omega$ have positive values of the coordinate y_n. With the new coordinates consider the boundary value problem:

$$P_0(x_0, D)u = f(y) \quad \text{for} \quad y_n > 0, \tag{2}$$

and

$$B_{j0}(x_0, D)u = g_j(y) \quad \text{for} \quad y_n = 0, \, j = 1, \ldots, k, \tag{3}$$

where $B_{j0}(x, D)$ is the principal homogeneous part of order m_j of the operator $B_j(x, D)$ and $P_0(x, D)$ is the principal homogeneous part of order m of the operator $P(x, D)$. If this problem is elliptic for each point $x_0 \in \Gamma$, then the original problem in Ω is called *elliptic*.

Remark 3. The ellipticity condition is often called the *Lopatinsky* or *Shapiro-Lopatinsky* condition. Since the class of transformations under consideration preserves the principal parts of the symbols of the operators P and B_j up to some non-degenerate transformations in the ξ-space, the ellipticity condition is independent of the transformation used.

Remark 4. If $n \geq 3$ and the considered boundary value problem is elliptic, then the number k is defined uniquely for each connected component of Γ, since then the set $S^*(\Gamma)$, the set of unit vectors cotangent to Γ, is connected and the imaginary parts of the roots of the characteristic equation cannot vanish.

The following example shows that the condition $n \geq 3$ is essential.

4.2.2　Examples

1. (A.V. Bizadze) Let $n = 2, \Omega = \{(x_1, x_2) : x_1^2 + x_2^2 < 1\}$ and

$$Pu = (\frac{\partial}{\partial x_1} + i\frac{\partial}{\partial x_2})^2 u.$$

The general solution of the equation $Pu = 0$ has the form

$$u = \bar{z}f(z) + g(z),$$

where $z = x_1 + ix_2$. Each function of the form

$$u = (x_1^2 + x_2^2 - 1)f(x_1 + ix_2),$$

where f is a holomorphic in Ω function, satisfies the equation $Pu = 0$ and vanishes on Γ. Therefore, the Dirichlet problem for the elliptic operator P has an infinite-dimensional kernel. It is easy to see that in this case the number k is different for $\xi' > 0$ and $\xi' < 0$.

2. The Dirichlet problem for an elliptic operator of the second order with real coefficients is always elliptic. Let $n > 2$. Then *the oblique derivative problem* with the boundary condition

$$\sum_{j=1}^{n} a_j(x)D_j u + a_0(x)u = g(x) \quad \text{on} \quad \Gamma$$

for such equation is elliptic if and only if

$$\sum_{j=1}^{n} a_j(x)\nu_j \neq 0$$

on Γ, where $(\nu_1, \ldots, \nu_n)$ is the normal vector to Γ.

4.2.3　Construction of a parametrix

Put as above

$$\mathcal{H} = L_2(\Omega) \times H^{m-m_1-1/2}(\Gamma) \times \ldots \times H^{m-m_k-1/2}(\Gamma),$$

$$\mathcal{H}' = H^1(\Omega) \times H^{m-m_1+1/2}(\Gamma) \times \ldots \times H^{m-m_k+1/2}(\Gamma).$$

Theorem 5. *If a boundary value problem is elliptic, then there exists a left and a right parametrix.*

Proof. **A.** Let $\varepsilon > 0$ be a small number and $\delta = \delta(\varepsilon)$ be so small that for $x \in \Omega$, $y \in \Omega$, $|x - y| < \delta$ the inequalities

$$|a_\alpha(x) - a_\alpha(y)| < \varepsilon, \text{ for } |\alpha| = m$$

are true and, besides, for $x \in \Gamma$, $y \in \Gamma$, $|x - y| < \delta$ the ineqalities for the higher order coefficients of the operators $B_j(x, D)$

$$|b_{j,\alpha}(x) - b_{j,\alpha}(y)| < \varepsilon, \text{ for } j = 1, \ldots, k, \ |\alpha| = m_j$$

are also true. Let us assume also that in a δ-neighbourhood of each point of Γ we can construct the above-indicated transformation of the boundary to the part of the plane $y_n = 0$. Let us construct a partition of the unity $\sum \psi_j(x) = 1$, such that $diam \operatorname{supp} \psi_j < \delta/2$, $\psi_j \in C^\infty(\overline{\Omega})$, $0 \le \psi_j \le 1$.

If $\operatorname{supp} \psi_j$ has an empty intersection with Γ, then we construct a parametrix for the operator $P(x_j, D)$, where x_j is a point of $\operatorname{supp} \psi_j$, i.e. an operator $E_j^0 \in \mathcal{L}(L_2(\Omega), H^m(\Omega))$ such that

$$E_j^0 P(x_j, D)(\psi_j u) = \psi_j u + T_j u$$

and

$$P(x_j, D)E_j^0(\psi_j u) = \psi_j u + T_j' u,$$

where

$$T_j \in \mathcal{L}(H^m(\Omega), \ H^{m+1}(\Omega)); \ T_j' \in \mathcal{L}(L_2(\Omega), \ H^1(\Omega)).$$

As we have seen above, it is possible, for example, to put

$$E_j^0 v = (2\pi)^{-n} \int h(\xi)\tilde{v}(\xi)P_0(x_j, \xi)^{-1}e^{ix\xi}d\xi,$$

where $h \in C^\infty(\mathbb{R}^n)$, $0 \le h \le 1, h(\xi) = 0$ for $|\xi| \le 1/2, h(\xi) = 1$ for $|\xi| > 1$.

If the $\operatorname{supp} \psi_j$ does intersect with Γ, then we apply the above transformation of the boundary and construct a parametrix $E_j : \mathcal{L}(\mathcal{H}, H^m(\Omega))$ for the operator

$$\mathcal{A} = (P_0(x_j, D), B_{10}(x_j, D), \ldots, B_{k0}(x_j, D)),$$

where x_j is a point of $\Gamma \cup \operatorname{supp} \psi_j$. Then we have

$$E_j \mathcal{A}(\psi_j u) = \psi_j u + T_j u \text{ for } u \in H^m(\Omega)$$

and

$$\mathcal{A}E_j(\psi_j F) = \psi_j F + T_j' F \text{ for } F \in \mathcal{H},$$

where

$$T_j \in \mathcal{L}(H^m(\Omega), \ H^{m+1}(\Omega)), \ T_j' \in \mathcal{L}(\mathcal{H}(\Omega), \ \mathcal{H}'(\Omega)).$$

B. Now let

$$E = \sum_j \varphi_j(x)E_j\psi_j(x),$$

where $\varphi_j, \psi_j \in C_0^\infty(\mathbb{R}^n)$, the diameter of the $\operatorname{supp} \varphi_j$ is less than δ and $\varphi_j(x) = 1$ in $\operatorname{supp} \psi_j$. Let $u_j = E_j^0(\psi_j f, \psi_j g_1, \ldots, \psi_j g_k)$ if $\operatorname{supp} \psi_j$ intersects Γ, and $u_j = E_j(\psi_j f)$ if it does not.

We have for $F = (f, g_1, \ldots, g_k) \in \mathcal{H}$,

$$PEF = P \sum_j \varphi_j u_j = \sum_j \varphi_j P u_j + T_1 F$$

$$= \sum_j \varphi_j P_0(x_j, D) u_j + \sum_j [P_0(x, D) - P_0(x_j, D)] \varphi_j u_j + T_2 F$$

$$= \sum_j \varphi_j \psi_j f(x) + T_3 F + T_4 F = f(x) + T_3 F + T_4 F,$$

where T_1, T_2, T_3 are bounded linear operators from $\mathcal{H}$ in $H^1(\Omega)$, and T_4 is a bounded linear operator from $\mathcal{H}$ in $L_2(\Omega)$ such that $\|T_4\| \le C_1 \delta$ and the constant C_1 is independent of δ.

Analogously,

$$B_s EF = B_s \sum_j \varphi_j u_j = \sum_j \varphi_j B_s u_j + T_{1s} F$$

$$= \sum_j \varphi_j B_{s0}(x_j, D) u_j + \sum_j [B_{s0}(x, D) - B_{s0}(x_j, D)] \varphi_j u_j + T_{2s} F$$

$$= \sum_j \varphi_j \psi_j g_s(x) + T_{3s} F + T_{4s} F = f(x) + T_{3s} F + T_{4s} F,$$

where T_{1s}, T_{2s}, T_{3s} are bounded linear operators from the space $\mathcal{H}$ to $H^{m-m_s+1/2}(\Gamma)$, and T_{4s} is a bounded linear operator from $\mathcal{H}$ in $H^{m-m_s-1/2}(\Gamma)$ such that $\|T_{4s}\| \le C_{1s} \delta$ and the constant C_{1s} is independent of δ.

Thus, we have obtained that

$$\mathcal{A} EF = F + T' F + T'' F,$$

where T' is a bounded linear operator from $\mathcal{H}$ in $\mathcal{H}'$, and $T'' \in \mathcal{L}(\mathcal{H}, \mathcal{H})$ with its norm $\|T''\| < C\delta$, and the constant C is independent of δ. If δ is so small that $C\delta < 1$, then the operator $I + T''$ is invertible and

$$\mathcal{A} E (I + T'')^{-1} = I + T'(I + T'')^{-1}.$$

Since the operator $T'(I + T'')^{-1}$ belongs to $\mathcal{L}(\mathcal{H}, \mathcal{H}')$, we see that the operator

$$E_r = E(I + T'')^{-1}$$

is a right parametrix.

C. The left parametrix is constructed in the same way. By the construction,

$$E_j \mathcal{A}_0(x_j, D)(\psi_j u) = \psi_j u + T_j' \psi_j u,$$

where $u \in H^m(\Omega)$,

$$\mathcal{A}_0(x_j, D) = (P_0(x, D), B_{10}(x, D), \ldots, B_{k0}(x, D)),$$
$$T_j \in \mathcal{L}(H^m(\mathbb{R}^n_+), H^{m+1}(\mathbb{R}^n_+)).$$

As above, let $E = \sum_j \varphi_j E_j \psi_j$. Put $v = E\mathcal{A}u$. Then

$$v = \sum_j \varphi_j E_j \psi_j \mathcal{A}u = \sum_j \varphi_j E_j \mathcal{A}\psi_j u + T_1 = \sum_j \varphi_j E_j \mathcal{A}_0(x_j, D)\psi_j u$$

$$+ \sum_j \varphi_j E_j [\mathcal{A}_0(x, D) - \mathcal{A}_0(x_j, D)]\psi_j u + T_2 u$$

$$= \sum_j \varphi_j \psi_j u + T_3 u + T_4 u = u + T_3 u + T_4 u,$$

where T_1, T_2, T_3 are linear operators from $\mathcal{L}(H^m(\Omega)$ to $H^{m+1}(\Omega))$, and $T_4 \in \mathcal{L}(H^m(\Omega),\ H^m(\Omega))$, whose norm does not exceed $C\delta$. The latter follows from the fact that

$$\|\varphi_j E_j [\mathcal{A}_0(x, D) - \mathcal{A}_0(x_j, D)]\psi_j u\|_m \le C_1 (\|[P_0(x, D) - P_0(x_j, D)]\psi_j u\|_0$$

$$+ \sum_{j=1}^{k} \|[B_{s0}(x, D) - B_{s0}(x_j, D)]\psi_j u\|_{m-m_j-1/2}^{\Gamma}).$$

If δ is so small that $C\delta < 1/2$, then the operator $I + T_4$ is invertible. Putting $E_l = (I + T_4)^{-1}E$, we get the relation

$$E_l \mathcal{A} = T + (I + T_4)^{-1}T_3,$$

showing that E_l is a left parametrix. $\qquad\square$

4.2.4 Continuity of the parametrices

Theorem 6. *The operators E_l and E_r are bounded as operators from*

$$\mathcal{H}^{(s)} = H^s(\Omega) \times H^{m-m_1+s-1/2}(\Gamma) \times \ldots \times H^{m-m_k+s-1/2}(\Gamma)$$

in the space $H^{m+s}(\Omega)$ for any integer $s \ge 0$. For any integer $s \ge 0$ there exists such a constant $C = C(s)$ that for all $u \in H^{s+m}(\Omega)$

$$\|u\|_{s+m} \le C(\|Pu\|_s + \sum_{j=1}^{k} \|B_j u\|_{s+m-m_j-1/2} + \|u\|_0).$$

Proof. Using a partition of the unity as above, we can reduce the proof to the case, when $\Omega = \mathbb{R}_+^n$. Let $u = E_r(f, g_1, \ldots, g_k)$. In the proof of Theorem 2 of 4.1.2 we obtained the inequality

$$\sum_{j=0}^{m} |\xi'|^{2m-2j} \int_0^\infty |D_n^j \tilde{u}(\xi', x_n)|^2 dx_n$$

$$\le C_1 \Big(\int_0^\infty |\tilde{f}(\xi', x_n)|^2 dx_n + \sum_{j=1}^{k} |\xi'|^{2m-2m_j-1} |\tilde{g}_j(\xi')|^2 \Big).$$

Multiplying both sides of this inequality by $|\xi'|^{2s}$, integrating with respect to ξ' and adding $\|u\|_0^2$ to both sides, we get the estimate

$$\|u\|_{s+m}^2 \leq C(\|f\|_s^2 + \sum_{j=1}^{k} \|g_j\|_{s+m-m_j-1/2}^2 + \|u\|_0^2).$$

It is easy to see that the operator $\mathcal{A}E_r - I$ is a bounded operator from $\mathcal{H}^{(s)}$ in $\mathcal{H}^{(s+1)}$ for any s and the same is true for the operator

$$E_l\mathcal{A} - I : H^s(\mathbb{R}_+^n) \to H^{s+1}(\mathbb{R}_+^n).$$

In order to prove our theorem it is sufficient to repeat the proof of Theorem 5 of 4.2.3 with the substitution of the spaces $\mathcal{H}$ and $H^m(\Omega)$ by $\mathcal{H}^{(m)}(\Omega)$ and $H^{s+m}(\Omega)$.

$\square$

4.2.5 Fredholm property

Theorem 7. *If the boundary value problem $\mathcal{A}u = F$ is elliptic, then $\ker \mathcal{A}$ is a finite-dimensional subspace in the space $C^\infty(\overline{\Omega})$. The set $\operatorname{im}\mathcal{A}$ for the operator $\mathcal{A} : H^{m+s}(\Omega) \to \mathcal{H}^s$ is closed for any real $s \geq 0$.*

Proof. If $u \in \ker \mathcal{A}$, i.e. $\mathcal{A}u = 0$, then $E_l\mathcal{A}u = u + Tu = 0$, where $T \in \mathcal{L}(H^{s+m}(\Omega), H^{s+m+1}(\Omega))$ for all integers $s \geq 0$. If $\|u\|_m \leq 1$, then $\|Tu\|_{m+1} \leq C$, and hence $\|u\|_{m+1} \leq C$. Therefore, the unit ball in the space $\ker \mathcal{A}$, with the the same norm as in the space $H^m(\Omega)$, is compact, which is possible in a finite-dimensional space only.

If $u \in \ker \mathcal{A} \cap H^m(\Omega)$, then $u = -Tu \in H^{m+1}(\Omega)$. However, then $u = -Tu \in H^{m+2}(\Omega)$, and so on. We see that $u \in H^{m+s}(\Omega)$ for all $s \geq 0$, i.e. $u \in C^\infty(\overline{\Omega})$.

On the other hand the a priori estimate obtained in the Theorem 6 of 4.2.4 implies that the image of the operator $\mathcal{A}$ is closed. $\square$

Theorem 8. *If u is a solution of the elliptic boundary value problem $\mathcal{A}u = F$ from the class $H^m(\Omega)$, and $f \in H^s(\Omega)$, $g_j \in H^{m+s-m_j-1/2}(\Gamma)$ for some integer $s > 0$, then $u \in H^{m+s}(\Omega)$. If $f \in C^\infty(\overline{\Omega})$, $g_j \in C^\infty(\Gamma)$, $j = 1, \dots, k$, then $u \in C^\infty(\overline{\Omega})$.*

Proof. It is evident that $F = (f, g_1, \dots, g_k) \in \mathcal{H}^{(s)}$. Since the operator E_l is bounded as an operator from $\mathcal{H}^{(s)}$ in $H^{s+m}(\Omega)$, we see that

$$E_lF = u + Tu \in H^{s+m}(\Omega).$$

Since the operator

$$T : H^{m+t}(\Omega) \to H^{m+t+1}(\Omega)$$

is bounded for all integers $t \geq 0$, it follows that $u \in H^{m+s}(\Omega)$. $\square$

Theorem 9. *Let E_r be a right parametrix of the elliptic operator $\mathcal{A} : H^m(\Omega) \to \mathcal{H}$, so that*

$$\mathcal{A}E_r = I + T.$$

Let

$$\mathcal{N} = \{\Phi \in \mathcal{H} : (I + T^*)\Phi = 0\},$$

where T^* is the operator, adjoint to T in $\mathcal{H}$ with respect to some fixed Hilbert structure in $\mathcal{H}$. The space $\mathcal{N}$ is finite-dimensional and the corresponding boundary value problem has a solution in $H^m(\Omega)$ iff $F = (f, g_1, \ldots, g_k) \in \mathcal{H}$ and $(F, \Phi)_{\mathcal{H}} = 0$ for all Φ from $\mathcal{N}$.

Proof. Since $\mathcal{A}E_rG = G + TG$, solving the boundary value problem is reduced to solving the equation

$$G + TG = F,$$

which is a Fredholm type equation in $\mathcal{H}$. Thus the statements of the theorem follow from the Fredholm theorem. $\qquad\square$

4.2.6 Necessity of the ellipticity condition

Theorem 10. *If the operators $P, B_1, \ldots, B_k$ are such that the a priori estimate*

$$\|u\|_m \leq C(\|Pu\|_0 + \sum_{j=1}^{k} \|B_j u\|_{m-m_j-1/2} + \|u\|_0)$$

is true for all $u \in C^\infty(\overline{\Omega})$, then the problem is elliptic.

Proof. Let $x_0 \in \Gamma$ and ω be a neighbourhood of this point so small that it is possible to introduce local coordinates in ω in which the point x_0 corresponds to the origin, Γ is defined by the equation $y_n = 0$, and ω lies in the domain $y_n > 0$. It is easy to see that the estimate is true for all functions $u \in C^\infty(\overline{\omega})$, vanishing in a neighbourhood of $\partial\omega \setminus \Gamma$, after the substitution of the operators P and B_j by the operators $P_0(x, D)$ and $B_{j0}(x, D)$ in the new coordinates. Furthermore, if ω is small enough, it is possible to replace these operators by their values at the point x_0.

Assume now that the problem

$$P_0(0, D)u = f \text{ for } y_n > 0,$$

$$B_{j0}(0, D)u = g_j \text{ for } y_n = 0, \ j = 1, \ldots, k,$$

is not elliptic, i.e. there exists for $y_n > 0$ a function $v(y_n)$, exponentially decreasing as $y_n \to +\infty$, not vanishing identically and satisfying the relations

$$P_0(0, \xi_0', D_n)v(y_n) = 0 \text{ for } y_n > 0,$$

$$B_{j0}(0, \xi_0', D_n)v(y_n) = 0 \text{ for } y_n = 0, \ j = 1, \ldots, k$$

for a vector $\xi_0' \neq 0$.

Put
$$u_\lambda(y) = \varphi_1(y')\varphi_2(y_n)e^{i\lambda y'\xi_0'}v(\lambda y_n),$$

where $\varphi_1(y')$ and $\varphi_2(y_n) \in C_0^\infty$. Furthermore, $0 \le \varphi_1(y') \le 1, 0 \le \varphi_2(y_n) \le 1, \varphi_1(y')\varphi_2(y_n) = 0$ outside ω and $\varphi_1(y')\varphi_2(y_n) = 1$ in a neighbourhood ω' of the origin. The parameter λ has large positive values. By the condition we have

$$\|u_\lambda\|_m^2 \le C_1(\|P_0(0, D)u_\lambda\|_0^2 + \sum_{j=1}^{k} \|B_j(0, D)u_\lambda\|_{m-m_j-1/2}^2 + \|u_\lambda\|_0^2).$$

If $\xi_{0i} \ne 0$ for some $i, 1 \le i \le n - 1$, then

$$\|u_\lambda\|_m^2 \ge c_0\|D_i^m u_\lambda\|_0^2$$

$$\ge c_0\lambda^{2m} \int |\varphi_1(y')|^2 dy' \int |v(\lambda y_n)|^2 dy_n - C_2\lambda^{2m-2} \int |v(\lambda y_n)|^2 dy_n$$

$$\ge c_0'\lambda^{2m-1} - C_3\lambda^{2m-3}, \ c_0' = \text{const} > 0.$$

Furthermore,

$$\|P_0(0, D)u_\lambda\|_0^2 \le \|P_0(0, \lambda\xi_0', D_n)\varphi_1(y')v(\lambda y_n)\|_0^2 + C_4\lambda^{2m-3},$$

$$\|B_j(0, D)u_\lambda(y', 0)\|_{m-m_j-1/2}^2$$

$$\le \|B_j(0, \lambda\xi_0', D_n)\varphi_1(y')v(y_n)\|_0^2 + C_5\lambda^{2m-3} = C_5\lambda^{2m-3}.$$

Finally,
$$\|u_\lambda\|_0^2 \le C_6\lambda^{-1}.$$

Thus we come to the inequality

$$c_0'\lambda^{2m-1} \le C_2[(C_3 + C_4 + C_5)\lambda^{2m-3} + C_6\lambda^{-1}],$$

with constants C_j independent of λ. Since this inequality cannot be valid for all $\lambda > 0$, our assumption is not true, i.e. the problem is elliptic. $\qquad\square$

Chapter 5

Kondratiev's theory

5.1 A model problem

Consider at first the simplest model problem. It can serve as a good illustration of the general approach for the study of boundary value problems in conical domains.

Let u be a solution of the Poisson equation $\Delta u = f$ in the angle domain

$$\Omega = \{(x, y) : 0 < \varphi < \theta, \ 0 < r < \infty\}$$

where r, φ are polar coordinates. Let u satisfy the boundary conditions

$$u(r, 0) = 0, \ u(r, \theta) = 0.$$

After the change $r = e^{-t}$ the domain Ω passes in the strip

$$C = \{(t, \varphi) : 0 < \varphi < \theta, \ -\infty < t < \infty\},$$

and the Poisson equation takes the form

$$\frac{\partial^2 u}{\partial t^2} + \frac{\partial^2 u}{\partial \varphi^2} = f(x, y)e^{-2t} = f_1(t, \varphi).$$

The Fourier transform with respect to t,

$$v(\lambda, \varphi) = \int_{-\infty}^{+\infty} u(t, \varphi)e^{-it\lambda}dt, \ g_1(\lambda, \varphi) = \int_{-\infty}^{+\infty} f_1(t, \varphi)e^{-it\lambda}dt$$

yields an ordinary differential equation

$$\frac{\partial^2 v}{\partial \varphi^2} - \lambda^2 v = g_1(\lambda, \varphi)$$

with the boundary conditions

$$v(\lambda, 0) = 0, \ v(\lambda, \theta) = 0.$$

This problem has a unique solution for all complex λ except for the points $\lambda = \pm in\pi/\theta$, $n = 1, 2, \ldots$ on the imaginary axis. Its resolvent R_λ has simple poles at these points. The solution is represented in the form $v(\lambda, \varphi) = R_\lambda g_1$ so that

$$u(t, \varphi) = \frac{1}{2\pi} \int_{-\infty}^{+\infty} R_\lambda g_1(\lambda, \varphi) e^{it\lambda} d\lambda.$$

Note that the change $r = e^{-t}$ and the Fourier transform with respect to t in fact generate the Mellin transform

$$v(\lambda, \varphi) = \int_0^\infty u(r, \varphi) r^{i\lambda - 1} dr.$$

The inverse transform takes the form

$$u(r, \varphi) = \frac{1}{2\pi} \int_{-\infty}^{+\infty} v(\lambda, \varphi) r^{-i\lambda} d\lambda.$$

Assume that f satisfies the condition

$$\sum_{|\alpha| \leq k} \int_{\mathbb{R}^2} |D^\alpha f(x, y)|^2 r^{2(|\alpha| - k - \gamma)} dx dy < \infty$$

with a positive γ and a natural k. This condition means that

$$\sum_{|\alpha| \leq k} \int_{-\infty}^{+\infty} \int_0^\theta |D^\alpha_{t,\varphi} f_1(t, \varphi)|^2 e^{2(k+1+\gamma)t} d\varphi dt < \infty.$$

By virtue of the Parseval equality it is equivalent to the following:

$$\int_{ih-\infty}^{ih+\infty} |\lambda|^{2j} \int_0^\theta |D^l_\varphi g_1(\lambda, \varphi)|^2 d\varphi d\lambda < \infty, \; j + l \leq k,$$

where $h = k + \gamma + 1$. This implies that for $h \neq \pm \pi n/\theta$, $n = 1, 2, \ldots$ the problem has a unique solution v such that

$$\sum_{j+l \leq k+2} \int_{ih-\infty}^{ih+\infty} |\lambda|^{2j} \int_0^\theta |D^l_\varphi v(\lambda, \varphi)|^2 d\varphi d\lambda < \infty,$$

i.e.

$$\sum_{|\alpha| \leq k+2} \int_{\mathbb{R}^2} |D^\alpha u(x, y)|^2 r^{2(|\alpha| - k - \gamma)} dx dy < \infty.$$

If $f \in H^k$, we may apply similar formulas with $\gamma = 0$. If $0 < \theta < \pi/(k+1)$ and $f \in H^k$, then $u \in H^{k+2}$. Moreover, the solution is represented in the form

$$u = \sum_{0 < j\pi < (k+1)\theta} C_j r^{\frac{j\pi}{\theta}} \sin\frac{j\pi}{\theta}\varphi + w,$$

where $w \in H^{k+2}$, and the coefficients C_j can be found after computation of the residues of the function $R_\lambda g_1(\lambda, \varphi) r^{-i\lambda}$ at the points $\lambda = ij\pi/\theta$ for $j = 1, \ldots, k$.

5.2 The general problem

5.2.1 The conditions on a domain and differential operators

Let Ω be a bounded domain in $\mathbb{R}^n$ and

$$P(x, D) = \sum_{|\alpha| \leq 2m} a_\alpha(x) D^\alpha$$

be an elliptic operator with smooth coefficients of order m defined in $\overline{\Omega}$. The boundary Γ of Ω is assumed to be an infinitely differentiable surface everywhere except at the origin, and in a neighbourhood of the origin it coincides with the cone

$$x_n^{2p} = \sum_{|\alpha|=2p, \ \alpha_n=0} h_\alpha x^\alpha + h(x),$$

where $p > 0$, $\sum_{|\alpha|=2p} h_\alpha x^\alpha > 0$ for $x \neq 0$ and h is a smooth function such that

$$h(x) = o(|x|^{2p}) \text{ as } |x| \to 0.$$

On the boundary Γ outside the origin there are defined m differential operators

$$B_l(x, D) = \sum_{|\beta| \leq m_l} b_{l\beta}(x) D^\beta, \ l = 1, \ldots, m$$

with smooth coefficients. Let us put

$$P^0(D) = \sum_{|\alpha|=2m} a_\alpha(0) D^\alpha, \ B_l^0(D) = \sum_{|\beta|=m_l} b_{l\beta}(0) D^\beta.$$

5.2.2 Functional spaces

We will use the following functional spaces:

1. The Sobolev space $H^s(\Omega)$, where $s \in \mathbf{N}$, with the norm

$$\|u\|_s = \Big(\sum_{|\alpha| \leq s} \int_\Omega |D^\alpha u(x)|^2 dx \Big)^{1/2};$$

2. The space $H^{s-1/2}(\Gamma)$ of functions φ defined on the boundary Γ, with the norm

$$\|\varphi\|_{s-1/2} = \inf \|\varphi\|_s,$$

where inf is taken over all $\varphi \in H^s(\Omega)$ such that $\varphi|_\Gamma = \varphi$;

3. $H^{s,\gamma}(\Omega)$, where $s \in \mathbf{N}, \gamma \geq 0$, is the space of functions u defined in Ω with the norm

$$\|u\|_{s,\gamma}^2 = \sum_{|\alpha| \leq s} \int_\Omega |D^\alpha u(x)|^2 r^{2(|\alpha|-\gamma-s)} dx, \quad r = |x|;$$

4. The space $H^{s-1/2,\gamma}(\Gamma)$ of functions φ defined on the boundary Γ, with the norm

$$\|\varphi\|_{s-1/2,\gamma} = \inf \|\Phi\|_{s,\gamma}, \ \Phi|_\Gamma = \varphi.$$

Put also

$$\mathcal{H}^{s,\gamma} = H^{s,\gamma}(\Omega) \times H^{s+2m-m_1-1/2,\gamma}(\Gamma) \times \ldots \times H^{s+2m-m_m-1/2,\gamma}(\Gamma).$$

5.2.3 The statement of the general boundary value problem and main results

Now let Ω be a domain described in 5.2.1 with a boundary Γ, and

$$P(x,D) = \sum_{|\alpha| \leq 2m} a_\alpha(x) D^\alpha$$

be an elliptic differential operator of order m with smooth coefficients in $\overline{\Omega}$. We are looking for a solution u of the boundary value problem

$$P(x,D)u = f(x) \text{ in } \Omega, \tag{1}$$

$$B_l(x,D)u = g_l(x) \text{ on } \Gamma, \ l = 1,\ldots,m, \tag{2}$$

where B_l are differential operators satisfying the ellipticity conditions on $\Gamma \setminus 0$.

One can show (see, for instance [Kon]) that there exists an infinitely differentiable transformation of coordinates after which the surface Γ takes the form

$$x_n^{2p} = \sum_{|\alpha|=2p, \ \alpha_n=0} h_\alpha x^\alpha$$

in a neighbourhood of the origin. Since the ellipticity conditions are not violated by this transformation, we will assume in what follows that Γ has this form in a neighbourhood of the origin.

Theorem 1. *There exists a sequence $\mu_1,\ldots,\mu_j,\ldots$ of real numbers such that for $(f,g_1,\ldots,g_m) \in \mathcal{H}^{s,\gamma}$ and $s+\gamma \neq \mu_j, \ j=1,2,\ldots$ the boundary value problem has a unique solution $u \in \mathcal{H}^{s+m,\gamma}$, provided that the vector $(f,g_1,\ldots,g_m)$ satisfies a finite number of conditions of the form*

$$\Phi_j[f,g_1,\ldots,g_m] = 0, \ \Phi_j \in (\mathcal{H}^{s,\gamma})^*, \ j=1,\ldots,N.$$

Moreover, there is a constant C independent of $(f,g_1,\ldots,g_m)$ such that

$$\|u\|_{s+2m,\gamma} \leq C(\|f\|_{s,\gamma} + \sum_{l+1}^{m} \|g_l\|_{s+2m-m_l-1/2,\gamma} + \|u\|_{s+2m-1,\gamma}).$$

Other results on the asymptotics of the solution as $|x| \to 0$ and on the solvability of the boundary value problem in the usual Sobolev spaces can be found below, in Section 5.6.

In order to prove Theorem 1 we will consider in the next section the model problem for operators with constant coefficients in an infinite cone and in Section 5.4 the problem in an infinite cone for operators with variable coefficients. The proof of Theorem 1 will be given in 5.5.

5.3   The boundary value problem in an infinite cone for operators with constant coefficients

5.3.1   The statement of the problem and its transformations

Now we consider an auxiliary problem

$$P^0(D_x)u = f(x)$$

in the domain K, whose boundary S is the infinite cone

$$x_n^{2p} = \sum_{|\alpha|=2p,\ \alpha_n=0} h_\alpha x^\alpha,$$

with the boundary conditions

$$B_l^0(D_x)u = g_l(x) \text{ on } S,\ l = 1,\ldots,m.$$

In the spheric coordinates $(r, \omega_1, \ldots, \omega_{n-1})$ the problem takes the form

$$\sum_{j=0}^{2m} \sum_{|\alpha|\leq j} \frac{a_{\alpha,j}(\omega)}{r^j} D_r^{2m-j} D_\omega^\alpha u = f(x) \text{ in } K,$$

$$\sum_{j=0}^{m_l} \sum_{|\beta|\leq j} \frac{b_{\beta,j}^{(l)}(\omega)}{r^j} D_r^{m_l-j} D_\omega^\beta u = g_l(x) \text{ on } S,\ l = 1,\ldots,m.$$

Let us use the change of variables $t = \ln(1/r)$. As a result the domain K will become an infinite cylinder C, the boundary of which is a closed smooth surface

$$M = \{(t,\omega):\ h(\omega_1,\ldots,\omega_{n-1}) = 1,\ -\infty < t < \infty\}.$$

The problem will take the form

$$L(\omega, D_t, D_\omega)u = \sum_{j+|\alpha|\leq 2m} a_{j\alpha} D_t^j D_\omega^\alpha u = fe^{-2mt} = F(t,\omega) \text{ in } C,$$

$$T_l(\omega, D_t, D_\omega)u = \sum_{j+|\beta|\leq m_l} b_{j\beta}^{(l)} D_t^j D_\omega^\beta u = g_l e^{-m_l t} = G_l(t,\omega) \text{ on } M,$$

$$l = 1,\ldots,m,$$

where $a_{j\alpha}$, $b_{j\beta}^{(l)}$ are smooth functions of ω. The first equation is elliptic of order $2m$ and the boundary conditions satisfy the ellipticity conditions, since we have applied only non-singular smooth changes of coordinates, which did not affect the ellipticity conditions.

Assume that

$$f \in H^{k,\gamma}(K), \ g_l \in H^{k+2m-m_l-1/2,\gamma}(S), \ l = 1,\ldots,m,$$

that implies the boundedness of the integrals

$$\sum_{j+|\alpha|\leq k} \int_C |D_t^j D_\omega^\alpha F(t,\omega)|^2 e^{-nt+2(\gamma+k+2m)t} d\omega dt \leq C\|f\|_{k,\gamma}^2,$$

$$\sum_{j+|\alpha|\leq k+2m-m_l} \int_C |D_t^j D_\omega^\alpha v_l(t,\omega)|^2 e^{-nt+2(\gamma+k+2m)t} d\omega dt$$

$$\leq C\|g_l\|_{k+2m-m_l-1/2,\gamma}^2, \ l = 1,\ldots,m,$$

where v_l is a function such that $v_l|_S = g_l$. The existence of v_l follows from the
definition of the space $H^{s,\gamma}(S)$.

Let $\psi_l = v_l e^{-m_l t}$ and $G_l = \psi_l|_M$,

$$\tilde{F}(\lambda,\omega) = \int_{-\infty}^{\infty} F(t,\omega)e^{-i\lambda t} dt, \quad \tilde{\psi}_l(\lambda,\omega) = \int_{-\infty}^{\infty} \psi_l(t,\omega)e^{-i\lambda t} dt,$$

$$\tilde{G}_l(\lambda,\omega) = \int_{-\infty}^{\infty} G_l(t,\omega)e^{-i\lambda t} dt, \ l = 1,\ldots,m.$$

Then the Parseval identity implies the inequalities

$$\sum_{s=0}^{k} \int_{-\infty+i\tau}^{\infty+i\tau} |\lambda|^{2s}\|\tilde{F}(\lambda,\omega)\|_{k-s}^2 d\lambda \leq C\|f\|_{k,\gamma}^2, \ \tau = 2m+k+\gamma-n/2;$$

$$\sum_{s=0}^{k+2m-m_l} \int_{-\infty+i\tau}^{\infty+i\tau} |\lambda|^{2s}\|\tilde{\psi}_l(\lambda,\omega)\|_{k+2m-m_l-s}^2 d\lambda \leq C\|g_l\|_{k+2m-m_l-1/2,\gamma}^2,$$

where the norms in the left-hand side are the norms of the spaces $H^t(X)$ for the
domain $X \subset \mathbb{R}^{n-1}$ bounded by the surface Y:

$$h(\omega_1,\ldots,\omega_{n-1}) = 1.$$

Since ψ_l is a continuation of G_l from the boundary, one can consider $\tilde{\psi}_l$ a continuation of $\tilde{G}_l$ and

$$|\lambda|^{2s}\|\tilde{G}_l(\lambda,\omega)\|_{k+2m-m_l-s-1/2}^2 = \inf_{\tilde{\psi}_l} |\lambda|^{2s}\|\tilde{\psi}_l(\lambda,\omega)\|_{k+2m-m_l-s}^2, \tilde{\psi}_l|_Y = \tilde{G}_l,$$

where inf is taken over all such continuations. Hence

$$\int_{-\infty+i\tau}^{\infty+i\tau} [|\lambda|^{2k+4m-2m_l-2}\|\tilde{G}_l\|_{1/2}^2 + \|\tilde{G}_l\|_{k+2m-m_l-1/2}^2] d\lambda \leq C\|g_l\|_{k+2m-m_l-1/2,\gamma}^2.$$

5.3.2 The resolution of the model boundary value problem

After the Fourier transform with respect to t the boundary value problem takes the form

$$L(\omega, i\lambda, D_\omega)\tilde{u} = \tilde{F} \quad \text{in } X,$$

$$T_l(\omega, i\lambda, D_\omega)\tilde{u} = \tilde{G}_l \quad \text{on } Y, \; l = 1, \ldots, m.$$

For each λ this problem is elliptic in the variables ω, and therefore there exists an operator $R_\lambda : \mathcal{H}^{k,\gamma} \to H^{k+2m}(X)$, where

$$\mathcal{H}^{k,\gamma} = H^k(X) \times H^{k+2m-m_1-1/2}(Y) \times \ldots \times H^{k+2m-m_m-1/2}(Y),$$

which is a meromorphic function of λ such that

$$AR_\lambda = I, \text{ if } \mathcal{A}u = (Lu, T_1 u, \ldots, T_m u).$$

For every $C_1 > 0$ there exists a real C_2 such that for

$$|\operatorname{Im}\lambda| < C_1, \; |\operatorname{Re}\lambda| > C_2$$

the function R_λ has no singularities and

$$|\lambda|^{2k+4m}\|R_\lambda[\tilde{F}, \tilde{G}_1, \ldots, \tilde{G}_m]\|_0^2 + \|R_\lambda[\tilde{F}, \tilde{G}_1, \ldots, \tilde{G}_m]\|_{k+2m}^2$$

$$\leq C[|\lambda|^{2k}\|\tilde{F}\|_0^2 + \|\tilde{F}\|_k^2 + \sum_{l=1}^{m}(\|\tilde{G}_l\|_{k+2m-m_l-1/2}^2 + |\lambda|^{2k+4m-2m_l-2}\|\tilde{G}_l\|_{1/2}^2)].$$

Theorem 2. *If the function R_λ has no poles on the line $\operatorname{Im}\lambda = k + 2m + \gamma - n/2$, then for any functions*

$$f \in H^{k,\gamma}(K), \; g_l \in H^{k+2m-m_l-1/2,\gamma}(S), \; l = 1, \ldots, m$$

the boundary problem has a unique solution belonging to the space $H^{k+2m,\gamma}(K)$. Moreover,

$$\|u\|_{k+2m,\gamma} \leq C[\|f\|_{k,\gamma} + \sum_{l=1}^{m}\|g_l\|_{k+2m-m_l-1/2,\gamma}].$$

Proof. Integrating the above estimate of R_λ along the straight line $\operatorname{Im}\lambda = \tau = k + 2m + \gamma - n/2$ we obtain that

$$\int_{-\infty+i\tau}^{\infty+i\tau} [|\lambda|^{2k+4m}\|R_\lambda[\tilde{F}, \tilde{G}_1, \ldots, \tilde{G}_m]\|_0^2 + \|R_\lambda[\tilde{F}, \tilde{G}_1, \ldots, \tilde{G}_m]\|_{k+2m}^2]d\lambda$$

$$\leq C \int_{-\infty+i\tau}^{\infty+i\tau} [|\lambda|^{2k}\|\tilde{F}\|_0^2 + \|\tilde{F}\|_k^2]d\lambda$$

$$+ C \sum_{l=1}^{m} \int_{-\infty+i\tau}^{\infty+i\tau} (\|\tilde{G}_l\|_{k+2m-m_l-1/2}^2 + |\lambda|^{2k+4m-2m_l-2}\|\tilde{G}_l\|_{1/2}^2)d\lambda.$$

The right-hand side is not greater than

$$C_1(\|f\|_{k,\gamma}^2 + \sum_{l=1}^{m} \|g_l\|_{k+2m-m_l-1/2,\gamma}^2).$$

Put

$$v(\lambda,\omega) = R_\lambda[\tilde{F},\tilde{G}_1,\ldots,\tilde{G}_m].$$

The integral $\int_{-\infty+i\tau}^{\infty+i\tau} v(\lambda,\omega)e^{i\lambda t}d\lambda$ converges in the space L_2 and determines a function $u(t,\omega)$ such that

$$\sum_{j+|\alpha|\leq k+2m} \int_C |D_t^j D_\omega^\alpha u|^2 e^{-nt+2(k+2m+\gamma)t}dtd\omega$$

$$\leq C_2 \int_{-\infty+i\tau}^{\infty+i\tau} [|\lambda|^{2k+4m}\|v\|_0^2 + \|v\|_{k+2m}^2]d\lambda$$

$$\leq C_3[\|f\|_{k,\gamma}^2 + \sum_{l=1}^{m} \|g_l\|_{k+2m-m_l-1/2,\gamma}^2].$$

Using the definition of R_λ, we see that

$$Lu = F \text{ in } C, \quad T_l u = G_l \text{ on } M, \ l = 1,\ldots,m.$$

Returning to the variables x we can see that u is a solution of the original problem. It is clear that it is unique. Indeed, if u is a solution of the homogeneous problem from the class $H^{k+2m,\gamma}$, then the function $\tilde{u}$ is a solution of the other homogeneous problem:

$$L(\omega, i\lambda, D_\omega)\tilde{u} = 0 \text{ in } X; \ T_l(\omega, i\lambda, D_\omega)\tilde{u} = 0 \text{ on } Y, \ l = 1,\ldots,m.$$

But this problem has a unique solution for $\text{Im}\,\lambda = h$. Therefore, $\tilde{u} = 0$, and by the uniqueness of the Fourier transform, $u \equiv 0$. $\qquad\square$

5.3.3   The asymptotics of the solution

Now we will study the behavior of the solution $u \in H^{k+2m,\gamma}$ in a neighbourhood of the conical point.

Theorem 3. *Let*

$$u \in H^{k+2m,\gamma}(K), \ f = P^0 u \in H^{k_1,\gamma_1}(K),$$

$$g_l = B_l^0 u|_S \in H^{k_1+2m-m_l-1/2,\gamma_1}(S), \ l = 1,\ldots,m,$$

and the function R_λ have no poles on the straight line $\text{Im}\,\lambda = h_1$, where the numbers h_1, k_1 are such that

$$k_1 \geq k, \ h_1 := k_1 + 2m + \gamma_1 - n/2 > h := k + 2m + \gamma - n/2.$$

Then

$$u = \sum_j \sum_{s=0}^{\mu_j-1} a_{js} r^{-i\lambda_j} \ln^s r \cdot \varphi_{sj}(\omega) + w(\omega),$$

where

$$\|w\|_{k_1+2m,\gamma_1} \le C(\|f\|_{k_1,\gamma_1} + \sum_{l=1}^m \|g_l\|_{k_1+2m-m_l-1/2,\gamma_1} + \|u\|_{k+2m}),$$

g_{js} *are infinitely differentiable functions independent of u and λ_j is the pole of multiplicity μ_j of the function R_λ such that $h < \operatorname{Im}\lambda_j < h_1$.*

5.3.4   Lemmas

The proof of the above Theorem 3 is based on the following lemmas.

Lemma 4. *There exists a constant C such that for all functions $\varphi \in H^{l-1/2,\gamma}(S)$, the estimate*

$$\int_S |\varphi|^2 r^{1-2\gamma-2l} ds \le C\|\varphi\|_{l-1/2,\gamma}^2$$

holds.

Proof. Let v be a function from $H^{l,\gamma}(K)$ such that $v|_S = \varphi$ and

$$\|v\|_{l,\gamma} \le 2\|\varphi\|_{l-1/2,\gamma}.$$

Then

$$\int_K |v|^2 r^{-2\gamma-2l} dx < \infty, \quad \int_K |\nabla v|^2 r^{-2\gamma-2l+2} dx < \infty.$$

After the change $t = \ln(1/r)$, we obtain

$$\int_C |v|^2 e^{-nt+2(l+\gamma)t} dt d\omega < \infty, \quad \int_C |\nabla v|^2 e^{-nt+2(l+\gamma+2)t} dt d\omega < \infty.$$

This implies that there exists a number β such that $0 < \beta < 1/2$ and

$$\int_{M_\beta} \int_{-\infty}^{\infty} |v|^2 e^{-nt+2(l+\gamma)t} dt d\omega \le C \int_C |v|^2 e^{-nt+2(l+\gamma)t} dt d\omega,$$

where M_β is the surface

$$1 - \beta = h(\omega_1, \ldots \omega_{n-1}).$$

Now we can estimate the integral over M in the following way:

$$\int_{-\infty}^{\infty} \int_M |v|^2 e^{-nt+2(l+\gamma)t} dt d\omega \le 2[\int_C |\nabla v|^2 e^{-nt+2(l+\gamma)t} dt d\omega$$

$$+ \int_{M_\beta} |v|^2 e^{-nt+2(l+\gamma)t} dt d\omega] \le C\|\varphi\|_{l-1/2,\gamma}^2.$$

Passing to the variables $x_1, \ldots, x_n$, we obtain the estimate

$$\int_S |v|^2 r^{1-2\gamma-2l} dx = \int_S |\varphi|^2 r^{1-2\gamma-2l} dx \leq C\|\varphi\|_{l-1/2,\gamma}^2. \qquad \square$$

Lemma 5. *If* $\varphi \in H^{l-1/2,\gamma_1}(S) \cap H^{l-1/2,\gamma_2}(S)$, $\gamma_2 < \gamma_1$, *then*

$$\|\varphi\|_{l-1/2,\gamma} \leq C(\|\varphi\|_{l-1/2,\gamma_1} + \|\varphi\|_{l-1/2,\gamma_2})$$

for all $\gamma \in [\gamma_1, \gamma_2]$. *The constant* C *is independent of* γ.

Proof. Let $\theta(r) \in C^\infty(\mathbb{R})$, $\theta(r) = 0$ for $r > 1$, $\theta(r) = 1$ for $r < 1/2$. Then

$$\varphi = \theta(r)\varphi + [1 - \theta(r)]\varphi = \theta(r)\varphi + \theta_1(r)\varphi.$$

Let

$$v_1|_S = \varphi, \ \|v_1\|_{l,\gamma_1} \leq 2\|\varphi\|_{l-1/2,\gamma_1};$$
$$v_2|_S = \varphi, \ \|v_2\|_{l,\gamma_2} \leq 2\|\varphi\|_{l-1/2,\gamma_2}.$$

The function θv_1 vanishes when $r > 1$ and $\theta v_1 \in H^{l,\gamma_1}(K)$. Therefore,

$$\|\theta v_1\|_{l,\gamma} \leq C\|v_1\|_{l,\gamma_2} \leq 2C\|\varphi\|_{l-1/2,\gamma_2}$$

for $\gamma \geq \gamma_2$. The function $\theta_1 v_2$ vanishes when $r < 1/2$, and for $\gamma \leq \gamma_1$ we have

$$\|\theta_1 v_2\|_{l,\gamma} \leq C\|v_2\|_{l,\gamma_1} \leq 2C\|\varphi\|_{l-1/2,\gamma_1}.$$

Thus

$$\|\theta v_1 + \theta_1 v_2\|_{l,\gamma} \leq 2C(\|\varphi\|_{l-1/2,\gamma_1} + \|\varphi\|_{l-1/2,\gamma_2}),$$
$$(\theta v_1 + \theta_1 v_2)|_S = \theta\varphi + \theta_1\varphi = \varphi.$$

Therefore,

$$\|\varphi\|_{l-1/2,\gamma} = \inf \|v\|_{l,\gamma} \leq \|\theta v_1 + \theta_1 v_2\|_{l,\gamma} \leq 2C(\|\varphi\|_{l-1/2,\gamma_1} + \|\varphi\|_{l-1/2,\gamma_2}). \quad \square$$

5.3.5 The proof of Theorem 3

As above, put

$$F(t,\omega) = fe^{-2mt}, \ G_l(t,\omega) = g_l e^{-m_l t}, \ l = 1, \ldots, m.$$

Since $f \in H^{k_1,\gamma_1}(K) \cap H^{k,\gamma}(K)$, we have

$$\sum_{|\alpha|+j\leq k_1} \int_C |D_t^j D_\omega^\alpha F|^2 e^{-nt+2(k_1+2m+\gamma_1)t} d\omega dt < \infty,$$

$$\sum_{|\alpha|+j\leq k} \int_C |D_t^j D_\omega^\alpha F|^2 e^{-nt+2(k+2m+\gamma)t} d\omega dt < \infty,$$

implying that $\tilde{F}(\lambda,\omega)$ is an analytic function in the strip $\tau < \operatorname{Im}\lambda < \tau_1$, and on each straight line $\operatorname{Im}\lambda = \beta$ with $\tau < \beta < \tau_1$ we have

$$\int_{-\infty+i\beta}^{\infty+i\beta} (|\lambda|^{2k}\|\tilde{F}\|_0^2 + \|\tilde{F}\|_k^2)d\lambda$$

$$\leq C[\int_{-\infty+i\tau}^{\infty+i\tau} (|\lambda|^{2k}\|\tilde{F}\|_0^2 + \|\tilde{F}\|_k^2)d\lambda + \int_{-\infty+i\tau_1}^{\infty+i\tau_1} (|\lambda|^{2k}\|\tilde{F}\|_0^2 + \|\tilde{F}\|_k^2)d\lambda],$$

where $\|\cdot\|_t$ is the norm in the space $H^t(X)$. Analogously using Lemmas 4, 5, we obtain that

$$\int_{-\infty+i\beta}^{\infty+i\beta} [|\lambda|^{2k+4m-2m_l-2}\|\tilde{G}_l\|_{1/2}^2 + \|\tilde{G}_l\|_{k+2m-m_l-1/2}^2)]d\lambda$$

$$\leq C(\|g_l\|_{k+2m-m_l-1/2,\gamma}^2 + \|g_l\|_{k_1+2m-m_l-1/2,\gamma_1}^2)$$

uniformly with respect to all β from the interval $[\tau_1,\tau_2]$. The function $\tilde{u} = R_\lambda[\tilde{F},\tilde{G}_1,\ldots,\tilde{G}_m]$ is a meromorphic function in the strip $\tau < \operatorname{Im}\lambda < \tau_1$.

Using the Cauchy formula, we obtain

$$u = \lim_{N\to\infty} \int_{-N+i\tau}^{N+i\tau} e^{i\lambda t}\tilde{u}(\lambda,\omega)d\lambda$$

$$= \int_{-N+i\tau}^{-N+i\tau_1} + \int_{-N+i\tau_1}^{N+i\tau_1} + \int_{N+i\tau_1}^{N+i\tau} + \sum_j \operatorname{Res} e^{i\lambda t} R_\lambda[\tilde{F},\tilde{g}_1,\ldots,\tilde{g}_m].$$

The summation in the right-hand side of this equality is taken over all residues of the integrated function, lying inside the rectangle with vertices $-N+i\tau, -N+i\tau_1, N+i\tau_1, N+i\tau$. As $N\to\infty$, the first and the third integrals in the right-hand side tend to zero.

Passing to the limit we obtain the equality

$$u = w + \sum_{\tau<\operatorname{Im}\lambda_j<\tau_1} \operatorname{Res} e^{i\lambda_j t} R_{\lambda_j}[\tilde{F},\tilde{G}_1,\ldots,\tilde{G}_m],$$

where

$$w = \int_{-\infty+i\tau_1}^{\infty+i\tau_1} R_\lambda[\tilde{F},\tilde{G}_1,\ldots,\tilde{G}_m]e^{i\lambda t}d\lambda.$$

From the Parseval identity and Theorem 2 it follows that

$$\sum_{j+|\alpha|\leq 2m+k_1} \int_C e^{2h_1 t}|D_t^j D_\omega^\alpha w|^2 dtd\omega \leq C(\|f\|_{k_1,\gamma_1}^2 + \sum_{l=1}^m \|g_l\|_{k_1+2m-m_l-1/2,\gamma_1}^2),$$

or in variables x

$$\sum_{|\alpha|\leq 2m+k_1}\int r^{-2\gamma_1-2k_1-4m+2|\alpha|}|D^\alpha w(x)|^2 dx$$

$$\leq C(\|f\|^2_{k_1,\gamma_1}+\sum_{l=1}^{m}\|g_l\|^2_{k_1+2m-m_l-1/2,\gamma_1}).$$

The residue of the function $e^{i\lambda t}R_\lambda[\tilde{F},\tilde{G}_1,\ldots,\tilde{G}_m]$ at the pole λ_j of multiplicity μ_j is equal to

$$\sum_{s=0}^{\mu_j-1}a_{js}t^s e^{i\lambda_j t}\psi_{sj}(\omega_1,\ldots,\omega_{n-1}),$$

where $\psi_{sj}(\omega_1,\ldots,\omega_{n-1})$ are the characteristic or the associated functions of the homogeneous problem

$$L(\omega,i\lambda,D_\omega)v=0 \ \text{ in } \ X, \ \ T_l(\omega,i\lambda,D_\omega)v=0 \ \text{ on } \ Y, \ l=1,\ldots,m. \qquad \Box$$

Remark 6. If the straight line

$$\text{Im }\lambda = 2m+k+\gamma_1-n/2$$

contains a pole of the function R_λ, then Theorem 3 cannot be applied directly. However, we can apply it in a somewhat different form.

Let $M(D)$ be a differential operator of order $m_0\leq k+2m$ vanishing on all functions

$$r^{-i\lambda_0}ln^s r\cdot\psi_{sj}(\omega_1,\ldots,\omega_{n-1}), \ 0\leq s\leq\mu_j-1,$$

where λ_0 is a pole of R_λ of multiplicity μ_j such that $\text{Im }\lambda_0 = 2m+k+\gamma_1-n/2$ and ψ_{sj} are the characteristic and the associated functions of the indicated homogeneous problem.

After the change of variables and the Fourier transform the operator $M(D)$ is transformed into a differential operator $M^*(\lambda,D_\omega)$ of order m_0. The function M^*R_λ will be a meromorphic function as is R_λ, but it will be regular at the point λ_0, and from the equality

$$\int_{-N+i\tau}^{N+i\tau}e^{i\lambda t}M^*R_\lambda[\tilde{F},\tilde{G}_1,\ldots,\tilde{G}_m]d\lambda = \int_{-N+i\tau}^{-N+i\tau_1}$$

$$+\int_{-N+i\tau_1}^{N+i\tau_1}+\int_{-N+i\tau_1}^{N+i\tau}+\sum_{\tau<\text{Im }\lambda_j<\tau_1}Res\ e^{i\lambda_j t}M^*R_{\lambda_j}[\tilde{F},\tilde{G}_1,\ldots,\tilde{G}_m],$$

we obtain, passing to the limit as N tends to infinity, that

$$Mu = \sum_{\tau < \operatorname{Im} \lambda_j < \tau_1} Res \; e^{i\lambda_j t} M^* R_{\lambda_j} [\tilde{F}, \tilde{G}_1, \ldots, \tilde{G}_m]$$

$$+ \int_{-\infty + i\tau_1}^{\infty + i\tau_1} e^{i\lambda t} M^* R_\lambda [\tilde{F}, \tilde{G}_1, \ldots, \tilde{G}_m] d\lambda + Mw$$

$$= \sum_{\tau < \operatorname{Im} \lambda_j < \tau_1} \sum_{s=0}^{\mu_j - 1} M(D) r^{-i\lambda_j} ln^s r \cdot \psi_{js}(\omega) + Mw,$$

where $Mw \in H^{k_1 + 2m - m_0, \gamma_1}(K)$.

5.4 Equations with variable coefficients in an infinite cone

5.4.1 Conditions on the coefficients

Assume now that the operators P and B_l are homogeneous but with variable coefficients. Suppose that

$$r^{|\alpha| - 1} |D^\alpha [a_\beta(x) - a_\beta(0)]| \le M,$$

$$r^{|\alpha| - 1} |D^\alpha [b_{l\beta'}(x) - b_{l\beta'}(0)]| \le M,$$

where $|\alpha| \le k$, $|\beta| = 2m$, $|\beta'| = m_l$, $l = 1, \ldots, m$.

5.4.2 Lemmas

Lemma 7. *If $u \in H^{k,\gamma}(K)$ and the function $a(x)$ is such that*

$$r^{|\alpha| - \gamma_1} |D^\alpha a(x)| \le C_1, \; |\alpha| \le k,$$

then

$$\|au\|_{k, \gamma + \gamma_1} \le C C_1 \|u\|_{k, \gamma},$$

where the constant C depends neither on a nor on u.

Proof. Let $|\beta| \le k$ and $v_\beta = D^\beta(au)$. The function v_β is a linear combination of the functions $D^{\beta - \alpha} u \cdot D^\alpha a$. Therefore,

$$|v_\beta| \le C C_1 \sum_{\alpha \le \beta} r^{\gamma_1 - |\alpha|} |D^{\beta - \alpha} u|.$$

But then

$$\int_K |v_\beta|^2 r^{2(|\beta|-k-\gamma-\gamma_1)} dx$$

$$\leq C^2 C_1^2 \sum_{\alpha \leq \beta} \int_K |D^{\beta-\alpha} u(x)|^2 r^{2(|\beta|-|\alpha|-\gamma-k)} dx \leq C^2 C_1^2 \|u\|_{k,\gamma}^2.$$

By taking the sum of these inequalities over β we find that

$$\|au\|_{k,\gamma+\gamma_1} \leq CC_1 \|u\|_{k,\gamma}. \qquad \square$$

Lemma 8. *If $\varphi \in H^{k-1/2,\gamma}(S)$ and the function $b(x)$ defined on the boundary S is such that*

$$r^{|\alpha|}|D^\alpha b(x)| \leq C_1, \ |\alpha| \leq k,$$

where the derivatives are taken tangentially to the boundary, then

$$\|r^{\gamma_1} b\varphi\|_{k-1/2,\gamma+\gamma_1} \leq CC_1 \|\varphi\|_{k-1/2,\gamma}.$$

Proof. Let us apply the change $t = \ln(1/r)$ in the spheric coordinates. The function $b(x)$ will turn into the function $b^*(t,\omega)$, defined on the boundary surface $M : g(\omega) = 0$ of the cylinder C. By the condition, the derivatives of the function b^* up to the order k along directions tangential to M are bounded by the constant CC_1. Let φ^* be the image of φ. By the definition of the space $H^{k-1/2,\gamma}(K)$ there exists a function v such that $v|_S = \varphi$ and

$$\int_K |D^\alpha v(x)|^2 r^{2(-\gamma-k+|\alpha|)} dx \leq 2\|\varphi\|_{k-1/2,\gamma}^2.$$

After the transformation the function v becomes a function v^* satisfying the inequality

$$\int_C |D_{t,\omega}^\alpha v^*|^2 e^{-nt+2(k+\gamma)t} dt d\omega \leq 2\|\varphi\|_{k-1/2,\gamma}^2.$$

Thus the function $v_1^* = v^* e^{(k+\gamma-n/2)t}$ belongs to $H^k(C)$ and

$$\|v_1^*\|_k \leq C\|\varphi\|_{k-1/2,\gamma}.$$

The function v_1^* coincides on the boundary M with the function $g_1 = \varphi^* e^{(k+\gamma-n/2)t}$. Therefore

$$\|g_1\|_{k-1/2,\gamma} \leq C\|v_1^*\|_k \leq C\|\varphi\|_{k-1/2,\gamma}.$$

Since $g_1 \in H^{k-1/2}$ and b^* has bounded derivatives up to order k, we have $b^* g_1 \in H^{k-1/2}(M)$ and

$$\|b^* g_1\|_{k-1/2} \leq CC_1 \|g_1\|_{k-1/2}.$$

In the variables x it means that the function $bg_1 = b\varphi r^{n/2-\gamma-k}$ admits a continuation ψ onto the domain K such that $\psi|_S = bg_1$ and

$$\int_K |D^\alpha \psi|^2 r^{2(|\alpha|-k-\gamma-\gamma_1)} dx < \infty, \quad |\alpha| \leq k,$$

i.e. $bg_1 \in H^{k-1/2,\gamma+\gamma_1}(S)$, and the proof is complete. $\qquad \square$

5.4.3 The existence of the solution

Let

$$\mathcal{A}u = (P(0,D)u, B_1(0,D)u, \ldots, B_m(0,D)u),$$

$$\mathcal{A}_1 u = ([P(x,D) - P(0,D)]u, [B_1(x,D) - B_1(0,D)]u, \ldots,$$

$$[B_m(x,D) - B_m(0,D)]u).$$

Theorem 9. *Let the following conditions hold:*

1) The function R_λ has no poles on the line $\operatorname{Im}\lambda = k + 2m + \gamma - n/2$;

2) $(f, g_1, \ldots, g_m) \in \mathcal{H}^{k,\gamma}$;

3) $r^{|\alpha|}|D^\alpha a_\beta(x) - D^\alpha a_\beta(0)| \le \delta$, for $|\alpha| \le k$, $|\beta| = 2m$;

$$r^{|\alpha|}|D^\alpha b_{l,\beta'}(x) - D^\alpha b_{l,\beta'}(0)| \le \delta \text{ for } |\alpha| \le k, \ |\beta'| = m_l, \ l = 1, \ldots, m.$$

If δ is small enough, then there exists a solution of the boundary value problem $(\mathcal{A} + \mathcal{A}_1)u = (f, g_1, \ldots, g_m)$ of the class $H^{k+2m,\gamma}(K)$. This solution is unique and satisfies the inequality

$$\|u\|_{k+2m,\gamma} \le C(\|f\|_{k,\gamma} + \|g_1\|_{k+m_1-2m-1/2,\gamma} + \ldots + \|g_m\|_{k+m_m-2m-1/2,\gamma}),$$

where the constant C is independent of $f, g_1, \ldots, g_m$.

Proof. By Theorem 2 there exists an operator $\mathcal{A}^{-1}$. The operator $\mathcal{A} + \mathcal{A}_1$ is invertible if the norm of the operator $\mathcal{A}_1 : H^{k+2m,\gamma}(K) \to \mathcal{H}^{k,\gamma}$ is sufficiently small, for example, if $\|\mathcal{A}_1\| \le \|\mathcal{A}^{-1}\|^{-1}/2$. Using Lemmas 7, 8 we can see that

$$\|\mathcal{A}_1 u\|_{k,\gamma} \le C_k \delta \|u\|_{k+2m,\gamma},$$

where C_k is a constant independent of the coefficients of P and B_l. Thus if $C_k\delta < \|\mathcal{A}^{-1}\|^{-1}/2$, the operator $\mathcal{A} + \mathcal{A}_1$ is invertible and the proof is complete. $\square$

5.4.4 The smoothness of the solution

Theorem 10. *Let u be the solution of the problem in an infinite cone K of the class $H^{k+2m,\gamma}(K)$ and let the functions f, g_l be such that*

$$(f, g_1, \ldots, g_m) \in \mathcal{H}^{k,\gamma_1},$$

where $\gamma < \gamma_1 \le \gamma + 1$, provided that $f = 0$ and $g_l = 0$ for $r \ge 1$. Suppose that R_λ has no poles on the line $\operatorname{Im}\lambda = k + 2m + \gamma_1 - n/2$. Then the solution $u(x)$ has the form

$$u = \sum_j \sum_{s=0}^{\mu_j - 1} a_{js} r^{-i\lambda_j} \ln^s r \cdot \psi_{sj}(\omega) + w(x),$$

where λ_j is the pole of multiplicity μ_j of the function $R(\lambda)$ such that

$$2m + k + \gamma - n/2 < \operatorname{Im}\lambda_j < 2m + k + \gamma_1 - n/2,$$

ψ_{js} *are the normed characteristic and associated functions of the problem*

$$P(\omega, i\lambda_j, D_\omega)v = 0, \ B_l(\omega, i\lambda_j, D_\omega)v = 0, \ l = 1, \ldots, m$$

and w satisfies the inequality

$$\|w\|_{k+2m,\gamma_1} \leq C(\|f\|_{k,\gamma_1} + \sum_{l=1}^{m} \|g_l\|_{k+2m-m_l-1/2,\gamma_1} + \|u\|_{k+2m,\gamma}).$$

Proof. Since $u \in H^{k+2m,\gamma}(K)$, we find using Lemmas 7 and 8 (with $\gamma_1 = 1$ there) that

$$f_1 = [P(x,D) - P(0,D)]u \in H^{k,\gamma+1}(K),$$

$$g_l^{(1)} = [B_l(x,D) - B_l(0,D)]u \in H^{k+2m-m_l-1/2,\gamma+1}(S).$$

This means that the function u satisfies the equation

$$P(0,D)u = f - f_1 = f_2 \in H^{k,\gamma_1}(K)$$

and the boundary conditions

$$B_l(0,D)u = g_l - g_l^{(1)} = g_l^{(2)} \in H^{k+2m-m_l-1/2,\gamma_1}(S).$$

Applying Theorem 3 we obtain the result. $\qquad\square$

5.5　The boundary value problem in a bounded domain

5.5.1　Lemmas

Lemma 11. *If $\varphi \in H^{k-1/2,\gamma}(\Gamma)$, then $\varphi \in H^{k-1/2,\gamma_1}(\Gamma)$ for all $\gamma_1 \leq \gamma$ and*

$$\|\varphi\|_{k-1/2,\gamma_1}(\Gamma) \leq C\|\varphi\|_{k-1/2,\gamma},$$

where the constant C is independent of φ.

Proof. Let $v \in H^{k,\gamma}(\Omega)$ and $v|_\Gamma = \varphi$. Then, obviously, $v \in H^{k,\gamma_1}(\Omega)$, if $\gamma_1 \leq \gamma$. Therefore,

$$\|\varphi\|_{k-1/2,\gamma_1} = \inf_{v|_\Gamma=\varphi} \|v\|_{k,\gamma_1} \leq \inf_{v|_\Gamma=\varphi} \|v\|_{k,\gamma} = \|\varphi\|_{k-1/2,\gamma}. \qquad\square$$

The proofs of the three following lemmas are rather obvious.

Lemma 12. *If $u \in H^{k,\gamma}(\Omega)$, $l \in \mathbf{N}$, then*

$$\|u\|_{k-l,\gamma+l} \leq C\|u\|_{k,\gamma},$$

where the constant C is independent of M.

Lemma 13. *If $u \in H^{k,\gamma}(\Omega)$ and the function $a(x)$ has derivatives up to order k bounded by a constant M, then*

$$\|au\|_{k,\gamma} \leq CM\|u\|_{k,\gamma},$$

where the constant C is independent of u. If in addition $a(0) = 0$, then

$$\|au\|_{k,\gamma+1} \leq CM\|u\|_{k,\gamma}.$$

Lemma 14. *If $\varphi \in H^{k-1/2,\gamma}(\Gamma)$, then $\varphi \in H^{k-l-1/2,\gamma-l}(\Gamma)$ for $l \in \mathbf{N}$ and*

$$\|\varphi\|_{k-l-1/2,\gamma-l} \leq C\|\varphi\|_{k-1/2,\gamma},$$

where the constant C is independent of φ.

Lemma 15. *The operator of embedding $H^{k,\gamma}(\Omega) \subset H^{k_1,\gamma_1}(\Omega)$ is compact if $k_1 < k$, $\gamma_1 + k_1 < \gamma + k$.*

Proof. Let B be a bounded set in $H^{k,\gamma}(\Omega)$, i.e. for $u \in B$ we have

$$\sum_{|\alpha| \leq k} \int_\Omega |D^\alpha u(x)|^2 r^{2(|\alpha|-k-\gamma)} dx \leq C_1.$$

Given an $\varepsilon > 0$, choose a $\delta > 0$ so that

$$\sum_{|\alpha| \leq k} \int_{r < \delta} |D^\alpha u(x)|^2 r^{2(|\alpha|-k_1-\gamma_1)} dx \leq \varepsilon/2.$$

To this end it is sufficient to take δ such that

$$\delta^{2(\gamma-\gamma_1+k-k_1)} C_1 = \varepsilon/2.$$

In the domain Ω_δ, obtained from Ω by removing a ball of radius δ, the set of functions $u \in B$ is bounded in $H^{k_1}(\Omega_\delta)$. Since the inclusion $H^k(\Omega_\delta) \subset H^{k_1}(\Omega_\delta)$ is compact, we can choose on the set B an $\varepsilon/2$-net in the space $H^{k_1}(\Omega_\delta)$. This $\varepsilon/2$-net is an ε-net in the space $H^{k_1,\gamma_1}(\Omega_\delta)$. Therefore, the set B is compact in $H^{k_1,\gamma_1}(\Omega)$. $\square$

Lemma 16. *The operator of inclusion $H^{k-1/2,\gamma}(\Gamma) \subset H^{k_1-1/2,\gamma_1}(\Gamma)$ is compact if $k_1 < k$, $\gamma_1 + k_1 < \gamma + k$.*

Proof. Let B be a bounded set in $H^{k-1/2,\gamma}(\Gamma)$. For $\varphi \in B$ there exists a function $v \in H^{k,\gamma}(\Omega)$ such that $v|_\Gamma = \varphi$. The set B_1 of such v is bounded in $H^{k,\gamma}(\Omega)$. By Lemma 15 the set B_1 contains a subsequence v_j converging in $H^{k_1,\gamma_1}(\Omega)$. Since $\|\varphi\|_{k_1-1/2,\gamma_1} \leq \|v\|_{k_1,\gamma_1}$, the sequence g_j converges in $H^{k_1-1/2,\gamma_1}(\Gamma)$. $\square$

Lemma 17. *If $\varphi \in H^{k-1/2,\gamma}(\Gamma)$ and the function $b(x)$ is such that*

$$r^{|\alpha|+\gamma_1}|D^\alpha b(x)| \leq M \text{ for } |\alpha| \leq k,$$

then

$$\|b\varphi\|_{k-1/2,\gamma+\gamma_1} \leq CM\|\varphi\|_{k-1/2,\gamma},$$

where the constant C is independent of M.

Proof. Let $\theta(r)$ be an infinitely differentiable function which is equal to one in some neighbourhood of the origin and equal to zero outside that neighbourhood of the origin, in which the boundary surface has the form

$$x_n^{2p} = \sum_{|\alpha|=2p,\alpha_n=0} h_\alpha x^\alpha.$$

It is clear that

$$(1-\theta)\varphi \in H^{k-1/2,\gamma}(\Gamma), \theta\varphi \in H^{k-1/2,\gamma}(\Gamma),$$

and by Lemma 8

$$\|b(x)\theta\varphi\|_{k-1/2,\gamma+\gamma_1} \leq C_1 M\|\varphi\|_{k-1/2,\gamma}.$$

The function $b(x)(1-\theta)\varphi$ vanishes outside some neighbourhood of the origin, and therefore

$$\|b(x)(1-\theta)\varphi\|_{k-1/2,\gamma+\gamma_1} \leq C_2 M\|\varphi\|_{k-1/2,\gamma}.$$

Thus

$$\|b(x)\varphi\|_{k-1/2,\gamma+\gamma_1} \leq \|b(x)\theta\varphi\|_{k-1/2,\gamma+\gamma_1} + \|b(x)(1-\theta)\varphi\|_{k-1/2,\gamma+\gamma_1}$$

$$\leq CM\|\varphi\|_{k-1/2,\gamma}. \qquad \square$$

Lemma 18. *If $u \in H^{k+2m,\gamma}(\Omega), Pu = f \in H^{k,\gamma}(\Omega)$,*

$$B_l u = g_l \in H^{k+2m-m_l-1/2,\gamma}(\Gamma), \ l = 1,\ldots,m,$$

then

$$\|P^0(0,D)u - f\|_{k,\gamma+1} \leq C\|u\|_{k+2m,\gamma},$$

$$\sum_{l=1}^m \|B_l^0(0,D)u - g_l\|_{k+2m-m_l-1/2,\gamma+1} \leq C\|u\|_{k+2m,\gamma}.$$

Proof. The function $P_0(0,D)u - f$ can be represented as

$$\sum_{|\alpha|=2m} [a_\alpha(0) - a_\alpha(x)]D^\alpha u - \sum_{|\beta|<2m} a_\beta(x)D^\beta u.$$

By Lemma 12 the function $D^\beta u$ for $|\beta| < 2m$ belongs to $H^{k,\gamma+2m-|\beta|}(\Omega)$ and therefore to $H^{k,\gamma+1}(\Omega)$. Hence by Lemma 13 we see that

$$\|a_\beta(x)D^\beta u\|_{k,\gamma+1} \le C\|u\|_{k+2m,\gamma}.$$

By Lemma 13

$$[a_\alpha(0) - a_\alpha(x)]D^\alpha u \in H^{k,\gamma+1}(\Omega) \text{ for } |\alpha| = 2m.$$

Thus all the summands in the expression $P^0(0,D)u - f$ belongs to $H^{k,\gamma+1}$. In an analogous way we consider the function

$$B_l^0(0,D)u - g_l = \sum_{|\alpha|=m_l} [b_{l,\alpha}(0) - b_{l,\alpha}(x)]D^\alpha u - \sum_{|\beta|<m_l} b_{l,\beta}(x)D^\beta u.$$

Since $u \in H^{k+2m,\gamma}(\Omega)$, we see that by Lemma 7, for $|\beta| < m_l$

$$b_{l,\beta}(x)D^\beta u \in H^{k+2m-m_l-1/2,\gamma+1}(\Gamma).$$

Also by Lemma 7 we have for $|\alpha| = m_l$

$$[b_{l,\alpha}(0) - b_{l,\alpha}(x)]D^\alpha u \in H^{k+2m-|\beta|-1/2,\gamma+1}(\Gamma). \qquad \Box$$

5.5.2 The construction of the parametrix

Let us fix a partition of unity $\sum \psi_j(x) = 1$, where $\psi_j \in C_0^\infty(\mathbb{R}^n)$ and the diameters of the supports of ψ_j are smaller than $\varepsilon/2$. The number $\varepsilon > 0$ is supposed to be so small that in a 2ε-neighbourhood of the origin the boundary is described by the equation

$$x_n^{2p} = \sum_{|\alpha'|=2p,\alpha_n=0} h_\alpha x^\alpha,$$

and for admissible values of β the inequalities

$$r^{|\alpha|}|D^\alpha[a_\beta(x) - a_\beta(0)]| \le \delta/2, \ r^{|\alpha|}|D^\alpha[b_{l\beta}(x) - b_{l,\beta}(0)]| \le \delta/2$$

are valid for $|\alpha| \le k$, where δ was indicated in Theorem 9. On the other hand, the variations of coefficients of the operators P, B_l in the ε-neighbourhood of the origin are so small that it is possible to construct a parametrix in such a neighbourhood, acting as in 5.3.2.

Let G_j be the support of the function ψ_j. If G_j does not intersect the boundary Γ, then there exists a function u_j satisfying the conditions

$$P_0(x, D_x)u_j = \psi_j f + T_j f,$$

where T_j is an operator such that

$$\|T_j f\|_{k+1} \le C\|f\|_k.$$

If G_j intersects the boundary Γ, but not the $\varepsilon/2$-neghbourhood of the origin, then we can find a function $u_j = R[\psi_j f, \psi_j g_1, \ldots, \psi_j g_m]$ such that

$$P^0(x, D_x)u_j = \psi_j f + T_{j0}[\psi_j f, \psi_j g_1, \ldots, \psi_j g_m] \text{ in } \Omega,$$

$$B_l^0(x, D_x)u_j = \psi_j g_l + T_{jl}[\psi_j f, \psi_j g_1, \ldots, \psi_j g_m] \text{ on } \Gamma,$$

where

$$\|T_{jl}[\psi_j f, \psi_j g_1, \ldots, \psi_j g_m]\|_{k+1,\gamma} \leq C(\|f\|_{k,\gamma} + \sum_{l=1}^{m} \|g_l\|_{k+2m-m_l-1/2,\gamma})$$

for $l = 0, 1, \ldots, m$.

Finally, let G_j intersect the $\varepsilon/2$-neighbourhood of the origin. Taking as ψ_0 the sum of all ψ_j satisfying this condition, continuing the coefficients of the operators P^0, B_l^0 onto the infinite cone so that the conditions of Theorem 2 are satisfied, and applying the construction of Section 5.3.2, we get the function u_0 such that

$$P'(x, D)u_0 = \psi_0 f, \ B_l' u_0 = \psi_0 g_l, \ l = 1, \ldots, m,$$

where P', B_l' are the operators obtained as the result of the indicated continuation of the coefficients of P^0 and B_l^0. The existence of the function u_0 follows from Theorem 2, if R_λ has no poles on the straight line $\operatorname{Im} \lambda = 2m + k + \gamma - n/2$.

Now if we put $v = R[f, g_1, \ldots, g_m]$, where

$$v = \sum_j \psi_j u_j + \psi_0 u_0,$$

then we have

$$P(x, D)v = f + \sum_{j=0}^{N} \sum_{|\alpha|<m} c_\alpha(x)D^\alpha u_j,$$

$$B_l(x, D)v = g_l + \sum_{j=0}^{N} \sum_{|\alpha|<m_l} c_{l\alpha}(x)D^\alpha u_j.$$

The functions $c_\alpha(x)D^\alpha u_j$ with $j \neq 0$ belong to $H^{k+1}(\Omega)$ and have supports at a positive distance from the origin. Therefore, if $\|f\|_{k,\gamma} \leq c_0$, $\|g_l\|_{k+2m-m_l,\gamma} \leq c_0$, then the set of the functions $c_\alpha(x)D^\alpha u_j$ with $|\alpha| < 2m, j > 0$ is compact in $H^{k,\gamma}(\Omega)$ and the set of the functions $c_{l\alpha}(x)D^\alpha u_j$ with $|\alpha| \leq m_l - 1, j > 0$ is also compact in $H^{k,\gamma}(\Omega)$.

The compactness of the sets of the functions $c_\alpha(x)D^\alpha u_0$ with $|\alpha| < 2m$ and of the functions $c_{l\alpha}(x)D^\alpha u_0$ with $|\alpha| \leq m_l - 1$ follows from Lemmas 15 and 16.

So putting $F = (f, g_1, \ldots, g_m)$, we obtain the relation

$$\mathcal{A}RF = F + TF,$$

where T is a compact operator in $H^{k,\gamma}(\Omega)$.

We can now prove the main result.

5.5.3 Proof of Theorem 1

If R is the parametrix which has been constructed above, then we have

$$\mathcal{A}RF = F + TF, \ F = (f, g_1, \ldots, g_m).$$

The Fredholm theory says that the conditions of Theorem 1 are sufficient to solve the equation

$$(I + T)F_1 = F$$

for a given F. But then the function $u = RF_1$ is a solution of the boundary value problem, since

$$\mathcal{A}RF_1 = F_1 + TF_1 = F. \qquad \square$$

5.5.4 Smoothness of solutions

Theorem 19. *Let $u \in H^{k+2m,\gamma}(\Omega)$ and*

$$f \in H^{k+1,\gamma-1}(\Omega), \quad g_l \in H^{k+2m-m_l+1/2,\gamma-1}(\Gamma) \text{ for } l = 1, \ldots, m$$

and the resolvent R_λ of the auxiliary problem have no poles on the straight line $\operatorname{Im} \lambda = k + 2m + \gamma - n/2$. *Then $u \in H^{k+2m+1,\gamma-1}(\Omega)$ and there exists a constant C independent of u such that*

$$\|u\|_{k+2m+1,\gamma-1} \leq C(\|f\|_{k+1,\gamma-1} + \sum_{l=1}^{m} \|g_l\|_{k+2m-m_l+1/2,\gamma-1} + \|u\|_{k+2m,\gamma}).$$

Proof. Let $\psi_j(x)$, $j = 0, 1, 2, \ldots$ be the functions introduced in Section 5.5.2. The function $u_0 = \psi_0 u$ vanishes outside a neighbourhood of the origin. It is clear that

$$P^0(x, D)u_0 = \psi_0 f + P'(x, D)u \text{ in } \Omega,$$

$$B_{l0}(x, D)u_0 = \psi_0 g_l + B_l'(x, D)u \text{ on } \Gamma, \ l = 1, \ldots, m,$$

where P' and B_l' are linear differential operators, whose orders are less than $2m$ and m_l, respectively. The coefficients of these operators depend on the choice of the function ψ_0 and vanish outside a neighbourhood of the origin. By Lemmas 12 and 14 we have

$$\|P'(x, D)u\|_{k+1,\gamma-1} \leq M\|u\|_{k+2m,\gamma},$$

$$\|B_l'(x, D)u\|_{k+2m-m_l+1/2,\gamma-1} \leq M\|u\|_{k+2m,\gamma}.$$

Therefore

$$P_0(x, D)u_0 = f_0 \text{ in } \Omega, \ B_{l0}(x, D)u_0 = g_{l0} \text{ on } \Gamma, \ l = 1, \ldots, m,$$

where f_0 and g_{l0} vanish outside some neighbourhood of the origin and satisfy the inequalities

$$\|f_0\|_{k+1,\gamma-1} \leq C(\|f\|_{k+1,\gamma+1} + \|u\|_{k+2m,\gamma}),$$

$$\|g_{l0}\|_{k+2m-m_l+1/2,\gamma-1} \leq C(\|g_l\|_{k+2m-m_l+1/2,\gamma+1} + \|u\|_{k+2m,\gamma}).$$

One can continue the coefficients of the operator P_0 on the whole space and the coefficients of the operators B_{l0} on the whole boundary of the cone $K = \{x : x_n^{2p} = \sum g_\alpha x^\alpha\}$ in such a way that these inequalities remain satisfied. Then we define the function u_0 to be zero outside the support of the function ψ_0 and consider it as a solution of the boundary value problem in the infinite cone.

It follows from Theorem 3 that the solution u_0 of this auxiliary problem is unique in the space $H^{k+2m+1,\gamma}(K)$ and

$$\|u_0\|_{k+2m+1,\gamma-1} \leq C(\|f\|_{k+1,\gamma-1} + \sum_{l=1}^{m} \|g_l\|_{k+2m-m_l+1/2,\gamma-1} + \|u_0\|_{k+2m,\gamma}).$$

Now we represent u in the form

$$u = \psi_0 u + \sum \psi_j u$$

and consider a function $u_j = \psi_j u$. This function vanishes in a neighbourhood of the conical point and

$$P(x,D)u_j = \psi_j f + P'(x,D)u = f_j \text{ in } \Omega,$$

$$B_l(x,D)u_j = \psi_j g_l + B_l'(x,D)u = g_{lj} \text{ on } \Gamma, \ l = 1,\ldots,m,$$

where P' and B_l' are linear differential operators, whose orders are smaller than $2m$ and m_l, respectively. The coefficients of these operators depend on the choice of the function ψ_0 and vanish in a neighbourhood of the origin. Therefore,

$$\|f_j\|_{k+1} \leq C(\|f\|_{k+1,\gamma+1} + \|u\|_{k+2m,\gamma}),$$

$$\|g_{lj}\|_{k+2m-m_l+1/2,\gamma+1} \leq C(\|g_l\|_{k+2m-m_l+1/2,\gamma+1} + \|u\|_{k+2m,\gamma}).$$

By the smoothness theorem for elliptic problems (see Theorem 8 of 4.2.5) we have $u_j \in H^{k+2m+1}(\Omega)$ and

$$\|u_j\|_{k+2m+1} \leq C(\|f\|_{k+1,\gamma+1} + \sum_{l=1}^{m} \|g_l\|_{k+2m-m_l+1/2,\gamma+1} + \|u\|_{k+2m,\gamma}).$$

Therefore, $u \in H^{k+2m+1,\gamma+1}(\Omega)$ and

$$\|u\|_{k+2m+1,\gamma+1} \leq C(\|f\|_{k+1,\gamma+1} + \sum_{l=1}^{m} \|g_l\|_{k+2m-m_l+1/2,\gamma+1} + \|u\|_{k+2m,\gamma}). \qquad \square$$

5.5.5 The solution of the boundary value problem in usual Sobolev spaces

Using the above construction of a parametrix in the space $H^{k,\gamma}(\Omega)$ it is possible to prove similar theorems for usual Sobolev spaces $H^k(\Omega)$. It is clear that the value $\gamma = 0$ is then essential. The main condition of Theorem 1 in this case is the absence of poles of the resolvent for an auxiliary problem on the line $\operatorname{Re}\lambda = k+2m-n/2$. If n is even and $k \geq n/2$, then it generates new conditions of compatibility, necessary for the solvability of the boundary value problem.

In order to understand this let us return for a moment to the model problem in Section 5.1. Let u be a solution of the equation $\Delta u = f$ in the domain $K = \{(x,y) : x > 0, y > 0\}$ and $u(0,y) = 0$, $u(x,0) = 0$. If $u \in C^2(\overline{K})$, then

$$\frac{\partial^2 u}{\partial x^2}(0,0) = 0, \quad \frac{\partial^2 u}{\partial y^2}(0,0) = 0$$

and the function f vanishes at the origin. Moreover, if $u \in C^k(\overline{K})$ and $k > 2$, then

$$D_x^i D_y^j f(0,0) = 0 \quad \text{for } i + j \leq k - 2,$$

since all derivatives $D^\alpha u(0,0)$ vanish for $|\alpha| \leq k$ except for $D_x D_y u(0,0)$. Therefore, there is a constant C such that the function $u_1 = u - Cxy$ is also a solution of the boundary value problem and satisfies the conditions

$$D_x^i D_y^j u_1(0,0) = 0 \text{ for } i + j \leq k.$$

If $u \in H^{k,0}$ and $k \geq 3$, then $f \in H^{k-2,0}$, and for all derivatives $D^\alpha f$ with $|\alpha| \leq k-3$ we have

$$\int r^{-2}|D^\alpha f|^2 dx < \infty.$$

In the general case the compatibility conditions mean that there is a function $v(x)$ of the class $H^{k+2m}(\Omega)$ such that

$$P(x,D)v(x) - f(x) \in H^k(\Omega),$$

$$B_l(x,D)v(x)|_\Gamma - g_l(x) \in H^{k+2m-m_l-1/2}(\Gamma).$$

Theorem 20. *There exist N linear continuous functionals $\Phi_1,\ldots,\Phi_N$ on the space $\mathcal{H}^k = H^k(\Omega) \times H^{k+2m-m_1-1/2}(\Gamma) \times \ldots \times H^{k+2m-m_m-1/2}(\Gamma)$ such that if $f \in H^k(\Omega), g_l(x) \in H^{k+2m-m_l-1/2}(\Gamma)$, the finite number of conditions*

$$\Phi_j[f,g_1,\ldots,g_m] = 0, \ j = 1,\ldots,N$$

are satisfied as well as the compatibility conditions and the line $\operatorname{Im}\lambda = 2m+k-n/2$ does not contain poles of $R(\lambda)$, then there exists a solution to the boundary value problem

$$Pu = f \text{ in } \Omega, \ B_l u = g_l \text{ on } \Gamma, \ l = 1,\ldots,m$$

belonging to the class $H^{k+2m}(\Omega)$ such that

$$\|u\|_{k+2m} \le C\Big(\|f\|_k + \sum_{l=1}^{m} \|g_l\|_{k+2m-m_l-1/2}\Big).$$

Proof. Let $v(x)$ be a function of the class $H^{k+2m,0}(\Omega)$ such that

$$Pv - f \in H^{k,0}(\Omega), \ B_l v - g_l \in H^{k+2m-m_l-n/2,0}(\Gamma), \ l = 1,\ldots,m.$$

Put $u_1 = u - v$. Then u_1 is a solution of the equation $Pu_1 = f_1 \in H^{k,0}(\Omega)$ and satisfies the boundary conditions

$$B_l u_1 = g_{l1} \in H^{k+2m-m_l-1/2,0}(\Gamma), \ l = 1,\ldots,m.$$

It follows from Theorem 1 that this problem is solvable when a finite number of conditions of the form $\Phi_j[f, g_1, \ldots, g_m] = 0$ are fulfilled. $\qquad\square$

Chapter 6

Non-elliptic operators; propagation of singularities

6.1 Canonical transformations and Fourier integral operators

6.1.1 Definitions

Let M be a smooth manifold of dimension $2n$, on which a smooth differential form ω^2 is defined, such that in every local coordinate system,

$$\omega^2 = \sum \omega_{ij} dx_i \wedge dx_j, \quad \omega_{ij} = -\omega_{ji}.$$

It is assumed that this form is non-degenerate, i.e. the matrix $\|\omega_{ij}(x)\|$ is non-degenerate at each point x, and closed, i.e. $d\omega^2 = 0$, where d is the exterior differentiation operator. Such a form ω^2 is called a *symplectic structure* on the manifold. A manifold having such a structure is called *symplectic*.

If Ω is a domain in $\mathbb{R}^n$ then the manifold $T^*\Omega$ has the dimension $2n$ and the symplectic structure can be defined by the form

$$\omega^2 = dx \wedge d\xi = \sum_{j=1}^{n} dx_j \wedge d\xi_j.$$

Indeed, this form is non-degenerate: $\omega_{ij} = -\delta_{i+n,j}$ for $i \leq n$, $\omega_{ij} = \delta_{i-n,j}$ for $n+1 \leq i \leq 2n$, so that $\det \|\omega_{ij}\| = 1$. Furthermore, $\omega^2 = d\omega^1$, where the form ω^1 is equal to $\omega^1 = -\xi dx$. For an arbitrary smooth manifold Ω a symplectic structure on $T^*\Omega$ can be defined with the help of the form ω^2, equal to $dx \wedge d\xi$ in every local coordinate neighbourhood. Moreover, $\omega^2 = d\omega^1$ and the form ω^1 is defined on $T^*\Omega$.

A transformation $\Phi : T^*\Omega \to T^*\Omega$ is called *canonical*, if it smooth and preserves the symplectic structure.

6.1.2　Examples

Example 1. The change of coordinates $y = F(x)$ in the domain Ω of $\mathbb{R}^n$ induces a transformation of the cotangent space $T^*\Omega$:

$$y = F(x), \ \eta = (F'(x)^t)^{-1}\xi,$$

where $F'(x)^t$ is the transposed Jacobi matrix. It is evident that

$$dy = F'(x)dx, \ d\eta = (F'(x)^t)^{-1}d\xi + ((F'(x)^t)^{-1})'\xi dx.$$

Therefore,

$$dx \wedge d\xi = F'(x)^{-1}dy \wedge d\xi,$$

$$dy \wedge d\eta = dy \wedge (F'(x)^t)^{-1}d\xi + dy \wedge ((F'(x)^t)^{-1})'\xi dx$$

$$= dy \wedge (F'(x)^t)^{-1}d\xi + F'(x)dx \wedge ((F'(x)^t)^{-1})'\xi dx$$

$$= F'(x)^{-1}dy \wedge d\xi,$$

and this transformation is canonical.

Example 2. Let $S(x,\xi)$ be a real-valued function defined on $T^*\Omega$ and

$$\det \|\partial^2 S/\partial x \partial \xi\| \neq 0.$$

Let $(y,\eta) = \Phi(x,\xi)$, where

$$\xi = \frac{\partial S(x,\eta)}{\partial x}, \ y = \frac{\partial S(x,\eta)}{\partial \eta}.$$

It is obvious that Φ is a diffeomorphism. Moreover,

$$dx \wedge d\xi = dx \wedge (\frac{\partial^2 S(x,\eta)}{\partial x^2}dx + \frac{\partial^2 S(x,\eta)}{\partial x \partial \eta}d\eta) = dx \wedge \frac{\partial^2 S}{\partial x \partial \eta}d\eta,$$

$$dy \wedge d\eta = (\frac{\partial^2 S(x,\eta)}{\partial \eta \partial x}dx + \frac{\partial^2 S(x,\eta)}{\partial \eta^2}d\eta) \wedge d\eta = dx \wedge \frac{\partial^2 S}{\partial x \partial \eta}d\eta,$$

so that Φ is a canonical transformation. The function S is called *generating* function of the transformation Φ. The transformation of Example 1 is the particular case corresponding to the function $S(x,\xi) = F(x) \cdot \xi$.

Example 3. Let $H(x,\xi)$ be a real-valued function defined on $T^*\Omega$ and $(x(t),\xi(t))$ be the solution of the Hamiltonian system

$$\dot{x} = \frac{\partial H(x,\xi)}{\partial \xi}, \ \dot{\xi} = -\frac{\partial H(x,\xi)}{\partial x},$$

satisfying the initial conditions $x(0) = x$, $\xi(0) = \xi$. Then

$$\frac{d}{dt}(dx(t) \wedge d\xi(t)) = d\dot{x}(t) \wedge d\xi(t) + dx \wedge d\dot{\xi} = \frac{\partial^2 H}{\partial \xi^2} d\xi(t) \wedge d\xi(t)$$

$$+ \frac{\partial^2 H}{\partial \xi \partial x} dx(t) \wedge d\xi(t) - dx(t) \wedge \frac{\partial^2 H}{\partial x^2} dx(t) - dx(t) \wedge \frac{\partial^2 H}{\partial x \partial \xi} d\xi(t) = 0.$$

Therefore, the map

$$\Phi_t : (x, \xi) \to (x(t), \xi(t))$$

is canonical for all t, i.e. the Hamiltonian flow preserves a symplectic structure.

The Poisson bracket $\{f, g\}$ of two functions from $C^1(T^*\Omega)$ is the derivative of the function f with respect to the phase Hamiltonian flow of the function g. In local coordinates

$$\{f, g\} = \frac{d}{dt} f(x(t), \xi(t)),$$

where $\dot{x}(t) = g_\xi(x, \xi)$, $\dot{\xi}(t) = -g_x(x, \xi)$, so that

$$\{f, g\} = \sum_{j=1}^{n} \left(\frac{\partial f}{\partial x_j} \frac{\partial g}{\partial \xi_j} - \frac{\partial f}{\partial \xi_j} \frac{\partial g}{\partial x_j} \right).$$

Theorem 4. *A transformation Φ is canonical if and only if it preserves the Poisson brackets for any two smooth functions defined on $T^*\Omega$.*

Proof. The form ω^2 allows us to find an isomorphism between $T^*(T^*\Omega)$ and $T(T^*\Omega)$ in the following way. Put in the correspondence to each element $\alpha \in T(T^*\Omega)$ an 1-form ω_α^1 so that

$$\omega_\alpha^1(\beta) = \omega^2(\beta, \alpha), \quad \beta \in T(T^*\Omega).$$

Since the form ω^2 is not degenerate, a unique element $\alpha \in T(T^*\Omega)$ satisfying this relation corresponds to each form ω^1. Therefore, putting $\alpha = J\omega^1$, we get the isomorphism $J : T^*(T^*\Omega) \to T(T^*\Omega)$. If $\omega^2 = dx \wedge d\xi$ in local coordinates, then

$$J = \begin{pmatrix} 0 & -I \\ I & 0 \end{pmatrix},$$

where I is the unit matrix of n-th order.

Let us note that

$$\{f, g\} = \omega^2(Jdf, Jdg),$$

and the invariance of Poisson brackets with respect to canonical transformations follows from the invariance of the form ω^2 and of the isomorphism J. Since the differentials of the functions x_j, ξ_k form a base in $T^*(T^*\Omega)$, the invariance of their Poisson brackets implies the invariance of values of the form ω^2. $\qquad\square$

6.1.3 Fourier integral operators and canonical transformations

The canonical transformations of the space $T^*\Omega$, preserving the symbol classes $S^m(\Omega)$, are of special interest for us. It is easy to see that a canonical transformation, defined by a generating function $S(x,\xi)$, satisfies this condition if

(1) $|D_\xi^\alpha D_x^\beta S(x,\xi)| \le C_{\alpha,\beta}(1+|\xi|)^{1-|\alpha|}$; $\forall\ \alpha,\ \beta$;

(2) $|\partial S(x,\xi)/\partial x| \ge C_1|\xi|$ for $|\xi| \ge C_2$, $C_1 = \mathrm{const} > 0$;

(3) $|\partial^2 S(x,\xi)/\partial x\partial\xi| \ge c_0$, $c_0 = \mathrm{const} > 0$ for $|\xi| \ge C_2$.

A real-valued function S satisfying these conditions is called a *phase function*. The operator, moving a function u in the function

$$(2\pi)^{-n}\int a(x,\xi)\widetilde{u}(\xi)e^{iS(x,\xi)}d\xi,$$

is called a *Fourier integral operator*, if $a \in S^k(\Omega)$ for some k. The function a is then called a *symbol of the Fourier integral operator*. The partial case $S(x,\xi) = x\cdot\xi$ corresponds to the usual pseudo-differential operators.

Theorem 5. *Let S be a phase function, $a \in S_0^0(\Omega)$ and P be a pseudo-differential operator with a symbol $p \in S_0^m(\Omega)$. Let*

$$\Phi u(x) = (2\pi)^{-n}\int a(x,\xi)\widetilde{u}(\xi)e^{iS(x,\xi)}d\xi,$$

Ω' be the image of Ω at the application, defined by the function S, and Q_0 be the operator with the symbol q_0, defined by the relation

$$q_0\Big(\frac{\partial S(x,\eta)}{\partial\eta},\eta\Big) = p\Big(x,\frac{\partial S(x,\eta)}{\partial x}\Big).$$

Then the operator $P\Phi - \Phi Q_0$ has the order $m-1$.

Moreover, if $|a(x,\xi)| \ge C_0 > 0$, then there is a pseudo-differential operator Q with a symbol $q \in S_0^m(\Omega')$, such that the operator $Q - Q_0$ has the order $m-1$ and the order of the operator $T = P\Phi - \Phi Q$ is equal to $-\infty$.

Proof. **A.** Consider first the operator $P\Phi$. We have

$$P\Phi u(x) = (2\pi)^{-2n}\int\int\int p(x,\eta)e^{ix\eta}d\eta e^{-iy\eta}dy\, a(y,\xi)\widetilde{u}(\xi)e^{iS(y,\xi)}d\xi$$

$$= (2\pi)^{-n}\int k(x,\xi)\widetilde{u}(\xi)e^{iS(x,\xi)}d\xi,$$

where we have put

$$k(x,\xi) = (2\pi)^{-n}\int\int a(y,\xi)p(x,\eta)e^{i[S(y,\xi)-S(x,\xi)+(x-y)\eta]}dyd\eta.$$

Let a function $h_1(x, y)$ from the space $C^\infty(\mathbb{R}^{2n})$ be such that $h_1(x, y) = 0$ for $|x - y| \leq \delta$; $h_1(x, y) = 1$ for $|x - y| > 2\delta$, where $\delta = \text{const} > 0$, and k_1 be the integral of the same form as k but with a supplementary factor $h_1(x, y)$ under the integral sign. Let us show that $k_1 \in S^{-\infty}(\Omega)$. Note that for any N

$$e^{i(x-y)\eta} = |x - y|^{-2N}(-\Delta_\eta)^N e^{i(x-y)\eta}.$$

Furthermore, we can see that for $|x - y| \geq \delta$ the inequalities

$$1 + |\eta - S_y(y, \xi)|^2 + |\eta|^2 |x - y|^2$$

$$\geq 1 + (1 + \delta^2)|\eta|^2 + |S_y(y, \xi)|^2 - 2|\eta| \cdot |S_y(y, \xi)|$$

$$\geq 1 + \frac{\delta^2}{1 + \delta^2}|S_y(y, \xi)|^2 \geq C_1(\delta)(1 + |\xi|^2);$$

$$|\Delta_y S(y, \xi)| \leq C(1 + |\xi|) \leq \frac{1}{2}C_1(\delta)(1 + |\xi|^2) + C_2(\delta)$$

are valid. Therefore, for large enough A_0 the inequality

$$F(x, y, \xi, \eta) := |A_0 + |\eta - S_y(y, \xi)|^2 + |\eta|^2 |x - y|^2$$

$$- i\Delta_y S(y, \xi)| \geq c_0(1 + |\xi|^2), \quad c_0 > 0$$

is fulfilled. From the equality

$$e^{i[S(y,\xi) - S(x,\xi) + (x-y)\eta]}$$

$$= [F(x, y, \xi, \eta)^{-1}(A_0 - \Delta_y - |\eta|^2 \Delta_\eta)]^M e^{i[S(y,\xi) - S(x,\xi) + (x-y)\eta]}$$

it follows that

$$k_1(x, \xi) = (2\pi)^{-n} \int \int [(A_0 - \Delta_y - |\eta|^2 \Delta_\eta)F^{-1}]^M \{|x - y|^{-2N}$$

$$\times h_1(x, y)a(y, \xi)(-\Delta_\eta)^N p(x, \eta)\} e^{i[S(y,\xi) - S(x,\xi) + (x-y)\eta]} dy d\eta.$$

This integral converges if $2N > |m| + n$. Moreover, then $|k_1(x, \xi)| \leq C(\delta, M)(1 + |\xi|^2)^{-M}$ for any δ and M. Similar inequalities can be obtained for the derivatives $D_\xi^\alpha D_x^\beta k_1(x, \xi)$, so that $k_1 \in S^{-\infty}(\Omega)$. Let $h_2(\xi, \eta) \in C^\infty(\mathbb{R}^{2n})$, $h_2(\xi, \eta) = 0$ for $|\xi| + |\eta| < \frac{1}{2}$;

$$h_2(\xi, \eta) = 1 \quad \text{for} \quad |\eta - \frac{\partial S(x, \xi)}{\partial x}| \geq 4\mu\delta(|\xi| + |\eta|);$$

$$h_2(\xi, \eta) = 0 \quad \text{for} \quad |\eta - \frac{\partial S(x, \xi)}{\partial x}| < 3\mu\delta(|\xi| + |\eta|),$$

where

$$\mu^2 = \sup \sum_{i,j} |\frac{\partial^2 S(x, \xi)}{\partial x_i \partial x_j}|^2 (1 + |\xi|^2)^{-1}.$$

Put

$$k_2(x,\xi) = (2\pi)^{-n} \int\int a(y,\xi)p(x,\eta)[1 - h_1(x,y)]h_2(\xi,\eta)$$

$$\times e^{i[S(y,\xi)-S(x,\xi)+(x-y)\eta]}dyd\eta.$$

Note that for $|x - y| \geq 2\delta$ the relation

$$\left|\frac{\partial S(x,\xi)}{\partial x} - \frac{\partial S(y,\xi)}{\partial y}\right| \geq 2\delta\mu|\xi|$$

holds, and therefore $h_2(\xi,\eta) = 0$, if $|\eta - \partial S(y,\xi)/\partial y| < \mu\delta(|\xi| + |\eta|)$, and $|x - y| \geq 2\delta$. Therefore, we have for points from $\operatorname{supp} h_2$ the inequality

$$|F(x,y,\xi,\eta)| \geq A_0 + [\mu\delta(|\xi| + |\eta|)]^2,$$

and we can repeat the arguments used for the estimates of the function k_1, so that $k_2 \in S^{-\infty}$. Let us consider the function

$$k_3(x,\xi) = (2\pi)^{-n} \int\int a(y,\xi)p(x,\eta)[1 - h_1(x,y)][1 - h_2(\xi,\eta)]$$

$$\times e^{i[S(y,\xi)-S(x,\xi)+(x-y)\eta]}dyd\eta.$$

It is obvious that

$$S(y,\xi) - S(x,\xi) = \frac{\partial S(x,\xi)}{\partial x}(y - x) + A(x,y,\xi)(y - x) \cdot (y - x),$$

where

$$A_{ij}(x,y,\xi) = \int_0^1 (1-t)\frac{\partial^2 S(x + t(y - x),\xi)}{\partial x_i\partial x_j}dt.$$

Let $\eta_1 = \eta - A(x,y,\xi)(y - x)$, $\xi_1 = \partial S(x,\xi)/\partial x$. Then

$$k_3(x,\xi) = (2\pi)^{-n} \int\int a(y,\xi)p(x,\eta_1 + A(x,y,\xi)(y - x))$$

$$[1 - h_1(x,y)]e^{i(x-y)(\eta_1-\xi_1)}dyd\eta_1.$$

If δ is small enough, then for

$$|\eta - \xi_1| < 4\mu\delta(|\xi| + |\eta|)$$

the inequality $|\eta_1 + A(x,y,\xi)(y - x)| \geq \frac{1}{4}(|\xi_1| + |\eta_1|)$ holds. But then we have

$$|D_\xi^\alpha D_{\eta_1}^{\alpha'} D_x^\beta D_y^{\beta'} a(y,\xi)p(x,\eta_1 + A(x,y,\xi)(y - x))[1 - h_1(x,y)]$$

$$\times [1 - h_2(\xi,\eta)]| \leq C_{\alpha,\alpha',\beta,\beta'}(1 + |\xi|)^{-|\alpha|}(1 + |\eta_1|)^{m-|\alpha'|}.$$

By Lemma 13 of 2.3.1 we get that $k_3 \in S^m(\Omega)$. Moreover, $k \in S^m(\Omega)$, $k - k_0 \in S^{m-1}(\Omega)$, where $k_0(x,\xi) = a(x,\xi)p(x,\partial S(x,\xi)/\partial x)$.

B. We consider the operator ΦQ in the same way. We have

$$\Phi Q u(x) = (2\pi)^{-2n} \int\int\int a(x,\eta)e^{iS(x,\eta)}d\eta \cdot e^{-iy\eta}dy \cdot q(y,\xi)\tilde{u}(\xi)e^{iy\xi}d\xi$$

$$= (2\pi)^{-n}\int b(x,\xi)\tilde{u}(\xi)e^{iS(x,\xi)}d\xi,$$

where

$$b(x,\xi) = (2\pi)^{-n}\int\int q(y,\xi)a(x,\eta)e^{i[S(x,\eta)-S(x,\xi)+y(\xi-\eta)]}dyd\eta.$$

As above, consider the integral $b_1(x,\xi)$ obtained from b by inserting the additional factor $h(\xi,\eta)$ under the integral sign, where $h \in C^\infty$, $h(\xi,\eta) = 0$ if $|\xi - \eta| \leq \delta(1 + |\xi| + |\eta|)$ and $h(\xi,\eta) = 1$ if $|\xi - \eta| \geq 2\delta(1 + |\xi| + |\eta|)$. Representing $e^{iy(\xi-\eta)}$ in the form $|\xi - \eta|^{-2N}(-\Delta_y)^N e^{iy(\xi-\eta)}$ and integrating by parts with respect to y, we obtain the integral of an expression majorized by

$$C_1(1 + |\xi| + |\eta|)^{-2N} \leq C_1(1 + |\xi|)^{-N}(1 + |\eta|)^{-N},$$

with a bounded domain of integration with respect to y. Since similar estimates are also valid for the derivatives $D_\xi^\alpha D_x^\beta b_1(x,\xi)$, we obtain that $b_1 \in S^{-\infty}$.

Let us now show that $b - b_1 \in S^m$. We have

$$S(x,\eta) - S(x,\xi) = B(x,\xi,\eta)(\eta - \xi),$$

where

$$B(x,\xi,\eta) = \int_0^1 \partial S(x,\xi + t(\eta - \xi))/\partial\xi dt,$$

so that $B(x,\xi,\xi) = \partial S(x,\xi)/\partial\xi$ and

$$|D_\xi^\alpha D_\eta^\beta D_x^\gamma B(x,\xi,\eta)| \leq C_{\alpha,\beta,\gamma}(1 + |\xi|)^{-|\alpha|}(1 + |\eta|)^{-|\beta|}(1 + |\xi - \eta|)^{|\alpha+\beta|}.$$

After the change of variables $y_1 = y + x - B(x,\xi,\eta)$ we obtain that

$$b(x,\xi) - b_1(x,\xi) = (2\pi)^{-n}\int\int q(y_1 + B(x,\xi,\eta) - x,\xi)$$

$$\times [1 - h(\zeta,\eta)]u(x,\eta)e^{i(\eta-\xi)(x-y_1)}d\eta dy_1,$$

and

$$|D_\xi^\alpha D_\eta^{\alpha'} D_x^\beta D_{y_1}^{\beta'} q(y_1 + B(x,\xi,\eta) - x,\xi)a(x,\eta)[1 - h(\xi,\eta)]|$$

$$\leq C_{\alpha,\alpha',\beta,\beta'}(1 + |\xi|)^{m-|\alpha|}(1 + |\eta|)^{-|\alpha'|}(1 + |\xi - \eta|)^{|\alpha+\alpha'|}.$$

Lemma 13 of 2.3.1 implies that $b - b_1 \in S^m$ and

$$b(x,\xi) - a(x,\xi)q(\partial S(x,\xi)/\partial\xi,\xi) \in S^{m-1}.$$

C. Thus, if

$$q_0(\partial S(x,\xi)/\partial \xi,\xi) = p(x,\partial S(x,\xi)/\partial x),$$

then the order of the operator $P\Phi - \Phi Q_0$ is equal to $m - 1$ and

$$(P\Phi - \Phi Q_0)u(x) = (2\pi)^{-n}\int p_1(x,\xi)\widetilde{u}(\xi)e^{iS(x,\xi)}d\xi,$$

where $p_1 \in S^{m-1}$. Now let

$$a(x,\xi)q_1(\partial S(x,\xi)/\partial \xi,\xi) = p_1(x,\xi).$$

Then
$$(P\Phi - \Phi Q_0 - \Phi Q_1)u(x) = (2\pi)^{-n}\int p_2(x,\xi)\widetilde{u}(\xi)e^{iS(x,\xi)}d\xi,$$

where $p_2 \in S^{m-2}$. Continuing the process we can find $q_j \in S^{m-j}$ such that

$$(P\Phi - \Phi(Q_0 + \ldots + Q_j))u = (2\pi)^{-n}\int p_{j+1}(x,\xi)\widetilde{u}(\xi)e^{iS(x,\xi)}d\xi,$$

where $p_{j+1} \in S^{m-j-1}$. Lemma 1 of 3.1.1 guarantees the existence of a pseudo-differential operator Q satisfying the conditions of Theorem 5 and such that $q(x,\xi) \in S^m$. $\qquad\qquad\qquad\qquad\qquad\qquad\qquad\qquad\qquad\qquad\qquad\qquad\quad \square$

6.1.4 Canonical transformations and quadratic forms

Theorem 6. *Let* $a \in S_0^0$, $p \in S_0^m$, *P be a pseudo-differential operator with the symbol p and*

$$\Phi u(x) = (2\pi)^{-n}\int a(x,\xi)\widetilde{u}(\xi)e^{iS(x,\xi)}d\xi.$$

Then the operator $Q = \Phi^* P\Phi$ *is pseudo-differential with the symbol* $q(x,\xi)$ *of the class* S_0^m *and*

$$q(y,\eta) - p(x,\xi)|a(x,\xi)|^2|\det S_{x\xi}|^{-1} \in S_0^{m-1},$$

where (y,η) *is the image of a point* (x,ξ) *under the canonical transformation defined by the generating function* S.

Proof. The symbol k of the operator $P\Phi$ was found in the proof of Theorem 5. Therefore,

$$Qu(x) = (2\pi)^{-2n}\int e^{ix\eta}d\eta \int \overline{a(y,\eta)}e^{-iS(y,\eta)}dy$$

$$\times \int k(y,\xi)e^{iS(y,\xi)}\widetilde{u}(\xi)d\xi = (2\pi)^{-n}\int q(x,\xi)\widetilde{u}(\xi)e^{ix\xi}d\xi,$$

where
$$q(x,\xi) = (2\pi)^{-n}\int\int k(y,\xi)\overline{a(y,\eta)}e^{i[S(y,\xi)-S(y,\eta)]+ix(\eta-\xi)}dyd\eta.$$

Let $S(y,\xi) - S(y,\eta) = B(y,\xi,\eta)(\xi - \eta)$, where

$$B(y,\xi,\eta) = \int_0^1 \frac{\partial S(y,\eta + t(\xi - \eta))}{\partial \xi}dt, \quad B(y,\xi,\xi) = \frac{\partial S(y,\xi)}{\partial \xi}.$$

Let us consider the symbol q_1 obtained from the integral q by writing the additional factor $h(\xi, \eta)$ under the integral sign, where $h \in C^\infty$, $h(\xi, \eta) = 0$ if $|\xi - \eta| < \delta(|\xi + |\eta|)$, $h(\xi, \eta) = 1$, if $|\xi - \eta| > 2\delta(|\xi| + |\eta|)$. Show that $q_1 \in S^{-\infty}$. By assumption, the equation $\partial S(x, \xi)/\partial x = \xi_1$ for ξ is solvable if $|\xi_1| \geq C$ and the solution $\xi = \varphi(x, \xi_1)$ is such that

$$|\xi - \eta| \geq C|\xi_1 - \eta_1| = |\frac{\partial S(x, \xi)}{\partial x} - \frac{\partial S(x, \eta)}{\partial x}|.$$

Therefore, there are positive constants A_0 and c_0 such that

$$|A_0 + |\frac{\partial S(y, \xi)}{\partial y} - \frac{\partial S(y, \eta)}{\partial y}|^2 + i\Delta_y[S(y, \xi) - S(y, \eta)]| \geq c_0(1 + |\xi - \eta|^2).$$

From the equality

$$q_1(x, \xi) = (2\pi)^{-n} \int \int h(\xi, \eta)\{(A_0 - \Delta_y)(A_0 + |S_y(y, \xi) - S_y(y, \eta)|^2$$
$$+ i\Delta_y[S(y, \xi) - S(y, \eta)])^{-1}\}^N[k(y, \xi)\overline{a(y, \eta)}]e^{i[S(y,\xi)-S(y,\eta)]+ix(\eta-\xi)}dy d\eta,$$

where the integral with respect to y is taken over a bounded domain and the module of the integrand is majorized by

$$C(\delta, N)(1 + |\xi|)^{m-N}(1 + |\eta|)^{-N}$$

and from the similar equalities for derivatives of q_1 with respect to x and ξ, we can see that $q_1 \in S^{-\infty}$.

Now let δ be so small that $|\det(\partial B(y, \xi, \eta)/\partial y)| \geq c_0/2$ in the support of $(1 - h)$. Put $z = B(y, \xi, \eta)$. Then

$$q(x, \xi) - q_1(x, \xi) = (2\pi)^{-n} \int \int [1 - h(\xi, \eta)]k(y, \xi)\overline{a(y, \eta)}e^{i(x-z)(\eta-\xi)}\frac{dz d\eta}{|\det(\partial B/\partial y)|}$$

and one can apply Lemma 1 of 3.1.1, so that $q \in S_0^m$, and $q - q_0 \in S_0^{m-1}$, if

$$q_0(\frac{\partial S(x, \xi)}{\partial \xi}, \xi) = |a(x, \xi)|^2 p(x, \frac{\partial S(x, \xi)}{\partial x})|\det \frac{\partial^2 S(x, \xi)}{\partial x \partial \xi}|^{-1}.$$

Corollary 7. *If the conditions of Theorem 6 are fulfilled and*

$$a(x, \xi) = |\det \partial^2 S/\partial x \partial \xi|^{1/2},$$

then $\Phi^*\Phi = I + T$, *where* T *is a pseudo-differential operator of order* -1. *In particular, for all* $s \in \mathbf{N}$ *this implies the inequalities*

$$\|\Phi u\|_s \leq C\|u\|_s, \quad \|u\|_s \leq C(\|\Phi u\|_s + \|u\|_{s-1}),$$

for $u \in C_0^\infty(K)$, *where* K *is a compact subset in* Ω.

Repeating the arguments of the proof of Theorem 5, one can find a symbol a such that the operator $\Phi^*\Phi - I$ will be smoothing.

6.1.5　Reduction of operators of principal type to canonical forms

Definition 8. If $p \in S^m$ and there exists

$$\lim_{t \to +\infty} t^{-m} p(x, t\xi) = p_0(x, \xi)$$

for $x \in \Omega$, $\xi \neq 0$, such that $p_0(x, \xi)$ satisfies the inequalities

$$|D_\xi^\alpha D_x^\beta p_0(x, \xi)| \leq C_{\alpha, \beta, K}(1 + |\xi|)^{m - |\alpha|},$$

for $|\xi| \geq 1$, then p_0 is called a *principal symbol.* An operator P with the principal symbol p_0 is an operator *of principal type*, if the form $dp_0(x, \xi)$ is not proportional to the form ξdx at any point of $T^*\Omega \setminus 0$.

The function p_0 is positively homogeneous in the variable ξ of degree m and therefore, the Euler identity

$$\sum_{j=1}^{n} \xi_j \partial p_0(x, \xi)/\partial \xi_j = m p_0(x, \xi)$$

is valid. Therefore, $d_\xi p_0(x, \xi) \neq 0$, if $p_0(x, \xi) \neq 0$. Hence P is an operator of principal type if and only if at each point $(x, \xi) \in T^*\Omega \setminus 0$ with $p_0(x, \xi) = 0$, $d_\xi p_0(x, \xi) = 0$, the vector $\operatorname{grad}_x p_0(x, \xi)$ is not collinear to the vector ξ.

If P_1 is an operator of principal symbol of order m, then $P_1 = \Lambda^{m-1} P$, where Λ is the operator with the symbol $(1 + |\xi^2|)^{1/2}$, and P is an operator of principal type of first order.

Theorem 9. *Let P be a pseudo-differential operator of principal type of first order, the function p_0 be real and $p_0(x_0, \xi_0) = 0$, $x_0 \in \Omega$, $\xi_0 \neq 0$. Then there exists a Fourier integral operator Φ, such that the operator $P\Phi - \Phi Q$ is smoothing, and Q is a pseudo-differential operator of first order, whose symbol is equal to ξ_k in a conical neighbourhood of the point (x_0, ξ_0) for some $k, 1 \leq k \leq n$.*

Proof. **A.** First let $\partial p_0(x_0, \xi_0)/\partial \xi_1 \neq 0$. Consider the following Cauchy problem:

$$p_0\left(x, \frac{\partial S(x, \eta)}{\partial x}\right) = \eta_1, \quad S = \sum_{j=2}^{n}(x_j - x_{0j})\eta_j \text{ at } x_1 = x_{01}.$$

Since the plane $x_1 = x_{01}$ is non-characteristic, this problem has a solution in a neighbourhood of the point (x_0, ξ_0). Since a solution of the Cauchy problem is unique and the function p_0 is homogeneous, the function $tS(x, \eta)$ is a solution of the same problem after replacing η by $t\eta$, so that $S(x, t\eta) = tS(x, \eta)$ for $t > 0$. Differentiating the equation for S with respect to η_1, we obtain that

$$\sum_{j=1}^{n} \frac{\partial p_0}{\partial \xi_j}\left(x, \frac{\partial S}{\partial x}\right) \frac{\partial^2 S}{\partial x_j \partial \eta_1} = 1.$$

But $\partial^2 S(x_0, \eta_0)/\partial x_j \partial \eta_1 = 0$ for $j = 2, \ldots, n$ by virtue of the initial conditions. Hence

$$\frac{\partial^2 S(x_0, \eta_0)}{\partial x_1 \partial \eta_1} = \left(\frac{\partial p_0(x_0, \xi_0)}{\partial \xi_1}\right)^{-1} \neq 0$$

and

$$\det \left\| \frac{\partial^2 S(x_0, \eta_0)}{\partial x_i \partial \eta_j} \right\| = \frac{\partial^2 S(x_0, \eta_0)}{\partial x_1 \partial \eta_1} \neq 0,$$

if $\partial S(x_0, \eta_0)/\partial x_1 = \xi_{01}$ and $\eta_{0j} = \xi_{0j}$, $j = 2 \ldots, n$.

By the implicit function theorem, the equation

$$\frac{\partial S(x, \eta)}{\partial x} = \xi$$

is solvable with respect to η in a neighbourhood of the point (x_0, ξ_0). Moreover, if $|\xi| \leq 1$, then $|\eta| \leq C$, so that the homogeneity implies the inequality $|\eta| \leq C|\partial S(x, \eta)/\partial x|$.

Therefore, if $h \in C^\infty$, $h(\xi) = 0$ for $|\xi| \leq 1/2$ and $h(\xi) = 1$, for $|\xi| \geq 1$, then the function $h(\xi)S(x, \xi)$ is a phase function. If Φ is an integral Fourier operator with this phase function and a symbol $a(x, \xi) \in S^0(\Omega)$, which is equal to 1 in a conical neighbourhood of the point (x_0, ξ_0) lying inside the neighbourhood V in which the function $S(x, \xi)$ was defined, and equal to 0 outside V, then by Theorem 5, the operator $P\Phi - \Phi Q$ has the order 0.

B. Now let

$$p_0(x_0, \xi_0) = 0, \operatorname{grad}_\xi p_0(x_0, \xi_0) = 0,$$

but the vector $\operatorname{grad}_x p_0(x_0, \xi_0)$ does not vanish and is not collinear to the vector ξ_0. Let for example, $\operatorname{grad}_x p_0(x_0, \xi_0) = (1, 0, \ldots, 0)$ and the vector $\xi_0 \neq 0$ have non-vanishing coordinate ξ_{01}.

Let us look for the generating function $S(y, \xi)$ as a solution of the Cauchy problem

$$p_0\left(\frac{\partial S}{\partial \xi}, \xi\right) = \frac{\partial S}{\partial y_1}; \quad S = \sum_{j=2}^{n} y_j \xi_j + \frac{1}{2|\xi|}\xi_1^2 \text{ at } y_1 = 0.$$

It is easy to see that such a solution exists in a conical neighbourhood of the point $(0, \xi_0)$ and $S(y, t\xi) = tS(y, \xi)$ for $t > 0$. At the point $(0, \xi_0)$ the equalities

$$\frac{\partial^2 S}{\partial \xi_1 \partial y_1} = |\xi_0|^{-1}; \quad \frac{\partial^2 S}{\partial y_i \partial \xi_j} = \delta_{ij} \text{ for } i, j = 2, \ldots, n$$

hold, so that $\det \|\partial^2 S/\partial y_i \partial \xi_j\| = |\xi_0|^{-1} \neq 0$. It follows that in a conical neighbourhood of the point $(0, \xi_0)$ the inequality $|\xi| \leq C|\partial S(y, \xi)/\partial y|$ is fulfilled.

Now extend S to a phase function. It is sufficient to do this for $|\xi| = 1$, thereafter to continue S as a homogeneous first-order function of ξ, and to multiply it by a function $\psi(\xi)$ vanishing for $|\xi| \leq 1/2$ and equal to 1 for $|\xi| \geq 1$.

Suppose that the solution S of the Cauchy problem exists for $|y| \leq \rho$, $|\xi - \xi_0| < \rho$ and $|\xi_0| = 1$. After computing the values of the first and second order derivatives of the function S at the point $(0, \xi_0)$, we can see that for $|\xi| = 1$,

$$S = S_0(y, \xi) + O(|y^3| + |\xi - \xi_0|^3),$$

where

$$S_0(y, \xi) = \sum_{j=2}^{n} y_j \xi_j + \frac{1}{2|\xi_0|} \xi_1^2 + \frac{1}{2|\xi_0|} y_1^2 + \frac{1}{|\xi_0|} \xi_1 y_1.$$

Let $h(y, \xi) = 1$ for $|y| + |\xi - \xi_0| \leq \rho/2$ and $h(y, \xi) = 0$ for $|y| + |\xi - \xi_0| \geq \rho$. Put for $|\xi| = 1$

$$S_1(y, \xi) = h(y, \xi) S(y, \xi) + [1 - h(y, \xi)] S_0(y, \xi).$$

Then for $i = 2, \ldots, n$; $j = 1, \ldots, n$, the inequalities

$$\left| \frac{\partial^2 S_1}{\partial y_1 \partial \xi_1} - \frac{1}{|\xi_0|} \right| \leq C\rho, \quad \left| \frac{\partial^2 S_1}{\partial y_i \partial \xi_j} - \delta_{ij} \right| \leq C\rho$$

hold. Therefore,

$$\left| \det \frac{\partial^2 S_1}{\partial y_i \partial \xi_j} \right| \geq c_0 > 0$$

for $|\xi| = 1$, $|y - y_0| \leq \rho$ and $|\partial S_1/\partial y| \geq C_1 > 0$. It is evident that $S_1 = S$ if $|y| + |\xi - \xi_0| \leq \rho/2$, $|\xi| = 1$.

By Theorem 5, if

$$q(y, \partial S_1(y, \xi)/\partial y) = p(\partial S_1(y, \xi)/\partial \xi, \xi)$$

and Q is the operator with the symbol q, then the operator $Q\Phi - \Phi P$ has the order 0, where Φ is the integral operator with the phase function S_1 and the symbol $a(x, \xi) = |\det \partial^2 S_1/\partial x \partial \xi|^{1/2}$. By virtue of Corollary 7 the order of the operator $\Phi^*(Q\Phi - \Phi P)\Phi^* = \Phi^* Q - P\Phi^* + T_0$ is zero and so is that of the operator T_0.

C. Thus we have shown that the operator P is equivalent to the operator Q whose symbol is equal to ξ_1 in a conical neighbourhood of the point (x_0, ξ_0) up to an operator T_0 of order 0. We now show that there is an elliptic pseudo-differential operator A of order 0, such that

$$B(D_1 + T_0)A = D_1 + T,$$

where B is a parametrix for A, and the operator T is smoothing.

The symbol of the operator AD_1 is equal to $\xi_1 a(x, \xi)$. The symbol of the operator $(D_1 + T_0)A$ is equal to

$$[\xi_1 + t_0(x, \xi)] a(x, \xi) + \frac{1}{i} \frac{\partial a}{\partial x_1} + \sum_{|\alpha|=1}^{N-1} \frac{i^{|\alpha|}}{\alpha!} D_\xi^\alpha t_0(x, \xi) \cdot D_x^\alpha a(x, \xi) + t_N(x, \xi),$$

where $t_N \in S^{-N}$. This formula enables us to find a. Let a_0 be a non-zero solution of the equation $\partial a_0/\partial x_1 + it_0 a_0 = 0$. For example, one can take $a_0 = exp(-i \int_0^{x_1} t_0 dx_1)$, so that $a_0 \in S^0$. Solving the equation

$$\frac{1}{i}\frac{\partial a_1}{\partial x_1} + t_0 a_1 + \sum_{|\alpha|=1} iD_\xi^\alpha t_0 \cdot D_x^\alpha a_0 = 0, \; a_1|_{x_1=0} = 0,$$

we can find the function a_1 of the class S^{-1}. We can then define the functions $a_j \in S^{-j}$ so that

$$\frac{1}{i}\frac{\partial a_j}{\partial x_1} + t_0 a_j + \sum_{|\alpha|+k=j} \frac{i^{|\alpha|}}{\alpha!}D_\xi^\alpha t_0 \cdot D_x^\alpha a_k = 0, \; a_j|_{x_1=0} = 0.$$

The symbol $a \in S^0$ is constructed then as an asymptotic sum of $\{a_j\}$, using Lemma 1 of 3.1.1. $\qquad\qquad\square$

6.2 Wave fronts of distributions

6.2.1 Definitions

Until now we have regarded the points as singular for a distribution u, if in every neighbourhood of them u was not a function of the class C^∞. The set of all such points is called a *singular support* and is denoted as $\operatorname{sing\,supp} u$. However, this notion is often insufficient to describe the singularities of u. For example, both the distributions $\delta(x_1)$ and $\delta(x_1, x_2)$ have a singular point $(0,0)$ on the plane $\mathbb{R}^2$. However, the distribution $\delta(x_1)$ is an infinitely differentiable function of the variable x_2, and for $\delta(x_1, x_2)$ all directions are equivalent. A generalization of this simple observation leads to the notion of the *wave front*, which is very useful for description of the singularities of solutions of differential equations.

The essence of this notion is in the representation of a distribution u as a sum of plane waves u_ω, where $|\omega| = 1$ and the directions of the plane waves, having singularities at a considered point, are called singular. The set of such directions can be characterized as a subset in $T^*\Omega \setminus 0$.

Definition 10. *A point (x_0, ξ_0) of $T^*\Omega \setminus 0$ does not belong to the wave front of a distribution u of $\mathcal{D}'(\Omega)$, if there exist two functions φ and ψ such that*
1) $\varphi \in C_0^\infty(\Omega)$, $\varphi(x) = 1$ in a neighbourhood of the point x_0;
2) $\psi \in C^\infty(\mathbb{R}^n)$, $\psi(t\xi) = \psi(\xi)$ for $t \geq 1$, $|\xi| \geq 1$; $\psi(\xi_0) \neq 0$;
3) $\psi(D)\varphi(x)u(x) \in C^\infty(\mathbb{R}^n)$.
Here, as usual,

$$\psi(D)v(x) = (2\pi)^{-n}\int \psi(\xi)\tilde{v}(\xi)e^{ix\xi}d\xi.$$

The wave front of a distribution u is a conical set and is denoted by $WF(u)$.

Theorem 11. *A point $(x_0, \xi_0) \in T^*\Omega \setminus 0$ does not belong to the wave front of a distribution u if and only if there exist a function φ of $C_0^\infty(\Omega)$ such that $\varphi(x_0) \neq 0$ and a cone Γ in $\mathbb{R}^n$ having its vertex at the origin and containing in interior the ray $\{\xi;\ \xi = t\xi_0,\ t > 0\}$ such that*

$$|\widetilde{\varphi u}(\xi)| \leq C_N (1 + |\xi|)^{-N}$$

for all $\xi \in \Gamma$ and all N.

Proof. Let $\psi \in C^\infty(\mathbb{R}^n), \operatorname{supp}\psi \subset \Gamma,\ \psi(\xi_0) = 1$ and $\psi(t\xi) = \psi(\xi)$ for $t \geq 1,\ |\xi| \geq |\xi_0|$. Then we have

$$|\psi(\xi)\widetilde{\varphi u}(\xi)| \leq C_N(1 + |\xi|)^{-N},$$

and therefore, $\psi(D)\varphi(x)u(x) \in C^\infty(\mathbb{R}^n)$.

On the other hand, let a point $(x_0, \xi_0) \in T^*\Omega \setminus 0$ not belong to $WF(u)$ and functions $\varphi,\ \psi$ satisfy the conditions of Definition 10. Let us show that $\psi(D)\varphi u \in \mathcal{S}(\mathbb{R}^n)$. This implies that $\psi(\xi)\widetilde{\varphi u}(\xi) \in \mathcal{S}$ and therefore the estimates of Theorem 11 for $\widetilde{\varphi u}(\xi)$ will hold in a cone Γ containing the point ξ_0.

Let $\varphi(x) = 0$ for $|x - x_0| \geq \rho$ and $h(x) = 1$ for $|x - x_0| \leq 2\rho$, $h(x) = 0$ for $|x - x_0| \geq 3\rho,\ h \in C_0^\infty$.

It is evident that $h(x)\psi(D)\varphi(x)u(x) \in C_0^\infty$. Hence it is sufficient to verify that for all α and N

$$|D^\alpha[(1 - h(x))\psi(D)v(x)]| \leq C_{\alpha,N}(1 + |x|)^{-N},\ v = \varphi u.$$

While the distribution v has a compact support, there exist an integer k and continuous functions w_α with compact supports such that $v = \sum_{|\alpha| \leq k} D^\alpha w_\alpha$. Therefore,

$$\psi(D)v(x) = (2\pi)^{-n} \int \int \psi(\xi)v(y)e^{i(x-y)\xi}\,dy\,d\xi$$

$$= (2\pi)^{-n} \int \int \psi(\xi) \sum_{|\alpha| \leq k} \xi^\alpha w_\alpha(y)e^{i(x-y)\xi}\,dy\,d\xi.$$

Since $(-\Delta_\xi)^N e^{i(x-y)\xi} = |x - y|^{2N} e^{i(x-y)\xi}$, we have

$$\psi(D)v(x) = (2\pi)^{-n} \int \int (-\Delta_\xi)^N \Big[\sum \xi^\alpha \psi(\xi)\Big]$$

$$\times w_\alpha(y)|x - y|^{-2N} e^{i(x-y)\xi}\,dy\,d\xi,$$

if $x \in \operatorname{supp}(1 - h),\ y \in \operatorname{supp} w_\alpha$, and therefore

$$|x - y| \geq c_0(1 + |x|),\ c_0 = \text{const} > 0.$$

If $2N > k + n + 1$, then the last integral converges absolutely and $|\psi(D)v(x)| \leq C_N(1 + |x|)^{-2N}$. In a similar way we can verify that $|D^\alpha \psi(D)v(x)| \leq C_{\alpha,N}(1 + |x|)^{-2N+|\alpha|}$, and since N is an arbitrary integer, it follows that $\psi(D)\varphi u \in \mathcal{S}$. $\square$

6.2.2 Examples

Example 12. If $u(x) = f(x_1)$, $n > 1$, then a point (x_0, ξ_0) cannot belong to $WF(u)$ if $\xi_{0j} \neq 0$ for some $j \neq 1$.

Indeed, if $\varphi \in C_0^\infty$, then

$$\widetilde{\varphi u}(\xi) = \widetilde{\varphi}(\xi_1) * \widetilde{f}(\xi_1)\delta(\xi') = \int \widetilde{\varphi}(\xi_1 - \eta_1, \xi')\widetilde{f}(\eta_1)d\eta_1.$$

Let Γ be a cone containing the point ξ_0 and the points of this cone satisfy the inequality $|\xi_1| \leq C|\xi'|$, $C = \mathrm{const} > 0$. Then for $\xi \in \Gamma$ we have

$$|\widetilde{\varphi}(\xi_1 - \eta_1, \xi')| \leq C_N(1 + |\xi_1 - \eta_1| + |\xi'|)^{-N}$$

$$\leq C_N(1 + |\xi_1 - \eta_1|)^{-N/3}(1 + |\xi'|)^{-2N/3}$$

$$\leq C_N(1 + |\xi_1|)^{N/3}(1 + |\eta_1|)^{-N/3}(1 + |\xi'|)^{-2N/3}$$

$$\leq C_N'(1 + |\eta_1|)^{-N/3}(1 + |\xi|)^{-N/3}.$$

Without loss of generality we can assume that $f(x_1) = 0$ for $|x_1 - x_{01}| \geq \delta > 0$. Therefore $|\widetilde{f}(\eta_1)| \leq C(1 + |\eta_1|)^k$ for some integer k. Thus if $N/3 > k + n$, then

$$|\widetilde{\varphi u}(\xi)| \leq C_N''(1 + |\xi|)^{-N/3},$$

and therefore, $(x_0, \xi_0) \notin WF(u)$. $\qquad\square$

Example 13. We will construct a continuous function u such that

$$WF(u) = \{(0, t\eta),\ \eta \in \mathbb{R}^n \setminus 0,\ t > 0\}.$$

Put

$$u(x) = \sum_{k=1}^\infty \frac{\varphi(kx)}{k^2} e^{ik^2 x\eta},$$

where $\varphi \in C_0^\infty(\mathbb{R}^n)$, $\int \varphi(x)dx = 1$, $\widetilde{\varphi}(\xi) \geq 0$. At each point $x \neq 0$ a finite number of terms does not vanish, so that $u \in C^\infty(\mathbb{R}^n \setminus 0)$. Note that $\widetilde{\varphi}(0) = 1$. Furthermore,

$$\widetilde{u}(\xi) = \sum_{k=1}^\infty k^{-n-2}\widetilde{\varphi}\left(\frac{\xi - k^2\eta}{k}\right),$$

so that $\widetilde{u}(k^2\eta) \geq k^{-n-2}$, i.e., $(0, \eta) \in WF(u)$.

Let a vector ξ_0 not be collinear to η so that in a conical neighbourhood Γ of this vector the inequality

$$|\xi - t\eta| \geq c(|\xi| + t|\eta|),\ c > 0,\ t > 0$$

holds. Then for $k \geq 1$

$$\left|\frac{\xi - k^2\eta}{k}\right| \geq \frac{|\xi| + k^2|\eta|}{k} \geq c_1\sqrt{|\xi|},\ c_1 > 0.$$

Therefore, for all N and $\xi \in \Gamma$ we have

$$|\widetilde{u}(\xi)| \leq \sum_{k=1}^{\infty} k^{-n-2} C_N (1 + \sqrt{|\xi|})^{-N} \leq C'_N (1 + |\xi|)^{-N/2},$$

i.e. $(0, \xi_0) \notin WF(u)$. $\square$

6.2.3   Properties of wave fronts

Theorem 14. *If* $h \in C_0^{\infty}(\mathbb{R}^n)$, $u \in \mathcal{D}'(\mathbb{R}^n)$, *then* $WF(hu) \subset WF(u)$.

Proof. Let $v = hu$ and $(x_0, \xi_0) \notin WF(u)$. The distribution theory implies that there are C and k such that $|\widetilde{\varphi u}(\xi)| \leq C(1 + |\xi|)^k$. By Theorem 11 there exists a cone Γ with its vertex at the origin such that $|\widetilde{\varphi u}(\xi)| \leq C_N(1 + |\xi|)^{-N}$ for $\xi \in \Gamma$ and any N with a function $\varphi \in C_0^{\infty}$ equal to 1 in a neighbourhood of the point x_0. It is obvious that

$$\widetilde{\varphi v}(\xi) = \widetilde{h\varphi u}(\xi) = (2\pi)^{-n} \int_{\Gamma} \widetilde{h}(\xi - \eta) \widetilde{\varphi u}(\eta) d\eta$$

$$+ (2\pi)^{-n} \int_{\mathbb{R}^n \setminus \Gamma} \widetilde{h}(\xi - \eta) \widetilde{\varphi u}(\eta) d\eta.$$

The first integral is not greater than

$$C_N \int (1 + |\xi - \eta|)^{-N} (1 + |\eta|)^{-N-n-1} d\eta \leq C_N (1 + |\xi|)^{-N} \int (1 + |\eta|)^{-n-1} d\eta.$$

For points $\eta \in \mathbb{R}^n \setminus \Gamma$ and $\xi \in \Gamma'$, where Γ' is a cone lying inside Γ and containing the point ξ_0, the inequality $|\xi - \eta| \geq c_0 |\xi|$, $c_0 = \text{const} > 0$ holds. Therefore, the second integral is majorated for $\xi \in \Gamma'$ by

$$B_N \int (1 + |\xi - \eta|)^{-N-k-n-1} (1 + |\eta|)^k d\eta$$

$$\leq B'_N (1 + |\xi|)^{-N+k} \int (1 + |\eta - \xi|)^{-n-1} d\eta.$$

Therefore, $|\widetilde{\varphi v}(\xi)| \leq C'_N (1 + |\xi|)^{-N}$ for all N and $\xi \in \Gamma'$. Thus the point (x_0, ξ_0) does not belong to $WF(v)$. $\square$

Theorem 15. *If* $\pi: \ T^*\Omega \to \Omega$ *is the natural projection, then the relation*

$$\pi WF(u) = \text{sing supp}\, u$$

holds for $u \in \mathcal{D}'(\Omega)$.

Proof. If $x_0 \notin \text{sing supp}\, u$, then $\varphi u \in C_0^{\infty}(\Omega)$, where φ is a function of $C_0^{\infty}(\Omega)$, $\varphi(x_0) \neq 0$. Therefore,

$$|\widetilde{\psi u}(\xi)| \leq C_N (1 + |\xi|)^{-N}$$

for all N and all $\xi \in \mathbb{R}^n$, so that $(x_0, \xi) \notin WF(u)$ for all $\xi \in \mathbb{R}^n \setminus 0$.

If $(x_0, \xi)) \notin WF(u)$ for all $\xi \in \mathbb{R}^n \setminus 0$, then for each point $\xi_0 \in \mathbb{R}^n$ with $|\xi_0| = 1$ there are a neighbourhood ω_0 on the sphere $|\xi| = 1$, such that $|\widetilde{\varphi u}(t\xi)| \leq C_N(1 + t)^{-N}$ for $\xi \in \omega_0$, and some function φ of $C_0^\infty(\mathbb{R}^n)$ such that $\varphi(x_0) \neq 0$. Let $\omega_1, \ldots, \omega_N$ be a finite covering of the sphere by such neighbourhoods and $\varphi_1, \ldots \varphi_N$ be the corresponding functions. Put $\varphi_0 = \varphi_1, \ldots \varphi_N$. Then $\varphi_0(x_0) \neq 0$, $\varphi_0 \in C_0^\infty(\mathbb{R}^n)$ and by Theorem 14 $|\widetilde{\varphi_0 u}(\xi)| \leq C_N(1 + |\xi|)^{-N}$ for all N and ξ. However then $\varphi_0 u \in C_0^\infty(\Omega)$, i.e. $x_0 \notin \operatorname{sing\,supp} u$. $\square$

Theorem 16. *Let $u \in \mathcal{E}'(\Omega)$. Then $WF(Au) \subset WF(u)$ for any pseudo-differential operator A.*

Proof. Let $(x_0, \xi_0) \notin WF(u)$. By Definition 10 there exist functions φ and ψ such that $\psi(D)\varphi(x)u \in C^\infty$. Moreover, we can assume that $\varphi(x) = 1$ in a neighbourhood of the point x_0 and $\psi(\xi) = 1$ in a conical neighbourhood of the point ξ_0, so that $\varphi(x)\psi(\xi) = 1$ in a conical neighbourhood ω of the point (x_0, ξ_0).

Let $B = \psi_1(D)\varphi_1(x)$ be a pseudo-differential operator with the symbol $\psi_1(\xi)\varphi_1(x)$, homogeneous with respect to ξ of zero degree for $|\xi| \geq 1$, having its support in ω and equal to 1 in a smaller conical neighbourhood ω' of the point (x_0, ξ_0).

Note that

$$BAu = ABu + [B, A]u = AB\psi(D)\varphi(x)u$$
$$- AB[\psi(D)\varphi(x) - I]u + [B, A]\psi(D)\varphi(x)u$$
$$+ [B, A]\,[I - \psi(D)\varphi(x)]u.$$

Since $\psi(D)\varphi(x)u \in \mathcal{S}$, the terms containing this function are infinitely differentiable. The other terms are infinitely differentiable, too, since they are obtained by applying the smoothing operator to the function u. The latter follows from the fact that the symbol of B vanishes in a neighbourhood of the support of the function $\psi(\xi)\varphi(x) - 1$. $\square$

Let X and Y be smooth manifolds and $f : X \to Y$ be a smooth map. If this map is *proper*, i.e. the pre-image of each compact is compact in X, then the map $f^* : C_0^\infty(Y) \to C_0^\infty(X)$ is defined such that $f^*\varphi(x) = \varphi(f(x))$. Using the duality, one can define the map $f_* : \mathcal{D}'(X) \to \mathcal{D}'(Y)$ such that $f_* u(\varphi) = u(f^*\varphi)$. In particular, the map $f_* : \mathcal{E}'(X) \to \mathcal{D}'(Y)$ is defined for any smooth map f.

Theorem 17. *Let Ω be a domain in $\mathbb{R}^n$, $u \in \mathcal{D}'(\Omega)$. Then a point $(x_0, \xi_0) \notin WF(u)$ if and only if there exist a function $\varphi \in C_0^\infty(\Omega)$, such that $\varphi(x_0) \neq 0$, and a real $\varepsilon > 0$, such that for any real-valued function f satisfying $|f'(x_0) - \xi_0| < \varepsilon$, the function $f_*(\varphi u)(t)$ belongs to the space $C_0^\infty(\mathbb{R})$.*

The proof is based on the following lemmas.

Lemma 18. *Theorem 17 is true for distributions of the form $u(x) = g((\omega, x))$, $g \in \mathcal{D}'(\mathbb{R})$, $\omega \in S^{n-1}$.*

Proof. We can assume that $u = g(x_1)$. Let f be a smooth function in Ω, such that $\operatorname{grad} f$ is not collinear to the x_1-axis. Using a smooth transformation we can reduce f to the function $f = x_2$. Then

$$f_* \varphi(x) g(x_1) = \int (\int g(x_1)\varphi(x)dx_1)dx_3 \ldots dx_n \in C_0^\infty(\mathbb{R}).$$

Analogously, if $f = x_1$, then

$$f_* \varphi(x) g(x_1) = \int g(x_1)\varphi(x)dx_2 \ldots dx_n,$$

and if $f_* \varphi(x) g(x_1) \in C^\infty(\omega)$ on some interval ω, then $g(x_1) \in C^\infty(\omega)$. As we have seen in Example 12,

$$WF(u) = \{(x,\xi);\ x_1 \in \text{sing supp } g,\ \xi_2 = \ldots = \xi_n = 0\}. \qquad \square$$

Lemma 19. *If* $u \in \mathcal{E}'(\mathbb{R}^n)$, *then* $\widetilde{(f_\omega)_* u}(r) = \widetilde{u}(\omega r)$, *where* $f_\omega(x) = (\omega, x)$ *for* $\omega \in S^{n-1}$.

Proof. For $u \in \mathcal{E}'(\mathbb{R}^n)$ we have $\widetilde{u}(r\omega) = \int u(y)e^{-ir\omega y}dy$. Substituting z for y so that $z_1 = \omega y$, we obtain that

$$\widetilde{u}(r\omega) = \int u(z)e^{-irz_1}dz = \int (f_\omega)_* u e^{-irz_1}dz_1 = \widetilde{(f_\omega)_* u}(r). \qquad \square$$

Lemma 20. (Radon) *If* $u \in \mathcal{E}'(\mathbb{R}^n)$, *then*

$$u(x) = (2\pi)^{-n+1} \int_{S^{n-1}} f_\omega^* I_{n-1}((f_\omega)_* u)d\omega,$$

where

$$I_{n-1}v(t) = \frac{1}{2\pi} \int_0^\infty r^{n-1}\widetilde{v}(r)e^{irt}dr.$$

Proof. Note that

$$u(x) = (2\pi)^{-n} \int \widetilde{u}(\xi)e^{ix\xi}d\xi = (2\pi)^{-n} \int_0^\infty dr \int_{S^{n-1}} \widetilde{u}(r\omega)r^{n-1}e^{ir\omega x}d\omega.$$

Since by Lemma 19 $\widetilde{u}(r\omega) = \widetilde{(f_\omega)_* u}(r)$, it follows that

$$u(x) = (2\pi)^{-n+1} \int_{S^{n-1}} f_\omega^* (I_{n-1}(f_\omega)_* u)(\omega x)d\omega,$$

i.e.

$$u(x) = (2\pi)^{-n+1} \int_{S^{n-1}} f_\omega^* I_{n-1}((f_\omega)_* u)d\omega. \qquad \square$$

Proof of Theorem 17. Let $(x_0, \omega_0) \notin WF(u)$. Then the function $u(t\omega)$ tends to 0 faster than any negative power of t as $t \to +\infty$ for $\omega \in S^{n-1}$ close to ω_0. By

Lemma 19, the function $(f_\omega)_* u$ is smooth, and hence the function $(f_\omega)^* I_{n-1}(f_\omega)_* u$ is smooth, too, if ω belongs to some neighbourhood of the point ω_0. If

$$u_1 = (2\pi)^{-n+1} \int f_\omega^* I_{n-1}(f_\omega)_* u)d\omega$$

and the integral is taken over this neighbourhood, then $u_1 \in C^\infty$ and therefore $f_*(\varphi u_1) \in C^\infty$. Put $u - u_1 = u_2$.

If the direction of the vector $f'(x_0)$ is near ω_0, then the function $f_*(\varphi u_2)$ belongs to C^∞ by virtue of Lemma 18. Therefore, the function $f_*(\varphi u)$ is infinitely differentiable.

Conversely, if $f_*(\varphi u) \in C_0^\infty$ for any function f such that $f'(x_0) = \omega_0$, then the function $(f_\omega)_* \varphi u \in C^\infty$ for ω close to ω_0. Lemma 19 implies that $\widetilde{\varphi u}(t\omega)$ is fast decreasing as $t \to +\infty$ for ω close to ω_0. By definition this means that $(x_0, \omega_0) \notin WF(u)$. $\qquad\qquad\square$

6.2.4 Wave fronts under push-forwards and pull-backs

A submersion is a map $f : X \to Y$ such that the pre-image of each point is a smooth submanifold, diffeomorphic to a fixed k-dimensional manifold. In this case, to each function $\varphi \in C_0^\infty(X)$ there corresponds a function $f_* \varphi \in C_0^\infty(Y)$ obtained by integrating the function φ along the fibers. Thereafter one can define the *pull-back* $f^* u$ of $u \in \mathcal{D}'(Y)$, putting $f^* u(\varphi) = u(f_* \varphi)$.

Example 21. Let $\pi : \mathbb{R}^n \to \mathbb{R}$ be the projection on the x_1-axis. If $u \in \mathcal{E}'(\mathbb{R})$, then $\pi^* u(x) = u(x_1) \otimes 1_{x_2,\ldots,x_n} \in \mathcal{D}'(\mathbb{R}^n)$. If $v \in \mathcal{E}'(\mathbf{R}^n)$, then

$$\pi_* v(x_1) = \int v(x)dx_2 \ldots dx_n \in \mathcal{D}'(\mathbb{R}).$$

Theorem 22. *If f is a submersion of X on Y and $u \in \mathcal{E}'(X)$, then $WF(f_* u)$ is contained in the set*

$$\{(f(x), \eta) : \ x \in X : (x, f'(x)^t \eta) \in WF(u) \text{ or } f'(x)^t \eta = 0\}.$$

Proof. Let $x_0 \in X$, $y_0 = f(x_0) \in Y$ and $f'(x_0)^t \eta_0 \neq 0$. Suppose that $(x_0, f'(x_0)^t \eta_0) \notin WF(u)$. Show that then $(y_0, \eta_0) \notin WF(f_* u)$.

Let $g \in C^\infty(Y)$ be a real-valued function and $g'(y_0) = \eta_0$. Then $g \circ f : X \to \mathbf{R}$ is a smooth map, and therefore $(g \circ f)'(x_0) = f'(x_0)^t \cdot \eta_0$. Then Theorem 17 implies that $(g \circ f)_* = g_*(f_* u) \in C_0^\infty(\mathbb{R})$ and thus by virtue of Theorem 17 we have $(y_0, \eta_0) \notin WF(f_* u)$. $\qquad\qquad\square$

If $f : X \to Y$ is a smooth map, we put

$$N_f = \{(y, \eta) \in T^* Y : \ \exists x \in X, \ f'(x)^t \eta = 0\}.$$

Theorem 23. *Let $u \in \mathcal{D}'(Y)$. If $WF(u) \cap N_f = \emptyset$, then one can define the distribution $f^* u \in \mathcal{D}'(X)$, and $WF(f^* u)$ is contained in the set*

$$\{(x, \xi) \in T^* X : \exists(y, \eta) \in WF(u), \ y = f(x), \xi = f'(x)^t \eta\}.$$

In particular, if $u \in C^\infty(Y)$, then $f^* u(x) = u(f(x)) \in C^\infty(X)$.

Proof. Since the statement is local, one can assume that X is a subset of $\mathbb{R}^n$, Y is a subset in $\mathbb{R}^m$ and Y is a small neighbourhood of a point y_0, such that $u(y_0) \neq 0$.

If $u \in C_0^\infty(Y)$, then

$$f^*u(x) = u(f(x)) = (2\pi)^{-n}\int \widetilde{u}(\eta)e^{if(x)\eta}d\eta.$$

Let $\varphi \in C_0^\infty(X)$, $\varphi(x_0) \neq 0$, $f(x_0) = y_0$. Then

$$f^*u(\varphi) = (2\pi)^{-n}\int \widetilde{u}(\eta)I_\varphi(\eta)d\eta,$$

where

$$I_\varphi(\eta) = \int \varphi(x)e^{if(x)\eta}dx.$$

We will justify these formulas for $u \in \mathcal{E}'(Y)$, if $\operatorname{supp}\varphi$ is lying in a small neighbourhood U of the point x_0. Put $V_0 = \{\eta \in \mathbb{R}^m :\ f'(x_0)^t\eta = 0\}$. If V is a small conical neighbourhood of the set V_0, then for $x \in U$, $\eta \notin V$ the inequality

$$|f'(x)^t\eta|^2 \geq C|\eta|^2,\ C = \text{const} > 0$$

is valid.

We have

$$\Delta_x e^{if(x)\eta} = [i\Delta f\eta - (f'(x)^t\eta)^2]e^{if(x)\eta}.$$

If $A(x,\eta) = [i\Delta f\eta - (f'(x)^t\eta)^2]^{-1}$, then for $\eta \notin V$ the inequality $|A(x,\eta)| \leq C_1|\eta|^{-2}$ holds and

$$I_\varphi(\eta) = \int (\Delta \cdot A(x,\eta))^N \varphi(x) \cdot e^{if(x)\eta}dx.$$

Therefore, if $\eta \notin V$, then $|I_\varphi(\eta)| \leq C_N|\eta|^{-2N}$. By assumption, V does not contain the directions η, along which the function $\widetilde{u}(\eta)$ is decreasing faster than any power of $|\eta|^{-1}$. Thus

$$|\widetilde{u}(\eta)| \cdot |I_\varphi(\eta)| \leq C_N'|\eta|^{-N}$$

for all η, and the distribution $f^*u(\varphi)$ is defined for all $\varphi \in C_0^\infty(V)$, i.e. $f^*u \in \mathcal{D}'(U)$.

Furthermore, for $u \in \mathcal{E}'(Y)$, $\varphi \in C_0^\infty(U)$, we have

$$\widetilde{\varphi f^*u}(\xi) = (2\pi)^{-n}\int \widetilde{u}(\eta)d\eta \int e^{if(x)\eta - ix\xi}\varphi(x)dx.$$

Let $V_1 = \{(\xi,\eta) :\ \exists x \in U,\ \xi = f'(x)^t\eta\}$. Outside a conical neighbourhood V_2 of the set V_1 the inequality

$$|\xi - f'(x)^t\eta| \geq \varepsilon(|\xi| + |\eta|),\ \varepsilon = \text{const} > 0$$

is true. So, integrating by parts as above, we obtain that for $(\xi, \eta) \notin V_2$ the inequality

$$\left| \int e^{if(x)\eta - ix\xi} \varphi(x) dx \right| \leq C_N (1 + |\xi| + |\eta|)^{-N}$$

holds. Since $|\widetilde{u}(\eta)| \leq C_1 (1 + |\eta|)^k$, it follows that for any N

$$\left| \int_{(\xi,\eta)\notin V_2} \widetilde{u}(\eta) d\eta \int e^{if(x)\eta - ix\xi} \varphi(x) dx \right| \leq C_N (1 + |\xi|)^{-N}.$$

On the other hand, if η is outside a conical neighbourhood of the set

$$\{ f'(x_0)^t \xi : \ (f(x_0), \xi) \in WF(u) \},$$

then $|\widetilde{u}(\eta)| = O(|\eta|^{-N})$ for all N, if U is sufficiently small. Besides, if $(\xi, \eta) \in V_2$, then $|\xi| \leq C_2 |\eta|$, and hence

$$\left| \int_{(\xi,\eta)\in V_2} \widetilde{u}(\eta) d\eta \int e^{if(x)\eta - ix\xi} \varphi(x) dx \right| \leq C'_N (1 + |\xi|)^{-N},$$

if $\operatorname{supp} \varphi$ and $\operatorname{supp} u$ are sufficiently small. Therefore, $WF(f^*u) \subset V_2$. $\square$

Example 24. Let M be a smooth k-dimensional submanifold in the space $\mathbb{R}^n$, $\rho \in C_0^\infty(M)$, and $u = \rho \otimes \delta(M)$ be a function of simple layer type. Then

$$WF(u) \subset N(M) \setminus 0,$$

where $N(M)$ is the bundle conormal to M in $T^*M \setminus 0$, and 0 is the zero section in T^*M. If $\rho(x_0) \neq 0$, then $(x_0, \xi) \in WF(u)$ if and only if the vector ξ is normal to M.

Theorem 25. *If $f : X \to \mathbb{R}$ is a smooth function, then for $u \in \mathcal{E}'(\mathbb{R})$ the distribution $f^*u \in \mathcal{D}'(X)$ is defined if $f'(x) \neq 0$ in* $\operatorname{sing supp} u$. *Moreover,*

$$WF(f^*u) \subset \{(x, \xi); \ \exists (y, \eta) \in WF(u), \ y = f(x), \ \xi = f'(x)^t \eta \}.$$

Proof. This is a particular case of Theorem 23. $\square$

6.2.5 Wave fronts and traces of distributions on a manifold of a lower dimension

Theorem 26. *Let M be a smooth k-dimensional submanifold in $\mathbb{R}^n$, defined by the equations $f_1(x) = 0, \ldots, f_{n-k}(x) = 0$; $u \in \mathcal{D}'(\mathbb{R}^n)$. If $WF(u) \cap N(M) = \emptyset$, then the restriction $u|_M \subset \mathcal{D}'(M)$ is defined such that*

$$WF(u|_M) \subset \{(x, \xi) \in T^*M \setminus 0 : \ \exists \eta \in \mathbb{R}^n), \ \alpha_1, \ldots, \alpha_{n-k} \in \mathbb{R},$$

$$(x, \eta) \in WF(u), \ \eta = \xi + \sum_{1}^{n-k} \alpha_j f'_j(x) \}.$$

Proof. Let $i : M \to \mathbb{R}^n$ be an embedding. Then the set N_i corresponding to N_f in Theorem 23 coincides with $N(M)$. By Theorem 23 $i^*u \in \mathcal{D}'(M)$ is defined if $WF(u) \cap N(M) = \emptyset$. $\square$

6.2.6 Products of distributions

Theorem 27. *If u_1, $u_2 \in \mathcal{D}'(\mathbb{R}^n)$ and $WF(u_1) + WF(u_2) \subset T^*\mathbb{R}^n \setminus 0$, then the product $u_1 \cdot u_2$ is a distribution, and*

$$WF(u_1 \cdot u_2) \subset \{(x, \xi + \eta) :\ (x, \xi) \in WF(u_1) \text{ or } \xi = 0,$$

$$(x, \eta) \in WF(u_2) \text{ or } \eta = 0,\ \xi + \eta \neq 0\}.$$

Proof. The distribution $u_1(x)u_2(y)$ is always defined in $\mathbb{R}^{2n}$. Let us consider the embedding operator $i : \mathbb{R}^n \to \mathbb{R}^n \times \mathbb{R}^n$, such that $i(x) = (x, x)$. Theorem 23 can be applied if $WF(u_1(x)u_2(y)) \cap N_i = \emptyset$, where $N_i = \{(x, x, \xi, -\xi) :\ x \in \mathbb{R}^n,\ \xi \in \mathbb{R}^n \setminus 0\}$, i.e. if $\xi + \eta \neq 0$, when $(x, \xi) \in WF(u_1)$, $(x, \eta) \in WF(u_2)$. Then Theorem 23 guarantees the existence of the distribution $i^* u_1 \otimes u_2 = u_1 u_2$ such that

$$WF(u_1 u_2) \subset \{(x, \xi + \eta) :\ (x, x, \xi, \eta) \in WF(u_1 \otimes u_2)\}. \qquad \square$$

6.3 Wave fronts and Fourier integral operators

6.3.1 Wave fronts and integral operators

Theorem 28. *Let $K \in \mathcal{D}'(\Omega_1 \times \Omega_2)$, $\Omega_1 \subset \mathbf{R}^n, \Omega_2 \subset \mathbb{R}^m$ and $A : C_0^\infty(\Omega_2) \to \mathcal{D}'(\Omega_1)$ be a map defined by the equality*

$$Au(\varphi) = K(\varphi \otimes u),\ u \in C_0^\infty(\Omega_2),\ \varphi \in C_0^\infty(\Omega_1).$$

If $u \in C_0^\infty(\Omega_2)$, then

$$WF(Au) \subset \{(x, \xi) \in T^*\Omega_1 \setminus 0 :\ \exists y \in \Omega_2,\ (x, y, \xi, 0) \in WF(K)\}.$$

Proof. Without loss of generality we can suppose that the support of K is compact in $\Omega_1 \times \Omega_2$ and $WF(K) \subset \Omega_1 \times \Omega_2 \times \Gamma$, where Γ is a closed cone in $\mathbb{R}^{n+m}$ with its vertex at the origin. It is sufficient to prove that $WF(Au) \subset \Omega_1 \times \Gamma_0$, where $\Gamma_0 = \{\xi :\ (\xi, 0) \in \Gamma\}$.

It is obvious that $\widetilde{Ku}(\xi) = (2\pi)^{-m} \int \widetilde{K}(\xi, -\eta)\widetilde{u}(\eta)d\eta$. If a closed cone Γ_1 in $\mathbb{R}^n$ with its vertex at the origin does not intersect Γ_0, then there is an $\varepsilon > 0$ such that

$$|\widetilde{K}(\xi, -\eta)| \leq C_N(1 + |\xi|)^{-N} \text{ for } N > 0,\ |\eta| < \varepsilon|\xi|,\ \xi \in \Gamma_1.$$

Since $|\widetilde{K}(\xi, -\eta)| \leq C(1 + |\xi| + |\eta|)^k$ with some C and k, then for $\xi \in \Gamma_1$ the inequality

$$|\widetilde{Ku}(\xi)| \leq C_N(1 + |\xi|)^{-N}\Big(\int_{|\eta|<\varepsilon|\xi|} |\widetilde{u}(\eta)|d\eta$$

$$+ \int_{|\eta|>\varepsilon|\xi|} (1 + |\eta|)^{k+N}|\widetilde{u}(\eta)|d\eta \Big) \leq C_N'(1 + |\xi|)^{-N}$$

holds, i.e. $\widetilde{Ku}(\xi)$ is fast decreasing as $|\xi| \to \infty$, $\xi \in \Gamma_1$. Therefore, $WF(Au) \cap \Gamma_1 = \emptyset$. $\qquad \square$

Theorem 29. *Let $K \in \mathcal{D}'(\Omega_1 \times \Omega_2)$, $\Omega_1 \subset \mathbf{R}^n, \Omega_2 \subset \mathbb{R}^m$, $u \in \mathcal{E}'(\Omega_2)$. Assume that the sets*

$$M_1 = \{(y,\eta) \in WF(u) : \ \exists x \in \Omega_1, \ (x,y,0,-\eta) \in WF(K)\}$$

and $\quad M_2 = \{(x,\xi) \in T^*\Omega_1 \setminus 0 : \ \exists y \in \Omega_2, \ (x,y,\xi,0) \in WF(K)\}$

are empty. Then the distribution $Au(x) = \int K(x,y)u(y)dy \in \mathcal{D}'(\Omega_1)$ is defined, and

$$WF(Au) = \{(x,\xi) : \ \exists (y,\eta) \in WF(u), \ (x,y,\xi,-\eta) \in WF(K)\}.$$

Proof. By virtue of Theorem 27 the product $K(x,y)u(y)$ is defined if $M_1 = \emptyset$, since

$$WF[1_x \otimes u(y)] = \{(x,y,0,\eta) : \ (y,\eta) \in WF(u)\}.$$

By Theorem 28 the condition $M_2 = \emptyset$ means that $Au \in C^\infty(\Omega_1)$, if $u_1 \in C_0^\infty(\Omega_2)$.

Without loss of generality we can assume that $\operatorname{supp} K$ is compact and

$$WF(K) \subset \Omega_1 \times \Omega_2 \times \Gamma, \ WF(u) \subset \Omega_2 \times \Gamma',$$

where Γ and Γ' are closed cones with their vertices at the origin in the spaces $\mathbb{R}^{n+m}$ and $\mathbb{R}^m$, respectively. Suppose that these cones are minimal, so that if $(\xi,\eta) \in \Gamma \setminus 0$, then $\eta \neq 0$ and if $\xi = 0$, then $-\eta \notin \Gamma'$. We have to prove that

$$WF(Au) \subset \Omega_1 \times \{\xi : \ \exists \eta \in \Gamma', \ (\xi,-\eta) \in \Gamma\}.$$

Let $\psi(\xi,\eta)$ and $\varphi(\eta)$ be homogeneous functions of degree zero, equal to 1 in neighbourhoods of the cones Γ and Γ' and having their supports in small neighbourhoods of these cones. Then

$$\widetilde{Ku}(\xi) = (2\pi)^{-m} \int \widetilde{K}(\xi,-\eta)\widetilde{u}(\eta)d\eta = (2\pi)^{-m}(I_1 + I_2 + I_3 + I_4),$$

where

$$I_1 = \int_{|\eta|<\varepsilon|\xi|} [1 - \varphi(\eta)]\widetilde{K}(\xi,-\eta)\widetilde{u}(\eta)d\eta;$$

$$I_2 = \int_{|\eta|>\varepsilon|\xi|} [1 - \varphi(\eta)]\widetilde{K}(\xi,-\eta)\widetilde{u}(\eta)d\eta;$$

$$I_3 = \int \varphi(\eta)[1 - \psi(\xi,-\eta)]\widetilde{K}(\xi,-\eta)\widetilde{u}(\eta)d\eta;$$

$$I_4 = \int \varphi(\eta)\psi(\xi,-\eta)\widetilde{K}(\xi,-\eta)\widetilde{u}(\eta)d\eta.$$

By our construction, $[1 - \psi(\xi,-\eta)]\widetilde{K}(\xi,-\eta) = 0$ for $(\xi,-\eta) \in \Gamma$, so that

$$|[1 - \psi(\xi,-\eta)]\widetilde{K}(\xi,-\eta)| \leq C_N(1 + |\xi| + |\eta|)^{-N}$$

and $|I_3| \leq C_N'(1 + |\xi|)^{-N}$ for all N. In order to estimate I_1, note that $\eta \notin \Gamma'$ and hence the point $(\xi,-\eta)$ lies outside Γ so that $|I_1| \leq C_N''(1 + |\xi|)^{-N}$. The integral I_2 is also taken over the domain where $\eta \notin \Gamma'$ and since then $|\xi|^k |\eta|^{-N} \le C|\xi|^{k-N/2}|\eta|^{-N/2}$, we have $|I_2| \le C_N^{(3)}(1+|\xi|)^{-N}$. Finally, in the domain of integration in I_4 we have $\eta \in \operatorname{supp}\varphi, (\xi, -\eta) \in \operatorname{supp}\psi$ so that

$$WF(Au) \subset \Omega_1 \times \{\xi : \exists \eta, (\xi, -\eta) \in \operatorname{supp}\psi, \ \eta \in \operatorname{supp}\varphi\}. \qquad \square$$

6.3.2   Wave fronts and pseudo-differential operators

Theorem 30. *Let P be a pseudo-differential operator with a symbol $p(x, \xi)$. Then*

$$WF(K) \subset \{(x, x, \xi, -\xi) : \ x \in \Omega, \ \xi \in \mathbb{R}^n \setminus 0\},$$

where K is the distributional kernel of the operator P.

Proof. We can assume that $p(x, \xi) = 0$ for $|x| \ge R$. Let $Pu(x) = \int K(x, y)u(y)dy$, where

$$K(x, y) = (2\pi)^{-n} \int p(x, \xi)e^{i(x-y)\xi}d\xi \in \mathcal{D}'(\Omega \times \Omega).$$

Then

$$\widetilde{K}(\xi, \eta) = \int p(x, -\eta)e^{-ix(\xi+\eta)}dx.$$

If $|\xi + \eta| > \varepsilon(|\xi| + |\eta|)$, then the equality

$$e^{-ix(\xi+\eta)} = (1 + |\xi + \eta|^2)^{-N}(1 - \Delta_x)^N e^{-ix(\xi+\eta)}$$

implies that

$$\widetilde{K}(\xi, \eta) = \int (1 + |\xi + \eta|^2)^{-N}[(1 - \Delta_x)^N p(x, -\eta)]e^{-ix(\xi+\eta)}dx,$$

and therefore

$$|\widetilde{K}(\xi, \eta)| \le C_N(1 + |\eta|)^m(1 + |\xi + \eta|^2)^{-N}$$

$$\le C_N'(1 + |\xi| + |\eta|)^{-2N+|m|},$$

i.e. $\widetilde{K}(\xi, \eta)$ is fast decreasing as $|\xi| + |\eta| \to \infty$ outside a conical neighbourhood of the set $\{(\xi, \eta) : \ \xi + \eta = 0\}$. $\qquad \square$

Theorem 31. *If P is a pseudo-differential operator and $u \in \mathcal{E}'(\Omega)$, then $WF(Pu) \subset WF(u)$.*

Proof. Theorem 29 implies that

$$WF(Pu) \subset \{(x, \xi) : \ \exists(y, \eta) \in WF(u), \ (x, y, \xi, -\eta) \in WF(K)\},$$

since now $M_1 = \emptyset$, $M_2 = \emptyset$. However, if $(x, y, \xi, -\eta) \in WF(K)$, then by Theorem 30 we have $y = x$, $\eta = \xi$. Therefore, $WF(Pu) \subset WF(u)$. $\qquad \square$

Theorem 32. *Let P be a pseudo-differential operator, $u \in \mathcal{E}'(\Omega)$ and*

$$\text{char } P = \{(x, \xi) \in T^*\Omega \setminus 0 : \; p_0(x, \xi) = 0\}.$$

Then $WF(u) \subset WF(Pu) \cup \text{char } P$.

Proof. Let $(x_0, \xi_0) \notin WF(Pu)$ and $p_0(x_0, \xi_0) \neq 0$. Let us show that $(x_0, \xi_0) \notin WF(u)$. By the definition of the wave front set there exist such functions $\varphi \in C_0^\infty(\Omega)$, $\psi \in C^\infty(\mathbb{R}^n \setminus 0)$, that $\varphi(x) = 1$ in a neighbourhood of the point x_0, $\psi(t\xi) = \psi(\xi)$ for $t > 1$, $\psi(\xi_0) \neq 0$ and $|\psi(\xi)\widetilde{\varphi Pu}(\xi)| \leq C_N(1 + |\xi|)^{-N}$ for all N. Let $Q = \psi(D)\varphi(x)P(x, D)$. Then $Qu \in C^\infty$ and $q_0(x_0, \xi_0) = \psi(\xi_0)\varphi(x_0)p_0(x_0, \xi_0) \neq 0$. One can construct a left parametrix A of the operator Q in a small conical neighbourhood V of the point (x_0, ξ_0), such that there

$$\sum \frac{1}{\alpha!}(iD_\xi)^\alpha a(x, \xi)D_x^\alpha q(x, \xi) = 1,$$

where a is the symbol of the operator A. Let $R = AQ$. The symbol of the operator R is equal to 1 in a conical neighbourhood V_1 of the point (x_0, ξ_0) and $Ru \in C^\infty$. Let functions φ' and ψ' be such that $\psi'(t\xi) = \psi'(\xi)$ for $t > 0$, $\psi'(\xi) = 0$ outside V_1, $\varphi'(x) = 1$ in a neighbourhood of the point x_0, and $\psi'(\xi) = 1$ in a neighbourhood of the point ξ_0, so that

$$\varphi'(x)\psi'(\xi)r(x, \xi) = \varphi'(x)\psi'(\xi).$$

Then $\psi'(D)\varphi'(x)Ru - \psi'(D)\varphi'(x)u \in C^\infty$, and hence $\psi'(D)\varphi'(x)u \in C^\infty$. Therefore, $(x_0, \xi_0) \notin WF(u)$. $\qquad\square$

Corollary 33. *If P is an elliptic pseudo-differential operator, then $WF(Pu) = WF(u)$ for all u of $\mathcal{E}'(\Omega)$.*

Definition 34. An operator P is called *microlocally hypoelliptic*, if $WF(Pu) \supset WF(u)$.

Theorem 35. *Let $a \in S^m(\Omega)$ and*

$$K(x, y) = (2\pi)^{-n} \int a(x, \xi)e^{iS(y,\xi)-iy\xi}d\xi \in \mathcal{D}'(\Omega \times \Omega).$$

Then

$$WF(K) \subset \{(x, y, \xi, -\eta) : \; y = \partial S(x, \eta)/\partial\eta, \; \xi = \partial S(x, \eta)/\partial x\}.$$

Proof. Using a partition of the unity, one can reduce the statement to the case when $K(x, y) = 0$ outside a small neighbourhood of the point (x_0, y_0). Put

$$\Gamma = \{(\xi, \eta) \in \mathbb{R}^{2n} : \; |y - \frac{\partial S(x, \eta)}{\partial\eta}|$$

$$+ |\xi - \frac{\partial S(x, \eta)}{\partial x}|(|\xi| + |\eta|)^{-1} \geq \varepsilon, \; \forall(x, y) \in U\}.$$

Let us show that $|\widetilde{K}(\xi,-\eta)| \le C_N(1+|\xi|+|\eta|)^{-N}$ for all N, if $(\xi,\eta) \notin \Gamma$. Note that

$$\widetilde{K}(\xi,-\eta) = (2\pi)^{-n}\int\int\int h(x,y)a(x,\zeta)e^{iS(y,\zeta)-iy\zeta-ix\xi+iy\eta}d\zeta\,dx\,dy,$$

where $h \in C_0^\infty(\mathbb{R}^{2n})$, $h = 1$ in U. Using the equalities

$$(1-\Delta_y)^N e^{iy(\eta-\zeta)} = (1+|\eta-\zeta|^2)^N e^{iy(\eta-\zeta)};$$

$$(1-\Delta_x)^N e^{iS(x,\zeta)-ix\zeta} = (1+|S_x(x,\zeta)-\xi|^2 - i\Delta_x S(x,\zeta))^N e^{iS(x,\zeta)-ix\zeta};$$

$$[1-(1+|\zeta|^2)\Delta_\zeta]^N e^{iS(x,\zeta)-iy\zeta}$$
$$= [1+(1+|\zeta|^2)|y-S_\zeta(x,\zeta)|^2 - i(1+|\zeta|^2)\Delta_\zeta S(x,\zeta)]^N e^{iS(x,\zeta)-iy\zeta},$$

and the partial integration, one can represent $\widetilde{K}$ in the form of the integral of a function whose module does not exceed

$$C_N(1+|\zeta|)^{-n-1}(1+|\xi|+|\eta|)^{-N}$$

for all N. $\qquad\qquad\qquad\qquad\qquad\qquad\qquad\qquad\qquad\qquad\qquad\qquad\square$

Theorem 36. *Let*

$$Au(x) = (2\pi)^{-n}\int a(x,\zeta)\widetilde{u}(\zeta)e^{iS(x,\zeta)}d\zeta$$

with $a \in S^m(\Omega)$, and let the set

$$\{(y,\xi) \in WF(u): \ \exists x \in \Omega, \ y = \partial S(x,\xi)/\partial\xi; \ \partial S(x,\xi)/\partial x = 0\}$$

be empty. If $u \in \mathcal{E}'(\Omega)$, then

$$WF(Au) \subset \{(x,\xi): \exists\eta \in \mathbb{R}^n, \ \xi = \frac{\partial S(x,\eta)}{\partial x}; (\frac{\partial S(x,\eta)}{\partial\eta},\eta) \in WF(u)\}.$$

Proof. The distributional kernel of the operator A has been considered in Theorem 35. Let us apply Theorem 29. For this we need the following condition: the sets

$$M_1 = \{(y,\xi) \in WF(u): \ \exists x \in \Omega, \ (x,y,0,-\xi) \in WF(K)\},$$

$$M_2 = \{(x,\xi) \in T^*\Omega \setminus 0: \ \exists y \in \Omega, \ (x,y,\xi,0) \in WF(K)\}$$

must be empty. The set M_1 is empty by our condition. The set M_2 is empty, too, since $\eta \ne 0$ in $WF(K)$. Therefore,

$$WF(Au) \subset \{(x,\xi): \ \exists(y,\eta) \in WF(u); \ (x,y,\xi,-\eta) \in WF(K)\}. \qquad\square$$

6.4 Propagation of singularities

6.4.1 Propagation of singularities for operators of real principal type

In this section we will prove Hörmander's remarkable theorem on the propagation of singularities. Let $Pu = f \in C^\infty(\Omega)$ and a point $(x_0, \xi_0) \in T^*\Omega \setminus 0$ belong to $WF(u)$. Which other points of $T^*\Omega \setminus 0$ will also belong to $WF(u)$? It turns out that in the case when the principal symbol p_0 of P is real-valued, these points form a set, invariant with respect to the Hamilton phase flow for the Hamiltonian $p_0(x, \xi)$, so that each bicharacteristic line, i.e. the integral curve $x = x(t)$, $\xi = \xi(t)$ of the system of equations

$$\frac{dx_j}{\partial t} = \frac{\partial p_0(x(t), \xi(t))}{\partial \xi_j}, \ \frac{d\xi_j}{dt} = -\frac{\partial p_0(x(t), \xi(t))}{\partial x_j}, \quad j = 1, \dots, n,$$

either does not intersect $WF(u)$ or lies in it. Of course, by virtue of Theorem 32, if $p_0(x(t), \xi(t)) \neq 0$, then the corresponding curve does not intersect $WF(u)$, so it is interesting to consider only those curves along which $p_0(x(t), \xi(t)) \equiv 0$.

Let us first consider the case when $P = D_1$.

Theorem 37. *Let $u \in \mathcal{D}'(\Omega)$, Γ be a segment of a line parallel to the x_1-axis, and $\Gamma \cap WF(D_1 u) = \emptyset$. Then either $\Gamma \subset WF(u)$ or $\Gamma \cap WF(u) = \emptyset$.*

Proof. Let $\Gamma = \{(x_{01} + t, \ x_0', \ \xi_0) : a \leq t \leq b\}$. If $\xi_{01} \neq 0$, then $\Gamma \cap WF(u) = \emptyset$ by Theorem 32. On the other hand, if $\xi_{01} = 0$, then Theorem 26 implies that $(x, \xi) \in WF(u)$ at $\xi_1 = \xi_{01}$ if and only if $(x, \xi') \in WF(u|_{x=x_{01}})$. Therefore, $\psi(D')\varphi(x')u \in C^\infty$ at $x_1 = x_{10}$, if $\psi(D')\varphi(x')u \in C^\infty$ at $x_1 = a$ and the segment of the bicharacteristic line $[a, x_{01}]$ does not intersect $WF(D_1 u)$. $\qquad\square$

Theorem 38. *Let P be a pseudo-differential operator of principal type with a real symbol p_0, let Γ be a connected segment of a bicharacteristic line and $\Gamma \cap WF(Pu) = \emptyset$. Then either $\Gamma \cap WF(u) = \emptyset$ or $\Gamma \subset WF(u)$.*

Proof. It is obvious that the statement and the conditions of Theorem 38 are invariant with respect to the composition of P with an elliptic operator. So we can suppose that $m = 1$. By Theorem 9 there is a Fourier integral operator Φ such that the operator $P - \Phi^* Q \Phi$ is smoothing and the symbol of the operator Q is equal to D_1 in a conical neighbourhood of the point (x_0, ξ_0). Put $\Phi u = v$. Then $D_1 v = \Phi f + Tv$, where T is a smoothing operator. Besides, the segments of bicharacteristic lines of the function p_0 are transformed under the associated canonical transformation to the segments of lines parallel to the x_1-axis, and the statement follows from Theorem 37. $\qquad\square$

6.4.2 Propagation of singularities in the Sobolev spaces

For applications it is useful to make the notion of the wave front set more precise, using the Sobolev spaces H^s.

Definition 39. A distribution u belongs to the space H^s at a point (x_0, ξ_0), if $u = u_1 + u_2$, where $u_1 \in H^s$ and $(x_0, \xi_0) \notin WF(u_2)$. The complement to the set of points (x, ξ) in $T^*\Omega \setminus 0$, at which $u \in H^s$, is denoted by $WF_s(u)$.

It is easy to see that Theorem 38 can be made precise in the following way.

Theorem 40. *Let $u \in \mathcal{D}'(\Omega)$ and Γ be a segment of a bicharacteristic line such that $\Gamma \cap WF_s(Pu) = \emptyset$, where P is a pseudo-differential operator of principal type of order m with a real principal symbol. Then either $\Gamma \subset WF_{s+m-1}(u)$ or $\Gamma \cap WF_{s+m-1}(u) = \emptyset$.*

Proof. It is sufficient to repeat the proof of Theorem 38, since in the case when $P = D_1$ Theorem 40 is easy to prove and a Fourier integral operator with non-vanishing symbol of the class $S^0(\Omega)$ and its parametrix are bounded in the space H^s. $\qquad\square$

6.4.3   Solvability of real principal type

Recall that a pseudo-differential operator is of principal type if the vector $\mathrm{grad}_{x,\xi}\, p_0(x, \xi)$ is not collinear to the vector $(\xi, 0)$. We will assume in this section that the function p_0 is real-valued. Let us first prove the local solvability of the equation $Pu = f$.

Theorem 41. *Let P be a pseudo-differential operator of principal type with real-valued p_0. Then for any point $x_0 \in \Omega$ there is a neighbourhood U such that if $f \in H^s(\Omega)$, then there exists a function $u \in H_0^{s+m-1}(\Omega)$ satisfying the equation $Pu = f$ in U, and such that $\|u\|_{s+m-1} \leq C\|f\|_s$ with a constant C depending on s only.*

Proof. Since P is an operator of principal type, the bicharacteristic lines of the function p_0 passing through the point (x_0, ξ_0) for each $\xi_0 \in S^{n-1}$ are such that their projections on the space of x-variables come on the boundary of the set U. This follows, for instance, from Theorem 6.

Note that P^* is also an operator of principal type, and by Theorem 40, if $P^*\varphi \in H^t(U)$ for some real t, $\varphi \in \mathcal{E}'(U)$, then $\varphi \in H^{t+m-1}(U)$. If the neighbourhood U is small enough, then each bicharacteristic line passing through points $(x, \xi) \in T^*U \setminus 0$ are such that their projections on the space of the x-variables exit in the domain where $\varphi \in C^\infty$. Therefore, $\varphi \in H^{t+m-1}$ at each point of $T^*U \setminus 0$. It follows then that $\varphi \in H_0^{t+m-1}(U)$. By the Banach open mapping theorem on an open map there is a constant C such that $\|\varphi\|_{t-m-1} \leq C\|P^*\varphi\|_t$ for $\varphi \in C_0^\infty(U)$. If $f \in H^s(\Omega)$, then this inequality shows that the integral $\int f(x)\varphi(x)dx$ is a bounded linear functional of $\overline{P^*\varphi} \in H^{-s-m+1}$. By the Hahn-Banach theorem, there is a function u of the space $H^{s+m-1}(\Omega)$, such that

$$\int f(x)\overline{\varphi(x)}dx = \int u(x)\overline{P^*\varphi}dx, \ \varphi \in C_0^\infty(U)$$

and $\|u\|_{s+m-1} \leq C\|f\|_s$. Therefore, $Pu = f$ in U. $\qquad\square$

In the case when the domain U is not small, the bicharacteristic lines of P can never exit on the boundary of U. The main danger here is in the presence of closed bicharacteristic lines.

Example 42. Let Ω be an annulus on the plane, defined in the polar coordinates by the inequalities $1 < r < 2$. The operator $D_\varphi = (xD_y - yD_x)r^{-1}$ is of principal type. However, the equation $D_\varphi u = f$ is solvable in Ω if and only if $\int_0^\pi f(r, \varphi)d\varphi = 0$ for $1 < r < 2$. Therefore, for solvability it is necessary that the function f satisfies an infinite set of orthogonality conditions.

Condition A. Each bicharacteristic line of the function $p_0(x, \xi)$ such that $p_0(x, \xi) = 0$ along it, intersecting the set $T^*K \setminus 0$, where K is a compact in Ω, has points whose projections on Ω lie outside K.

Theorem 43. *For some compact K let the condition A be fulfilled and P be an operator of principal type with a real principal type. If $\varphi \in \mathcal{E}'(K)$, $P^*\varphi = 0$ in Ω, then $\varphi \in C_0^\infty(K)$. The linear space $\mathcal{N}$ of functions φ of $C_0^\infty(K)$, satisfying the equation $P^*\varphi = 0$, is finite-dimensional.*

Proof. The smoothness of the functions φ follows from Theorem 40, since by condition A each bicharacteristic line passing through characteristic points, contains points in which $\varphi \in C^\infty$. If a bicharacteristic line does not contain characteristic points, then by Theorem 32 $\varphi \in C^\infty$ at its points.

Let us introduce in $\mathcal{N}$ a topology by the norm

$$\|u\|_{\mathcal{N}} = \left(\int |u|^2 dx\right)^{1/2}.$$

Each function of $\mathcal{N}$ belongs to $H^1(K)$. By the open mapping theorem the estimate $\|\varphi\|_1 \leq C\|\varphi\|_{\mathcal{N}}$ is valid. Hence the unit ball in the space $\mathcal{N}$ is compact. By Kolmogorov's theorem it follows that $\dim \mathcal{N}$ is finite. $\square$

Theorem 44. *For a compact subset K, let condition A be fulfilled and P be a pseudo-differential operator of principal type with a real principal symbol. The equation $Pu = f$ with $f \in H^s(\Omega)$ is solvable in K if and only if $\int f(x)\overline{\varphi(x)}dx = 0$ for all $\varphi \in \mathcal{N}$. If this condition holds, then there exists a solution $u \in H^{s+m-1}(\Omega)$ and $\|u\|_{s+m-1} \leq C\|f\|_s$, where C is independent of f.*

Proof. If $Pu = f$, then $u(\overline{P^*\varphi}) = \int f(x)\overline{\varphi(x)}dx$ for all $\varphi \in C_0^\infty(K)$. Therefore, for $\varphi \in \mathcal{N}$ we have $\int f(x)\overline{\varphi(x)}dx = 0$.

Let us show that for functions of $C_0^\infty(K)$ satisfying the condition $\int v(x)\overline{\varphi(x)}dx = 0$ for all $\varphi \in \mathcal{N}$, the inequality

$$\|v\|_{-s} \leq C\|P^*v\|_{-s-m+1}$$

is valid. Indeed, if it is not true, then there is a sequence v_k of functions belonging to $C_0^\infty(K)$ such that $\|v_k\|_{-s} = 1$,

$$\|P^*v_k\|_{-s-m+1} \leq k^{-1}, \quad \int v_k(x)\overline{\varphi(x)}dx = 0$$

for all $\varphi \in \mathcal{N}$. By virtue of Theorem 40 and the closed graph theorem, we then have the estimate

$$\|v\|_{-s} \le C(\|P^*v\|_{-s-m+1} + \|v\|_{-s-1}), \ v \in C_0^\infty(K).$$

Therefore $C\|v_k\|_{-s-1} \ge 1 - Ck^{-1} > 1/2$ if k is sufficiently large. The boundedness of the norms $\|v_k\|_{-s}$ implies the compactness of the sequence v_k in the space H^{-s-1}. Choosing a subsequence, we can assume that $v_k \to v$ in H^{-s-1}. However

$$\|v_k - v_j\|_{-s} \le C(\|P^*v_k\|_{-s-m+1} + \|P^*v_j\|_{-s-m+1} + \|v_k - v_j\|_{-s-1}).$$

Therefore, $v_k \to v$ in H^{-s}. Moreover, $P^*v = 0$ and $C\|v\|_{-s-1} > 1/2$, i.e. $v \in \mathcal{N}$, which contradicts the condition $\int v(x)\overline{\varphi(x)}dx = 0$ for $\varphi \in \mathcal{N}$. Hence the inequality

$$\|v\|_{-s} \le C\|P^*v\|_{-s-m+1}$$

is true. Now let $f \in H^s(\Omega)$. We can define a linear continuous functional $l(v) = \int \overline{f(x)}\varphi(x)dx$, on the linear space

$$\{v : \exists \varphi \in C_0^\infty(\Omega), \ v = P^*\varphi\}$$

with the norm $\|v\| = \|v\|_{-s-m+1}$, since the inequality

$$|l(v)| \le \|f\|_s \cdot \|\varphi\|_{-s} \le C\|f\|_s \cdot \|v\|$$

is true. By the Hahn-Banach theorem there is an element $u \in H^{s+m-1}$ such that

$$\int \overline{f(x)}\varphi(x)dx = \int \overline{u(x)}v(x)dx,$$

i.e.

$$\int f(x)\overline{\varphi(x)}dx = \int u(x)\overline{P^*\varphi(x)}dx$$

and $\|u\|_{s+m-1} \le C\|f\|_s$. Therefore the function u is the sought solution. $\qquad\square$

6.5 The Cauchy problem for a strongly hyperbolic equation

6.5.1 The Cauchy problem for the wave equation

We are going to show how the theory of Fourier integral operators can be applied to study the Cauchy problem for a strongly hyperbolic equation.

Let us consider first the Cauchy problem for the wave equation:

$$\frac{\partial^2 u}{\partial t^2} = a^2 \Delta u \text{ for } 0 < t < T,$$

$$u(0, x) = \varphi(x), \ \frac{\partial u}{\partial t}(0, x) = \psi(x).$$

If we put $v(t,\xi) = \int u(t,x)e^{-ix\xi}dx$, then

$$\frac{\partial^2 v}{\partial t^2} = -a^2|\xi|^2 v$$

for $0 < t < T$, and

$$v(0,\xi) = \tilde{\varphi}(\xi), \ \frac{\partial v}{\partial t}(0,\xi) = \tilde{\psi}(\xi).$$

Therefore,

$$v(t,\xi) = \tilde{\varphi}(\xi)\cos\ at|\xi| + \tilde{\psi}(\xi)\frac{\sin\ at|\xi|}{a|\xi|},$$

so that

$$u(t,x) = (2\pi)^{-n}\int \tilde{\varphi}(\xi) \cdot \frac{1}{2}(e^{iat|\xi|} + e^{-iat|\xi|})e^{ix\xi}d\xi$$

$$+ (2\pi)^{-n}\int \tilde{\psi}(\xi) \cdot \frac{1}{2ia|\xi|}(e^{iat|\xi|} - e^{-iat|\xi|})e^{ix\xi}d\xi.$$

Thus the solution of the Cauchy problem for the wave equation is represented as a sum of two Fourier integral operators applied to the initial data, i.e. to the functions φ and ψ. The phase functions of these operators are equal to $\pm at|\xi|+x\xi$. By Theorem 36 it follows that for $0 \le t \le T$,

$$WFu(t,\cdot) \subset \{(x \pm at\xi|\xi|^{-1},\xi) : (x,\xi) \in WF(\varphi) \cup WF(\psi)\}.$$

6.5.2 The Cauchy problem for a hyperbolic equation

Now let $P(t,x,D_t,D_x)$ be a *strongly hyperbolic* differential operator of order m in the strip $\Omega = \{(t,x) : \ 0 < t < T, \ x \in \mathbb{R}^n\}$. This means that the roots $\lambda_1,\ldots,\lambda_m$ of the characteristic equation $p_0(t,x,\lambda,\xi) = 0$ are real and different for $\xi \ne 0$, so that

$$|\lambda_j(t,x,\xi) - \lambda_k(t,x,\xi)| \ge c_0 = \text{const} > 0,$$

if $(t,x) \in \Omega$, $|\xi| = 1$, $j \ne k$. Suppose that the coefficients of this operator are smooth functions, bounded in Ω.

Let us consider the Cauchy problem

$$P(t,x,D_t,D_x)u = 0 \ \text{ for } \ (t,x) \in \Omega$$

$$D_t^j u(0,x) = \varphi_j(x) \ \text{ for } \ x \in \mathbb{R}^n, \ j = 0,1,\ldots,m-1.$$

We will reduce this problem to an equivalent Cauchy problem for a hyperbolic system of equations of first order. To this end we introduce new unknown functions

$$v_j = D_t^j \Lambda^{m-j-1}u, \ j = 0,1,\ldots,m-1,$$

where Λ is the pseudo-differential operator with the symbol $\Lambda(\xi) = (1 + |\xi|^2)^{1/2}$. Let

$$P(t, x, D_t, D_x) = \sum_{j+|\alpha|\leq m} a_{j,\alpha}(t, x)D_t^j D_x^\alpha = \sum_{j=0}^m H_j(t, x, D_x)D_t^j,$$

where $H_m(t, x, D_x) = 1$, and the order of the operator $H_j(t, x, D_x)$ is equal to $m - j$. Then

$$P(t, x, D_t, D_x)u = D_t v_{m-1} + \sum_{j=0}^{m-1} A_j(t, x, D_x)v_j,$$

where $A_j(t, x, D_x) = H_j(t, x, D_x)\Lambda^{j+1-m}$ is an operator of first order for $0 \leq j \leq m - 1$. Let $a_j(t, x, \xi)$ denote the principal symbol of $A_j(t, x, D_x)$. It is obvious that the Cauchy problem is equivalent to the following one:

$$D_t v_j = \Lambda v_{j+1}, \; j = 0, 1, \ldots, m - 2,$$

$$D_t v_{m-1} = -\sum_{j=0}^{m-1} A_j(t, x, D_x)v_j,$$

$$v_j(0, x) = \Lambda^{m-j-1}\varphi_j(x), \; j = 0, 1, \ldots, m - 1.$$

These equations are pseudo-differential with respect to x. It is important to note that the characteristic equation of the obtained system coincides with the characteristic equation for the operator P. Indeed, this equation has the form $\det|A(t, x, \xi) - \lambda I| = 0$, where

$$A(t, x, \xi) = \begin{pmatrix} 0 & |\xi| & 0 & \cdots & 0 \\ 0 & 0 & |\xi| & \cdots & 0 \\ \cdots & \cdots & \cdots & \cdots & \cdots \\ 0 & 0 & 0 & \cdots & |\xi| \\ -a_0 & -a_1 & -a_2 & \cdots & -a_{m-1} \end{pmatrix}.$$

Therefore, the roots of this equation are real and different for $|\xi| = 1$. They are positively homogeneous in ξ of degree 1.

Let $N(t, x, \xi)$ be a matrix of order m, whose columns are eigenvectors of the matrix A. The strong hyperbolicity of the operator P yields that the matrix N can be chosen in such a way that the elements of the matrices N and N^{-1} are uniformly bounded for $|\xi| = 1$ and are smooth functions. We can assume also that these elements are positively homogeneous in ξ of degree 1. By our construction,

$$A(t, x, \xi)N(t, x, \xi) = N(t, x, \xi)M(t, x, \xi),$$

where M is the diagonal matrix, having $\lambda_1, \ldots, \lambda_m$ as diagonal elements. Now put $v = N(t, x, D_x)w$, where v is a vector-column $v = (v_0, v_1, \ldots, v_{m-1})$.

Since the matrix N is non-degenerate, the operator $N(t, x, D)$ is elliptic. Therefore, there exists a linear elliptic operator $N_1(t, x, D)$ of order zero such that the operator $T = N_1 N - I$ is smoothing. The operator N_1 can be constructed as the operator Q in Theorem 3 of 3.1.2. The system will take the form

$$D_t w = M(t, x, D)w + B(t, x, D)w,$$

where B is a matrix of pseudo-differential operators of order zero. The initial conditions allows us to find the vector

$$w(0, x) = \Phi(x)$$

up to a smoothing operator, so that

$$\Phi(x) + T\Phi(x) = N_1 v(0, x).$$

6.5.3   The construction of the phase function and the symbol

The system obtained has the diagonal principal part and can be investigated using the same scheme as for the solution of one hyperbolic equation of first order. Let us first consider the Cauchy problem for such an equation:

$$D_t u = \lambda(t, x, D_x)u, \ \ u(0, x) = \varphi(x),$$

where λ is a smooth real-valued function such that $\lambda(t, x, r\xi) = r\lambda(t, x, \xi)$ for $r > 0$. As in the above example we shall look for the solution in the form

$$u(t, x) = (2\pi)^{-n} \int a(t, x, \xi)\tilde{\varphi}(\xi)e^{iS(t, x, \xi)}\, d\xi,$$

where $a \in S^0$. The initial condition will be satisfied if

$$a(0, x, \xi) = 1, \ \ S(0, x, \xi) = x\xi.$$

It is clear that

$$D_t u(t, x) = (2\pi)^{-n} \int (a\frac{\partial S}{\partial t} + \frac{1}{i}\frac{\partial a}{\partial t})\tilde{\varphi}(\xi)e^{iS(t, x, \xi)}\, d\xi.$$

On the other hand, as the proof of Theorem 5 shows, we have

$$\lambda(t, x, D_x)u = (2\pi)^{-n} \int k(t, x, \xi)\tilde{\varphi}(\xi)e^{iS(t, x, \xi)}\, d\xi,$$

where $k \in S^1$, $k - \sum_{j=0}^{N} k_j \in S^{1-N}$ where $k_j \in S^{1-j}$, and

$$k_0(t, x, \xi) = a(t, x, \xi)\lambda(t, x, \partial S(t, x, \xi)/\partial x).$$

It follows that the function S satisfies the equation

$$\frac{\partial S}{\partial t} = \lambda(t, x, \frac{\partial S}{\partial x}),$$

called the *iconal equation*. This term is taken from geometrical optics, where the value $S(t)$ of the function S means the distance travelled in time t by a light ray starting from the point x in the direction of the vector ξ. The function λ thus characterizes the optical properties of the medium. The solution of this non-linear equation with the initial condition $S(0, x, \xi) = x\xi$ exists on some, generally speaking, small time interval $[0, T]$. If T is small enough, then the homogeneous function S in ξ will be a phase function for $0 \leq t \leq T$.

Having defined the function S, we can find the functions $a_0, a_1, \ldots$, such that $a_j \in S^{-j}$,

$$\frac{1}{i}\frac{\partial a_0}{\partial t} = k_1, \ a_0(0, x, \xi) = 1;$$

$$\frac{1}{i}\frac{\partial a_j}{\partial t} + a_{j-1}\frac{\partial S}{\partial t} = k_{j+1}, \ a_j(0, x, \xi) = 0$$

for $j \geq 1$. Using Lemma 1 of 3.1.2 we can construct a function $a \in S^0$ having the given asymptotic expansion $\sum a_j$. It is obvious that $u(t, x) = A\varphi + T\varphi$, where

$$A\varphi(t, x) = (2\pi)^{-n} \int a(t, x, \xi)\tilde{\varphi}(\xi)e^{iS(t,x,\xi)}d\xi,$$

and the operator T is smoothing. As in the above example, using Theorem 36, we find that

$$WFu(t, \cdot) \subset \{(x, \partial S(t, x, \xi)/\partial x) : (\partial S(t, x, \xi)/\partial\xi, \xi) \in WF(\varphi)\}.$$

This means that $WFu(t, \cdot)$ is contained in the image of the set $WF(\varphi)$ under the canonical transformation, defined by the generating function $S(t, x, \xi)$.

6.5.4   The Cauchy problem for a hyperbolic system of first order

Now we can expand the stated method for studying the Cauchy problem for a hyperbolic system of first order. We will look for a solution of the Cauchy problem in the form

$$w_j(t, x) = (2\pi)^{-n} \int a_j(t, x, \xi)\tilde{\varphi}_j(\xi)e^{iS_j(t,x,\xi)}d\xi,$$

$$j = 1, \ldots, m,$$

so that

$$a_j(0, x, \xi) = 1, \ S_j(0, x, \xi) = x\xi.$$

Repeating the above arguments, we obtain that

$$\frac{\partial S_j}{\partial t} = \lambda_j(t, x, \frac{\partial S_j(t, x, \xi)}{\partial x}), \ j = 1, \ldots, m.$$

The asymptotic expansion of the functions $a_j \sim \sum a_{jk}$ is found in the same way as above, with the help of integration. The terms in the equations for a_{jk} corresponding to the lower order terms Bw depend only on $\{a_{l0}, \ldots, a_{lk}\}$, $l = 1, \ldots, m$.

Thus the singularities of the functions $w_j(t, x)$ can be found from the singularities of the functions Φ_j using the canonical transformation corresponding to the generating function $S_j(t, x, \xi)$.

Let us stress that the stated arguments are valid for a small segment $[0, T]$ only, since the iconal equation has a solution on a small segment only and the equation $\det(\partial^2 S_j / \partial x \partial \xi) \neq 0$ is satisfied in the general case also only for small values of t. The construction of the solution on a large interval of time requires the use of non-trivial topological notions.

Chapter 7

Pseudo-differential operators on manifolds with conical and edge singularities; motivation and technical preparations

7.1 The general background

7.1.1 The program of the analysis on manifolds with singularities

This section summarizes a number of results from the calculus of pseudo-differential operators with a view to formulating the program around the concept of ellipticity on manifolds with singularities. This is continued in Sections 7.1.2 and 7.1.3 where we give a survey of the results which ought to be valid in the case of conical singularities and edges. The details are elaborated in the remaining part of the exposition. It turns out that the way of obtaining algebras with adequate symbolic structures for the "elementary singularities" requires a number of suitably prepared variants of the classical pseudo-differential calculus, such as a parameter-dependent version, in particular, with holomorphic or meromorphic dependence on parameters, or the calculus in the Mellin operator convention. This material is also of independent interest and suggests analogous constructions, starting with more complicated operator stuctures than the pseudo-differential ones. The repeated application of the ideas is actually a method for treating higher singularities, and in particular, corners, cf. [Sc4]. Another remarkable aspect is that the wedge theory contains the pseudo-differential calculus of boundary value problems.

Elliptic partial differential equations in a bounded domain with C^∞ boundary or on a compact C^∞ manifold (closed or with C^∞ boundary) belong to the classical subjects of analysis. The pseudo-differential operators allow us to express the parametrices and to derive the elliptic regularity in Sobolev spaces.

The same kind of calculus is interesting for spaces with piecewise C^∞ geometry in the sense of conical singularities, edges, corners, ... , "higher" stratifications. The spaces (here called in short manifolds with singularities) are common in many applications in mathematical physics, mechanics, engineering, and also in branches

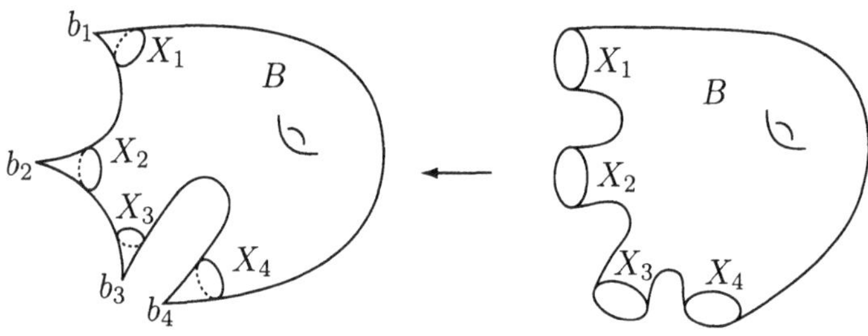

Fig. 1

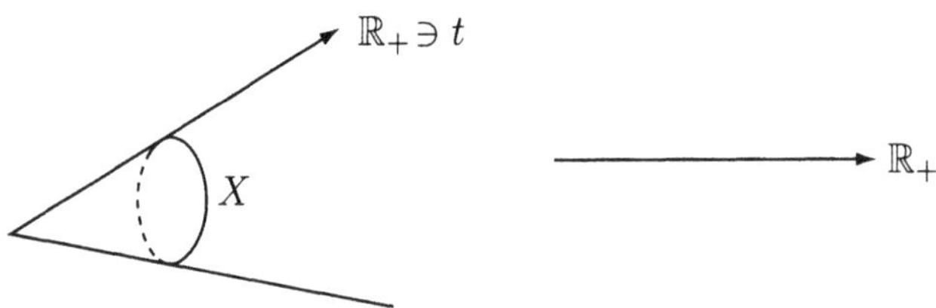

Fig. 2

of pure mathematics such as geometry and topology. The pictures give an idea of the geometric assumptions.

Fig. 1 shows a manifold B with conical singularities $\{b_1, b_2, b_3, b_4\}$. The cone bases X_1, X_2, X_3, X_4 are closed compact C^∞ manifolds. On the right we have the stretched manifold $\mathbb{B}$ associated with B where $\partial\mathbb{B} \cong X_1 \cup X_2 \cup X_3 \cup X_4$. In Fig. 2 we have an infinite cone with base X and the cone axis $\mathbb{R}_+$. The half axis where $\dim X = 0$ is a special case.

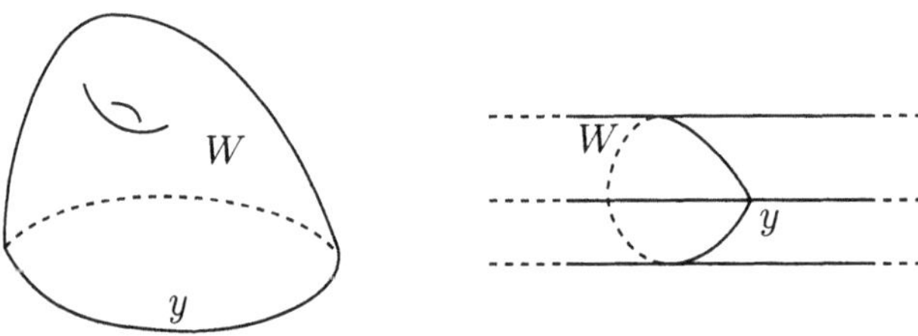

Fig. 3

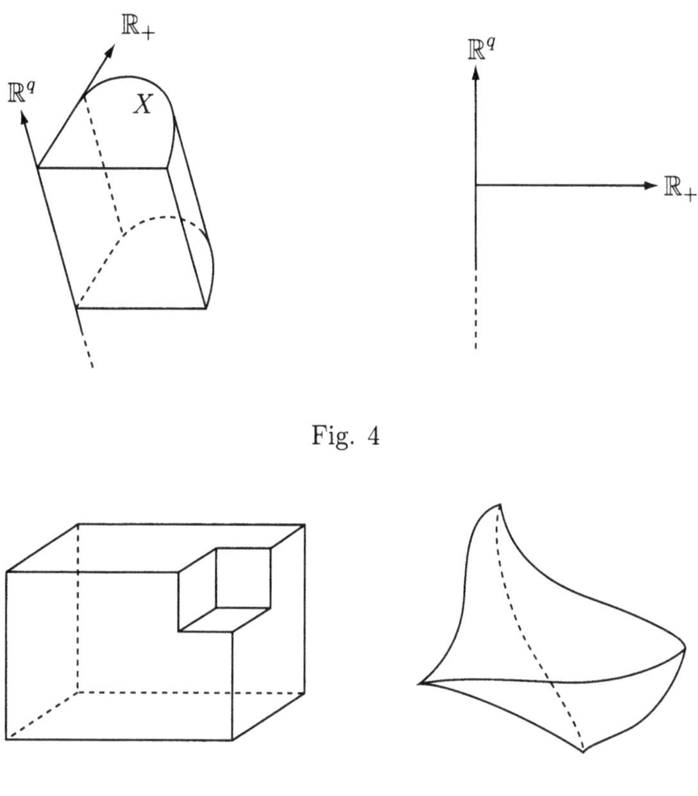

Fig. 4

Fig. 5

Fig. 3 contains a manifold W with edges Y; the right one is an infinite cylinder where the base has conical singularities. Fig. 4 shows an infinite wedge with $\mathbb{R}^q \ni y$ as edge and the infinite cone with base X as the "model cone". The half space $\{(r, y) \in \mathbb{R}^{q+1} : r \geq 0, \ y \in \mathbb{R}^q\}$ is a special case of such a wedge. In other words, domains with C^∞ boundary belong to the class of manifolds with edges, where the edge is the boundary, and the "inner normal" $\mathbb{R}_+$ is the model cone.

Figs. 5 and 6 show manifolds with corners of second order which locally are cones with bases with conical singularities. The system of one-dimensional edges including the corners on the left in Fig. 5 is an example of a one-dimensional manifold with conical singularities. In other words, our spaces with singularities are in general manifolds only outside the singular subsets. The definitions also allow fictitious singularities, e.g. fixed isolated points on a C^∞ manifold, interpreted as conical singularities. By iterating the procedure of forming cones and wedges we get the local forms of corresponding manifolds with higher corners and edges. C^∞ manifolds will also belong to the permitted cases. A basic requirement for

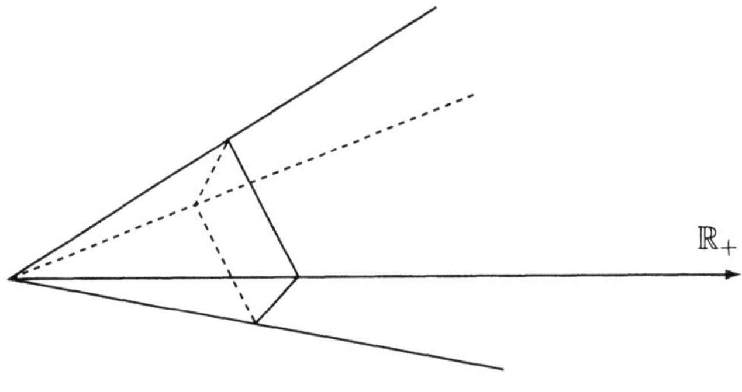

Fig. 6

manifolds with singularities will be that the analytical objects (e.g. operators or the analogues of the Sobolev spaces) coincide with the standard ones outside the singularities.

In order to discuss our program in more detail we will first look at the case of a closed compact C^∞ manifold X (when we speak here of manifolds we will always have in mind paracompact finite-dimensional ones).

We start with the space

$$Diff^m(X) \tag{7.1}$$

of all differential operators of order $m \in \mathbb{Z}_+ = \{0, 1, 2, \ldots\}$ on X (with C^∞ coefficients in local coordinates) and look at the scale

$$\{H^s(X)\}_{s \in \mathbb{R}} \tag{7.2}$$

of standard Sobolev spaces of smoothness s over X. Then every $A \in Diff^m(X)$ induces continuous operators

$$A: \ H^s(X) \longrightarrow H^{s-m}(X) \tag{7.3}$$

for all $s \in \mathbb{R}$. In local coordinates every $A \in Diff^m(X)$ is of the form

$$A = \sum_{|\alpha| \leq m} a_\alpha(x) D_x^\alpha \tag{7.4}$$

in the usual multi-index notation $\alpha = (\alpha_1, \ldots, \alpha_n)$, $n = \dim X$, $|\alpha| = \alpha_1 + \ldots + \alpha_n$, $D_x^\alpha = D_{x_1}^{\alpha_1} \cdot \ldots \cdot D_{x_n}^{\alpha_n}$, $D_{x_j} = \frac{-i\partial}{\partial x_j}$, $i = \sqrt{-1}$. Let us set

$$\sigma_\psi^m(A)(x, \xi) = \sum_{|\alpha| = m} a_\alpha(x) \xi^\alpha, \tag{7.5}$$

$\xi^\alpha = \xi_1^{\alpha_1} \cdot \ldots \cdot \xi_n^{\alpha_n}$. Then (7.5), called the homogeneous principal symbol of A of order m, is invariantly defined as a function

$$\sigma_\psi^m(A) \in C^\infty(T^*X \setminus 0),$$

where T^*X is the cotangent bundle of X, with the local coordinates (x, ξ), x in a coordinate neighbourhood on X, $\xi \in \mathbb{R}^n$. The zero section $\{\xi = 0\}$ of T^*X may be identified with X, and $T^*X \setminus 0$ means T^*X minus the zero section. $C^\infty(\,\cdot\,)$ is the space of all infinitely differentiable functions on the underlying C^∞ manifold. Let us set

$$S^{(m)}(T^*X \setminus 0) = \{\sigma \in C^\infty(T^*X \setminus 0) : \sigma(x, \lambda\xi) = \lambda^m \sigma(x, \xi) \text{ for all } \lambda > 0\}.$$

Here, for convenience, we employ (x, ξ) to indicate points in T^*X, and we use the fact that the multiplication of ξ by $\lambda \in \mathbb{R}_+$ is an invariant $\mathbb{R}_+$ action on T^*X. Then we have a principal symbolic map

$$\sigma_\psi^m : \ Diff^m(X) \longrightarrow S^{(m)}(T^*X \setminus 0). \tag{7.6}$$

Definition 1. $A \in Diff^m(X)$ is called elliptic if $\sigma_\psi^m(A)(x, \xi) \neq 0$ for all $(x, \xi) \in T^*X \setminus 0$.

Let us fix a Riemannian metric g on X. Then we get a measure dx on X and we can talk about the space of square integrable functions

$$L_2(X) \ \text{ with respect to } \ dx \tag{7.7}$$

on X, with the scalar product

$$(u, v)_{L_2(X)} = \int_X u(x)\overline{v(x)}dx. \tag{7.8}$$

Then, in particular, $H^0(X) \cong L_2(X)$.

The nondegenerate sesquilinear pairing

$$(\,\cdot,\,\cdot\,)_{L_2(X)} : \ C^\infty(X) \times C^\infty(X) \longrightarrow \mathbb{C}$$

has an extension to a nondegenerate sesquilinear pairing

$$H^s(X) \times H^{-s}(X) \longrightarrow \mathbb{C} \ \text{ for all } \ s \in \mathbb{R}.$$

By

$$(Au, v)_{L_2(X)} = (u, A^*v)_{L_2(X)}$$

for $u, v \in C^\infty(X)$ we define the formal adjoint A^* of an operator (7.3). If (7.3) is continuous for all $s \in \mathbb{R}$, then A^* has extensions to continuous operators

$$A^* : \ H^s(X) \longrightarrow H^{s-m}(X)$$

for all $s \in \mathbb{R}$.

Remark 2. If $A \in Diff^m(X)$ is elliptic, then so is the formal adjoint A^*.

Remember that g allows an identification between T^*X and the tangent bundle TX of X. Then g also induces metrics on the fibres of T^*X. In particular, we have the unit cosphere bundle

$$S^*X = \{(x,\xi) \in T^*X : |\xi| = 1\}, \tag{7.9}$$

where $|\xi|$ refers to the metric in the fibre over x.

Denote local coordinates by $x = (x_1, \ldots, x_n)$ and set

$$g_{ij} = g\left(\frac{\partial}{\partial x_i}, \frac{\partial}{\partial x_j}\right), \qquad G = (|\det(g_{ij})|)^{\frac{1}{2}}. \tag{7.10}$$

Define g^{jk} by $\sum_{j=1}^{n} g_{ij}g^{jk} = \delta_{ik}$. The Laplace-Beltrami operator associated with g is defined by

$$\Delta_X = G^{-1} \sum_{k=1}^{n} \frac{\partial}{\partial x_k} \left\{ \sum_{i=1}^{n} g^{ik} G \frac{\partial}{\partial x_i} \right\}. \tag{7.11}$$

Then

$$\sigma_\psi^2(\Delta_X)(x,\xi) = -|\xi|^2. \tag{7.12}$$

Thus (7.11) is an example of an elliptic operator.

Definition 3. An operator

$$P \in \bigcap_{s \in \mathbb{R}} \mathcal{L}(H^s(X),\, H^{s+m}(X)) \tag{7.13}$$

is called a *parametrix* of A if

$$AP - I, PA - I \in \bigcap_{s \in \mathbb{R}} \mathcal{L}(H^s(X), C^\infty(X)). \tag{7.14}$$

Here I indicates the identical operator. As usual $\mathcal{L}(E,F)$ is the space of all linear continuous operators $E \longrightarrow F$. For Banach spaces E, F we endow $\mathcal{L}(E,F)$ with the operator norm topology under which this is a Banach space. Note that

$$C^\infty(X) = \bigcap_{s \in \mathbb{R}} H^s(X) =: H^\infty(X), \tag{7.15}$$

where a countable intersection suffices, say over all $s \in \mathbb{Z} = \{\ldots, -2, -1, 0, 1, 2, \ldots\}$. Then $C^\infty(X)$ with the standard Fréchet topology[1] may be interpreted as the projective limit

$$C^\infty(X) = \varprojlim_{s \in \mathbb{Z}}(H^s(X)).$$

---

[1] We will often employ standard material from functional analysis. In particular, a Fréchet space is a metrizable complete topological vector space. In our exposition all occurring Fréchet spaces are locally convex. Their topology can be described by a countable system of semi-norms, cf. also [Sf].

The space

$$\mathcal{L}(H^s(X), C^\infty(X)) = \varprojlim_{t\in\mathbb{Z}} \mathcal{L}(H^s(X), H^t(X)) \tag{7.16}$$

is also endowed with a Fréchet topology. The space

$$H^{-\infty}(X) = \bigcup_{s\in\mathbb{R}} H^s(X)$$

coincides with the space $\mathcal{D}'(X)$ of all distributions on X (the dual of $C^\infty(X)$).

Theorem 4. *The following conditions are equivalent :*

(i) the operator A is elliptic,

(ii) (7.3) is a Fredholm operator for a certain fixed $s = s_0 \in \mathbb{R}$.

If A is elliptic, (7.3) is a Fredholm operator for all $s \in \mathbb{R}$. There exists a parametrix P of A. Furthermore

$$Au = f \in H^r(X) \text{ for some } r \in \mathbb{R}, \tag{7.17}$$

$$u \in H^{-\infty}(X) \tag{7.18}$$

implies

$$u \in H^{r+m}(X). \tag{7.19}$$

The implication (7.17), (7.18) $\Rightarrow$ (7.19), also called elliptic regularity, follows easily by using a parametrix P of A. In fact, from (7.17) and (7.18) we get

$$PAu = Pf \in H^{r+m}(X). \tag{7.20}$$

Set $G = PA - I$. Then (7.20) yields

$$(I + G)u = u + Gu \in H^{r+m}(X)$$

Now (7.14) and (7.15) actually imply $u \in H^{r+m}(X)$.

Remark 5. From the elliptic regularity it follows that

$$\ker A = \{u \in H^s(X) : Au = 0\}$$

is independent of $s \in \mathbb{R}$. In particular, ker A is a finite-dimensional subspace of $C^\infty(X)$. The index of A

$$\operatorname{ind} A = \dim \ker A - \dim \ker A^*$$

does not depend on s.

It is very fruitful to interpret the elements of $Diff^m(X)$, $m \in \mathbb{N}$, as elements of an algebra of operators contained in $\bigcap\limits_{s \in \mathbb{R}} \mathcal{L}(H^s(X), H^{s-m}(X))$, such that the parametrix of any elliptic differential operator is in that algebra. This will be the class $L^m_{cl}(X)$ of (classical) pseudo-differential operators on X of order $m \in \mathbb{R}$ (cf. Section 7.2.1 below). Then the concept of ellipticity has a natural generalization to $L^m_{cl}(X)$, including the notion of a parametrix, and Theorem 4 allows a corresponding extension to $L^m_{cl}(X)$.

A program of the analysis of manifolds with singularities should contain similar ideas, starting with corresponding *typical differential* operators. More precisely the following are required:

(*i*) an algebra of pseudo-differential operators,

(*ii*) symbolic structures,

(*iii*) adequate Sobolev spaces,

such that for an operator A the ellipticity in terms of (*ii*) is equivalent to the Fredholm property in (*iii*) and that then (*i*) contains the parametrices of elliptic operators. It turns out that (*iii*) will also contain

(*iv*) subspaces of distributions with asymptotics.

Then the analogue of the elliptic regularity is a natural problem not only in (*iii*) but also in (*iv*). Furthermore, the

(*v*) index theory on manifolds with singularities.

should be developed.

(By algebra we always mean that the algebra operations are allowed under certain natural restrictions, e.g., when the rows and columns of matrices to be composed fit together). Clearly, these questions are only a minimal program, compared with the huge variety of concrete results on partial differential equations on a C^∞ manifold. We shall content ourselves here with the aspects mentioned. It would be interesting, of course, to study also non-elliptic and nonlinear partial differential equations in a neighbourhood of the singularities of the underlying space. The literature contains many specific results on this. Let us finally note that it does not seem to be too hard to establish a rather general theory for parabolic equations in the singular set-up. The latter area of problems may also be a starting point for mathematicians who want to be active in this field.

7.1.2   Typical differential operators on manifolds with conical singularities

If X is a topological space we set

$$X^\Delta = (\overline{\mathbb{R}}_+ \times X)/(\{0\} \times X), \qquad (7.21)$$

interpreted as the cone with base X and

$$X^\wedge = \mathbb{R}_+ \times X \tag{7.22}$$

as the open stretched cone with base X, where $\mathbb{R}_+ = \{r \in \mathbb{R} : r > 0\}$.

A (finite-dimensional) *manifold B with conical singularities* is a topological space with a finite subset

$$B_0 = \{b_1, \ldots, b_M\} \subset B \tag{7.23}$$

of conical singularities, such that the following properties hold:

   (i) $B \setminus B_0$ is a C^∞ manifold,

  (ii) every $b \in B_0$ has an open neighbourhood U in B, such that there exists a diffeomorphism

$$\phi : U \setminus \{b\} \longrightarrow X^\wedge \tag{7.24}$$

for some closed compact C^∞ manifold $X = X(b)$, where (7.24) is extendible to a homeomorphism

$$\overline{\phi} : \ U \longrightarrow X^\Delta. \tag{7.25}$$

If

$$\psi : \ U \setminus \{b\} \longrightarrow X^\wedge \tag{7.26}$$

is another diffeomorphism, analogously extendible to U, we say that (7.24) and (7.26) are equivalent if

$$\phi\psi^{-1} : \ X^\wedge \longrightarrow X^\wedge$$

is the restriction of some diffeomorphism $\overline{\mathbb{R}}_+ \times X \longrightarrow \overline{\mathbb{R}}_+ \times X$ to $\mathbb{R}_+ \times X$.

The equivalence classes of the systems of maps (7.24) when b runs over B_0 are regarded as part of the "geometry" on $B \setminus B_0$ in any neighbourhood of the conical singularities, which is fixed once and for all.

If we keep the system of maps (7.24) fixed, then

$$\delta_\lambda u := \phi^{-1}(\lambda r, x) \ \text{ for } \ \phi(u) = (r, x), \ \lambda \in \mathbb{R}_+, \ u \in U \setminus \{b\},$$

induces an $\mathbb{R}_+$ action on $U \setminus \{b\}$. The manifold X is called the *base of the cone* (in the neighbourhood of b), and $\mathbb{R}_+$ the *cone axis*.

Note that B is not a manifold in a neighbourhood of $b \in B_0$ unless the base is a sphere of dimension $n = \dim (B \setminus B_0) - 1$. We hope our notation will not lead to confusion; the analysis takes place on $B \setminus B_0$ anyway.

We will assume from now on that B is paracompact.

It follows easily from the assumption that there is a C^∞ manifold $\mathbb{B}$ with compact C^∞ boundary

$$\partial\mathbb{B} \cong \bigcup_{b \in B_0} X(b)$$

such that there is a diffeomorphism

$$B \setminus B_0 \cong \mathbb{B} \setminus \partial\mathbb{B}$$

the restriction of which to $U_1 \setminus B_0$ is a diffeomorphism

$$U_1 \setminus B_0 \cong V_1 \setminus \partial\mathbb{B}$$

for an open neighbourhood $U_1 \subset B$ of B_0 and a collar neighbourhood $V_1 \subset \mathbb{B}$ of $\partial\mathbb{B}$

$$V_1 \cong \bigcup_{b \in B_0} \{[0,1) \times X(b)\}.$$

$\mathbb{B}$ is also called the *stretched manifold* with "conical singularities" associated with B.

Note that $\mathbb{B}/\partial\mathbb{B}$ is a manifold with conical singularities, with only one conical point b_0. The stretched manifold is $\mathbb{B}$, again. More generally, if B is a manifold with conical singularities B_0, and if B_0' is a subset of B_0, then $\widetilde{B} := B/B_0'$ is a manifold with conical singularities $\widetilde{B}_0 := B_0/B_0'$. We then always have $\widetilde{\mathbb{B}} \cong \mathbb{B}$.

Example 1. Let X be an arbitrary closed compact C^∞ manifold. Then there is an N and a C^∞ submanifold $\tilde{X}$ of $S^{N-1} = \{\tilde{x} \in \mathbb{R}^N : |\tilde{x}| = 1\}$ which is diffeomorphic to X. The set

$$B := \left\{ \tilde{x} \in \mathbb{R}^N \setminus \{0\} : \frac{\tilde{x}}{|\tilde{x}|} \in \tilde{X} \right\} \cup \{0\}$$

is an infinite cone with base $\tilde{X}$ and $B_0 = \{0\}$. Then there is a homeomorphism

$$X^\Delta \cong B$$

which is a diffeomorphism outside the conical point, commuting with the canonical $\mathbb{R}_+$ actions on X^Δ and B, respectively.

Note, in particular, that polar coordinates in $\mathbb{R}^N \setminus \{0\}$

$$\tilde{x} \longrightarrow (r, x), \qquad r = |\tilde{x}|, \qquad x = \frac{\tilde{x}}{|\tilde{x}|} \tag{7.27}$$

allow the interpretation

$$\mathbb{R}^N \cong (S^{N-1})^\Delta \quad \text{or} \quad \mathbb{R}^N \setminus \{0\} = (S^{N-1})^\wedge. \tag{7.28}$$

We now pass to the definition of the typical differential operators on manifolds with conical singularities.

The space

$$Diff^m(X) \quad \text{for} \quad m \in \mathbb{Z}_+ \tag{7.29}$$

of all differential operators on X of order m, with C^∞ coefficients in local coordinates, is a Fréchet space in a natural way.

An operator $A \in Diff^m(X^\wedge)$, expressed in $(r,x) \in \mathbb{R}_+ \times X = X^\wedge$, is said to be of *Fuchs type* if A is in a neighbourhood of $r = 0$ of the form

$$A = r^{-m} \sum_{k=0}^{m} a_k(r) \left(-r \frac{\partial}{\partial r} \right)^k \tag{7.30}$$

with coefficients

$$a_k(r) \in C^\infty(\overline{\mathbb{R}}_+, Diff^{m-k}(X)). \tag{7.31}$$

r^{-m} is also called a *weight factor*, with weight $-m$.

Definition 2. Let B be a manifold with conical singularities B_0. Then $A \in Diff^m(B \setminus B_0)$ is said to be of Fuchs type if every $b \in B_0$ has a neighbourhood such that A in local coordinates (r,x) is of Fuchs type in the sense of (7.30). Similarly we say that an $A \in Diff^m(\text{int } \mathbb{B})$ is of Fuchs type if it is in a neighbourhood of $\partial \mathbb{B}$ of form (7.30).

It is an easy exercise to verify the following.

Remark 3. Definition 2 is correct in the sense that it only depends on the equivalence class of (7.24).

The operators of Fuchs type are regarded as the typical operators over a manifold with conical singularities.

Example 4. Let $g_X(r)$ be an r-dependent family of Riemannian metrics on a closed compact C^∞ manifold X, which is infinitely differentiable in $r \in [0, \infty)$. Then

$$g := dr^2 + r^2 g_X(r) \tag{7.32}$$

is a Riemannian metric on $X^\wedge = \mathbb{R}_+ \times X$. The Laplace-Beltrami operator corresponding to (7.32) is then of the form

$$\Delta = r^{-2} \sum_{k=0}^{2} a_k(r) \left(-r \frac{\partial}{\partial r} \right)^k \tag{7.33}$$

with certain coefficients $a_k(r) \in C^\infty(\overline{\mathbb{R}}_+, Diff^{m-k}(X))$, in other words (7.33) is of Fuchs type in X^Δ.

In order to check this property we employ (7.11) and for now set $f = g$, $F = \{|\det(f_{ij})|\}^{\frac{1}{2}}$, where i, j run over $(0, \ldots, n)$ with $x_0 = r$, and local coordinates $x = (x_1, \ldots, x_n)$ on X. Then, according to (7.10)

$$(f_{ij}) = \begin{pmatrix} 1 & 0 \\ 0 & r^2(g_{ij}) \end{pmatrix}, \qquad (f^{ij}) = \begin{pmatrix} 1 & 0 \\ 0 & r^{-2}(g^{ij}) \end{pmatrix}.$$

Clearly, (g_{ij}), (g^{ij}) are taken in the meaning of (7.10), whereas 1 the entry corresponds to $i = j = 0$. It follows that $F = r^n G$, and we get

$$
\begin{aligned}
\Delta u &= F^{-1} \sum_{k=1}^{n} \frac{\partial}{\partial x_k} \left\{ \sum_{i=0}^{n} f^{ik} F \frac{\partial}{\partial x_i} \right\} u \\
&= r^{-n} G^{-1} \sum_{k=0}^{n} \frac{\partial}{\partial x_k} \left\{ \sum_{i=1}^{n} r^{-2} g^{ik} r^n G \frac{\partial}{\partial x_i} \right\} u + r^{-n} G^{-1} \frac{\partial}{\partial r} r^n G \frac{\partial}{\partial r} u \\
&= r^{-2} \Delta_X u + r^{-n} G^{-1} \left\{ n r^{n-1} G + r^n \left(\frac{\partial}{\partial r} G \right) + r^n G \frac{\partial}{\partial r} \right\} \frac{\partial}{\partial r} u.
\end{aligned}
$$

Using $\frac{\partial^2}{\partial r^2} u = r^{-2} \left\{ \left(r \frac{\partial}{\partial r} \right)^2 - r \frac{\partial}{\partial r} \right\} u$ we obtain

$$
\begin{aligned}
\Delta u &= r^{-2} \Delta_X u + n r^{-2} \left(r \frac{\partial}{\partial r} \right) u + r^{-1} G^{-1} \left(\frac{\partial}{\partial r} G \right) r \frac{\partial}{\partial r} u + \\
&\qquad\qquad\qquad\qquad + r^{-2} \left\{ \left(r \frac{\partial}{\partial r} \right)^2 - r \frac{\partial}{\partial r} \right\} u \qquad\qquad (7.34) \\
&= r^{-2} \left\{ \left(-r \frac{\partial}{\partial r} \right)^2 + \left\{ -n + 1 - r G^{-1} \left(\frac{\partial}{\partial r} G \right) \right\} \left(-r \frac{\partial}{\partial r} \right) + \Delta_X \right\} u.
\end{aligned}
$$

Example 5. Let

$$
\widetilde{A} = \sum_{|\alpha| \leq m} a_\alpha(\tilde{x}) D_{\tilde{x}}^\alpha \in \mathit{Diff}^m(\mathbb{R}^{n+1}).
$$

We pass to polar coordinates (r, x) in $\mathbb{R}^{n+1} \setminus \{0\}$, with $r = |\tilde{x}|$, $x = \frac{\tilde{x}}{|\tilde{x}|} \in S^n$. We then have a diffeomorphism

$$
\chi : \mathbb{R}^{n+1} \setminus \{0\} \longrightarrow \mathbb{R}_+ \times S^n,
$$

and the operator

$$
A = \chi_* \{ \widetilde{A} |_{\mathbb{R}^{n+1} \setminus \{0\}} \} \qquad\qquad (7.35)
$$

is of the form (7.30) with (7.31) for $X = S^n$, i.e. A is of Fuchs type.

To prove this we first observe that $a_\alpha(\tilde{x}(r, x)) \in C^\infty(\overline{\mathbb{R}}_+ \times S^n)$. Furthermore we have in local coordinates $(x_1, \ldots, x_n)$ on S^n

$$
\frac{\partial u}{\partial \tilde{x}_k} = \frac{\partial u}{\partial r} \frac{\partial r}{\partial \tilde{x}_k} + \sum_{j=1}^{n} \frac{\partial u}{\partial x_j} \frac{\partial x_j}{\partial \tilde{x}_k}, \qquad k = 1, \ldots, n+1.
$$

Now let $f(\tilde{x}) \in C^\infty(\mathbb{R}^{n+1} \setminus \{0\})$ be homogeneous of order $\varrho \in \mathbb{R}$, i.e. $f(\lambda \tilde{x}) = \lambda^\varrho f(\tilde{x})$ for all $\lambda > 0$, $\tilde{x} \neq 0$. Then $(\frac{\partial f}{\partial \tilde{x}_k})(\tilde{x})$ is homogeneous of order $\varrho - 1$. In particular, since $r(\tilde{x})$ is of order 1, $x_j(\tilde{x})$ of order 0, we have

$$
\frac{\partial r}{\partial \tilde{x}_k}(\lambda \tilde{x}) = \frac{\partial r}{\partial \tilde{x}_k}(\tilde{x}), \qquad \frac{\partial x_j}{\partial \tilde{x}_k}(\lambda \tilde{x}) = \lambda^{-1} \frac{\partial x_j}{\partial \tilde{x}_k}(\tilde{x})
$$

for all $\lambda > 0$. It follows for $\lambda = |\tilde{x}|^{-1}$

$$\frac{\partial r}{\partial \tilde{x}_k}\left(\frac{\tilde{x}}{|\tilde{x}|}\right) = \frac{\partial r}{\partial \tilde{x}_k}(\tilde{x}) =: \tau_k(x), \qquad \frac{\partial x_j}{\partial \tilde{x}_k}\left(\frac{\tilde{x}}{|\tilde{x}|}\right) = |\tilde{x}|\frac{\partial x_j}{\partial \tilde{x}_k}(\tilde{x}) =: \xi_{jk}(x)$$

with functions $\tau_k(x), \xi_{jk}(x) \in C^\infty(S^n)$. We get

$$\frac{\partial u}{\partial \tilde{x}_k} = \left\{\tau_k(x)\frac{\partial}{\partial r} + r^{-1}\sum_{j=1}^{n}\xi_{jk}(x)\frac{\partial}{\partial x_j}\right\}u,$$

i.e.

$$\frac{\partial}{\partial \tilde{x}_k} = r^{-1}\left\{\tau_k(x)r\frac{\partial}{\partial r} + \sum_{j=1}^{n}\xi_{jk}(x)\frac{\partial}{\partial x_j}\right\}.$$

This immediately yields

$$D_{\tilde{x}}^\alpha = r^{-|\alpha|}\sum_{p+|\beta|\leq|\alpha|} d_{p\beta}(r,x)\left(-r\frac{\partial}{\partial r}\right)^p D_x^\beta$$

with coefficients $d_{p\beta}(r,x) \in C^\infty(\overline{\mathbb{R}}_+ \times S^n)$, in other words, (7.35) is of the asserted form.

In particular, we have

$$r\frac{\partial}{\partial r} = \sum_{k=1}^{n+1}\tilde{x}_k\frac{\partial}{\partial \tilde{x}_k}. \tag{7.36}$$

Remark 6. Let B be a manifold with conical singularities and $\mathbb{B}$ the associated stretched manifold. Then every operator A of Fuchs type on B of order m is the pull-back under the canonical projection $\mathbb{B} \longrightarrow B$ of an operator of the form

$$r^{-m}\sum_j \phi_j \sum_{|\alpha|\leq m} a_{j\alpha}v_j^\alpha. \tag{7.37}$$

Here $\{\phi_j\}$ is a partition of unity corresponding to a locally finite open covering of $\mathbb{B}$ and $a_{j\alpha} \in C^\infty(\mathbb{B})$, $v_j^\alpha = v_{j1}^{\alpha_1} \cdot \ldots \cdot v_{j,n+1}^{\alpha_{n+1}}$, with vector fields v_{jk} on $\mathbb{B}$ that are tangent to $\partial\mathbb{B}$. Conversely, every operator of the form (7.37) gives rise under push-forward to an operator of Fuchs type over B.

The proof is an easy exercise.

Corollary 7. *The operators of Fuchs type over a manifold B with conical singularities form an algebra.*

Note that the space of differential operators of Fuchs type on $\mathbb{R}^{n+1} \setminus \{0\}$ is much larger than the space of operators of the form (7.35) when $\tilde{A}$ runs through $Diff^m(\mathbb{R}^{n+1})$. For instance, the operator of multiplication by arbitrary $f \in C^\infty(\mathbb{R}^{n+1} \setminus \{0\})$ with $f(\lambda \tilde{x}) = f(\tilde{x})$ for all $\lambda > 0$, $\tilde{x} \neq 0$, is of Fuchs type, of order zero. But those f are only in exceptional cases restrictions of functions $\tilde{f} \in C^\infty(\mathbb{R}^{n+1})$ to $\tilde{x} \neq 0$ (namely when $\tilde{f}$ is a constant). In other words, operators of Fuchs type are much more singular in neighbourhoods of fictitious conical singularities than the corresponding restrictions of operators with C^∞ coefficients.

In contrast to (7.36) the minus sign at $r\frac{\partial}{\partial r}$ in (7.30) is motivated by properties of the Mellin transform

$$Mu(z) = \int_0^\infty r^{z-1} u(r)\,dr, \qquad (7.38)$$

that satisfies $-r\frac{\partial}{\partial r} = \mathcal{M}^{-1} z \mathcal{M}$. If we apply (7.38) first on $u \in C_0^\infty(\mathbb{R}_+)$, then $\mathcal{M}u(z) \in \mathcal{A}(\mathbb{C})$. Here $\mathcal{A}(U)$ for any open $U \subseteq \mathbb{C}$ is the space of holomorphic functions in U. Then

$$(\mathcal{M}^{-1}g)(r) = \frac{1}{2\pi i} \int_{\Gamma_\beta} r^{-z} g(z)\,dz, \qquad (7.39)$$

$$\Gamma_\beta = \{z \in \mathbb{C} : \text{Re } z = \beta\}. \qquad (7.40)$$

The basic properties of the Mellin transform (which is a classical integral transform) may be found in [Sc1] (cf. also [Sc5]). Note that

$$\mathcal{M}(r^\beta u)(z) = (\mathcal{M}u)(z + \beta). \qquad (7.41)$$

As we shall see below $C_0^\infty(\mathbb{R}_+) \ni u \longrightarrow \mathcal{M}(r^{-\gamma}u)(z + \gamma)$ has an extension by continuity to an isomorphism

$$\mathcal{M}_\gamma : r^\gamma L_2(\mathbb{R}_+) \xrightarrow{\cong} L_2(\Gamma_{\frac{1}{2}-\gamma}). \qquad (7.42)$$

Here $L_2(\mathbb{R}_+)$ is the space of square integrable functions with respect to dr, whereas $L_2(\Gamma_{\frac{1}{2}-\gamma})$ refers to $d\varrho$, $\varrho = \text{Im } z$. In particular, for $\gamma = 0$ we also write $\mathcal{M} = \mathcal{M}_0$, and then

$$\mathcal{M} : L_2(\mathbb{R}_+) \longrightarrow L_2(\Gamma_{\frac{1}{2}}). \qquad (7.43)$$

In order to obtain (7.42) we reduce the situation to the one-dimensional Fourier transform

$$\mathcal{F} : v(t) \longrightarrow \int_{-\infty}^\infty e^{-it\varrho} v(t)\,dt. \qquad (7.44)$$

$\mathcal{F}$ induces an isomorphism $L_2(\mathbb{R}_t) \longrightarrow L_2(\mathbb{R}_\varrho)$ with inverse

$$\mathcal{F}^{-1} : f(\varrho) \longrightarrow \frac{1}{2\pi} \int_{-\infty}^\infty e^{it\varrho} f(\varrho)\,d\varrho. \qquad (7.45)$$

Now let $u(r) \in C_0^\infty(\mathbb{R}_+)$ and set

$$(S_\gamma u)(t) = e^{-\left(\frac{1}{2}-\gamma\right)t} u(e^{-t}). \tag{7.46}$$

Then $S_\gamma : C_0^\infty(\mathbb{R}_+) \longrightarrow C_0^\infty(\mathbb{R})$ is an isomorphism and we have

$$\int\limits_0^\infty |r^{-\gamma} u(r)|^2 dr = \int\limits_{-\infty}^\infty |(S_\gamma u)(t)|^2 dt.$$

Thus S_γ is extendible to an isomorphism

$$S_\gamma : \; r^\gamma L_2(\mathbb{R}_+) \longrightarrow L_2(\mathbb{R}_t).$$

Hence

$$\mathcal{F}S_\gamma : \; r^\gamma L_2(\mathbb{R}_+) \longrightarrow L_2(\mathbb{R}_\varrho)$$

is also an isomorphism. Then (7.42) is a consequence of

$$(\mathcal{M}_\gamma u)\left(\frac{1}{2} - \gamma + i\varrho\right) = (\mathcal{F}S_\gamma u)(\varrho). \tag{7.47}$$

At the same time (7.39) follows easily from (7.45).

Definition 8. $\mathcal{H}^{s,\gamma}(\mathbb{R}_+)$ for $s, \gamma \in \mathbb{R}$ is the closure of $C_0^\infty(\mathbb{R}_+)$ with respect to the norm

$$u \longrightarrow \left\{ \frac{1}{2\pi i} \int\limits_{\Gamma_{\frac{1}{2}-\gamma}} (1 + |z|^2)^s |\mathcal{M}_\gamma u(z)|^2 dz \right\}^{\frac{1}{2}}.$$

From (7.41) it follows immediately that

$$\mathcal{H}^{s,\gamma}(\mathbb{R}_+) = r^\gamma \mathcal{H}^{s,0}(\mathbb{R}_+). \tag{7.48}$$

Furthermore, (7.42) implies

$$\mathcal{H}^{0,\gamma}(\mathbb{R}_+) = r^\gamma L_2(\mathbb{R}_+).$$

We shall also write $\mathcal{H}^s(\mathbb{R}_+) = \mathcal{H}^{s,0}(\mathbb{R}_+)$.

Proposition 9. *For* $s \in \mathbb{Z}_+$ *we have*

$$\mathcal{H}^s(\mathbb{R}_+) = \left\{ u \in L_2(\mathbb{R}_+) : \; \left(-r\frac{\partial}{\partial r}\right)^k u \in L_2(\mathbb{R}_+), \; k = 0, \ldots, s \right\}.$$

The proof is an easy exercise. Let $H^s(\mathbb{R})$ denote the standard Sobolev space on $\mathbb{R}$ of smoothness $s \in \mathbb{R}$, i.e.

$$H^s(\mathbb{R}) = \{u \in \mathcal{S}'(\mathbb{R}) :\ u \in \mathcal{F}^{-1}[(1 + |\varrho|^2)^{-\frac{s}{2}} L^2(\mathbb{R})]\}$$

with the one-dimensional Fourier transform $\mathcal{F} = \mathcal{F}_{t\to\varrho}$

$$\mathcal{F}:\ \mathcal{S}(\mathbb{R}) \xrightarrow{\cong} \mathcal{S}(\mathbb{R}), \qquad \mathcal{F}:\ \mathcal{S}'(\mathbb{R}) \xrightarrow{\cong} \mathcal{S}'(\mathbb{R}),$$

$\mathcal{S}(\mathbb{R})$ being the Schwartz space on $\mathbb{R}$, $\mathcal{S}'(\mathbb{R})$ the space of temperate distributions on $\mathbb{R}$. Then

$$\mathcal{M}\{\mathcal{H}^s(\mathbb{R}_+)\} = (1 + |\varrho|^2)^{-\frac{s}{2}} L_2(\Gamma_{\frac{1}{2}}), \tag{7.49}$$

$\varrho = \mathrm{Im}\ z$. In other words the Mellin image of $\mathcal{H}^s(\mathbb{R}_+)$ coincides on $\Gamma_{\frac{1}{2}}$ with the Fourier image of $H^s(\Gamma_{\frac{1}{2}})$ (under the natural identification $\mathbb{R} \cong \Gamma_{\frac{1}{2}}$, $\varrho \to \frac{1}{2} + i\varrho$).

Using (7.47) for $\gamma = 0$ we get

$$\begin{aligned}
\mathcal{H}^s(\mathbb{R}_+) &= S_0^{-1} \mathcal{F}_{\varrho\to t}^{-1}\{(1 + |\varrho|^2)^{-\frac{s}{2}} L_2(\Gamma_{\frac{1}{2}})\} \\
&= S_0^{-1} H^s(\mathbb{R}_t),
\end{aligned}$$

in other words

$$\mathcal{H}^s(\mathbb{R}_+) = \{r^{-\frac{1}{2}} u(\ln r) :\ u \in H^s(\mathbb{R})\}. \tag{7.50}$$

In view of the fact that for any open $\Omega \subseteq \mathbb{R}$ the space

$$H^s_{\mathrm{loc}}(\Omega) := \{u \in \mathcal{D}'(\Omega) :\ \phi u \in H^s(\mathbb{R})\ \text{ for every }\ \phi \in C_0^\infty(\Omega)\}$$

remains invariant under diffeomorphisms $\chi{:}\Omega{\to}\tilde{\Omega}$ (more precisely $\chi^*{:}C^\infty(\tilde{\Omega}){\to}C^\infty(\Omega)$ has an extension to an isomorphism $\chi^* :\ H^s_{\mathrm{loc}}(\tilde{\Omega}) \to H^s_{\mathrm{loc}}(\Omega)$), we obtain from (7.50), (7.48)

$$\mathcal{H}^{s,\gamma}(\mathbb{R}_+) \subset H^s_{\mathrm{loc}}(\mathbb{R}_+). \tag{7.51}$$

Similarly to standard pseudo-differential operators we can form pseudo-differential operators by using the Mellin transform. Let

$$h(r, z) \in S^m(\mathbb{R}_+ \times \Gamma_{\frac{1}{2}})$$

(which means by definition that $h(r, \frac{1}{2} + i\varrho)$ is in $S^m(\mathbb{R}_+ \times \mathbb{R})$ with ϱ as covariable to r). Then we can define

$$op_M(h)u(r) := \frac{1}{2\pi i} \int_{\Gamma_{\frac{1}{2}}} r^{-z} h(r, z) \left\{ \int_0^\infty r'^{z-1} u(r')dr' \right\} dz \tag{7.52}$$

for every $u \in C_0^\infty(\mathbb{R}_+)$. More generally, if $h(r, z) \in S^m(\mathbb{R}_+ \times \Gamma_{\frac{1}{2}-\gamma})$ (with analogous interpretation of $\mathrm{Im}\ z = \varrho$ as covariable) then we get

$$op_M^\gamma(h)u(r) = r^\gamma op_M(T^{-\gamma}h)r'^{-\gamma} u(r'),$$

with $(T^{-\gamma}h)(r,z) := h(r, z - \gamma)$. Let us consider (7.30) and set $h(r,z) = \sum\limits_{k=0}^{m} a_k(r) z^k$.
This is a $Diff^m(X)$-valued symbol and (7.30) can be written as $A = r^{-m} op_M(h)$.
In other words it makes sense to employ operator-valued symbols in the Mellin
pseudo-differential calculus on manifolds with singularities. This will be systematically employed in Chapter 8 below.

Let us now pass to the weighted Sobolev spaces

$$\mathcal{H}^{s,\gamma}(B), \qquad s, \gamma \in \mathbb{R} \tag{7.53}$$

on any compact manifold B with conical singularities. Their definition will follow
from a corresponding definition of the stretched manifold $\mathbb{B}$.

If Ω is a C^∞ manifold, $n = \dim \Omega$, then

$$H^s_{\text{loc}}(\Omega), \qquad s \in \mathbb{R},$$

will denote the space of all $u \in \mathcal{D}'(\Omega)$ such that for every chart $\chi : U \longrightarrow \mathbb{R}^n$ and
every $\phi \in C_0^\infty(U)$ we have $\chi_*(\phi u) \in H^s(\mathbb{R}^n)$. The space $H^s(\mathbb{R}^n)$ is defined as

$$H^s(\mathbb{R}^n) = \mathcal{F}_{\xi \to x}^{-1}\{(1 + |\xi|^2)^{-\frac{s}{2}} L_2(\mathbb{R}^n_\xi)\}$$

with the Fourier transform in $\mathbb{R}^n$

$$\mathcal{F}v(\xi) = \int e^{-ix\xi} v(x) dx, \qquad x\xi := \sum_{i=1}^{n} x_i \xi_i,$$

and the inverse

$$(\mathcal{F}^{-1}g)(x) = \int e^{ix\xi} g(\xi) d\xi, \qquad d\xi := (2\pi)^{-n} d\xi.$$

We also denote by

$$H^s_{\text{comp}}(\Omega), \qquad s \in \mathbb{R},$$

the subspace of all $u \in H^s_{\text{loc}}(\Omega)$ with compact support.

Definition 10. $\mathcal{H}^{s,\gamma}(\mathbb{B})$ for $s, \gamma \in \mathbb{R}$ and compact $\mathbb{B}$ is the space of all $u \in H^s_{\text{loc}}(\text{int } \mathbb{B})$
such that for every $\phi \in C_0^\infty(\mathbb{B})$ supported by a coordinate neighbourhood of a point
of $\partial\mathbb{B}$ with the local coordinates $(r,x) \in \overline{\mathbb{R}}_+ \times \mathbb{R}^n$ we have

$$\left\{ \frac{1}{2\pi i} \int\limits_{\Gamma_{\frac{n+1}{2} - \gamma}} \int\limits_{\mathbb{R}^n} (1 + |z|^2 + |\xi|^2)^s |\mathcal{M}_{\frac{n}{2} - \gamma, r \to z} \mathcal{F}_{x \to \xi}(\phi u)(z, \xi)|^2 d\xi\, dz \right\}^{\frac{1}{2}} < \infty. \tag{7.54}$$

$\gamma \in \mathbb{R}$ is called the weight of the space $\mathcal{H}^{s,\gamma}(\mathbb{B})$. It would be more complete to write everywhere

$$\mathcal{H}^{s,\gamma}(\text{int } \mathbb{B})$$

instead of $\mathcal{H}^{s,\gamma}(\mathbb{B})$. We hope that our shorter notation will not cause confusion. In the case of the corresponding space on the open stretched infinite cone $X^\wedge = \mathbb{R}_+ \times X$ we shall actually write

$$\mathcal{H}^{s,\gamma}(X^\wedge), \qquad s, \gamma \in \mathbb{R} \tag{7.55}$$

which means the space of all $u \in H^s_{\text{loc}}(X^\wedge)$ such that (7.54) holds in a neighbourhood of $r = 0$ for every ϕ as mentioned and that $v(r,x) = r^{-n-1}u(r^{-1}, x)$ satisfies the analogous conditions with $-\gamma$ instead of γ.

Note that

$$\mathcal{H}^{s,\gamma}(X^\wedge) = r^\gamma \mathcal{H}^s(X^\wedge),$$

with $\mathcal{H}^s(X^\wedge) := \mathcal{H}^{s,0}(X^\wedge)$ (cf. (7.48)). Moreover, we have

$$\mathcal{H}^0(X^\wedge) = r^{-\frac{n}{2}} L_2(\mathbb{R}_+ \times X)$$

with $n = \dim X$ and the space $L_2(\mathbb{R}_+ \times X)$ of square integrable functions with respect to $drdx$, dx associated with a fixed Riemannian metric on X.

Remark 11. $\mathcal{H}^{s,\gamma}(\mathbb{B})$ is a Banach space, and the norm can be generated by a Hilbert space scalar product. This is an easy consequence of Definition 10. An analogous remark holds for (7.55).

A further simple observation is that the nondegenerate sesquilinear pairing

$$(\cdot, \cdot)_{\mathcal{H}^0(\mathbb{B})} : \ C_0^\infty(\text{int } \mathbb{B}) \times C_0^\infty(\text{int } \mathbb{B}) \longrightarrow \mathbb{C}$$

can be extended to a nondegenerate sesquilinear pairing

$$\mathcal{H}^{s,\gamma}(\mathbb{B}) \times \mathcal{H}^{-s,-\gamma}(\mathbb{B}) \longrightarrow \mathbb{C} \ \text{ for all } \ s, \gamma \in \mathbb{R}.$$

By

$$(Au, v)_{\mathcal{H}^0(\mathbb{B})} = (u, A^* v)_{\mathcal{H}^0(\mathbb{B})}$$

for $u, v \in C_0^\infty(\text{int } \mathbb{B})$ we define the formal adjoint A^* of an operator (7.30). If

$$A : \ \mathcal{H}^{s,\gamma}(\mathbb{B}) \longrightarrow \mathcal{H}^{s-m,\gamma-\nu}(\mathbb{B})$$

is continuous for all $s \in \mathbb{R}$ and fixed reals m, γ, ν, then A^* is extendible to a continuous operator

$$A^* : \ \mathcal{H}^{s,-\gamma+\nu}(\mathbb{B}) \longrightarrow \mathcal{H}^{s-m,-\gamma}(\mathbb{B})$$

for all $s \in \mathbb{R}$.

Remark 12. There is a straightforward generalization of Definition 10 to spaces with weights $\gamma = (\gamma_1, \ldots, \gamma_M) \in \mathbb{R}^M$, where every $\gamma_j \in \mathbb{R}$ corresponds to one conical singularity $b_j \in B_0$ (cf. (7.23)). This generalization may be interesting for applications, but here we content ourselves for simplicity with the case of one weight.

The spaces of Definition 10 deserve more functional analytic consideration. We shall return to further details below. Let us only remark for the moment that the canonical embedding

$$\mathcal{H}^{s',\gamma'}(\mathbb{B}) \longrightarrow \mathcal{H}^{s,\gamma}(\mathbb{B}) \tag{7.56}$$

for $s' \geq s, \gamma' \geq \gamma$ is continuous and compact for $s' > s, \gamma' > \gamma$.

Exercise 13. If $A \in Diff^m(\text{int } \mathbb{B})$ is of Fuchs type (cf. Definition 2) then A induces continuous operators

$$A: \ \mathcal{H}^{s,\gamma}(\mathbb{B}) \longrightarrow \mathcal{H}^{s-m,\gamma-m}(\mathbb{B}) \tag{7.57}$$

for all $s, \gamma \in \mathbb{R}$.

The next problem will be to define an adequate notion of ellipticity under which (7.57) is a Fredholm operator. This will be done in terms of the symbolic structure.

As in (7.6) we first have the map

$$\sigma_\psi^m: \ Diff^m(\Omega) \longrightarrow S^{(m)}(T^*\Omega \setminus 0)$$

with $S^{(m)}(T^*\Omega \setminus 0)$ being the space of all C^∞ functions on $T^*\Omega \setminus 0$ which are (positively) homogeneous of order m with respect to the canonical $\mathbb{R}_+$ actions on the fibres of $T^*\Omega$. This can be applied, in particular, to $\Omega = \text{int } \mathbb{B}$. Let us express the homogeneous principal symbol of order m of a differential operator of Fuchs type written in $(r, x) \in \mathbb{R}_+ \times U$, U being a coordinate neighbourhood on X. We obtain

$$\sigma_\psi^m(A)(r, x, \varrho, \xi) = r^{-m} \sum_{k+|\alpha|=m} a_{k\alpha}(r, x)(-ir\varrho)^k \xi^\alpha \tag{7.58}$$

with coefficients $a_{k\alpha}(r, x) \in C^\infty(\overline{\mathbb{R}}_+ \times U)$. Here $(\varrho, \xi) \in \mathbb{R}^{n+1}$ are the covariables to (r, x). Let

$$C^\infty(T_b^*\mathbb{B} \setminus 0)$$

denote the space of all $p \in C^\infty(T^*(\text{int } \mathbb{B}) \setminus 0)$ which can be written in a neighbourhood of $\partial\mathbb{B}$ in the variables (r, x, ϱ, ξ) as $p(r, x, \varrho, \xi) = p_b(r, x, r\varrho, \xi)$ for some $p_b(r, x, \tilde{\varrho}, \xi) \in C^\infty([0, 1) \times U \times (\mathbb{R}^{n+1} \setminus \{0\}))$. The notation T_b^* is motivated by the interpretation of such functions on a "compressed cotangent bundle" $T_b^*\mathbb{B}$ of $\mathbb{B}$. We will not go into further details here. Let us restrict ourselves to indicating the form of C^∞ sections in $T_b^*\mathbb{B}$ locally in a neighbourhood of $\partial\mathbb{B}$, namely

$$r^{-1}\tilde{\alpha}_0(r, x)dr + \sum_{j=1}^{n} \alpha_j(r, x)dx_j$$

with $\tilde{\alpha}_0, \alpha_j \in C^\infty([0, 1) \times U)$. Note that a C^∞ section in the dual bundle $T_b\mathbb{B}$ can be interpreted as a vector field which is tangent to $\partial\mathbb{B}$ (cf. also Remark 6 above).

Every such vector field in a neighbourhood of $\partial\mathbb{B}$ will be of the form

$$r\tilde{v}_0(r,x)\frac{\partial}{\partial r} + \sum_{j=1}^{n} v_j(r,x)\frac{\partial}{\partial x_j}$$

with $\tilde{v}_0, v_j \in C^\infty([0,1) \times U)$.

Let

$$S^{(m)}(T_b^*\mathbb{B} \setminus 0) \tag{7.59}$$

denote the subspace of all $\sigma \in C^\infty(T_b^*\mathbb{B} \setminus 0)$ which are (positively) homogeneous of order m with respect to the canonical $\mathbb{R}_+$ actions on the fibres of $T_b^*\mathbb{B}$. This means, in particular, that in a neighbourhood of $\partial\mathbb{B}$

$$\sigma(r,x,\lambda\tilde{\varrho},\lambda\xi) = \lambda^m \sigma(r,x,\tilde{\varrho},\xi)$$

for all $\lambda \in \mathbb{R}_+$, $(\tilde{\varrho},\xi) \neq 0$, $(r,x) \in [0,1) \times U$, with $\tilde{\varrho} = r\varrho$. For every $\beta \in \mathbb{R}$ we choose a strictly positive function $k^\beta \in C^\infty(\text{int }\mathbb{B})$ with

$$k^\beta = r^\beta \quad \text{in a neighbourhood of } \partial\mathbb{B} \tag{7.60}$$

in the coordinates (r,x). Any such function will also be called a weight factor, of weight β.

In view of (7.58) the homogeneous principal symbol of a Fuchs operator allows this to be interpreted as a function

$$\sigma_{\psi,b}^m(A) \in S^{(m)}(T_b^*\mathbb{B} \setminus 0) \tag{7.61}$$

(equal to the homogeneous principal symbol of $p_b(r,x,\tilde{\varrho},\xi)$).

Furthermore, an operator (7.30) gives rise to a polynomial in $z \in \mathbb{C}$ with $Diff^m(X)$-valued coefficients

$$\sigma_M^m(A)(z) = \sum_{k=0}^{m} a_k(0)z^k, \tag{7.62}$$

called the conormal symbol of A of order m. If A is of Fuchs type on $\mathbb{B}$ we can also define (7.62) for $X = \partial\mathbb{B}$.

(7.62) is a family of continuous operators

$$\sigma_M^m(A)(z): \ H^s(X) \longrightarrow H^{s-m}(X) \tag{7.63}$$

for all $s \in \mathbb{R}$, parametrized by z.

Let us denote by

$$Diff^m(\mathbb{B})_{Fuchs} \tag{7.64}$$

the space of all differential operators of order m of Fuchs type on $\mathbb{B}$.

Definition 14. $A \in Diff^m(\mathbb{B})_{Fuchs}$ is called *elliptic* (of order m, with respect to a weight $\gamma \in \mathbb{R}$) if

(i) $\sigma_{\psi,b}^m(A) \neq 0$ on $T_b^*\mathbb{B} \setminus 0$,

(ii) the operators (7.63) are isomorphisms for some $s = s_0 \in \mathbb{R}$ and all $z \in \Gamma_{\frac{n+1}{2}-\gamma}$.

Note that condition (i) implies that $\sigma_M^m(A)(z)$ is elliptic on X for every fixed $z \in \mathbb{C}$ (cf. 7.1.1 Definition 1). According to 7.1.1 Theorem 4 $\sigma_M^m(A)(z)$ is a holomorphic family of Fredholm operators for all $s \in \mathbb{R}$. From 7.1.1 Remark 5, it follows that

$$D := \{ z \in \mathbb{C} : \ \sigma_M^m(A)(z) : \ H^s(X) \longrightarrow H^{s-m}(X) \ \text{ is no isomorphism } \} \quad (7.65)$$

is independent of $s \in \mathbb{R}$.

In order to give an appropriate generalization of 7.1.1 Definition 3 we need an adequate notion of smoothing operators over a manifold with conical singularities. The precise definition will be postponed to Section 8.1.1, where these operators will be called Green operators. But to formulate a result more simply we shall employ for the moment a more general notion of smoothing operators.

Definition 15. An operator

$$P \in \bigcap_{s \in \mathbb{R}} \mathcal{L}(\mathcal{H}^{s,\gamma-m}(\mathbb{B}), \mathcal{H}^{s+m,\gamma}(\mathbb{B})) \quad (7.66)$$

is called a *parametrix* of $A \in Diff^m(\mathbb{B})_{Fuchs}$ with respect to a fixed weight $\gamma \in \mathbb{R}$, if there is an $\varepsilon > 0$ such that

$$\begin{aligned}
AP - I \ &\in \ \bigcap_{s \in \mathbb{R}} \mathcal{L}(\mathcal{H}^{s,\gamma-m}(\mathbb{B}), \mathcal{H}^{\infty,\gamma-m+\varepsilon}(\mathbb{B})), \\
PA - I \ &\in \ \bigcap_{s \in \mathbb{R}} \mathcal{L}(\mathcal{H}^{s,\gamma}(\mathbb{B}), \mathcal{H}^{\infty,\gamma+\varepsilon}(\mathbb{B})).
\end{aligned} \quad (7.67)$$

Theorem 16. *The following conditions for some* $A \in Diff^m(\mathbb{B})_{Fuchs}$ *are equivalent:*

(i) $A : \mathcal{H}^{s,\gamma}(\mathbb{B}) \longrightarrow \mathcal{H}^{s-m,\gamma-m}(\mathbb{B})$ *is a Fredholm operator for a certain fixed* $s = s_0 \in \mathbb{R}$,

(ii) A *is elliptic with respect to the weight* γ.

If A *is elliptic with respect to* γ, *(7.57) is a Fredholm operator for all* $s \in \mathbb{R}$. *There exists a parametrix of* A. *Furthermore*

$$Au = f \in \mathcal{H}^{s,\gamma-m}(\mathbb{B}) \quad \text{for some} \ \ s \in \mathbb{R}, \quad (7.68)$$

$$u \in \mathcal{H}^{-\infty,\gamma}(\mathbb{B}) \quad (7.69)$$

implies

$$u \in \mathcal{H}^{s+m,\gamma}(\mathbb{B}). \quad (7.70)$$

This result as well as the other theorems of this preliminary discussion will be proved in Chapter 8. They will play the role of key results of the calculus and will then be extended to pseudo-differential operators.

We will also need a variant of weighted Sobolev spaces on the infinite open stretched cone $X^\wedge \ni (r, x)$ where the behaviour for $r \to \infty$ is modelled by the standard Sobolev spaces on the infinite cylinder

$$H^s(X^\wedge) := H^s(\mathbb{R} \times X)|_{\mathbb{R}_+ \times X} . \tag{7.71}$$

By definition we have

$$H^s(\mathbb{R} \times X) \subset H^s_{\mathrm{loc}}(\mathbb{R} \times X).$$

Then the only point to take into account is the behaviour of functions for $r \to \pm\infty$. Because of the symmetry when the sign of r changes it suffices to look at $r \to +\infty$. Choose an open covering of X by coordinate neighbourhoods $U_1, \ldots, U_N$ and let $\phi_1, \ldots, \phi_N$ be a subordinate partition of unity. If we fix diffeomorphisms

$$\kappa_j : U_j \longrightarrow V_j$$

to open subsets $V_j \subset S^n = \{y \in \mathbb{R}^{n+1} : |y| = 1\}$, then for $U_j^\wedge = \mathbb{R}_+ \times U_j$, $V_j^\wedge = \{y \in \mathbb{R}^{n+1} \setminus \{0\} : \frac{y}{|y|} \in V_j\}$ we get diffeomorphisms

$$\kappa_j^\wedge : U_j^\wedge \longrightarrow V_j^\wedge, \quad \kappa_j^\wedge(r, x) = r\kappa_j(x), \quad r > 0.$$

Now $u \in \mathcal{D}'(\mathbb{R} \times X)$ with supp $u \subseteq [\varepsilon, \infty) \times X$ for any $\varepsilon > 0$ belongs to $H^s(\mathbb{R} \times X)$ if

$$(\kappa_j^\wedge)_*(\phi_j u) \in H^s(\mathbb{R}_y^{n+1}) \quad \text{for all} \quad j = 1, \ldots, N.$$

Here subscript $*$ indicates the push-forward of distributions under which dt is transformed to $d|y|$. The resulting space $H^s(\mathbb{R} \times X)$ then has a natural Banach space structure. A norm may be defined by

$$H^s(\mathbb{R} \times X) \ni u \longrightarrow \left\{ \|\psi u\|_s^2 + \sum_{j=1}^N \|(\kappa_j^\wedge)_*(\phi_j u_+)\|_{H^s(\mathbb{R}^{n+1})}^2 \right.$$

$$\left. + \sum_{j=1}^N \|(\kappa_j^\wedge)_*(\phi_j \tilde{u}_-)\|_{H^s(\mathbb{R}^{n+1})}^2 \right\}^{\frac{1}{2}},$$

where $\psi(r) \in C_0^\infty(\mathbb{R})$ is a function with $\psi(r) \equiv 1$ in an open neighbourhood of $r = 0$, $(1 - \psi)u = u_+ + u_-$, where $u_+ = (1 - \psi)u|_{\mathbb{R}_+ \times X}$, $u_- = (1 - \psi)u|_{\mathbb{R}_- \times X}$, $\tilde{u}_-(r, x) := u_-(-r, x)$, and $\|\cdot\|_s$ is the norm in the Sobolev space $H^s((-c, c) \times X) = H^s_{\mathrm{loc}}(\mathbb{R} \times X)|_{(-c,c) \times X}$ with $c > 0$, supp $\psi \subset (-c, c)$.

Exercise 17. Check that the space $H^s(\mathbb{R} \times X)$ is independent of the choice of data such as the open covering of X, the partition of unity, the diffeomorphisms κ_j and of ψ. Moreover the norm in $H^s(\mathbb{R} \times X)$ can be obtained by a Hilbert space scalar product $(\cdot, \cdot)_s$, i.e. $\| \cdot \|_{H^s(\mathbb{R} \times X)} = (\cdot, \cdot)_s^{\frac{1}{2}}$.

We are not so much interested in the space (7.71) as in

$$\mathcal{K}^{s,\gamma}(X^\wedge) = \omega\mathcal{H}^{s,\gamma}(X^\wedge) + (1-\omega)H^s(X^\wedge), \tag{7.72}$$

where $\omega(r) \in C_0^\infty(\overline{\mathbb{R}}_+)$ is a cut-off function, i.e. $\omega(r) \equiv 1$ for $r < c_0$ with some $c_0 > 0$. Again, the space (7.72) is Banach, and a norm can be obtained by a Hilbert space scalar product. Clearly (7.72) is independent of the choice of ω.

Remark 18. The space $\mathcal{K}^{s,\gamma}(X^\wedge)$ will play a fundamental role below in the theory of pseudo-differential operators on manifolds with edges, where $X^\wedge$ has the meaning of the (open stretched) model cone of the wedge.

Exercise 19. The family of operators

$$(\kappa_\lambda u)(r,x) := \lambda^{\frac{n+1}{2}}u(\lambda r, x), \quad \lambda \in \mathbb{R}_+, \quad u \in \mathcal{K}^{s,\gamma}(X^\wedge)$$

depends continuously on λ in the strong operator topology of $\mathcal{L}(\mathcal{K}^{s,\gamma}(X^\wedge))$ for every $s,\gamma \in \mathbb{R}$.

Exercise 20. The space $C_0^\infty(X^\wedge)$ is dense in $\mathcal{K}^{s,\gamma}(X^\wedge)$ for every $s,\gamma \in \mathbb{R}$.

The operator-valued edge symbolic calculus below will make it necessary to extend the concept of ellipticity to the infinite cone $X^\wedge \ni (t,x)$ such that the associated operators

$$A: \mathcal{K}^{s,\gamma}(X^\wedge) \longrightarrow \mathcal{K}^{s-m,\gamma-m}(X^\wedge) \tag{7.73}$$

are Fredholm. This leads to special precautions near the *exit to infinity* $r \to \infty$. We write A in the form

$$A = \omega A_0\omega_0 + (1-\omega)A_1(1-\omega_1)$$

with differential operators $A_0, A_1 \in Diff^m(X^\wedge)$ and cut-off functions $\omega, \omega_0, \omega_1$ satisfying $\omega\omega_0 = \omega$, $\omega\omega_1 = \omega_1$. The operator A_0 will be assumed to be of the form (7.30) whereas A_1 looks like

$$A_1 = \sum_{k=0}^m b_k(r)D_r^k$$

with coefficients $b_k(r) \in C^\infty(\mathbb{R}_+, Diff^{m-k}(X))$ which behave as classical symbols in $r \to \infty$. More precisely, we assume $b_k(r) \in S_{cl}^0(\mathbb{R}_+) \otimes_\pi Diff^{m-k}(X)$. Here $\mathbb{R}_+$ is interpreted as a conical set in $\mathbb{R}$, the space of the one-dimensional cone axis variable r, and $\otimes_\pi$ is the completed projective tensor product between the spaces. (Note that $Diff^{m-k}(X)$ is a nuclear Fréchet space.) In order to describe the conditions for $r \to \infty$ we refer to the local coordinates $x \in \mathbb{R}^n$ in a coordinate neighbourhood U on X. Then

$$b_k(r) = \sum_{|\beta|\leq m-k} b_{k\beta}(r,x)D_x^\beta$$

with coefficients $b_{k\beta}(r,x) \in C^\infty(U, S^0_{cl}(\mathbb{R}_+))$. Let us pass to the homogeneous principal part of $b_{k\beta}(r,x)$ of order zero in r, denoted by $\sigma^0_e(b_{k\beta})_U(x)$. We then set

$$\sigma^0_e(b_k)_U(x,\xi) = \sum_{|\beta| \leq m-k} \sigma^0_e(b_{k\beta})_U(x)\xi^\beta$$

and

$$\sigma^0_e(A)_U(x,\varrho,\xi) = \sum_{k=0}^m \sigma^0_e(b_k)_U(x,\xi)\varrho^k. \tag{7.74}$$

Clearly we also have $\sigma^0_e(A) \in S^m(U \times \mathbb{R}^{n+1}_{\varrho,\xi})$ since this is a polynomial in (ϱ,ξ) of order m. Then we define the homogeneous principal part of $\sigma^0_e(A)$ of order m in (ϱ,ξ) as

$$\sigma^{m,0}_{\psi,e}(A)(x,\varrho,\xi). \tag{7.75}$$

The latter object may be regarded as an invariantly defined function on $T^*X^\wedge \setminus 0$ which is independent of r, whereas (7.74) is invariant as a function on $T^*X^\wedge$ (which is in fact independent of r, cf. also Section 8.2.4 below).

Definition 21. An operator $A \in Diff^m(X^\wedge)$ which is of Fuchs type in a neighbourhood of $r = 0$ and with the properties mentioned for $r \to \infty$ is called *elliptic* (with respect to a weight γ) if it satisfies the conditions of Definition 14 and in addition

$$\sigma^0_e(A) \neq 0 \quad \text{on} \quad T^*X^\wedge, \qquad \sigma^{m,0}_{\psi,e}(A) \neq 0 \quad \text{on} \quad T^*X^\wedge \setminus 0.$$

Theorem 22.   *Let $A \in Diff^m(X^\wedge)$ be of Fuchs type near $r = 0$ and with the properties mentioned for $r \to \infty$. Then the following conditions are equivalent:*

(i) (7.73) is a Fredholm operator for an $s = s_0$,

(ii) A is elliptic (with respect to the weight γ).

If A is elliptic then (7.73) is Fredholm for all $s \in \mathbb{R}$. Moreover,

$$Au = f \in \mathcal{K}^{s,\gamma-m}(X^\wedge) \quad \text{for some} \ \ s \in \mathbb{R},$$

and

$$u \in \mathcal{K}^{-\infty,\gamma}(X^\wedge) \quad \text{implies} \quad u \in \mathcal{K}^{s+m,\gamma}(X^\wedge).$$

The proof of the latter result will follow from a corresponding calculus in Section 8.2.5, where we also describe the nature of parametrices.

7.1.3 The typical differential operators on manifolds with edges and corners

A *manifold W with edge singularities* is a topological space with a subspace Y consisting of the edges, such that the following properties hold:

(i) Y is a C^∞ manifold of dimension $q \geq 1$,

(ii) $W \setminus Y$ is a C^∞ manifold of dimension $n + 1 + q$,

(iii) every $y \in Y$ has an open neighbourhood V in W, such that there is a diffeomorphism

$$\chi :\ V \setminus Y \longrightarrow X^\wedge \times \Omega \tag{7.76}$$

for some closed compact C^∞ manifold $X = X(y)$, $n = \dim X$, and an open subset $\Omega \subseteq \mathbb{R}^q$, that is extendible to a homeomorphism

$$\overline{\chi} : V \longrightarrow X^\Delta \times \Omega \tag{7.77}$$

inducing a diffeomorphism

$$\chi_0 : V \cap Y \longrightarrow \Omega. \tag{7.78}$$

If

$$\kappa : V \setminus Y \longrightarrow X^\wedge \times \tilde{\Omega} \tag{7.79}$$

is another diffeomorphism of that kind, with the associated maps

$$\overline{\kappa} : V \longrightarrow X^\Delta \times \tilde{\Omega}, \qquad \kappa_0 : V \cap Y \longrightarrow \tilde{\Omega},$$

we will say that χ and κ are equivalent if

$$\chi\kappa^{-1} : X^\wedge \times \tilde{\Omega} \longrightarrow X^\wedge \times \Omega$$

is the restriction of some diffeomorphism

$$\overline{\mathbb{R}}_+ \times X \times \tilde{\Omega} \longrightarrow \overline{\mathbb{R}}_+ \times X \times \Omega \ \text{ to } \ \mathbb{R}_+ \times X \times \tilde{\Omega}.$$

A further condition on the system of maps (7.76) for varying $y \in Y$ and different choices of V is that for every pair of neighbourhoods $V, \tilde{V}$ with $V \cap \tilde{V} \neq \emptyset$ and the corresponding maps

$$\chi : V \setminus Y \longrightarrow X^\wedge \times \Omega, \qquad \tilde{\chi} : \tilde{V} \setminus Y \longrightarrow X^\wedge \times \tilde{\Omega},$$

the restrictions of χ and $\tilde{\chi}$ to $(V \cap \tilde{V}) \setminus Y$ define equivalent diffeomorphisms

$$\chi' : (V \cap \tilde{V}) \setminus Y \longrightarrow X^\wedge \times \Omega', \qquad \tilde{\chi}' : (V \cap \tilde{V}) \setminus Y \longrightarrow X^\wedge \times \tilde{\Omega}'$$

with corresponding open subsets $\Omega' \subseteq \Omega$ and $\tilde{\Omega}' \subseteq \tilde{\Omega}$, respectively.

The equivalence class of the system of maps (7.76) when y runs over Y is regarded as part of the structure of $W \setminus Y$ in a neighbourhood of Y, and this is kept fixed. We shall in fact fix a system of representatives that leads on $V \setminus Y$ to an $\mathbb{R}_+$ action if we set

$$\delta_\lambda v = \chi^{-1}(\lambda r, x, y) \quad \text{for} \quad \chi(v) = (r, x, y), \tag{7.80}$$

$v \in V \setminus Y$, $(r, x, y) \in \mathbb{R}_+ \times X \times \Omega$, $\lambda \in \mathbb{R}_+$.

We shall also call $X^\wedge \times \Omega$ an open stretched wedge with edge Ω and the (open stretched) model cone $X^\wedge$.

Note that W is in a neighbourhood of $y \in Y$ no manifold unless the base of the model cone is a sphere of dimension n. Nevertheless we will talk about manifolds with edges because the analysis takes place on $W \setminus Y$.

It was assumed for simplicity that all connected components of Y have the same dimension q. We may easily generalize the notion of a manifold with edges by allowing Y to be a disjoint union of components $Y = Y_1 \cup Y_2 \cup \ldots \cup Y_M$ of different dimensions. This causes trivial modifications of our considerations. For similar reasons we suppose that Y is connected. Then the cone bases $X(y_1)$, $X(y_2)$ for different $y_1, y_2 \in Y$ are diffeomorphic. Therefore we simply talk about X.

It follows from the assumptions that there is a C^∞ manifold $\mathbb{W}$ with C^∞ boundary which is an X bundle over Y, such that there is a diffeomorphism

$$W \setminus Y \longrightarrow \mathbb{W} \setminus \partial \mathbb{W}$$

which has a restriction to a diffeomorphism

$$V_1 \setminus Y \cong T_1 \setminus \partial \mathbb{W}$$

for an open neighbourhood $V_1 \subset W$ of Y and a collar neighbourhood $T_1 \subset \mathbb{W}$ of $\partial \mathbb{W}$,

$$T_1 \cong [0, 1) \times \partial \mathbb{W}.$$

$\mathbb{W}$ is also called the *stretched manifold with edges* associated with W.

Example 1. Let B be a manifold with conical singularities, and $\Omega \subseteq \mathbb{R}^q$ be open. Then $W = B \times \Omega$ is a manifold with edges, and we have $\mathbb{W} = \mathbb{B} \times \Omega$.

Let X be a paracompact C^∞ manifold. An operator $A \in Diff^m(X^\wedge \times \Omega)$ for open $\Omega \subseteq \mathbb{R}^q$, written $(r, x, y) \in \mathbb{R}_+ \times X \times \Omega$, is said to be edge-degenerate if it is in a neighbourhood of $r = 0$ of the form

$$A = r^{-m} \sum_{k+|\alpha| \leq m} a_{k\alpha}(r, y) \left(-r \frac{\partial}{\partial r} \right)^k (r D_y)^\alpha \tag{7.81}$$

with coefficients

$$a_{k\alpha}(r, y) \in C^\infty(\overline{\mathbb{R}}_+ \times \Omega, Diff^{m-(k+|\alpha|)}(X)). \tag{7.82}$$

Note that the order of differentiation in (7.81) is not essential for the described class of operators. A can always be written in the form

$$\widetilde{A} = r^{-m} \sum_{k+|\alpha|\leq m} \tilde{a}_{k\alpha}(r,y)(rD_y)^\alpha \left(-r\frac{\partial}{\partial r}\right)^k$$

with coefficients $\tilde{a}_{k\alpha}(r,y) \in C^\infty(\overline{\mathbb{R}}_+ \times \Omega, Diff^{m-(k+|\alpha|)}(X))$, and conversely.

Definition 2. Let W be a manifold with edges Y. Then $A \in Diff^m(W\setminus Y)$ is said to be *edge-degenerate* if every $y \in Y$ has a neighbourhood where A in local coordinates (r,x,y) has the form (7.81). Similarly we say that an $A \in Diff^m(\text{int } \mathbb{W})$ is edge-degenerate on $\mathbb{W}$ if it is of the form (7.81) in a neighbourhood of $\partial\mathbb{W}$.

Example 3. Let $g_X(r,y)$ be an (r,y)-dependent family of Riemannian metrics on a closed compact C^∞ manifold X which is of the class C^∞ with respect to $(r,y) \in [0,\infty) \times \Omega$ with an open set $\Omega \subseteq \mathbb{R}^q$. Then

$$g = dr^2 + r^2 g_X(r,y) + dy^2 \tag{7.83}$$

is a Riemannian metric on the open stretched wedge $X^\wedge \times \Omega$. The metric (7.83) describes geometrically a proper wedge where the "angle of the model cone" varies along Ω. The Laplace-Beltrami operator related to (7.83) equals

$$\Delta = r^{-2} \sum_{k=0}^{2} a_k(r,y) \left(-r\frac{\partial}{\partial r}\right)^k + \sum_{j=1}^{q} r^2 \frac{\partial^2}{\partial y_j^2} \tag{7.84}$$

with coefficients $a_k(r,y) \in C^\infty(\overline{\mathbb{R}}_+ \times \Omega, Diff^{2-k}(X))$ (cf. 7.1.2 Example 4). The operator (7.84) is actually edge-degenerate.

Example 4. Let

$$\widetilde{A} = \sum_{|\alpha|+|\beta|\leq m} a_{\alpha\beta}(\tilde{x},y)D_{\tilde{x}}^\alpha D_y^\beta \in Diff^m(\mathbb{R}_{\tilde{x},y}^{n+1+q})$$

and introduce polar coordinates $\mathbb{R}_{\tilde{x}}^{n+1}\setminus\{0\} \ni \tilde{x} \to (r,x)$. Then we get a diffeomorphism

$$\chi: \ (\mathbb{R}_{\tilde{x}}^{n+1}\setminus\{0\}) \times \mathbb{R}_y^q \longrightarrow \mathbb{R}_+ \times S^n \times \mathbb{R}_y^n.$$

The operator

$$A = \chi_*\{\widetilde{A}|_{(\mathbb{R}^{n+1}\setminus\{0\})\times\mathbb{R}^q}\}$$

is then edge-degenerate.

This follows easily from the calculations in 7.1.2 Example 5.

As noted at the beginning, a paracompact C^∞ manifold Ω with C^∞ boundary belongs to the class of manifolds with edges, whose boundary is the edge and whose inner normal (with respect to a Riemannian metric) is the model cone of the "wedge". Let $A \in \textit{Diff}^m(\Omega)$ be an operator with C^∞ coefficients up to $\partial\Omega$. In local coordinates of any neighbourhood of a point of $\partial\Omega$ we may assume that A on the half space $\mathbb{R}_+^{q+1} = \{(r,y) : r \in \mathbb{R}_+, y \in \mathbb{R}^q\}$ is given by

$$A = \sum_{k+|\alpha|\leq m} a_{k\alpha}(r,y) D_r^k D_y^\alpha, \qquad a_{k\alpha}(r,y) \in C^\infty(\overline{\mathbb{R}}_+ \times \mathbb{R}^q). \tag{7.85}$$

For every k there are constants c_{jk} such that

$$r^k D_r^k = \sum_{j=0}^{k} c_{jk} \left(-r\frac{\partial}{\partial r}\right)^j.$$

Thus

$$\begin{aligned}
A &= r^{-m} \sum_{k+|\alpha|\leq m} a_{k\alpha}(r,y) r^{m-(k+|\alpha|)} \left\{\sum_{j=0}^{k} c_{jk} \left(-r\frac{\partial}{\partial r}\right)^j\right\} (rD_y)^\alpha \\
&= r^{-m} \sum_{j+|\beta|\leq m} b_{j\beta}(r,y) \left(-r\frac{\partial}{\partial r}\right)^j (rD_y)^\beta
\end{aligned}$$

with coefficients $b_{j\beta}(r,y) \in C^\infty(\overline{\mathbb{R}}_+ \times \mathbb{R}^q)$. Hence (7.85) may be interpreted as an edge-degenerate operator. This is, of course, a modification of Example 4.

Remark 5. Let W be a manifold with edges and $\mathbb{W}$ the associated stretched manifold. Then every edge-degenerate operator on $\mathbb{W}$ of order m is of the form

$$r^{-m} \sum_{j} \phi_j \sum_{|\alpha|\leq m} a_{j\alpha} w_j^\alpha.$$

Here $\{\phi_j\}$ is a partition of unity belonging to a locally finite open covering of $\mathbb{W}$, $a_{j\alpha} \in C^\infty(\mathbb{W})$, $w_j^\alpha = w_{j_1}^{\alpha_1} \cdot \ldots \cdot w_{j_{n+1+q}}^{\alpha_{n+1+q}}$, with vector fields w_j^k on $\mathbb{W}$ that are in a neighbourhood of $\partial\mathbb{W}$ of the form

$$r\widetilde{b}_0(r,x,y)\frac{\partial}{\partial r} + \sum_{l=1}^{n} b_l(r,x,y)\frac{\partial}{\partial x_l} + r\sum_{k=1}^{q} \widetilde{b}_k(r,x,y)\frac{\partial}{\partial y_k}$$

with $\widetilde{b}_k, b_l \in C^\infty(\overline{\mathbb{R}}_+ \times U \times \Omega)$, U being a coordinate neighbourhood on X, with local coordinates $x = (x_1, \ldots, x_n)$.

The proof is an easy exercise.

Corollary 6. *The edge-degenerate differential operators over a manifold W with edges form an algebra.*

For brevity,

$$Diff^m(\mathbb{W})_{ed} \tag{7.86}$$

will denote the space of all edge-degenerate differential operators on $\mathbb{W}$ of order m. We introduce the space

$$C^\infty(T_b^*\mathbb{W} \setminus 0) \tag{7.87}$$

of all $p \in C^\infty(T^*(\text{int } \mathbb{W}) \setminus 0)$ which can be written in a neighbourhood of $\partial\mathbb{W}$ in the variables $(r, x, y, \varrho, \xi, \eta)$ as

$$p(r, x, y, \varrho, \xi, \eta) = p_b(r, x, y, r\varrho, \xi, r\eta)$$

for some $p_b(r, x, y, \tilde{\varrho}, \xi, \tilde{\eta}) \in C^\infty(\overline{\mathbb{R}}_+ \times U \times \Omega \times (\mathbb{R}^{n+1+q} \setminus \{0\}))$. The subspace of those elements of (7.87) which are (positively) homogeneous with respect to the canonical $\mathbb{R}_+$ action on the fibres of $T^*(\text{int } \mathbb{W}) \setminus 0$ will be called

$$S^{(m)}(T_b^*\mathbb{W} \setminus 0) \tag{7.88}$$

(cf. analogously 7.1.2(7.59)). The evaluation of the homogeneous principal symbol of order m of an operator $A \in Diff^m(\mathbb{W})_{ed}$ yields a map

$$\sigma_{\psi,b}^m : \ Diff^m(\mathbb{W})_{ed} \longrightarrow S^{(m)}(T_b^*\mathbb{W} \setminus 0) \tag{7.89}$$

(i.e. $\sigma_{\psi,b}^m(A)$ equals the homogeneous principal symbol of $p_b(r, x, y, \tilde{\varrho}, \xi, \tilde{\eta})$, locally near $\partial\mathbb{W}$).

Here k^β for any $\beta \in \mathbb{R}$ is a strictly positive function in $C^\infty(\text{int } \mathbb{W})$ with

$$k^\beta = r^\beta \ \text{ in a neighbourhood of } \ \partial\mathbb{W}. \tag{7.90}$$

The map (7.89) allows us to define a notion of ellipticity of A. The operator A is said to be elliptic with respect to $\sigma_{\psi,b}^m(\cdot)$ if it is elliptic in the standard sense on int $\mathbb{W}$ and if in addition $\sigma_{\psi,b}^m(A)(r, x, y, \tilde{\varrho}, \xi, \tilde{\eta})$ does not vanish for all $(\tilde{\varrho}, \xi, \tilde{\eta}) \neq 0$ and all r, x, y including $r = 0$.

If $A \in Diff^m(\mathbb{W})_{ed}$ is written in a neighbourhood of $\partial\mathbb{W}$ in the form (7.81), then, in local coordinates $(r, x, y) \in \mathbb{R}_+ \times U \times \Omega$, U being a coordinate neighbourhood on X, Ω a coordinate neighbourhood on Y, we obtain

$$\sigma_{\psi,b}^m(A)(r, x, y, \tilde{\varrho}, \xi, \tilde{\eta}) = \sum_{k+|\beta|+|\alpha|=m} a_{k\beta\alpha}(r, x, y)\tilde{\varrho}^k \xi^\beta \tilde{\eta}^\alpha \tag{7.91}$$

with certain coefficients $a_{k\beta\alpha} \in C^\infty(\overline{\mathbb{R}}_+ \times U \times \Omega)$. The ordinary homogeneous principal symbol $\sigma_\psi^m(A)$ as a C^∞ function on $T^*(\text{int } \mathbb{W}) \setminus 0$ takes the form

$$\sigma_\psi^m(A)(r, x, y, \varrho, \xi, \eta) = r^{-m} \sum_{k+|\beta|+|\alpha|=m} a_{k\beta\alpha}(r, x, y)r^{k+|\alpha|}\varrho^k \xi^\beta \eta^\alpha. \tag{7.92}$$

Let us now set $E = \mathcal{K}^{s,\gamma}(X^\wedge)$, $\quad \widetilde{E} = \mathcal{K}^{s-m,\gamma-m}(X^\wedge)$ with a fixed weight $\gamma \in \mathbb{R}$ and arbitrary $s \in \mathbb{R}$. Remember (cf. 7.1.2 Exercise 19) that on E we have the action of a one-parameter group $\{\kappa_\lambda\}_{\lambda\in\mathbb{R}_+}$ of isomorphisms, namely $(\kappa_\lambda u)(r,x) = \lambda^{\frac{n+1}{2}} u(\lambda r, x)$ for all $\lambda \in \mathbb{R}_+$, $u \in E$. An analogous action holds on $\tilde{E}$, also denoted by κ_λ. Then let

$$S^{(m)}(T^*Y \setminus 0; E, \tilde{E}) \tag{7.93}$$

be the space of all operator functions

$$a_{(m)}(y,\eta) \in C^\infty(T^*Y \setminus 0, \mathcal{L}(E, \tilde{E}))$$

satisfying

$$a_{(m)}(y,\lambda\eta) = \lambda^m \kappa_\lambda a_{(m)}(y,\eta)\kappa_\lambda^{-1} \tag{7.94}$$

for all $(y,\eta) \in T^*Y \setminus 0$, $\lambda \in \mathbb{R}_+$. We now define the homogeneous principal edge symbol of order m of an $A \in Diff^m(\mathbb{W})_{ed}$ written as (7.81) in a neighbourhood of $\partial\mathbb{W}$, namely

$$\sigma_\wedge^m(A)(y,\eta) = r^{-m} \sum_{k+|\alpha|\leq m} a_{k\alpha}(0,y) \left(-r\frac{\partial}{\partial r}\right)^k (r\eta)^\alpha. \tag{7.95}$$

It can easily be verified that (7.95) gives rise to a map

$$\sigma_\wedge^m : \; Diff^m(\mathbb{W})_{ed} \longrightarrow S^{(m)}(T^*Y \setminus 0; \mathcal{K}^{s,\gamma}(X^\wedge), \mathcal{K}^{s-m,\gamma-m}(X^\wedge)), \tag{7.96}$$

for every $s \in \mathbb{R}$. (7.95) represents an operator-valued principal symbol for the operator A which will appear below together with $\sigma_{\psi,b}^m(A)$ in the adequate definition of ellipticity. Since ellipticity should allow a parametrix construction by inverting symbols, we have to demand, in particular, that $\sigma_\wedge^m(A)$ is invertible for every $(y,\eta) \in T^*Y \setminus 0$. However it turns out that the latter condition is too restrictive. We can only hope that

$$\sigma_\wedge^m(A)(y,\eta) : \; \mathcal{K}^{s,\gamma}(X^\wedge) \longrightarrow \mathcal{K}^{s-m,\gamma-m}(X^\wedge) \tag{7.97}$$

is a Fredholm operator for fixed (y,η). The Fredholm property follows from the concept of ellipticity on the infinite stretched cone $X^\wedge$ (cf. 7.1.2 Definition 21) with the symbol maps

$$\sigma_{\psi,b}^m, \qquad \sigma_M^m, \quad \sigma_e^0,$$

applied to $\sigma_\wedge^m(A)(y,\eta)$, regarded as a cone operator for every fixed (y,η).

Note that $\sigma_{\psi,b}^m$, applied to $\sigma_\wedge^m(A)(y,\eta)$ in the sense of the cone theory, yields in the coordinates $(r,x,y) \in \mathbb{R}_+ \times U \times \Omega$ the relation

$$\sigma_{\psi,b}^m(\sigma_\wedge^m(A)(y,\eta))(r,x,\tilde\varrho,\xi) = \sum_{k+|\beta|=m} a_{k\beta0}(0,x,y)\tilde\varrho^k\xi^\beta$$

(cf. (7.91)). The latter expression is independent of η. Furthermore

$$\sigma_e^0(\sigma_\wedge^m(A)(y,\eta))(x,\xi,\varrho) = \sum_{k+|\alpha|=m} a_{k0\alpha}(0,x,y)\varrho^k\eta^\alpha.$$

The ellipticity of $A \in Diff^m(\mathbb{W})_{ed}$ with respect to (7.89) (cf. Definition 8, below) will imply the ellipticity of $\sigma_\wedge^m(A)(y,\eta)$, $\eta \neq 0$ with respect to $\sigma_{\psi,b}^m$ and σ_e^0. For the Fredholm property of (7.97) we also need there to be a $\gamma \in \mathbb{R}$ such that

$$\sigma_M^m(\sigma_\wedge^m(A)(y,\eta))(z) = \sum_{k=0}^m a_{k0}(0,y)z^k \; : \; H^s(X) \longrightarrow H^{s-m}(X) \qquad (7.98)$$

is an isomorphism for all $z \in \Gamma_{\frac{n+1}{2}-\gamma}$ and for all $y \in Y$. We shall see in Section 7.2.5 that there is at most a discrete subset $D(y) \subset \mathbb{C}$ of points z where (7.98) is violated.

Since

$$\sigma_\wedge^m(A)(y,\lambda\eta) = \lambda^m \kappa_\lambda \sigma_\wedge^m(A)(y,\eta)\kappa_\lambda^{-1}$$

for all $\lambda \in \mathbb{R}_+$, it follows that (7.97) is a family of Fredholm operators parametrized by $T^*Y \setminus 0$ once $\sigma_\wedge^m(A)\,|_{S^*Y}$ is a family of Fredholm operators, S^*Y being the cosphere bundle induced by T^*Y and a fixed Riemannian metric on Y. We assume here that Y is compact. Then S^*Y is also compact. We will denote by $\mathrm{Vect}(Y)$ the set of all complex finite-dimensional vector bundles over Y.

We now assume that there are vector bundles $J^-, J^+ \in \mathrm{Vect}(Y)$ and operator families

$$\sigma_\wedge^m(T)(y,\eta) \quad : \quad \mathcal{K}^{s,\gamma}(X^\wedge) \longrightarrow J_y^+, \qquad (7.99)$$

$$\sigma_\wedge^m(K)(y,\eta) \quad : \quad J_y^- \longrightarrow \mathcal{K}^{s-m,\gamma-m}(X^\wedge), \qquad (7.100)$$

$$\sigma_\wedge^m(Q)(y,\eta) \quad : \quad J_y^- \longrightarrow J_y^+ \qquad (7.101)$$

(subscript y indicates the fibres over y) such that

$$\begin{pmatrix} \sigma_\wedge^m(A) & \sigma_\wedge^m(K) \\ \sigma_\wedge^m(T) & \sigma_\wedge^m(Q) \end{pmatrix}(y,\eta): \begin{matrix} \mathcal{K}^{s,\gamma}(X^\wedge) \\ \oplus \\ J_y^- \end{matrix} \longrightarrow \begin{matrix} \mathcal{K}^{s-m,\gamma-m}(X^\wedge) \\ \oplus \\ J_y^+ \end{matrix} \qquad (7.102)$$

are isomorphisms for all $(y,\eta) \in S^*Y$. By

$$\sigma_\wedge^m(T)(y,\eta) \;=\; |\eta|^m \sigma_\wedge^m(T)\left(y,\frac{\eta}{|\eta|}\right)\kappa_{|\eta|}^{-1},$$

$$\sigma_\wedge^m(K)(y,\eta) \;=\; |\eta|^m \kappa_{|\eta|}\sigma_\wedge^m(K)\left(y,\frac{\eta}{|\eta|}\right),$$

$$\sigma_\wedge^m(Q)(y,\eta) \;=\; |\eta|^m \sigma_\wedge^m(Q)\left(y,\frac{\eta}{|\eta|}\right)$$

we get an extension of the isomorphism (7.102) to $T^*Y \setminus 0$. The resulting matrix-valued function (7.102), now being defined on $T^*Y \setminus 0$, will be interpreted as the homogeneous principal part of order m of an operator-valued symbol. The elliptic regularity stated in 7.1.2 Theorem 22 allows the replacement of (7.99), (7.100) by the maps

$$\sigma_\wedge^m(T)(y,\eta) \quad : \quad \mathcal{K}^{-\infty,\gamma}(X^\wedge) \longrightarrow J_y^+, \tag{7.103}$$

$$\sigma_\wedge^m(K)(y,\eta) \quad : \quad J_y^- \longrightarrow \mathcal{K}^{\infty,\gamma-m}(X^\wedge). \tag{7.104}$$

If (7.102) is an isomorphism for some $s = s_0$, then it is an isomorphism for all $s \in \mathbb{R}$.

The next idea is to interprete A as an operator between adequate weighted Sobolev spaces $\mathcal{W}^{s,\gamma}(\mathbb{W})$ over the stretched manifold $\mathbb{W}$ with edges. Further, let $H^s(Y,J)$ for $J \in \mathrm{Vect}(Y)$ be the space of distributional sections in J, of Sobolev smoothness $s \in \mathbb{R}$. The spaces $\mathcal{W}^{s,\gamma}(\mathbb{W})$ have yet to be defined, whereas the meaning of $H^s(Y,J)$ is standard (cf. also [P]). We then pass to the matrix of operators

$$\mathcal{A} = \begin{pmatrix} A & K \\ T & Q \end{pmatrix} : \quad \begin{matrix} \mathcal{W}^{s,\gamma}(\mathbb{W}) \\ \oplus \\ H^s(Y,J^-) \end{matrix} \quad \longrightarrow \quad \begin{matrix} \mathcal{W}^{s-m,\gamma-m}(\mathbb{W}) \\ \oplus \\ H^{s-m}(Y,J^+) \end{matrix} \quad . \tag{7.105}$$

Let us first define T, K, Q. The operator Q is none other than a classical pseudo-differential operator of order m on Y, acting between the distributional sections of J^- and J^+, respectively (cf. also [P]). For T and K it suffices to give local descriptions. The global operators are then formulated in terms of an open covering of a collar neighbourhood of $\partial \mathbb{W}$ and a subordinated partition of unity. Thus we may restrict ourselves to a neighbourhood of $\partial \mathbb{W}$ of the form

$$[0,1) \times U \times \Omega \ni (r,x,y)$$

such that the bundles J^-, J^+ are trivial over Ω, of fibre dimensions N_- and N_+, respectively. Choose an excision function $\chi(\eta)$ in $\mathbb{R}^q \ni \eta$ (i.e. $\chi \in C^\infty(\mathbb{R}^q)$, $\chi = 0$ for $|\eta| < c_0$, $\chi = 1$ for $|\eta| > c_1$ with constants $c_0 < c_1$), and let $\mathcal{F} = \mathcal{F}_{y \to \eta}$ be the Fourier transform in $\mathbb{R}^q$. Distributions over $(0,1) \times U \times \Omega$ will for now be interpreted as vector-valued, in the sense that for $\mathcal{D}'((0,1) \times U \times \Omega) = \mathcal{D}'(\Omega, \mathcal{D}'((0,1) \times U))$. Accordingly they will be denoted by $u(y)$, $y \in \Omega$. We begin with some $u(y) \in \mathbb{C}^{N_+} \otimes C_0^\infty(\Omega, C_0^\infty((0,1) \times U))$. Further, let $\omega(r)$ be a cut-off function, supported by $[0,1)$. Then we set

$$(Tu)(y) = \mathcal{F}_{\eta \to y}^{-1}\{\chi(\eta)\sigma_\wedge^m(T)(y,\eta)\omega(r')\mathcal{F}_{y' \to \eta}u(y',r',x')\}.$$

Similarly, for $v(y) \in C_0^\infty(\Omega) \otimes \mathbb{C}^{N_-}$, we define

$$(Kv)(y,r,x) = \omega(r)\mathcal{F}_{\eta \to y}^{-1}\{\chi(\eta)\sigma_\wedge^m(K)(y,\eta)\mathcal{F}_{y' \to \eta}v(y')\}.$$

Since the present discussion serves as a motivation we shall postpone explanations of how the operators allow extensions to the Sobolev spaces.

The left-hand side of (7.102) will also be denoted by

$$\sigma_\wedge^m(\mathcal{A})(y,\eta), \tag{7.106}$$

called the *homogeneous principal edge* symbol of $\mathcal{A}$ of order m. Moreover,

$$\sigma_{\psi,b}^m(\mathcal{A}) := \sigma_{\psi,b}^m(A) \tag{7.107}$$

is called the *homogeneous principal interior* symbol of $\mathcal{A}$ of order m (in the compressed sense).

Next we look at weighted Sobolev spaces

$$\mathcal{W}^{s,\gamma}(\mathbb{W}), \qquad s,\gamma \in \mathbb{R}. \tag{7.108}$$

If E is a Banach space, we denote by

$$\mathcal{S}(\mathbb{R}^q, E) \tag{7.109}$$

the space of all E-valued Schwartz functions on $\mathbb{R}^q$. This is the subspace of all $u(y) \in C^\infty(\mathbb{R}^q, E)$ for which

$$\sup_{y \in \mathbb{R}^q} \|y^\alpha D_y^\beta u(y)\|_E < \infty \tag{7.110}$$

for all multi-indices $\alpha, \beta \in \mathbb{R}^q$. The expressions in (7.110) form a countable semi-norm system on $\mathcal{S}(\mathbb{R}^q, E)$ under which $\mathcal{S}(\mathbb{R}^q, E)$ is a Fréchet space.

Definition 7. $\mathcal{W}^s(\mathbb{R}^q, \mathcal{K}^{s,\gamma}(X^\wedge))$ for $s,\gamma \in \mathbb{R}$ is the closure of $\mathcal{S}(\mathbb{R}_y^q, \mathcal{K}^{s,\gamma}(X^\wedge))$ with respect to the norm

$$\left\{ \int [\eta]^{2s} \|\kappa(\eta)^{-1}(\mathcal{F}_{y\to\eta}u)(\eta)\|_{\mathcal{K}^{s,\gamma}(X^\wedge)}^2 \, d\eta \right\}^{\frac{1}{2}}.$$

Here $\eta \to [\eta]$ is a strictly positive function in $C^\infty(\mathbb{R}_\eta^q)$ with $[\eta] = |\eta|$ for $|\eta| > \text{const}$, and

$$\kappa(\eta) := \kappa_{[\eta]}, \tag{7.111}$$

with $(\kappa_\lambda v)(r,x) = \lambda^{\frac{n+1}{2}} v(\lambda r, x)$, $\lambda > 0$, $n = \dim X$.

Instead of $[\eta]$ we could take the function $\langle\eta\rangle = (1 + |\eta|^2)^{\frac{1}{2}}$. This would lead to an equivalent norm.

Now (7.108) is the subspace

$$\mathcal{W}^{s,\gamma}(\mathbb{W}) \subset H_{\text{loc}}^s(\text{int } \mathbb{W}) \tag{7.112}$$

of all u such that for an arbitrary coordinate neighbourhood $U \subset X$ and $\Omega \subset Y$ with local coordinates x and y, respectively, and the identification of a corresponding neighbourhood of a point in $\partial \mathbb{W}$ with $[0,1) \times U \times \Omega$ and every $\phi(r,x,y) \in C_0^\infty([0,1) \times U \times \Omega)$, it follows that

$$\phi u \in \mathcal{W}^s(\mathbb{R}^q, \mathcal{K}^{s,\gamma}(X^\wedge)).$$

For brevity we have dropped the obvious pull-backs under corresponding charts.

It belongs to the precise theory, of course, to verify that this is a correct definition. Other equivalent definitions also are of interest. We shall postpone those questions and pass to the concept of ellipticity for differential operators $A \in Diff^m(\mathbb{W})_{ed}$.

Definition 8. The operator (7.105) is called *elliptic* (of order m with respect to a weight $\gamma \in \mathbb{R}$) if

    (i) $\sigma^m_{\psi,b}(\mathcal{A}) \neq 0$ on $T^*_b \mathbb{W} \setminus 0$ (cf. the notations (7.90), (7.107)),

    (ii) (7.102) is an isomorphism for all $(y,\eta) \in T^*Y \setminus 0$ and some fixed $s = s_0 \in \mathbb{R}$.

It is a consequence of the theory for the (infinite open stretched) cone $X^\wedge$ that when (ii) is satisfied for $s = s_0$, it follows for all $s \in \mathbb{R}$, cf. 7.1.2 Theorem 22.

Definition 9. An operator

$$\mathcal{P} \in \bigcap_{s \in \mathbb{R}} \mathcal{L}(\mathcal{W}^{s,\gamma-m}(\mathbb{W}) \oplus H^s(Y, J^+), \mathcal{W}^{s+m,\gamma}(\mathbb{W}) \oplus H^{s+m}(Y, J^-))$$

is called a *parametrix* of (7.105) if there is an $\varepsilon > 0$ such that

$$\mathcal{AP} - \mathcal{I} \in \bigcap_{s \in \mathbb{R}} \mathcal{L}(\mathcal{W}^{s,\gamma-m}(\mathbb{W}) \oplus H^s(Y, J^+), \mathcal{W}^{\infty,\gamma-m+\varepsilon}(\mathbb{W}) \oplus H^\infty(Y, J^+)),$$

$$\mathcal{PA} - \mathcal{I} \in \bigcap_{s \in \mathbb{R}} \mathcal{L}(\mathcal{W}^{s,\gamma}(\mathbb{W}) \oplus H^s(Y, J^-), \mathcal{W}^{\infty,\gamma+\varepsilon}(\mathbb{W}) \oplus H^\infty(Y, J^-)).$$

The notion of a parametrix depends on the nature of the smoothing operators $\mathcal{AP} - \mathcal{I}$ and $\mathcal{PA} - \mathcal{I}$. Here we have chosen a version which does not refer to the more concrete behaviour of distributional kernels in the sense of asymptotics in a neighbourhood of the edge. The asymptotics will be described below in Chapter 9.

Theorem 10. *Let A be given in the form (7.105), in particular, $A \in Diff^m(\mathbb{W})_{ed}$. Then the following conditions are equivalent:*

    (i) *A is elliptic of order m, with respect to the weight $\gamma \in \mathbb{R}$,*

    (ii) *(7.105) is a Fredholm operator for a certain fixed $s = s_0 \in \mathbb{R}$.*

If $\mathcal{A}$ is elliptic of order m, with respect to γ, (7.105) is a Fredholm operator for all $s \in \mathbb{R}$. There exists a parametrix of $\mathcal{A}$. Furthermore,

$$\mathcal{A}u = f \; \in \; \mathcal{W}^{r,\gamma-m}(\mathbb{W}) \oplus H^r(Y, J^+) \;\; \text{for some } \; r \in \mathbb{R}, \qquad (7.113)$$

$$u \; \in \; \mathcal{W}^{-\infty,\gamma}(\mathbb{W}) \oplus H^{-\infty}(Y, J^-) \qquad (7.114)$$

implies

$$u \in \mathcal{W}^{r+m,\gamma}(\mathbb{W}) \oplus H^{r+m}(Y, J^-). \qquad (7.115)$$

This result is obtained below in Section 9.3.4, where we show that elliptic regularity holds also in the more precise version with asymptotics.

Finally, let us give an idea of what corner-degenerate differential operators look like. First, the local model of a corner (of second order) is a cone, where the base again has conical singularities. Denote the new "corner axis variable" by $t \in \mathbb{R}_+$ and let (r, x) be the points on the base with the inner cone axis variable $r \in \mathbb{R}_+$ and x varying along the corresponding base X that is a closed, compact, C^∞ manifold. Then the corner-degenerate operators have the form

$$A = r^{-m}t^{-m} \sum_{j+k \leq m} a_{jk}(r, t) \left(-tr\frac{\partial}{\partial t} \right)^j \left(-r\frac{\partial}{\partial r} \right)^k \qquad (7.116)$$

with coefficients

$$a_{jk}(r, t) \in C^\infty(\overline{\mathbb{R}}_+ \times \overline{\mathbb{R}}_+, \mathit{Diff}^{m-(j+k)}(X)).$$

The analysis of those operators is much more complicated than the cone and edge theory. We have in (7.116) two singular directions and may expect an interaction between the edge theory along the one-dimensional edge $\mathbb{R}_+ \ni t$, emanating from the corner, and the "higher" cone effects for $t \to 0$. A comprehensive calculus for this case including ellipticity and parametrix constructions may be found in [Sc3].

7.2 Parameter-dependent pseudo-differential operators and operator-valued Mellin symbols

7.2.1 Additional material on pseudo-differential operators on closed compact C^∞ manifolds

The role of this section is to remind the reader of some basic material on pseudo-differential operators that is needed below in several variants. In particular, we will have to reformulate certain items from Chapter 2 to enable us to arrange the corresponding theory near singularities.

As we have seen, a starting observation for introducing pseudo-differential operators is that a differential operator

$$A(x, D) = \sum_{|\alpha| \leq m} a_\alpha(x)D_x^\alpha \qquad (7.117)$$

with coefficients $a_\alpha(x) \in C^\infty(\Omega)$ for an open set $\Omega \subseteq \mathbb{R}^n$ can be written in the form

$$A = \mathcal{F}^{-1} a(x, \xi) \mathcal{F}. \tag{7.118}$$

Here $\mathcal{F}$ is the Fourier transform in $\mathbb{R}^n$, first applied to $u(x) \in C_0^\infty(\Omega)$ in the Schwartz space $\mathcal{S}(\mathbb{R}^n)$, cf. Sections 1.1.1, 1.1.2. The complete symbol

$$a(x, \xi) = \sum_{|\alpha| \leq m} a_\alpha(x) \xi^\alpha \tag{7.119}$$

of A is a polynomial in $\xi = (\xi_1, \ldots, \xi_n)$. Since

$$\mathcal{F} : \ \mathcal{S}(\mathbb{R}^n_x) \longrightarrow \mathcal{S}(\mathbb{R}^n_\xi)$$

is an isomorphism, and $\xi^\alpha \mathcal{S}(\mathbb{R}^n_\xi) \subset \mathcal{S}(\mathbb{R}^n_\xi)$, we obtain

$$D_x^\alpha = \mathcal{F}^{-1} \xi^\alpha \mathcal{F}$$

as a map $\mathcal{S}(\mathbb{R}^n) \longrightarrow \mathcal{S}(\mathbb{R}^n)$. The operator (7.117) shall be regarded as a map

$$A : \ C_0^\infty(\Omega) \longrightarrow C^\infty(\Omega)$$

(the image belongs in fact to $C_0^\infty(\Omega)$). This is compatible with (7.118), where we tacitly assume that (7.118) is composed from the left by the restriction to Ω.

To extend the discussion to pseudo-differential operators we refer once again to the symbol spaces of Section 2.2.1. If $U \subseteq \mathbb{R}^N$ is an open set and K a compact subset of U we shall also write $K \subset\subset U$.

Definition 1. Let $U \subseteq \mathbb{R}^N$ be an open set and $m \in \mathbb{R}$. Then

$$S^m(U \times \mathbb{R}^n) \tag{7.120}$$

is defined as the set of all $a(x, \xi) \in C^\infty(U \times \mathbb{R}^n)$ such that

$$|D_x^\alpha D_\xi^\beta a(x, \xi)| \leq c(1 + |\xi|)^{m - |\beta|} \tag{7.121}$$

for all $\alpha \in \mathbb{Z}_+^N$, $\beta \in \mathbb{Z}_+^n$, $x \in K$, $\xi \in \mathbb{R}^n$, with arbitrary $K \subset\subset U$, with constants $c = c(\alpha, \beta, K) > 0$. For $a \in S^m(U \times \mathbb{R}^n)$ we shall also write

$$m = \operatorname{ord} a. \tag{7.122}$$

Remark 2. The best constants c in (7.121) for fixed $a \in S^m(U \times \mathbb{R}^n)$ form a semi-norm system on the space $S^m(U \times \mathbb{R}^n)$,

$$a \to \sup_{\substack{x \in K \\ \xi \in \mathbb{R}^n}} (1 + |\xi|)^{-m + |\beta|} |D_x^\alpha D_\xi^\beta a(x, \xi)|,$$

parametrized by $\alpha \in \mathbb{Z}_+^N$, $\beta \in \mathbb{Z}_+^n$, $K \subset\subset U$. This turns (7.120) into a Fréchet space. It suffices to take a suitable countable system of compact sets K.

Note that

$$S^m(\mathbb{R}^n),$$

defined as the subspace of all elements of $S^m(U \times \mathbb{R}^n)$ which are independent of x, is a closed subspace in the induced topology. Then, it is an easy exercise to verify that

$$S^m(U \times \mathbb{R}^n) = C^\infty(U, S^m(\mathbb{R}^n)).$$

Here

$$C^\infty(U, E)$$

for any Fréchet space E is the space of E-valued C^∞ functions over U.

In particular, we have

$$S^{-\infty}(U \times \mathbb{R}^n) := \bigcap_{m \in \mathbb{R}} S^m(U \times \mathbb{R}^n) = C^\infty(U, \mathcal{S}(\mathbb{R}^n)).$$

We shall mainly need

$$U = \Omega \quad \text{or} \quad U = \Omega \times \Omega$$

for some open set $\Omega \subseteq \mathbb{R}^n$. In the latter case we will also write (x, x') for the variable in $\Omega \times \Omega$. Set

$$\text{diag } \Omega \times \Omega = \{(x, x) : \ x \in \Omega\}.$$

An *excision function* is a $\chi(\xi) \in C^\infty(\mathbb{R}^n)$ with

$$\chi(\xi) = \begin{cases} 0 & \text{for} & |\xi| < c_0, \\ 1 & \text{for} & |\xi| > c_1 \end{cases}$$

with constants $0 < c_0 < c_1 < \infty$. Let

$$S^{(m)}(U \times (\mathbb{R}^n \setminus \{0\})) \tag{7.123}$$

be the space of all $f(x, \xi) \in C^\infty(U \times (\mathbb{R}^n \setminus \{0\}))$ with

$$f(x, \lambda \xi) = \lambda^m f(x, \xi) \tag{7.124}$$

for all $\lambda \in \mathbb{R}_+$ and all $x \in U, \xi \in \mathbb{R}^n \setminus \{0\}$. Then

$$\chi(\xi) S^{(m)}(U \times (\mathbb{R}^n \setminus \{0\})) \subset S^m(U \times \mathbb{R}^n).$$

Note that for every $f_{(m)}(x, \xi) \in (7.123)$, the function $f(x, \xi) := \chi(\xi) f_{(m)}(x, \xi)$ satisfies (7.124) for all $\lambda \geq 1$, $|\xi| \geq$ const, for some constant > 0.

Lemma 3. *Let $a_j \in S^{m_j}(U \times \mathbb{R}^n)$, $j \in \mathbb{Z}_+$, be an arbitrary sequence, with $m_j \to -\infty$ as $j \to \infty$. Then there exists an $a \in S^m(U \times \mathbb{R}^n)$ for $m = \max\{m_j\}$ such that*

$$\text{ord} \left(a - \sum_{j=0}^{N} a_j \right) \to -\infty \quad \text{as} \ \ N \to \infty.$$

If $\tilde{a} \in S^m(U \times \mathbb{R}^n)$ is another symbol with this property, then $a - \tilde{a} \in S^{-\infty}(U \times \mathbb{R}^n)$.

We shall also write

$$a(x,\xi) \sim \sum_{j=0}^{\infty} a_j(x,\xi) \tag{7.125}$$

for the element a of Lemma 3, called the asymptotic sum of the a_j.

Lemma 3 is a slight extension of the corresponding 3.1.1 Lemma 1. Details of the proof are left as an exercise for the reader.

Definition 4. Let $U \subseteq \mathbb{R}^N$ be open and $m \in \mathbb{R}$. Then

$$S_{cl}^m(U \times \mathbb{R}^n) \tag{7.126}$$

is defined as the subspace of all $a \in S^m(U \times \mathbb{R}^n)$ for which there are a_j with

$$a_j(x, \lambda\xi) = \lambda^{m-j} a_j(x,\xi)$$

for all $\lambda \geq 1$, $x \in U$, $|\xi| \geq$ const, for all $j \in \mathbb{Z}_+$ such that (7.125) holds. The elements of (7.126) are called *classical symbols*.

The space (7.123) has a natural Fréchet topology from the isomorphism

$$C^\infty(U \times S^{n-1}) \cong S^{(m)}(U \times (\mathbb{R}^n \setminus \{0\})), \tag{7.127}$$

$S^{n-1} = \{\xi \in \mathbb{R}^n : |\xi| = 1\}$. We obtain (7.127) by

$$h(x,\xi) = |\xi|^m h\left(x, \frac{\xi}{|\xi|}\right). \tag{7.128}$$

(7.128) is also called the extension by homogeneity m of $h\left(x, \frac{\xi}{|\xi|}\right) \in C^\infty(U \times S^{n-1})$.

It is clear that for every $a \in S_{cl}^m(U \times \mathbb{R}^n)$ there is a unique

$$a_{(m)}(x,\xi) \in S^{(m)}(U \times (\mathbb{R}^n \setminus \{0\})) \tag{7.129}$$

with $a - \chi a_{(m)} \in S_{cl}^{m-1}(U \times \mathbb{R}^n)$ for every excision function $\chi(\xi)$. $a_{(m)}(x,\xi)$ is called the *homogeneous principal symbol* of a of order m. Now $a \to a_{(m)}$ defines a linear map

$$\sigma_\psi^m : S_{cl}^m(U \times \mathbb{R}^n) \longrightarrow S^{(m)}(U \times (\mathbb{R}^n \setminus \{0\})). \tag{7.130}$$

For any fixed excision function $\chi(\xi)$ we can form $a - \chi\sigma_\psi^m(a) \in S_{cl}^{m-1}(U \times \mathbb{R}^n)$. Then

$$\sigma^{m-1}(a) := \sigma_\psi^{m-1}(a - \chi\sigma_\psi^m(a))$$

gives us a linear map

$$\sigma^{m-1} : S_{cl}^m(U \times \mathbb{R}^n) \longrightarrow S^{(m-1)}(U \times (\mathbb{R}^n \setminus \{0\})). \tag{7.131}$$

This construction can be iterated and we obtain the maps

$$\sigma^{m-j} : S_{cl}^m(U \times \mathbb{R}^n) \longrightarrow S^{(m-j)}(U \times (\mathbb{R}^n \setminus \{0\})) \tag{7.132}$$

for all $j \in \mathbb{Z}_+$, $\sigma^m = \sigma_\psi^m$. Moreover, $a \to a - \sum_{j=0}^{k} \chi \sigma^{m-j}(a)$ yields a map

$$S_{cl}^m(U \times \mathbb{R}^n) \longrightarrow S^{m-(k+1)}(U \times \mathbb{R}^n) \tag{7.133}$$

for every $k \in \mathbb{Z}_+$. If we endow the space (7.126) with the topology of the projective limit under the countable system of maps (7.132), (7.133), then (7.126) will become a Fréchet space. It can easily be verified that this is independent of the concrete choice of χ.

The embeddings

$$S_{cl}^m(U \times \mathbb{R}^n) \longrightarrow S^m(U \times \mathbb{R}^n)$$

are continuous for all $m \in \mathbb{R}$, and the topology in $S_{cl}^m(U \times \mathbb{R}^n)$ is stronger than (not equal to) the topology induced by $S^m(U \times \mathbb{R}^n)$.

Definition 5. For every we set $m \in \mathbb{R} \cup \{-\infty\}$

$$L^m(\Omega) = \{Op(a) : \ a(x,x',\xi) \in S^m(\Omega \times \Omega \times \mathbb{R}^n)\}, \tag{7.134}$$
$$L_{cl}^m(\Omega) = \{Op(a) : \ a(x,x',\xi) \in S_{cl}^m(\Omega \times \Omega \times \mathbb{R}^n)\}. \tag{7.135}$$

The elements of $L^m(\Omega)$ are called pseudo-differential operators, and those in $L_{cl}^m(\Omega)$ are called classical.

In Section 2.2.4 it was proved that the singular support of the distributional kernel of every $A \in L^m(\Omega)$ is contained in diag $\Omega \times \Omega$. A relatively closed subset $K \subset \Omega \times \Omega$ is called proper if $\pi_i^{-1}(K_i) \cap K$ is compact for every compact $K_i \subset \Omega$, $i = 1, 2$, with the canonical projection $\pi_i : \ \Omega \times \Omega \longrightarrow \Omega$ to the i-th factor, $i = 1, 2$.

$A \in L^m(\Omega)$ is called *properly supported* if its distributional kernel $\in \mathcal{D}'(\Omega \times \Omega)$ has a proper support.

Let us choose a cut-off function $\omega(x,x')$, i.e. an $\omega \in C^\infty(\Omega \times \Omega)$ with proper support and $\omega(x,x') = 1$ for all (x,x') in some open neighbourhood of diag $\Omega \times \Omega$. Then, if $a(x,x',\xi) \in S^m(\Omega \times \Omega \times \mathbb{R})$ is arbitrary, we can write

$$Op(a) = Op(\omega a) + Op((1 - \omega)a).$$

Both items on the right are pseudo-differential operators. Theorem 1 of Section 2.2.4 shows that $Op((1 - \omega)a) \in L^{-\infty}(\Omega)$. Moreover $Op(\omega a)$ is obviously properly supported. In other words we have obtained the following:

Remark 6. Every $A \in L^m(\Omega)$ can be written as $A = A_0 + C$ with properly supported A_0 and some $C \in L^{-\infty}(\Omega)$.

If we set $K = \text{supp } \omega$ and denote by $L^m(\Omega)_K$ the set of all $A \in L^m(\Omega)$ with distributional kernel supported by K, then we can write

$$L^m(\Omega) = L^m(\Omega)_K + L^{-\infty}(\Omega) \tag{7.136}$$

in the sense of a non-direct sum of vector spaces.

Note that the properly supported operators allow a more precise version of 2.2.3 Theorem 1, namely, they induce continuous operators

$$A:\ C_0^\infty(\Omega) \longrightarrow C_0^\infty(\Omega), \quad A:\ C^\infty(\Omega) \longrightarrow C^\infty(\Omega). \tag{7.137}$$

In particular, we can form Au_ξ for $u_\xi := e^{ix\xi}$.

Exercise 7. Denote by $S^m(\Omega\times\mathbb{R}^n)_K$ the subspace of all $a(x,\xi) \in S^m(\Omega\times\mathbb{R}^n)$ which are of the form $\sigma(A)$ for some $A \in L^m(\Omega)_K$. Then $A \to u_{-\xi}Au_\xi =: \sigma(A)(x,\xi)$ induces a bijective map

$$\sigma:\ L^m(\Omega)_K \longrightarrow S^m(\Omega \times \mathbb{R}^n)_K. \tag{7.138}$$

Moreover, $S^m(\Omega \times \mathbb{R}^n)_K$ is a closed subspace of $S^m(\Omega \times \mathbb{R}^n)$. An analogous result holds for the classical pseudo-differential operators.

We shall frequently employ the notion of a non-direct sum of Fréchet spaces E, F which are vector subspaces of some topological vector Hausdorff space. First set

$$E + F = \{e + f :\ e \in E, f \in F\}, \tag{7.139}$$

first in the sense of a sum of vector spaces. If $\Delta := \{(e, -e) :\ e \in E \cap F\}$, then we have the algebraic isomorphism

$$E + F \cong E \oplus F/\Delta. \tag{7.140}$$

Now Δ is a Fréchet space in a canonical way. The same is true for $E \oplus F$. Then (7.140) allows us to introduce in $E + F$ the quotient topology, and it can easily be proved that then $E + F$ is also Fréchet. In particular, the non-direct sum of Banach spaces is also a Banach space. Furthermore, if E, F are Hilbert spaces, then Δ and $E \oplus F$ also have natural Hilbert space structures. Δ is closed in $E \oplus F$ and $E + F$ may be identified with the orthogonal complement of Δ. This turns $E + F$ into a Hilbert space.

Another useful notation is the following. Let a Fréchet space E be a (left) module over an algebra A. Then we set

$$[a]E =\ \text{closure of }\ \{ae :\ e \in E\} \tag{7.141}$$

for any fixed $a \in A$. An example is $E = H^s(\mathbb{R}^n)$ with $A = C_0^\infty(\mathbb{R}^n)$. In an analogous sense we can form $E[b]$ for $b \in A$, when E is a right A-module, or $[a]E[b]$ for a two-sided A-module E, $a, b \in A$.

By (7.136) we can define a Fréchet topology in $L^m(\Omega)$ by using the Fréchet structure of $L^m(\Omega)_K$ from the bijection (7.138) and the Fréchet structure of $L^{-\infty}(\Omega) \cong C^\infty(\Omega \times \Omega)$. This does not depend on the concrete choice of K (exercise!).

If $\chi:\ \Omega \longrightarrow \widetilde{\Omega}$ is a diffeomorphism, then there is a push-forward

$$\chi_*:\ L^m(\Omega) \longrightarrow L^m(\widetilde{\Omega}) \tag{7.142}$$

of pseudo-differential operators, and the same for classical operators, cf. Section 2.4.1. The map (7.142) is defined by

$$\chi_* A = (\chi^{-1})^* A \chi^*.$$

This enables the definition of the operator classes

$$L^m(\Omega), \qquad L^m_{cl}(\Omega)$$

on any C^∞ manifold Ω, cf. Section 2.4.2.

Let us look, in particular, at a closed compact C^∞ manifold X. If $U \subset X$ is a coordinate neighbourhood and $\kappa : U \longrightarrow \Omega$ a chart, $\Omega \subseteq \mathbb{R}^n$ open, then we obtain $L^m(U)$ with the Fréchet topology from the bijection $\kappa_* : L^m(U) \longrightarrow L^m(\Omega)$. Now, if

$$\mathcal{U} = \{U_1, \ldots, U_N\}$$

is a finite open covering of X by coordinate neighbourhoods and $\{\phi_1, \ldots, \phi_N\}$ a subordinate partition of unity, moreover $\{\psi_1, \ldots, \psi_N\}$ is another system of functions $\psi_j \in C_0^\infty(U_j)$ with $\psi_j \phi_j = \phi_j$ for all j, then we can write

$$L^m(X) = \sum_{j=1}^{N} \phi_j L^m(U_j) \psi_j + L^{-\infty}(X)$$

in the sense of a non-direct sum of vector spaces. $L^{-\infty}(X) \cong C^\infty(X \times X)$ has a Fréchet structure. According to the above notation we can form the Fréchet space $[\phi_j]L^m(U_j)[\psi_j]$ for all j. Then

$$L^m(X) = \sum_{j=1}^{N} [\phi_j] L^m(U_j) [\psi_j] + L^{-\infty}(X)$$

is a Fréchet space with the topology of the non-direct sum. This is independent of the concrete choice of the open covering $\mathcal{U}$ and of the system of ϕ_j, ψ_j.

From time to time we shall use some simple constructions in terms of projective tensor products of Fréchet spaces. Let E and F be Fréchet spaces with the semi-norm systems $\{p_j\}_{j \in \mathbb{Z}_+}$ and $\{q_k\}_{k \in \mathbb{Z}_+}$, respectively. Denote by $E \otimes F$ the algebraic tensor product, i.e. the space of all finite sums

$$h = \sum_{i=1}^{N} e_i \otimes f_i \quad \text{with arbitrary} \quad e_i \subset E, \ f_i \in F, \ N \in \mathbb{N},$$

where the $\otimes$ sign is understood in the standard algebraic sense. In particular, $\otimes$ satisfies the rules

$$\begin{aligned}
\lambda(e \otimes f) &= (\lambda e) \otimes f = e \otimes (\lambda f), \\
e \otimes (f_1 + f_2) &= e \otimes f_1 + e \otimes f_2, \quad (e_1 + e_2) \otimes f = e_1 \otimes f + e_2 \otimes f,
\end{aligned}$$

for arbitrary $e, e_1, e_2 \in E, \ f, f_1, f_2 \in F, \ \lambda \in \mathbb{C}$.

This shows, in particular, that every $h \in E \otimes F$ can be written in many different ways as a finite sum.

To every pair of semi-norms p_j and q_k we introduce their projective tensor product $p_j \otimes_\pi q_k$ as a semi-norm on $E \otimes F$ as follows: We set for $h \in E \otimes F$

$$(p_j \otimes_\pi q_k)(h) = \inf \sum_i p_j(e_i) q_k(f_i),$$

where the infimum is taken over all representations of h as a finite sum $h = \sum_i e_i \otimes f_i$.

It is an easy exercise to check that $p_j \otimes q_k$ is actually a semi-norm on $E \otimes F$.

Now $E \otimes_\pi F$, the so-called projective tensor product of E and F, is defined as the completion of $E \otimes F$ with respect to the semi-norm system $\{p_j \otimes_\pi q_k\}_{j,k \in \mathbb{Z}_+}$. Then again $E \otimes_\pi F$ is a Fréchet space.

Example 8. If $\Omega \subseteq \mathbb{R}^n$ is an open set, then $C^\infty(\Omega)$ is a Fréchet space with the semi-norm system

$$\varphi \longrightarrow \sup_{x \in K} | \, D^\alpha \varphi(x) \, |, \quad \alpha \in \mathbb{Z}_+^n, \quad K \subset\subset \Omega.$$

Let $\Omega \subseteq \mathbb{R}^n$, $\tilde{\Omega} \subseteq \mathbb{R}^{\tilde{n}}$ be open sets. Then there is a canonical isomorphism

$$C^\infty(\Omega \times \tilde{\Omega}) \cong C^\infty(\Omega) \otimes_\pi C^\infty(\tilde{\Omega})$$

under which $\varphi(x)\tilde{\varphi}(\tilde{x}) \in C^\infty(\Omega \times \tilde{\Omega})$ for $\varphi(\tilde{x}) \in C^\infty(\Omega)$, $\tilde{\varphi}(x) \in C^\infty(\tilde{\Omega})$ corresponds to $\varphi(x) \otimes \tilde{\varphi}(\tilde{x})$.

Exercise 9. If $S^m(\mathbb{R}^n)$ is the subspace of all $a(x,\xi) \in S^m(U \times \mathbb{R}^n)$ that are independent of x then there is a canonical isomorphism

$$S^m(U \times \mathbb{R}^n) \cong C^\infty(U) \otimes_\pi S^m(\mathbb{R}^n)$$

with $\varphi(x)a(\xi)$ corresponding to $\varphi(x) \otimes a(\xi)$ (cf. also Remark 2 and the subsequent notation). An analogous relation holds for classical symbols.

Theorem 10. *Let E, F be Fréchet spaces. Then every $h \in E \otimes_\pi F$ can be written as a convergent sum*

$$h = \sum_{j=0}^{\infty} \lambda_j e_j \otimes f_j \tag{7.143}$$

with $e_j \in E$, $f_j \in F$, $\lambda_j \in \mathbb{C}$, $j \in \mathbb{Z}_+$, and $e_j \to 0$ in E, $f_j \to 0$ in F as $j \to \infty$, $\sum_{j=0}^{\infty} | \, \lambda_j \, | < \infty$.

A proof of this result may be found in [Sf]. It has an extension to finitely many factors, i.e. Fréchet spaces E, F, G. The projective tensor product satisfies $(E \otimes_\pi F) \otimes_\pi G = E \otimes_\pi (F \otimes_\pi G)$ (exercise!).

We can then write $E \otimes_\pi F \otimes_\pi G$. Every h in that space can be written as a convergent sum

$$h = \sum_{j=0}^{\infty} \lambda_j e_j \otimes f_j \otimes g_j \tag{7.144}$$

with $e_j \in E$, $f_j \in F$, $g_j \in G$, $\lambda_j \in \mathbb{C}$, $j \in \mathbb{Z}_+$, and $e_j \to 0$, $f_j \to 0$, $g_j \to 0$ in the corresponding spaces, $\sum_{j=0}^{\infty} | \lambda_j | < \infty$.

Exercise 11. Let E, F, G be Fréchet spaces. Then there is a canonical isomorphism

$$(E + F) \otimes_\pi G \cong E \otimes_\pi G + F \otimes_\pi G,$$

where $+$ denotes the non-direct sum of Fréchet spaces, introduced above. Analogously we have

$$G \otimes_\pi (E + F) \cong G \otimes_\pi E + G \otimes_\pi F.$$

Hint: Employ the fact that for every two Fréchet spaces E, F the following is true: If $A : E \to F$ is a continuous and surjective linear map for which A^{-1} exists in the algebraic sense, then A is an isomorphism of Fréchet spaces.

A typical concrete realization of Theorem 10 is the following. Let H_1, H_2 be Banach spaces and $\mathcal{L}(H_1, H_2)$ be the space of all linear continuous operators $H_1 \to H_2$ in the norm topology and set $\mathcal{L}(H_1) = \mathcal{L}(H_1, H_1)$. Furthermore, let E, F be Fréchet spaces and

$$A \quad : \quad E \to \mathcal{L}(H_1), \tag{7.145}$$
$$B \quad : \quad F \to \mathcal{L}(H_1, H_2) \tag{7.146}$$

be continuous linear maps. Then, to every $h \in E \otimes_\pi F$ there corresponds an element $C(h) \in \mathcal{L}(H_1, H_2)$ as follows: We write h as a convergent sum (7.143). Then

$$\sum_{j=0}^{\infty} \lambda_j B(e_j) A(f_j) \tag{7.147}$$

converges in $\mathcal{L}(H_1, H_2)$. In fact, the continuities of A, B imply

$$\|A(f_j)\|_{\mathcal{L}(H_1)} \longrightarrow 0, \quad \|B(e_j)\|_{\mathcal{L}(H_1, H_2)} \longrightarrow 0$$

for $j \to \infty$. Then

$$\|\sum_{j=0}^{\infty} \lambda_j B(e_j) A(f_j)\|_{\mathcal{L}(H_1, H_2)} \leq \sum_{j=0}^{\infty} | \lambda_j | \, \|B(e_j)\|_{\mathcal{L}(H_1, H_2)} \|A(f_j)\|_{\mathcal{L}(H_1)} < \infty.$$

7.2.2 The parameter-dependent calculus; reductions of orders

The calculus of pseudo-differential operators permits to introduce a parameter-dependent variant, where the parameter λ varies along a conical subset Λ of some finite-dimensional vector space. We will mainly need the case

$$\Lambda = \mathbb{R}^l. \tag{7.148}$$

The analogous formulations for more general conical sets Λ are then straightforward and will be dropped. We shall see that the basic elements of the "usual" pseudo-differential calculus remain true in the parameter-dependent case. Most of this material may be regarded as a useful exercise in pseudo-differential operators. So we shall drop explanations in so far as the results are immediate modifications in the λ-independent case.

First let us introduce the space of smoothing parameter-dependent pseudo-differential operators $G(\lambda)$ on a paracompact C^∞ manifold Ω, namely

$$L^{-\infty}(\Omega; \Lambda) := \mathcal{S}(\Lambda, L^{-\infty}(\Omega)). \tag{7.149}$$

$L^{-\infty}(\Omega)$ is endowed with the Fréchet topology from the preceding section.

In order to define the classes of parameter-dependent pseudo-differential operators

$$L^m(\Omega; \Lambda), \quad L^m_{cl}(\Omega; \Lambda) \quad \text{for} \quad m \in \mathbb{R}, \tag{7.150}$$

we shall first assume that Ω is an open set in $\mathbb{R}^n$. Here we have the symbol spaces

$$S^m(\Omega \times \Omega \times \mathbb{R}^{n+l}_{\xi,\lambda}), \quad S^m_{cl}(\Omega \times \Omega \times \mathbb{R}^{n+l}_{\xi,\lambda}), \tag{7.151}$$

where (ξ, λ) is treated as a covariable. The space $S^m(\Omega \times \Omega \times \mathbb{R}^{n+l}_{\xi,\lambda})$ is described by the symbol estimates

$$|D^\alpha_{x,x'} D^\beta_{\xi,\lambda} a(x, x', \xi, \lambda)| \leq c(1 + |\xi| + |\lambda|)^{m-|\beta|} \tag{7.152}$$

for all $\alpha \in \mathbb{Z}^{2n}_+$, $\beta \in \mathbb{Z}^{n+l}_+$, all $(x, x') \in K$ for arbitrary $K \subset\subset \Omega \times \Omega$, $(\xi, \lambda) \in \mathbb{R}^{n+l}$, with constants $c = c(\alpha, \beta, K) > 0$. The notions and results from the "ordinary" symbol spaces remain valid, since we have only changed the notation. For classical parameter-dependent symbols, indicated by the subscript cl, the homogeneities of components refer to (ξ, λ). As usual, we also have the spaces of x'-independent symbols, where we write Ω instead of $\Omega \times \Omega$ in the corresponding cases.

Now $L^m(\Omega; \Lambda)$ for open $\Omega \subseteq \mathbb{R}^n$ is defined as the set of all $A(\lambda)$ with

$$A(\lambda)u(x) = \iint e^{i(x-x')\xi} a(x, x', \xi, \lambda) u(x')\, dx'\, d\xi$$

for some $a(x, x', \xi, \lambda) \in S^m(\Omega \times \Omega \times \mathbb{R}^{n+l}_{\xi,\lambda})$, for all $u \in C^\infty_0(\Omega)$. In an analogous manner we define $L^m_{cl}(\Omega; \Lambda)$ by requiring $a(x, x', \xi, \lambda) \in S^m_{cl}(\Omega \times \Omega \times \mathbb{R}^{n+l}_{\xi,\lambda})$.

For every $A(\lambda) \in L^m_{cl}(\Omega; \Lambda)$ we can define the λ-dependent principal symbol of order m

$$\sigma^m_{\psi,\lambda}(A)(x,\xi,\lambda) \in S^{(m)}(\Omega \times (\mathbb{R}^{n+l}_{\xi,\lambda} \setminus \{0\})).$$

Remark 1. $A(\lambda) \in L^m(\Omega; \Lambda)$ $(\in L^m_{cl}(\Omega; \Lambda))$ implies $A(\lambda_0) \in L^m(\Omega)$ $(\in L^m_{cl}(\Omega))$ for every fixed $\lambda_0 \in \Lambda$.

The global definition of (7.150) employs the following.

Proposition 2. *Let $\Omega, \widetilde{\Omega} \subset \mathbb{R}^n$ be open, and $\chi : \Omega \longrightarrow \widetilde{\Omega}$ be a diffeomorphism. Then the push-forward of pseudo-differential operators under χ, first defined for every fixed $\lambda \in \Lambda$, in fact defines a bijection*

$$\chi_* : L^m(\Omega; \Lambda) \longrightarrow L^m(\widetilde{\Omega}; \Lambda).$$

An analogous statement holds for the spaces of classical parameter-dependent pseudo-differential operators.

The proof follows easily by extending the technique of Section 2.4.1 to the parameter-dependent case in a straightforward manner (exercise!).

Now let Ω be a paracompact C^∞ manifold and $\kappa : U \longrightarrow \mathbb{R}^n$ be a chart on Ω. Then we define

$$L^m(U; \Lambda) = \{\kappa^* A(\lambda)(\kappa^*)^{-1} : A(\lambda) \in L^m(\mathbb{R}^n; \Lambda)\}$$

and analogously $L^m_{cl}(U; \Lambda)$. Every element of $L^m(U; \Lambda)$ is then an operator family $C^\infty_0(U) \longrightarrow C^\infty(U)$.

$L^m(\Omega; \Lambda)$ $(L^m_{cl}(\Omega; \Lambda))$ is defined as the space of all operator families

$$A(\lambda) : \ C^\infty_0(\Omega) \longrightarrow C^\infty(\Omega)$$

that allow a decomposition $A(\lambda) = A_0(\lambda) + G(\lambda)$ with $G(\lambda) \in L^{-\infty}(\Omega; \Lambda)$ and $A_0(\lambda)\,|_U \in L^m(U; \Lambda)$ $(L^m_{cl}(U; \Lambda))$ for every coordinate neighbourhood U on Ω.

Exercise 3. Show that, if $A(\lambda) \in L^m(\Omega; \Lambda)$ for any paracompact C^∞ manifold Ω and $\phi, \psi \in C^\infty_0(\Omega)$, supp $\phi \cap$ supp $\psi = \emptyset$ then

$$\phi A(\lambda)\psi \in L^{-\infty}(\Omega; \Lambda).$$

Exercise 4. Every $A(\lambda) \in L^m(\Omega; \Lambda)$ allows a decomposition

$$A(\lambda) = A_0(\lambda) + C(\lambda)$$

where $A_0(\lambda) \in L^m(\Omega; \Lambda)$ is properly supported for every $\lambda \in \Lambda$, and $C(\lambda) \in L^{-\infty}(\Omega; \Lambda)$.

Exercise 5. Let $A_j(\lambda) \in L^{m_j}(\Omega; \Lambda)$, $j \in \mathbb{Z}_+$, be an arbitrary sequence, $m_j \to -\infty$ as $j \to \infty$. Then there is an $A(\lambda) \in L^m(\Omega; \Lambda)$ for $m = \max\{m_j\}$ such that for every $M \in \mathbb{Z}_+$ there is an $N \in \mathbb{Z}_+$ with

$$A(\lambda) - \sum_{j=0}^{N} A_j(\lambda) \in L^{m-M}(\Omega; \Lambda).$$

$A(\lambda)$ is unique mod $L^{-\infty}(\Omega; \Lambda)$. An analogous result holds for classical operators.

Points in $T^*\Omega$ will be denoted by (x, ξ) though this refers to local coordinates. We set

$$T^*\Omega \times \Lambda \setminus 0 = \{(x, \xi, \lambda) \in T^*\Omega \times \Lambda : (\xi, \lambda) \neq 0\}.$$

Moreover,

$$S^{(m)}(T^*\Omega \times \Lambda \setminus 0)$$

will denote the space of all $p(x, \xi, \lambda) \in C^\infty(T^*\Omega \times \Lambda \setminus 0)$ such that

$$p(x, \varrho\xi, \varrho\lambda) = \varrho^m p(x, \xi, \lambda)$$

for all $\varrho \in \mathbb{R}_+$, $(x, \xi, \lambda) \in T^*\Omega \times \Lambda \setminus 0$. We then have a unique map

$$\sigma^m_{\psi,\lambda} : \ L^m_{cl}(\Omega; \Lambda) \longrightarrow S^{(m)}(T^*\Omega \times \Lambda \setminus 0) \tag{7.153}$$

which assigns to every $A(\lambda)$ the *parameter-dependent homogeneous principal* symbol of order m. This follows from the corresponding obvious construction in local coordinates and the invariance of the homogeneous principal symbol under coordinate diffeomorphisms, regarded as a function over $T^*\Omega \times \Lambda \setminus 0$.

Exercise 6. The symbolic map $\sigma^m_{\psi,\lambda}$ is surjective and

$$\ker \ \sigma^m_{\psi,\lambda} = L^{m-1}_{cl}(\Omega; \Lambda). \tag{7.154}$$

Remark 7. The spaces $L^m(\Omega; \Lambda)$ and $L^m_{cl}(\Omega; \Lambda)$ can be endowed with canonical Fréchet topologies, similar to the corresponding operator spaces without parameters, cf. Section 7.2.1.

Let X be a closed compact C^∞ manifold and $A(\lambda) \in L^m(X; \Lambda)$. We then have a family of continuous operators

$$A(\lambda) : \ H^s(X) \longrightarrow H^{s-m}(X),$$

$\lambda \in \Lambda$, $s \in \mathbb{R}$. Because of the continuous embeddings $H^{s-m}(X) \longrightarrow H^{s-r}(X)$ for all $r \geq m$ the operator family can also be regarded as

$$A(\lambda) : \ H^s(X) \longrightarrow H^{s-r}(X), \quad r \geq m.$$

Denote by $\| \cdot \|_{s,r}$ the operator norm in $\mathcal{L}(H^s(X), H^r(X))$.

Theorem 8. *$A(\lambda) \in L^m(X; \Lambda)$ implies for $r \geq m$*

$$\|A(\lambda)\|_{s,s-r} \leq \begin{cases} c(1+|\lambda|)^m & \text{for} \quad r \geq 0, \\ c(1+|\lambda|)^{m-r} & \text{for} \quad r \leq 0 \end{cases}$$

with constants $c = c(s, r)$.

Proof. First we want to show an analogue of the statement of the theorem for $A(\lambda) \in L^m(\mathbb{R}^n; \Lambda)$ where $A(\lambda) = Op(a)(\lambda)$ with $a(\xi, \lambda) \in S^m(\mathbb{R}^n \times \Lambda)$ being independent of x, x'. We have in this case

$$\begin{aligned} \|A(\lambda)u\|_{s-r}^2 &= \int (1+|\xi|^2)^{s-r} |a(\xi, \lambda)\mathcal{F}u(\xi)|^2 d\xi \\ &\leq \sup_{\xi \in \mathbb{R}^n} (1+|\xi|^2)^{-r} |a(\xi, \lambda)|^2 \int (1+|\xi|^2)^s |\mathcal{F}u(\xi)|^2 d\xi \\ &= \sup_{\xi \in \mathbb{R}^n} (1+|\xi|^2)^{-r} |a(\xi, \lambda)|^2 \|u\|_s^2. \end{aligned}$$

From (7.152) it follows that

$$(1+|\xi|^2)^{-r} |a(\xi, \lambda)|^2 \leq c(1+|\xi|^2)^{-r} (1+|\xi|^2 + |\lambda|^2)^m$$

with a constant $c = c(a) > 0$. Now we use the inequality

$$\sup_{\xi} (1+|\xi|^2)^{-r} (1+|\xi|^2 + |\lambda|^2)^m \leq p(\lambda, m, r)^2$$

with

$$p(\lambda, m, r) = c_1 \begin{cases} (1+|\lambda|)^m & \text{for} \quad r \geq 0, \\ (1+|\lambda|)^{m-r} & \text{for} \quad r \leq 0 \end{cases}$$

with a constant $c_1 > 0$. Here it was supposed that $r \geq m$. This yields

$$\|A(\lambda)\|_{s,s-r} \leq c(a)p(\lambda, m, r)$$

with a new constant $c(a) > 0$, where $c(a) \to 0$ as $a \to 0$ in the symbol space $S^m(\mathbb{R}^n \times \Lambda)$. Next we assume $u(x, x', \xi, \lambda) \in S^m(\mathbb{R}^{2n}_{x,x'} \times \mathbb{R}^n \times \Lambda)$ and $a(x, x', \xi, \lambda) = 0$ for $|x|, |x'| > c$ with some $c > 0$. Then $a(x, x', \xi, \lambda)$ can be regarded as an element of $C_0^\infty(K) \otimes_\pi C_0^\infty(K) \otimes_\pi S^m(\mathbb{R}^n \times \Lambda)$ where K is a compact set in $\mathbb{R}^n$ containing $|x| \leq c$ and $C_0^\infty(K)$ is the space of all $u \in C_0^\infty(\mathbb{R}^n)$ supported by K.

Applying 7.2.1 Theorem 10 to $a(x, x', \xi, \lambda)$ we get

$$a(x, x', \xi, \lambda) = \sum_{j=0}^{\infty} \beta_j \phi_j(x) \psi_j(x') a_j(\xi, \lambda)$$

with $\phi_j, \psi_j \in C_0^\infty(K)$, $a_j \in S^m(\mathbb{R}^n \times \Lambda)$, tending to zero as $j \to \infty$, $\sum_j |\beta_j| \leq \infty$. Now

$$
\begin{aligned}
\|Op(a)(\lambda)\|_{s,s-r} &= \|\sum_{j=0}^{\infty} \beta_j \mathcal{M}_{\phi_j} Op(a_j)(\lambda) \mathcal{M}_{\psi_j}\|_{s,s-r} \\
&\leq \sum_{j=0}^{\infty} |\beta_j| \|\mathcal{M}_{\phi_j}\|_{s-r,s-r} \|Op(a_j)(\lambda)\|_{s,s-r} \|\mathcal{M}_{\psi_j}\|_{s,s} \quad (7.155)
\end{aligned}
$$

Here $\mathcal{M}_\phi$ is the operator of multiplication by ϕ in the corresponding Sobolev space. From the first part of the proof we know that

$$\|Op(a_j)(\lambda)\|_{s,s-r} \leq c_j p(\lambda, m, r)$$

with constants $c_j \to 0$ as $j \to \infty$. We further know that

$$\|\mathcal{M}_{\phi_j}\|_{s-r,s-r} \to 0, \quad \|\mathcal{M}_{\psi_j}\|_{s,s} \to 0 \quad \text{for} \quad j \to \infty.$$

This gives us the inequality

$$\|Op(a)(\lambda)\|_{s,s-r} \leq cp(\lambda, m, r)$$

with some $c > 0$. Finally, every $A(\lambda) \in L^m(X; \Lambda)$ can be written as

$$A(\lambda) = \sum_{j=1}^{N} \phi_j A_j(\lambda) \psi_j + A_0(\lambda),$$

where $\{\phi_1, \ldots, \phi_N\}$ is a partition of unity belonging to an open covering of X by coordinate neighbourhoods $\{U_1, \ldots, U_N\}$, $\psi_j \in C_0^\infty(U_j)$, and

$$A_j(\lambda) = (\kappa_j)_*^{-1} Op(a_j)(\lambda)$$

with $\kappa_j : U_j \longrightarrow \mathbb{R}^n$ being a chart, $a_j(x, x', \xi, \lambda) \in S^m(\mathbb{R}_{x,x'}^{2n} \times \mathbb{R}^n \times \Lambda)$ and $A_0(\lambda) \in L^{-\infty}(X; \Lambda)$. Clearly ϕ_j, ψ_j can be regarded as a part of a_j, such that the above results concerning symbol functions can be applied with compact supports in x and x'. Thus it follows that

$$\|\phi_j A_j(\lambda) \psi_j\|_{s,s-r} \leq c_j p(\lambda, m, r)$$

and hence

$$
\begin{aligned}
\|A(\lambda)\|_{s,s-r} &\leq \sum_{j-1}^{N} \|\phi_j A_j(\lambda) \psi_j\|_{s,s-r} + \|A_0(\lambda)\|_{s,s-r} \\
&\leq cp(\lambda, m, r)
\end{aligned}
$$

with some constant $c > 0$. Here we have also employed the corresponding obvious estimate for $A_0(\lambda)$. $\qquad\square$

Corollary 9. $A(\lambda) \in L^m(X; \Lambda)$ *and* $m \le 0$ *implies*

$$\|A(\lambda)\|_{0,0} \le c(1 + |\lambda|)^m$$

with a constant $c > 0$.

Theorem 10. *Let* $A(\lambda) \in L^m(\Omega; \Lambda)$, $B(\lambda) \in L^{m'}(\Omega; \Lambda)$ *and* $A(\lambda)$ *or* $B(\lambda)$ *be uniformly properly supported in* $\lambda \in \Lambda$. *Then* $A(\lambda)B(\lambda) \in L^{m+m'}(\Omega; \Lambda)$. *If* $A(\lambda)$ *and* $B(\lambda)$ *are both classical, then the same is true for the composition, and we have*

$$\sigma_{\psi,\lambda}^{m+m'}(AB) = \sigma_{\psi,\lambda}^m(A)\sigma_{\psi,\lambda}^{m'}(B). \tag{7.156}$$

This result follows in the same manner as the corresponding one in the case without parameters. The simple details are left to the reader as an exercise.

We now turn to the concept of parameter-dependent ellipticity. Our applications will concern classical pseudo-differential operators, though the essential assertions allow immediate generalizations to non-classical operators.

Definition 11. An operator $A \in L_{cl}^m(\Omega; \Lambda)$ is called *parameter-dependent elliptic* if

$$\sigma_{\psi,\lambda}(A) \neq 0 \quad \text{on} \quad T^*\Omega \times \Lambda \setminus 0.$$

Example 12. Let $\{U_j\}_{j \in \mathbb{N}}$ be a locally finite open covering of Ω by coordinate neighbourhoods, $\{\phi_j\}_{j \in \mathbb{N}}$ a subordinate partion of unity, and $\{\psi_j\}_{j \in \mathbb{N}}$ a second system of functions in $C_0^\infty(U_j)$ with $\phi_j\psi_j = \phi_j$ for all j. Let us form

$$A(\lambda) = \sum_j \phi_j A_j(\lambda)\psi_j,$$

where for a corresponding chart $\kappa_j : U_j \longrightarrow \mathbb{R}^n$ the operator $A_j(\lambda)$ is the pull-back of

$$u \to \iint e^{i(x-x')\xi}(c + |\xi|^2 + |\lambda|^2)^{\frac{m}{2}} u(x')dx'd\xi, \tag{7.157}$$

with a $c > 0$. This is required for every j. Then $A(\lambda)$ is parameter-dependent elliptic and

$$\sigma_{\psi,\lambda}^m(A) = (|\xi|^2 + |\lambda|^2)^{\frac{m}{2}}.$$

Example 13. Let $A \in Diff^m(\mathbb{B})_{Fuchs}$ satisfy condition (i) of 7.1.2 Definition 14. Then

$$\sigma_M^m(A)(\varrho + i\lambda) \in L_{cl}^m(X; \mathbb{R}) \tag{7.158}$$

is parameter-dependent elliptic on X for every fixed $\varrho \in \mathbb{R}$.

We will also use the notation

$$L_{cl}^m(X; \Gamma_\varrho) \quad \text{for} \quad \Gamma_\varrho := \{z : \operatorname{Re} z = \varrho\} \tag{7.159}$$

for the class $L_{cl}^m(X; \mathbb{R})$ where $\mathbb{R}$ is identified with Γ_ϱ via $\lambda \to \varrho + i\lambda$.

Definition 14. An operator family $P(\lambda) \in L_{cl}^{-m}(\Omega; \Lambda)$ is called a *parameter-dependent parametrix* of $A(\lambda) \in L_{cl}^m(\Omega; \Lambda)$ if

$$P(\lambda)A(\lambda) - 1, \ A(\lambda)P(\lambda) - 1 \in L^{-\infty}(\Omega; \Lambda).$$

It is assumed here that $A(\lambda)$ or $P(\lambda)$ are properly supported for every fixed $\lambda \in \Lambda$.

Theorem 15. *Let* $A(\lambda) \in L_{cl}^m(\Omega; \Lambda)$ *be parameter-dependent elliptic. Then there is a parameter-dependent parametrix* $P(\lambda) \in L_{cl}^{-m}(\Omega; \Lambda)$ *of* $A(\lambda)$.

Proof. By assumption there exists

$$\sigma_{\psi,\lambda}^m(A)^{-1} \in S^{(-m)}(T^*\Omega \times \Lambda \setminus 0).$$

According to the surjectivity of (7.153) for all $m \in \mathbb{R}$ (cf. Exercise 6) there exists a $P_1(\lambda) \in L_{cl}^{-m}(\Omega; \Lambda)$ with $\sigma_{\psi,\lambda}^{-m}(P_1) = \sigma_{\psi,\lambda}^m(A)^{-1}$. Then, in view of Exercise 4 we find a $P_0(\lambda) \in L_{cl}^{-m}(\Omega; \Lambda)$ which is properly supported such that $P_0(\lambda) - P_1(\lambda) \in L^{-\infty}(\Omega, \Lambda)$. We obtain

$$C(\lambda) := A(\lambda)P_0(\lambda) - 1 \in L_{cl}^{-1}(\Omega; \Lambda). \tag{7.160}$$

Here we have employed (7.156) and (7.154). Choose a properly supported $C_0(\lambda) \in L_{cl}^{-1}(\Omega; \Lambda)$ with $C_0(\lambda) - C(\lambda) \in L^{-\infty}(\Omega; \Lambda)$. According to Exercise 5 there exists the asymptotic sum

$$D(\lambda) \sim \sum_{j=0}^{\infty} (-1)^j C_0(\lambda)^j \in L_{cl}^0(\Omega; \Lambda).$$

This can also chosen to be properly supported. We then obtain

$$(1 + C_0(\lambda))D(\lambda) = 1 \bmod L^{-\infty}(\Omega; \Lambda).$$

Since

$$C_0(\lambda) + 1 = A(\lambda)P_0(\lambda) \bmod L_{cl}^{-\infty}(\Omega; \Lambda),$$

we obtain

$$A(\lambda)\{P_0(\lambda)D(\lambda)\} = 1 \bmod L_{cl}^{-\infty}(\Omega; \Lambda).$$

Thus $P(\lambda) := P_0(\lambda)D(\lambda)$ is the desired parametrix. $\qquad\square$

Exercise 16. Let X be a closed compact C^∞ manifold and $C(\lambda) \in L^{-\infty}(X; \Lambda)$. Assume that

$$1 + C(\lambda): \ H^s(X) \longrightarrow H^s(X) \tag{7.161}$$

is invertible for all $\lambda \in \Lambda$ and fixed $s = s_0 \in \mathbb{R}$. Then (7.161) is invertible for all $s \in \mathbb{R}$ and

$$(1 + C(\lambda))^{-1} = 1 + C_1(\lambda)$$

with some $C_1(\lambda) \in L^{-\infty}(X; \Lambda)$.

Theorem 17. *Let X be a closed compact C^∞ manifold and $A(\lambda) \in L^m_{cl}(X; \Lambda)$ be a parameter-dependent elliptic operator. Then there is a $c_1 > 0$ such that*

$$A(\lambda): \ H^s(X) \longrightarrow H^{s-m}(X) \tag{7.162}$$

induces isomorphisms for all $|\lambda| \geq c_1$ and all $s \in \mathbb{R}$. Moreover, there exists for every $m \in \mathbb{R}$ a parameter-dependent elliptic $A(\lambda) \in L^m_{cl}(X; \Lambda)$ for which (7.162) are isomorphisms for all $\lambda \in \Lambda$, $s \in \mathbb{R}$. In that case $A^{-1}(\lambda) \in L^{-m}_{cl}(X; \Lambda)$.

Proof. Let $A(\lambda) \in L^m_{cl}(X; \Lambda)$ be parameter-dependent elliptic. Then, in view of Theorem 15 there exists a parameter-dependent parametrix $P(\lambda) \in L^{-m}_{cl}(X; \Lambda)$ which is properly supported since X is compact. Then

$$C(\lambda) = P(\lambda)A(\lambda) - 1 \in L^{-\infty}(X; \Lambda)$$

satisfies

$$\|C(\lambda)\|_{0,0} \leq c(1 + |\lambda|)^{-1}, \tag{7.163}$$

cf. Corollary 9. The isomorphisms in $\mathcal{L}(L_2(X))$ form an open set in the norm topology. Thus there exists an $\varepsilon > 0$ such that for every $C \in \mathcal{L}(L_2(X))$ with $\|C(\lambda)\|_{0,0} < \varepsilon$ the operator $1 + C: \ L_2(X) \longrightarrow L_2(X)$ is an isomorphism. From (7.163) it follows that

$$1 + C(\lambda): \ L_2(X) \longrightarrow L_2(X)$$

is an isomorphism for all $|\lambda| \geq c_0$ when c_0 is sufficiently large. Let $\psi(\lambda) \in C^\infty(\Lambda)$ be an excision function with $0 \leq \psi(\lambda) \leq 1$ and $\psi(\lambda) = 0$ for $|\lambda| < 2c_0$, $\psi(\lambda) = 1$ for $|\lambda| > 3c_0$. Then

$$1 + \psi(\lambda)C(\lambda): \ L_2(X) \longrightarrow L_2(X)$$

is an isomorphism for all $\lambda \in \Lambda$, since $\|\psi(\lambda)C(\lambda)\|_{0,0} < \varepsilon$ for all λ. Using the result of Exercise 16 we obtain a $C_1(\lambda) \in L^{-\infty}(X, \Lambda)$ with $(1 + \psi(\lambda)C(\lambda))^{-1} = 1 + C_1(\lambda)$. Thus

$$(1 + C_1(\lambda))(1 + C(\lambda)) = (1 + C_1(\lambda))P(\lambda)A(\lambda) = 1$$

once $|\lambda| \geq 3c_0$. It follows that $(1 + C_1(\lambda))P(\lambda)$ is a left inverse of $A(\lambda)$ for $|\lambda| \geq 3c_0$. In an analogous manner we can construct a right inverse, i.e. we can set

$$A(\lambda)^{-1} = (1 + C_1(\lambda))P(\lambda) \ \text{ for } \ |\lambda| =: c_1$$

for sufficiently large c_1. Hence $A(\lambda) : L_2(X) \longrightarrow H^{-m}(X)$ is invertible for all $|\lambda| \geq c_1$. Since kernel and cokernel of elliptic pseudo-differential operators are independent of s, the invertibility of (7.162) also follows for $|\lambda| \geq c_1$ and all $s \in \mathbb{R}$. The second statement follows from Example 12. It suffices after the above arguments to choose the constant c in (7.157) sufficiently large, since this has the same effect as to enlarge $|\lambda|$. Then we will obtain that $A(\lambda)$ is invertible for all λ. Our construction has provided $A(\lambda)^{-1}$ where this time the excision function $\psi(\lambda)$ may be replaced by 1. Because of Theorem 10 we therefore have $A(\lambda)^{-1} \in L_{cl}^{-m}(X; \Lambda)$. $\qquad\qquad\square$

7.2.3 Mellin pseudo-differential operators with operator-valued symbols

Next we shall establish the elements of the pseudo-differential calculus on

$$X^{\wedge} := \mathbb{R}_+ \times X \qquad\qquad (7.164)$$

with a closed compact C^{∞} manifold X, where we employ the Mellin transform along $\mathbb{R}_+ \ni r$ instead of the Fourier transform. $X^{\wedge}$ is regarded as an open stretched cone associated with a corresponding manifold with conical singularities. The manifold X is interpreted as the base of the cone and $r \in \mathbb{R}_+$ as the cone axis variable. The role of the calculus on $X^{\wedge}$ is to obtain a framework which allows the construction of parametrices of elliptic differential operators of Fuchs type, cf. 7.1.2 Definition 14.

The parameter-dependent pseudo-differential operators described in the preceding section will now serve as the operator-valued symbol for the Mellin pseudo-differential operators

$$op_M^{\gamma}(a) \quad \text{with symbols} \quad a(r, r', z) \quad \text{and weights} \quad \gamma \in \mathbb{R}.$$

We assume
$$a(r, r', z) \in C^{\infty}(\mathbb{R}_+ \times \mathbb{R}_+, L^m(X; \Gamma_{\frac{1}{2}-\gamma})), \qquad\qquad (7.165)$$

with $L^m(X; \Gamma_\beta) = L^m(X; \mathbb{R}_\varrho)$ under the identification $\Gamma_\beta \to \mathbb{R}, \beta + i\varrho \to \varrho$, cf. (7.148). Remember that these spaces of parameter-dependent pseudo-differential operators are endowed with adequate Fréchet topologies, cf. 7.2.2 Remark 7. Below we will employ the symbol spaces $C^{\infty}(\mathbb{R}_+ \times \mathbb{R}_+, L_{cl}^m(X; \Gamma_{\frac{1}{2}-\gamma}))$ as well as the subspaces

$$C^{\infty}(\overline{\mathbb{R}}_+ \times \overline{\mathbb{R}}_+, L^m(X; \Gamma_{\frac{1}{2}-\gamma})), \quad C^{\infty}(\overline{\mathbb{R}}_+ \times \overline{\mathbb{R}}_+, L_{cl}^m(X; \Gamma_{\frac{1}{2}-\gamma})).$$

For the moment we shall deal with (7.165).

For any $\gamma \in \mathbb{R}$ let us set

$$M_\gamma L^m(X^{\wedge}) = \{op_M^{\gamma}(a) : a(r, r', z) \in C^{\infty}(\mathbb{R}_+ \times \mathbb{R}_+, L^m(X; \Gamma_{\frac{1}{2}-\gamma}))\}.$$

Analogously we define $M_\gamma L_{cl}^m(X^{\wedge})$. Moreover, we have the spaces $L^m(X^{\wedge})$, $L_{cl}^m(X^{\wedge})$ of pseudo-differential operators on $X^{\wedge}$ from Section 7.2.1.

Proposition 1. *We have* $M_\gamma L^m(X^\wedge) = L^m(X^\wedge)$ *and* $M_\gamma L^m_{cl}(X^\wedge) = L^m_{cl}(X^\wedge)$ *for all* $m, \gamma \in \mathbb{R}$.

Proof. In view of

$$op_M^\gamma(a) = r^\gamma op_M(T^{-\gamma}a)r^{-\gamma}$$

with $(T^{-\gamma}a)(r, r', z) = a(r, r', z - \gamma)$ and $r^\gamma L^m(X^\wedge)r^{-\gamma} = L^m(X^\wedge)$, it suffices to restrict the consideration to a convenient weight. We shall choose $\gamma = \frac{1}{2}$. First it is obvious that

$$M_{\frac{1}{2}} L^{-\infty}(X^\wedge) = L^{-\infty}(X^\wedge) \cong C^\infty(X^\wedge \times X^\wedge).$$

Functions u in $C_0^\infty(X^\wedge)$ will also be written as $u(r)$, i.e. r-dependent with values in $C^\infty(X)$. Then

$$op_M^{\frac{1}{2}}(a)u(r) = \int\limits_{-\infty}^{\infty}\int\limits_{0}^{\infty} \left(\frac{r}{r'}\right)^{-i\varrho} a(r, r', i\varrho)u(r')\frac{dr'}{r'}\, d\varrho. \tag{7.166}$$

We will substitute the diffeomorphism

$$\kappa:\ \mathbb{R} \longrightarrow \mathbb{R}_+, \qquad \kappa(t) = r = e^{-t}. \tag{7.167}$$

Then, for $v = \kappa^* u$, i.e. $v(t) = u(e^{-t})$, we get

$$op_M^{\frac{1}{2}}(a)u(r) = (\kappa^*)^{-1} \iint e^{i(t-t')\varrho}p(t, t', \varrho)v(t')dt'd\varrho \tag{7.168}$$

with

$$p(t, t', \varrho) = a(e^{-t}, e^{-t'}, i\varrho). \tag{7.169}$$

The right-hand side of (7.168) defines the operator

$$Op(p) = \mathcal{F}^{-1}p\mathcal{F} \in L^m(X^\wedge).$$

Thus

$$op_M^{\frac{1}{2}}(a) = \kappa_* Op(p) \tag{7.170}$$

with the push-forward κ_* of pseudo-differential operators under (7.167). Since κ_* is a bijection

$$\kappa_*:\ L^m(\mathbb{R} \times X) \longrightarrow L^m(X^\wedge), \tag{7.171}$$

we obtain from (7.170) that $op_M^{\frac{1}{2}}(a) \in L^m(X^\wedge)$. Conversely, (7.171) shows that every $Q \in L^m(X^\wedge)$ is of the form $\kappa_* P$ with some $P \in L^m(\mathbb{R} \times X)$. Writing $P = Op(p)$ with some $p(t, t', \varrho) \in C^\infty(\mathbb{R} \times \mathbb{R}, L^m(X; \mathbb{R}_\varrho))$, which is always possible, we obtain from (7.169) an $a(r, r', z)$ with

$$Q = op_M^{\frac{1}{2}}(a). \tag{7.172}$$

Thus $Q \in M_{\frac{1}{2}} L^m(X^\wedge)$. The calculations remain true, of course, for classical symbols. Thus the assertion is proved. $\qquad\square$

Note that every $Q \in L^m(X^\wedge)$ is of the form $Q = Op(q)$ with a certain $q(r,r',\varrho) \in C^\infty(\mathbb{R}_+ \times \mathbb{R}_+, L^m(X;\mathbb{R}))$. Thus (7.172) may also be interpreted as a map $q(r,r',\varrho) \to a(r,r',z)$ such that the associated pseudo-differential operators, based on the Fourier transform and the Mellin transform, respectively, coincide mod $L^{-\infty}(X^\wedge)$. This holds analogously for arbitrary weights. Since the r' dependence of symbols may be removed at the expense of smoothing operators on the operator level we are led to the following

Remark 2. For every $\gamma \in \mathbb{R}$ there is a non-canonical map

$$m^\gamma : \ C^\infty(\mathbb{R}_+, L^m(X;\mathbb{R})) \longrightarrow C^\infty(\mathbb{R}_+, L^m(X;\Gamma_{\frac{1}{2}-\gamma})) \tag{7.173}$$

such that for $a = m^\gamma(q)$,

$$Op(q) = op^\gamma_M(a) \text{ mod } L^{-\infty}(X^\wedge).$$

Non-canonical means that m^γ is unique only mod $C^\infty(\mathbb{R}_+, L^{-\infty}(X;\Gamma_{\frac{1}{2}-\gamma}))$. The map (7.173) may be regarded as a *Mellin operator convention*. This will be discussed below in a more precise form when we control smoothness in r up to $r = 0$.

Let us now look at the weighted Sobolev spaces $\mathcal{H}^{s,\gamma}(X^\wedge)$, cf. (7.55). They have many properties analogous to the standard Sobolev spaces, based on the Fourier transform.

According to 7.2.2 Theorem 17, there exists for every $m \in \mathbb{R}$ an operator family

$$b^m(\lambda) \in L^m_{cl}(X;\mathbb{R})$$

which is parameter-dependent elliptic and induces isomorphisms

$$b^m(\lambda) : \ H^s(X) \longrightarrow H^{s-m}(X)$$

for all $s, \lambda \in \mathbb{R}$. Let us fix such a $b^m(\lambda)$ for every $m \in \mathbb{R}$. We may assume

$$b^m(\lambda)^{-1} = b^{-m}(\lambda).$$

Definition 3. The space $\mathcal{H}^{s,\gamma}(X^\wedge)$ for $s,\gamma \in \mathbb{R}$ is the closure of $C_0^\infty(X^\wedge)$ with respect to the norm

$$u \to \left\{ \frac{1}{2\pi i} \int_{\Gamma_{\frac{n+1}{2}-\gamma}} \|b^s(\text{Im } z)(\mathcal{M}_{r \to z}u)(z,\cdot)\|^2_{L_2(X)} dz \right\}^{\frac{1}{2}}. \tag{7.174}$$

Here $\mathcal{M} = \mathcal{M}_{r \to z}$ is the Mellin transform, applied with respect to the r-variable in $u(r,x)$.

Remark 4. Let $\tilde{b}^s(\lambda)$ be another choice of operator family with analogous properties to $b^s(\lambda)$. Then the corresponding norm with $\tilde{b}^s(\lambda)$ is equivalent to (7.174).

In fact, first we have $b^s(\varrho) \in L^s_{cl}(X;\mathbb{R})$, $\tilde{b}^{-s}(\varrho) \in L^{-s}(X;\mathbb{R})$, and hence $b^s(\varrho)\tilde{b}^{-s}(\varrho) \in L^0_{cl}(X;\mathbb{R})$ because of 7.2.2 Theorem 10. Using 7.2.2 Corrolary 9 we get

$$\sup_\varrho \|b^s(\varrho)\tilde{b}^{-s}(\varrho)\|_{0,0} < \infty.$$

Now we have for $z = \frac{n+1}{2} - \gamma + i\varrho$

$$\int \|b^s(\varrho)\mathcal{M}u(z,\cdot)\|^2_{L_2(X)}d\varrho \;=\; \int \|b^s(\varrho)\tilde{b}^{-s}(\varrho)\tilde{b}^s(\varrho)\mathcal{M}u(z,\cdot)\|^2_{L_2(X)}d\varrho$$

$$\leq \;\sup_\varrho \|b^s(\varrho)\tilde{b}^{-s}(\varrho)\|^2_{0,0} \int \|\tilde{b}^s(\varrho)\mathcal{M}u(z,\cdot)\|^2_{L_2(X)}d\varrho,$$

where the factor in front of the latter integral is finite. The estimate in the converse direction is analogous.

Exercise 5. Definition 3 is equivalent to (7.55).

For convenience we shall often write

$$u(r) = u(r,\cdot)$$

for functions on $X^\wedge \ni (r,x)$, in so far as they are treated as vector-valued on $\mathbb{R}_+$. For instance, $C_0^\infty(X^\wedge) = C_0^\infty(\mathbb{R}_+, C^\infty(X))$.

Note that there is a reasonable modification of Definition 3 if we replace the Mellin transform by the Fourier transform. Since

$$\mathcal{H}^{s,\gamma}(X^\wedge) = r^{-\varrho}\mathcal{H}^{s,\gamma+\varrho}(X^\wedge) \quad \text{for all} \quad s,\gamma \in \mathbb{R}, \tag{7.175}$$

we may restrict ourselves to convenient weights. From (7.47) we obtain

$$(\mathcal{M}_{r\to z}u)(i\varrho) = (\mathcal{F}S_{\frac{1}{2}}u)(\varrho), \qquad (S_{\frac{1}{2}}u)(t) = u(e^{-t}),$$

$\mathcal{F} = \mathcal{F}_{t\to\varrho}$ being the one-dimensional Fourier transform. Putting $v(t) = u(e^{-t})$ we get from (7.174) the norm

$$v \to \left\{ \frac{1}{2\pi} \int\limits_{-\infty}^{\infty} \|b^s(\varrho)(\mathcal{F}_{t\to\varrho}v)(\varrho)\|^2_{L_2(X)}d\varrho \right\}^{\frac{1}{2}}. \tag{7.176}$$

Set for a moment

$$\mathcal{H}^s(\mathbb{R} \times X)_{(F)} = \{ \text{ closure of } C_0^\infty(\mathbb{R} \times X) \text{ with respect to (7.176) } \}.$$

We then see that

$$\mathcal{H}^{s,\gamma}(X^\wedge) = \{r^{\gamma - \frac{n+1}{2}}v(-\ln r) : \; v \in \mathcal{H}^s(\mathbb{R} \times X)_{(F)}\} \tag{7.177}$$

(cf. (7.50)).

The operator of multiplication by a function ϕ will be denoted by

$$\mathcal{M}_\phi \quad \text{or likewise by} \quad \phi. \tag{7.178}$$

Throughout this exposition a cut-off function on $\mathbb{R}_+$ is an

$$\omega(r) \in C_0^\infty(\overline{\mathbb{R}}_+) \quad \text{with} \quad \omega(r) = 1 \quad \text{for} \quad 0 \leq r < c \tag{7.179}$$

with some $c > 0$.

Exercise 6. Show (by using (7.177)) that for every cut-off function $\omega(r)$ the operator $\mathcal{M}_\omega$ belongs to $\mathcal{L}(\mathcal{H}^{s,\gamma}(X^\wedge))$ for all $s, \gamma \in \mathbb{R}$.

Theorem 7. *$\phi(r) \in C_0^\infty(\overline{\mathbb{R}}_+)$ implies $\mathcal{M}_\phi \in \mathcal{L}(\mathcal{H}^{s,\gamma}(X^\wedge))$. The map $\phi \to \mathcal{M}_\phi$ induces continuous operators*

$$C_0^\infty(\overline{\mathbb{R}}_+) \longrightarrow \mathcal{L}(\mathcal{H}^{s,\gamma}(X^\wedge)) \tag{7.180}$$

for all $s, \gamma \in \mathbb{R}$.

Proof. In view of (7.175), it suffices to restrict the consideration to a convenient weight. Let us take, for instance, $\gamma = \frac{n}{2}$. Then the integrations in the complex plane will be taken along $\Gamma_{\frac{1}{2}}$. From Exercise 6 we know the result for $\phi = \omega$. Thus, we can assume that $\phi(r)$ is of the form $\phi(r) = \psi(r) - \psi(0)\omega(r)$ for given $\psi \in C_0^\infty(\overline{\mathbb{R}}_+)$, i.e. $\phi(0) = 0$. First let $u = u(r) \in C_0^\infty(X^\wedge)$. Then

$$
\begin{aligned}
\mathcal{M}(\phi u)(w) &= \mathcal{M}\left\{ \phi(r)(2\pi i)^{-1} \int_{\Gamma_{\frac{1}{2}}} r^{-z} \mathcal{M}u(z)\,dz \right\}(w) \\[2ex]
&= \int_0^\infty r^{w-1}\phi(r)\left\{ (2\pi i)^{-1} \int_{\Gamma_{\frac{1}{2}}} r^{-z} \mathcal{M}u(z)\,dz \right\}dr.
\end{aligned}
$$

Set $z = \frac{1}{2} + i\varrho$, $w = \frac{1}{2} + i\sigma$. Then, with obvious meaning of notation,

$$\mathcal{M}(\phi u)(\sigma) = \int b(\varrho - \sigma)\mathcal{M}u\left(\frac{1}{2} + i\varrho\right)d\varrho,$$

with

$$b(\varrho - \sigma) = \int_0^\infty r^{w-z-1}\phi(r)\,dr. \tag{7.181}$$

We have for $\phi \in C_0^N(\overline{\mathbb{R}}_+)$

$$(1+z)^N \int r^{z-1}\phi(r)\,dr = \int r^{z-1}\left(1 - r\frac{d}{dr}\right)^N \phi(r)\,dr.$$

Here we have used $\mathcal{M}(-r\frac{d}{dr}u) = z\mathcal{M}u$. Thus

$$\left| \int r^{z-1}\phi(r)dr \right| \leq \left| (1+z)^{-N}\int r^{z-1}\left(1 - r\frac{d}{dr}\right)^{N}\phi(r)dr \right|$$

$$\leq |(1+z)^{-N}| \int \left| r^{z-1}\left(1 - t\frac{d}{dr}\right)^{N}\phi(r) \right| dr$$

$$\leq |(1+z)^{-N}| \int r^{\mathrm{Re}\ z-1}\left| \left(1 - r\frac{d}{dr}\right)^{N}\phi(r) \right| dr.$$

In view of the fact that $\phi(0) = 0$, this estimate holds for $\mathrm{Re}\ z > -1$ and uniformly in ϱ in $-1 + \varepsilon^{-1} < \mathrm{Re}\ z < \varepsilon^{-1}$ for an $\varepsilon > 0$. In particular, for $w \in \Gamma_{\frac{1}{2}}$, $z \in \Gamma_{\frac{1}{2}}$ we may replace z by $z - w$. Using $|(1 + w - z)^{-N}| \leq c(1 + |\varrho - \sigma|)^{-N}$ for a $c > 0$, we obtain

$$|b(\varrho - \sigma)| \leq cc_N(\phi)(1 - |\varrho - \sigma|)^{-N} \tag{7.182}$$

with the constants

$$c_N(\phi) = \int \left| r^{-1}\left(1 - t\frac{d}{dr}\right)^{N}\phi(r) \right| dr, \tag{7.183}$$

dependent on ϕ. Next let $u,v \in C_0^\infty(X^\wedge)$ and set $g(\varrho)=(\mathcal{M}u)(\frac{1}{2}+i\varrho)$, $f(\sigma)=(\mathcal{M}v)(\frac{1}{2}+i\sigma)$. Then, if $(\cdot, \cdot)$ indicates the scalar product of $\mathcal{H}^{0,\frac{n}{2}}(X^\wedge) \cong L_2(\mathbb{R}_+, L_2(X))$, it follows that

$$\begin{aligned}
(\phi u, v) &= \int_{-\infty}^{\infty} (\mathcal{M}(\phi u)(\sigma), \mathcal{M}(v)(\sigma))_{L_2(X)}d\sigma \\
&= \int_{-\infty}^{\infty} \left(\int_{-\infty}^{\infty} b(\varrho - \sigma)g(\varrho)d\varrho, f(\sigma) \right)_{L_2(X)} d\sigma \\
&= \iint K(\varrho, \sigma)((1 + |\varrho|^s)g(\varrho), (1 + |\sigma|^{-s})f(\sigma))_{L_2(X)}d\varrho d\sigma
\end{aligned} \tag{7.184}$$

with $K(\varrho, \sigma) = b(\varrho - \sigma)(1 + |\varrho|)^{-s}(1 + |\sigma|)^s$. Using $(1 + |\alpha|)(1 + |\beta|)^{-1} \leq 1 + |\alpha - \beta|$ for $\alpha, \beta \in \mathbb{R}$, we get with (7.182)

$$\begin{aligned}
\int |K(\varrho, \sigma)|d\varrho &\leq \int |b(\varrho - \sigma)(1 + |\varrho - \sigma|)^{|s|}d\varrho \\
&\leq cc_N(\phi)\int (1 + |\varrho - \sigma|)^{-N+|s|}d\varrho \\
&\leq c_1c_N(\phi).
\end{aligned}$$

Here N was chosen so large that $-N + |s| < -1$. The constant c_1 is independent of σ. In an analogous manner it follows that

$$\int |K(\varrho, \sigma)| d\sigma \le c_2 c_N(\phi)$$

with a constant c_2, independent of ϱ. From (7.184) we thus obtain

$$
\begin{aligned}
|(\phi u, v)| \;\le\;& \iint |K(\varrho, \sigma)|(1 + |\varrho|)^{2s} \|g(\varrho)\|_{L_2(X)}^2 d\varrho d\sigma \;\cdot \\
& \quad \cdot \iint |K(\varrho, \sigma)|(1 + |\sigma|)^{-2s} \|f(\sigma)\|_{L_2(X)}^2 d\varrho d\sigma \\
\le\;& c c_N(\phi) \|u\|_{\mathcal{H}^{s,\frac{1}{2}}(X^\wedge)}^2 \|v\|_{\mathcal{H}^{-s,\frac{1}{2}}(X^\wedge)}^2
\end{aligned}
$$

with a constant $c > 0$. This yields

$$\sup \frac{|(\phi u, v)|}{\|v\|_{\mathcal{H}^{-s,\frac{1}{2}}(X^\wedge)}} \le c c_N(\phi) \|u\|_{\mathcal{H}^{s,\frac{1}{2}}(X^\wedge)} \tag{7.185}$$

with the supremum over all $v \in C_0^\infty(X^\wedge)$, $v \ne 0$. Since the expression on the left of (7.185) is up to a constant the norm of $\mathcal{M}_\phi u$ in $\mathcal{H}^{s,\frac{1}{2}}$, it follows that

$$\|\mathcal{M}_\phi\|_{\mathcal{L}(\mathcal{H}^{s,\frac{1}{2}}(X^\wedge))} \le c c_N(\phi). \tag{7.186}$$

This completes the proof of Theorem 7. $\qquad\qquad\qquad\qquad\qquad\square$

Remark 8. The proof of Theorem 7 in fact shows a stronger result, namely that for every $s_0 \in \mathbb{R}$ there is an $N \in \mathbb{N}$ such that $\phi(r) \in C_0^N(\overline{\mathbb{R}}_+)$ implies $\mathcal{M}_\phi \in \mathcal{L}(\mathcal{H}^{s,\gamma}(X^\wedge))$ for all $s \in \mathbb{R}$ with $|s| \le |s_0|$ and that $C_0^N(\overline{\mathbb{R}}_+) \longrightarrow \mathcal{L}(\mathcal{H}^{s,\gamma}(X^\wedge))$ is continuous for those s. This holds for all $\gamma \in \mathbb{R}$.

Theorem 9. *Let $a(r, r', z) \in C^\infty(\overline{\mathbb{R}}_+ \times \overline{\mathbb{R}}_+, L^m(X; \Gamma_{\frac{n+1}{2} - \gamma}))$, $m \in \mathbb{R}$, and let $\omega(r), \tilde{\omega}(r)$ be arbitrary cut-off functions. Then*

$$\omega \, op_M^{\gamma - \frac{n}{2}}(a) \tilde{\omega} : \; \mathcal{H}^{s,\gamma}(X^\wedge) \longrightarrow \mathcal{H}^{s-m,\gamma}(X^\wedge) \tag{7.187}$$

is continuous for all $s \in \mathbb{R}$, $\gamma \in \mathbb{R}$.

Proof. Let us first look at the case of (r, r')-independent a. Then, we can drop the cut-off factors and consider $A = op_M^{\gamma - \frac{n}{2}}(a)$ itself. We have according to (7.174) for $u \in C_0^\infty(X^\wedge)$

$$\|Au\|_{\mathcal{H}^{s-m,\gamma}(X^\wedge)}^2 = \left(\frac{1}{2\pi}\right)^2 \int \|b^{s-m}(\text{Im } z) a(z)(\mathcal{M}u)(z)\|_{L_2(X)}^2 dz.$$

The integrals with respect to z are always taken over $\Gamma_{\frac{n+1}{2}-\gamma}$. From 7.2.2 Theorem 8 we know that

$$\sup_{z} \|b^{s-m}(\mathrm{Im}\ z)a(z)b^{-s}(\mathrm{Im}\ z)\|_{\mathcal{L}(L_2(X))} =: c(a) < \infty$$

with the supremum over $z \in \Gamma_{\frac{n+1}{2}-\gamma}$. Thus

$$\|Au\|_{\mathcal{H}^{s-m,\gamma}(X^\wedge)} \leq cc(a)\|u\|_{\mathcal{H}^{s,\gamma}(X^\wedge)} \tag{7.188}$$

with some $c > 0$. Since $C_0^\infty(X^\wedge)$ is dense in our spaces, (7.188) follows for all $u \in \mathcal{H}^{s,\gamma}(X^\wedge)$. At the same time we obtain that $a \to op_M^{\gamma-\frac{n}{2}}(a)$ induces a continuous map

$$L^m(X;\Gamma_{\frac{n+1}{2}-\gamma}) \longrightarrow \mathcal{L}(\mathcal{H}^{s,\gamma}(X^\wedge)). \tag{7.189}$$

For $a(r,r',z)$ in general we can pass to $h(r,r',z) = \omega(r)a(r,r',z)\tilde{\omega}(r')$. Let $C_0^\infty([0,c))_0$ denote the subspace of all $\phi \in C_0^\infty(\overline{\mathbb{R}}_+)$ with $\phi(r) = 0$ for $r \geq c$.

The space $C_0^\infty([0,c))_0$ is Fréchet in the topology induced by the topology of $C_0^\infty(\overline{\mathbb{R}}_+)$. From the identification

$$C_0^\infty(\overline{\mathbb{R}}_+ \times \overline{\mathbb{R}}_+, E) = C_0^\infty(\overline{\mathbb{R}}_+) \otimes_\pi C_0^\infty(\overline{\mathbb{R}}_+) \otimes_\pi E$$

for a Fréchet space E, with $\otimes_\pi$ as the completed tensor product, we obtain for $E = L^m(X;\Gamma_{\frac{n+1}{2}-\gamma})$

$$\begin{aligned} h(r,r',z) &\in C_0^\infty(\overline{\mathbb{R}}_+ \times \overline{\mathbb{R}}_+, L^m(X;\Gamma_{\frac{n+1}{2}-\gamma})) \\ &= C_0^\infty(\overline{\mathbb{R}}_+) \otimes_\pi C_0^\infty(\overline{\mathbb{R}}_+) \otimes_\pi L^m(X;\Gamma_{\frac{n+1}{2}-\gamma}). \end{aligned}$$

The condition on the support of h in r and r' allows us to write for suitable $c > 0$

$$h(r,r',z) \in C_0^\infty([0,c))_0 \otimes_\pi C_0^\infty([0,c))_0 \otimes_\pi L^m(X;\Gamma_{\frac{n+1}{2}-\gamma}).$$

Now we can apply the above result on the representation of elements in a projective tensor product as a convergent series, cf. 7.2.1 Theorem 10. In our case we then get that $h(r,r',z)$ can be written as a convergent series

$$h(r,r',z) = \sum_{j=0}^{\infty} \lambda_j \phi_j(r) h_j(z) \psi_j(r')$$

with $\phi_j, \psi_j \in C_0^\infty([0,c))_0$, $h_j(z) \in L^m(X;\Gamma_{\frac{n+1}{2}-\gamma})$ tending to zero in corresponding spaces as $j \to \infty$, and $\lambda_j \in \mathbb{C}$, $\sum |\lambda_j| < \infty$. Now it suffices to show that

$$op_M^{\gamma-\frac{n}{2}}(h) = \sum_{j=0}^{\infty} \lambda_j \mathcal{M}_{\phi_j} op_M^{\gamma-\frac{n}{2}}(h_j) \mathcal{M}_{\psi_j}$$

converges in $\mathcal{L}(\mathcal{H}^{s,\gamma}(X^\wedge), \mathcal{H}^{s-m,\gamma}(X^\wedge))$. Since

$$\|op_M^{\gamma-\frac{n}{2}}(h)\|_{s,s-m} \leq \sum_{j=0}^{\infty} |\,\lambda_j\,|\,\|\mathcal{M}_{\varphi_j}\|_{s-m,s-m}\|op_m^{\gamma-\frac{n}{2}}(h_j)\|_{s,s-m}\|\mathcal{M}_{\psi_j}\|_{s,s}$$

with $\|\cdot\|_{s,t}$ indicating the norm in $\mathcal{L}(\mathcal{H}^{s,\gamma}(X^\wedge), \mathcal{H}^{t,\gamma}(X^\wedge))$, the result follows from (7.180), (7.189). $\square$

Remark 10. Analogously to Remark 8 we can show a sharper result, namely that for every $s_0 \in \mathbb{R}$ there is an $N \in \mathbb{N}$ such that $a(r, r', z) \in C^N(\overline{\mathbb{R}}_+ \times \overline{\mathbb{R}}_+, L^m(X; \Gamma_{\frac{n+1}{2}-\gamma}))$ implies the continuity of (7.187) for all $|s| \leq |s_0|$, $\gamma \in \mathbb{R}$.

7.2.4   Kernel cut-off and operator-valued holomorphic Mellin symbols

This section will study the aspects of the Mellin pseudo-differential calculus that are specific for the control of smoothness of symbols in a neighbourhood of $r = 0$. This will require the kernel cut-off constructions which are also useful for studying meromorphic operator-valued Mellin symbols below in the cone operator algebra with asymptotics. Since the weights γ play no major role for the moment we shall content ourselves with $\gamma = \frac{n+1}{2}$, $n = \dim X$.

Definition 1. $M_{\frac{n+1}{2}} L^m(\overline{\mathbb{R}}_+ \times X)$ is the subspace of all $A \in M_{\frac{n+1}{2}} L^m(X^\wedge)$ which are of the form

$$A = op_M^{\frac{1}{2}}(a) + C \tag{7.190}$$

with arbitrary

$$a(r, r', z) \in C^\infty(\overline{\mathbb{R}}_+ \times \overline{\mathbb{R}}_+, L^m(X; \Gamma_0))$$

and $C \in L^{-\infty}(X^\wedge)$ satisfying

$$\omega C \tilde{\omega}, \tilde{\omega} C^* \omega \in \bigcap_{s \in \mathbb{R}} \mathcal{L}(\mathcal{H}^{s, \frac{n+1}{2}}(X^\wedge), \mathcal{H}^{\infty, \frac{n+1}{2}}(X^\wedge)) \tag{7.191}$$

for arbitrary cut-off functions $\omega(r)$, $\tilde{\omega}(r)$. Analogously we define the corresponding classes with subscript cl.

The $*$ in (7.191) indicates the formal adjoint with respect to a scalar product in $\mathcal{H}^{0, \frac{n+1}{2}}(X^\wedge)$. We set

$$M_{\frac{n+1}{2}} L^{-\infty}(\overline{\mathbb{R}}_+ \times X) = \bigcap_{m \in \mathbb{R}} M_{\frac{n+1}{2}} L^m(\overline{\mathbb{R}}_+ \times X).$$

Remember that the interpretation of the Mellin action in (7.190) is that the operation on $u(r) \in C_0^\infty(\mathbb{R}_+, C^\infty(X))$ is carried out along X through the values of the symbol a and then the Mellin pseudo-differential operator is applied with respect to r.

According to (7.166) we set

$$k(a)(r, r', \beta) = \int\limits_{-\infty}^{\infty} \beta^{-i\varrho} a(r, r', i\varrho)\, \mathrm{d}\varrho,$$

interpreted in the distributional sense. The distributional kernel of $op_M^{\frac{1}{2}}(a)$ is obtained by inserting $\beta = \frac{r}{r'}$. Since

$$k(a)(r, r', \beta) = (\mathcal{M}_{\frac{1}{2}, z \to \beta}^{-1} a)(r, r', \beta), \tag{7.192}$$

we have

$$a(r, r', z) = (\mathcal{M}_{\frac{1}{2}, \beta \to z} k(a))(r, r', z). \tag{7.193}$$

Let us denote function or distribution spaces along $\Gamma_0 \ni i\varrho$ with respect to the variable ϱ analogously to the corresponding spaces on $\mathbb{R} \ni \varrho$, e.g., $\mathcal{S}(\Gamma_0) = \mathcal{S}(\mathbb{R})$ and so on. Set

$$\mathcal{T}(\mathbb{R}_+) = \mathcal{M}_{\frac{1}{2}, z \to \beta}^{-1} \mathcal{S}(\Gamma_0).$$

In view of $\mathcal{S}(\Gamma_0, L^{-\infty}(X)) = L^{-\infty}(X; \Gamma_0)$, it follows that

$$C^{\infty}(\overline{\mathbb{R}}_+ \times \overline{\mathbb{R}}_+, \mathcal{T}(\mathbb{R}_+, L^{-\infty}(X))) \tag{7.194}$$
$$= \{k(a)(r, r', \beta) = (\mathcal{M}_{\frac{1}{2}}^{-1} a)(r, r', \beta) : \ a(r, r', z) \in C^{\infty}(\overline{\mathbb{R}}_+ \times \overline{\mathbb{R}}_+, L^{-\infty}(X; \Gamma_0))\}$$

Theorem 2. *Let $\psi(\beta) \in C^{\infty}(\mathbb{R}_+)$ with $\psi(\beta) = 1$ for $|\beta - 1| \leq \varepsilon$ with some $0 < \varepsilon < \frac{1}{2}$. Then $a(r, r', z) \in C^{\infty}(\overline{\mathbb{R}}_+ \times \overline{\mathbb{R}}_+, L^m(X; \Gamma_0))$ implies*

$$(1 - \psi(\beta))k(a)(r, r', \beta) \in C^{\infty}(\overline{\mathbb{R}}_+ \times \overline{\mathbb{R}}_+, \mathcal{T}(\mathbb{R}_+, L^{-\infty}(X))). \tag{7.195}$$

Proof. The space $\mathcal{S}(\mathbb{R})$ can be characterized by

$$\mathcal{S}(\mathbb{R}) \ni a(\varrho) \iff \|D_\varrho^M \varrho^N a(\varrho)\|_{L_2(\mathbb{R})} < \infty \ \text{ for all } \ M, N \in \mathbb{N}.$$

From $\mathcal{M}\left(-\beta \frac{\partial}{\partial \beta} f(\beta)\right)(z) = z(\mathcal{M}f)(z)$, $\frac{d}{dz}(\mathcal{M}f(\beta))(z) = \mathcal{M}(\ln \beta f(\beta))(z)$ it follows that

$$\mathcal{T}(\mathbb{R}_+) \ni g(\beta) = \beta^{\frac{1}{2}} f(\beta) \iff \left\| \ln^M \beta \left(-\beta \frac{\partial}{\partial \beta}\right)^N f(\beta)\right\|_{L_2(\mathbb{R}_+)} < \infty \tag{7.196}$$
$$\text{for all } M, N \in \mathbb{Z}_+.$$

For $a(r, r', z) \in C^{\infty}(\overline{\mathbb{R}}_+ \times \overline{\mathbb{R}}_+, L^m(X; \Gamma_0))$ we also have

$$\ln^M \beta \left(-\beta \frac{\partial}{\partial \beta}\right)^N k(a)(r, r', \beta) = k(D_\varrho^M (i\varrho)^N a)(r, r', \beta). \tag{7.197}$$

Let us consider the case of (r, r')-independent symbols a. The general case is completely analogous and will be left to the reader as an exercise. From $D_\varrho^M(i\varrho)^N a(i\varrho) \in L^{m+N-M}(X; \Gamma_0)$ it follows for every $N \in \mathbb{N}$ and all $M \geq M_0$ for sufficiently large $M_0 = M_0(N)$ that

$$\|\ln^M \beta \left(-\beta \frac{\partial}{\partial \beta}\right)^N k(a)(\beta)\|_{L_2(\mathbb{R}_+ \times X \times X)} < \infty. \tag{7.198}$$

Here $k(a)(\beta)$ is regarded as a $\mathcal{D}'(X \times X)$-valued function which belongs to $L_2(X \times X)$ once ord a is negative enough, thus (7.198) holds for sufficiently large M_0. Clearly $D_\varrho^M(i\varrho)^N a(i\varrho)$ is identified in (7.198) with its kernel with respect to x-variables and interpreted as an element in $L_2(\Gamma_0 \times X \times X)$. In order to obtain (7.195) we have to show that

$$\|(\ln^M \beta) \left(-\beta \frac{\partial}{\partial \beta}\right)^N (1 - \psi(\beta)) k(a)(\beta)\|_{L_2(\mathbb{R}_+ \times X \times X)} < \infty \tag{7.199}$$

for all $M, N \in \mathbb{N}$. First observe that (7.198) implies

$$\ln^M \beta f(\beta) \left(-\beta \frac{\partial}{\partial \beta}\right)^N k(a)(\beta) \in L_2(\mathbb{R}_+ \times X \times X) \tag{7.200}$$

for every $f \in C^\infty(\mathbb{R}_+)$ with $\sup |f| < \infty$, $N \in \mathbb{Z}_+$, $M \geq M_0(N)$. If in addition $f = 0$ in a neighbourhood of $\beta = 1$, then (7.200) holds for all $M \in \mathbb{Z}_+$. In fact, (7.200) for $M \geq M_0$ may be multiplied by $\ln^{\widetilde{M}-M} \beta$ for every $\widetilde{M} \leq M_0$ without violating the L_2-property in β since $\sup\{\ln^{\widetilde{M}-M} \beta : \beta \in \text{supp } f\} < \infty$, whereas the L_2-property along $X \times X$ holds, since the singular support is concentrated on $\{\beta = 1\} \times \text{diag } X \times X$ and cut out by the multiplication by f. For (7.199) we now apply the Leibniz rule and obtain

$$\ln^M \beta \left(-\beta \frac{\partial}{\partial \beta}\right)^N (1 - \psi(\beta)) k(a)(\beta) = \sum_{j=0}^N (\ln^M \beta) f_j(\beta) \left(-\beta \frac{\partial}{\partial \beta}\right)^{N-j} k(a)(\beta)$$
$$\tag{7.201}$$

with certain $f_j \in C^\infty(\mathbb{R}_+)$, $\sup |f_j| < \infty$, $f_j = 0$ in a neighbourhood of $\beta = 1$, for all $j = 0, \ldots, N$. We have just seen that the items on the right of (7.201) belong to $L_2(\mathbb{R}_+ \times X \times X)$. Hence we have proved (7.199). $\qquad\square$

Remark 3. Let $\psi(\beta)$ be chosen as in Theorem 2. Then $a_{-\infty}(r, r', z) := \mathcal{M}_{\frac{1}{2}, \beta \to z}(1 - \psi(\beta)) k(a)(r, r', \beta) \in C^\infty(\overline{\mathbb{R}}_+ \times \overline{\mathbb{R}}_+, L^{-\infty}(X; \Gamma_0))$, cf. (7.194). Hence

$$a_0(r, r', z) \quad := \quad (a - a_{-\infty})(r, r', z)$$
$$= \quad \mathcal{M}_{\frac{1}{2}, \beta \to z} \psi(\beta) k(a)(r, r', \beta) \subset C^\infty(\overline{\mathbb{R}}_+ \times \overline{\mathbb{R}}_+, L^m(X; \Gamma_0)).$$

Since a_0 is the Mellin transform of a distribution on $\mathbb{R}_+ \ni \beta$ of compact support, it has a holomorphic extension $h(r, r', z)$ from Γ_0 to $z \in \mathbb{C}$. The map $H : a \to h$ will also be called a *kernel cut-off operator*.

Exercise 4. Let $\phi(\beta) \in C_0^\infty(\mathbb{R}_+)$ and set

$$(H(\phi)a)(r, r', z) = M_{\frac{1}{2}, \beta \to z}\{\phi(\beta)(M_{\frac{1}{2}, z \to \beta}^{-1} a)(r, r', \beta)\}.$$

Then $H(\phi)$ defines a continuous operator

$$H(\phi): \ C^\infty(\overline{\mathbb{R}}_+ \times \overline{\mathbb{R}}_+, L^m(X; \Gamma_0)) \longrightarrow C^\infty(\overline{\mathbb{R}}_+ \times \overline{\mathbb{R}}_+, L^m(X; \Gamma_0)) \qquad (7.202)$$

for all $m \in \mathbb{R}$. Moreover, for fixed $a(r, r', z) \in C^\infty(\overline{\mathbb{R}}_+ \times \overline{\mathbb{R}}_+, L^m(X; \Gamma_0))$ the map $\phi \to H(\phi)a$ induces a continuous operator

$$C_0^\infty(\mathbb{R}_+) \longrightarrow C^\infty(\overline{\mathbb{R}}_+ \times \overline{\mathbb{R}}_+, L^m(X; \Gamma_0)).$$

An analogous statement holds for the operator families with subscript *cl*.

Definition 5. Let $m \in \mathbb{R}$ and denote by

$$M_O^m(X) \qquad (7.203)$$

the space of all operator families $h(z) \in \mathcal{A}(\mathbb{C}_z, L_{cl}^m(X))$ with

$$h(\alpha + i\varrho) \in L_{cl}^m(X; \Gamma_\alpha)$$

for all $\alpha \in \mathbb{R}$, uniformly in $c \le \alpha \le c'$ for every $c < c'$. More generally for every $q \in \mathbb{N}$ we denote by

$$M_O^m(X; \mathbb{R}^q) \qquad (7.204)$$

the space of all $h(z, \eta) \in \mathcal{A}(\mathbb{C}_z, L_{cl}^m(X; \mathbb{R}_\eta^q))$ with

$$h(\alpha + i\varrho, \eta) \in L_{cl}^m(X; \Gamma_\alpha \times \mathbb{R}^q)$$

for all $\alpha \in \mathbb{R}$, uniformly in $c \le \alpha \le c'$ for every $c < c'$.

Exercise 6. Show that the kernel cut-off operator of Remark 3 induces a continuous map

$$H = H(\psi): \ C^\infty(\overline{\mathbb{R}}_+ \times \overline{\mathbb{R}}_+, L_{cl}^m(X; \Gamma_0)) \longrightarrow C^\infty(\overline{\mathbb{R}}_+ \times \overline{\mathbb{R}}_+, M_O^m(X)).$$

Remark 7. Let $\phi(\beta) \in C_0^\infty(\mathbb{R}_+)$, $a(r, r', z) \in C^\infty(\overline{\mathbb{R}}_+ \times \overline{\mathbb{R}}_+, L^m(X; \Gamma_0))$, and set $b(r, r', z) = \phi(\frac{r}{r'})a(r, r', z)$. Then $a_\phi(r, r', z) := H(\phi)a(r, r', z) \in C^\infty(\overline{\mathbb{R}}_+ \times \overline{\mathbb{R}}_+, L^m(X; \Gamma_0))$, and we have

$$\begin{aligned} k(a_\phi)(r, r', \beta) &= \phi(\beta)k(a)(r, r', \beta), \\ op_M^{\frac{1}{2}}(b) &= op_M^{\frac{1}{2}}(a_\phi). \end{aligned}$$

This is an immediate consequence of the fact that the Mellin operator with symbol a can always be expressed on the level of $k(a)$ by a Mellin convolution, where β is to be replaced by $\frac{r}{r'}$. In other words

$$op_M^{\frac{1}{2}}(a)u(r) = \left\langle k(a)\left(r,r',\frac{r}{r'}\right), u(r')(r')^{-1}\right\rangle,$$

where $\langle\cdot,\cdot\rangle$ is the bilinear pairing with respect to r' that extends $\langle f(r'),g(r')\rangle = \int f(r')g(r')dr'$ for $f(r') \in C^\infty(\mathbb{R}_+)$, $g(r') \in C_0^\infty(\mathbb{R}_+)$.

The spaces $M_O^m(X)$, $M_O^m(X;\mathbb{R}^q)$ have canonical Fréchet topologies. If $\{p_j\}_{j\in\mathbb{N}}$ is a semi-norm system for the Fréchet topology of $L_{cl}^m(X)$ and $\{q_l^\alpha\}_{l\in\mathbb{Z}_+}$ analogously a semi-norm system for $L_{cl}^m(X;\Gamma_\alpha)$, then

$$h(z) \quad \longrightarrow \quad \sup_{z\in K} p_j(h(z)), \ j \in \mathbb{Z}_+,$$

$$h(z) \quad \longrightarrow \quad \sup_{-k\leq\alpha\leq k} q_l^\alpha(h(z)), \ k,l \in \mathbb{Z}_+$$

is a semi-norm system for the Fréchet topology of $M_O^m(X)$. The construction for $M_O^m(X;\mathbb{R}^q)$ is completely analogous.

Exercise 8. For every sequence $a_j(r,r',z,\eta) \in C^\infty(\overline{\mathbb{R}}_+ \times \overline{\mathbb{R}}_+, M_O^{m-j}(X;\mathbb{R}^q))$, $j \in \mathbb{N}$, there exists an $a(r,r',z,\eta) \in C^\infty(\overline{\mathbb{R}}_+ \times \overline{\mathbb{R}}_+, M_O^m(X;\mathbb{R}^q))$ with

$$a(r,r',z,\eta) - \sum_{j=0}^N a_j(r,r',z,\eta) \in C^\infty(\overline{\mathbb{R}}_+ \times \overline{\mathbb{R}}_+, M_O^{m-(N+1)}(X;\mathbb{R}^q))$$

for all $N \in \mathbb{N}$, and a is unique mod $C^\infty(\overline{\mathbb{R}}_+ \times \overline{\mathbb{R}}_+, M_O^{-\infty}(X;\mathbb{R}^q))$.

Proposition 9. *Let $a(r,r',z) \in C^\infty(\overline{\mathbb{R}}_+ \times \overline{\mathbb{R}}_+, L^m(X;\Gamma_0))$ be a symbol with $(r - r')^{-N}a(r,r',z) \in C^\infty(\overline{\mathbb{R}}_+ \times \overline{\mathbb{R}}_+, L^m(X;\Gamma_0))$ for a certain $N \in \mathbb{Z}_+$. Then there is an $a_N(r,r',z) \in C^\infty(\overline{\mathbb{R}}_+ \times \overline{\mathbb{R}}_+, L^{m-N}(X;\Gamma_0))$ such that*

$$op_M^{\frac{1}{2}}(a) = op_M^{\frac{1}{2}}(a_N).$$

Proof. Let us write $f(r,r',z) = \left(\frac{r}{r'} - 1\right)^{-N} a(r,r',z) = (r')^N(r - r')^{-N}a(r,r',z)$ $\in C^\infty(\overline{\mathbb{R}}_+ \times \overline{\mathbb{R}}_+, L^m(X;\Gamma_0))$. We form the operator $H(\psi)$ with a cut-off function $\psi(\beta)$ as in Theorem 2 and write $H(1 - \psi) = id - H(\psi)$, $g := H(1 - \psi)a$, $b := H(\psi)a$. From Theorem 2 it follows that $g \in C^\infty(\overline{\mathbb{R}}_+ \times \overline{\mathbb{R}}_+, L^{-\infty}(X;\Gamma_0))$. Since $op_M^{\frac{1}{2}}(a) = op_M^{\frac{1}{2}}(b) + op_M^{\frac{1}{2}}(g)$, it suffices to look at b. We have

$$\begin{aligned}
k(b)(r,r',\beta) &= \psi(\beta)(\beta - 1)^N k(f)(r,r',\beta) \\
&= \phi(\beta)\ln^N \beta k(f)(r,r',\beta)
\end{aligned}$$

with $\phi(\beta) = (\beta - 1)^N \psi(\beta) \ln^{-N} \beta \in C_0^\infty(\mathbb{R}_+)$. Using Remark 7 we get

$$
\begin{aligned}
(\ln^N \beta)\phi(\beta)k(f)(r,r',\beta) &= \ln^N \beta k(f_\phi)(r,r',\beta) \\
&= k(D_\varrho^N f_\phi)(r,r',\beta)
\end{aligned}
$$

with $D_\varrho^N f_\phi(r,r',\beta) \in C^\infty(\overline{\mathbb{R}}_+ \times \overline{\mathbb{R}}_+, L^{m-N}(X;\Gamma_0))$. This yields $op_M^{\frac{1}{2}}(b) = op_M^{\frac{1}{2}}(D_\varrho^N f_\phi)$. Hence we may set $a_N = D_\varrho^N f_\phi + g$. $\qquad\square$

Theorem 10. *Let $A \in M_{\frac{n+1}{2}} L^m(\overline{\mathbb{R}}_+ \times X)$ be written in the form (7.190) with $a(r,r',z) \in C^\infty(\overline{\mathbb{R}}_+ \times \overline{\mathbb{R}}_+, L^m(X;\Gamma_0))$. Then there is an $\underline{a}(r,z) \in C^\infty(\overline{\mathbb{R}}_+, L^m(X;\Gamma_0))$ with $A = op_M^{\frac{1}{2}}(\underline{a}) + C_1$ with some $C_1 \in M_{\frac{n+1}{2}} L^{-\infty}(\overline{\mathbb{R}}_+ \times X)$.*

Proof. We may assume $A = op_M^{\frac{1}{2}}(a)$. Taylor's formula yields for every $N \in \mathbb{N}$, $N \geq 1$

$$
a(r,r',z) = \sum_{j=0}^{N-1} \frac{1}{j!}(r'-r)^j \left(\frac{\partial}{\partial r'}\right)^j a(r,r',z)\,|_{r'=r} + r_N(r,r',z) \tag{7.205}
$$

with a remainder r_N satisfying the assumptions of Proposition 9. Thus, we obtain an $a_N(r,r',z) \in C^\infty(\overline{\mathbb{R}}_+ \times \overline{\mathbb{R}}_+, L^{m-N}(X;\Gamma_0))$ with $op_M^{\frac{1}{2}}(r_N) = op_M^{\frac{1}{2}}(a_N)$. In other words, according to 7.2.3 Theorem 9,

$$
\omega op_M^{\frac{1}{2}}(r_N)\tilde{\omega} : \mathcal{H}^{s,\frac{n+1}{2}}(X^\wedge) \longrightarrow \mathcal{H}^{s+N,\frac{n+1}{2}}(X^\wedge)
$$

for all $s \in \mathbb{R}$. The same is true for the formal adjoints. For treating the items of the sum on the right of (7.205) we may argue as in the proof of Proposition 9. Write

$$
b_j(r,z) = \frac{1}{j!}r^j \left(\frac{\partial}{\partial r'}\right)^j a(r,r',z)\,|_{r'=r} .
$$

Then $f_j(r,r',z) := \frac{1}{j!}(r'-r)^j \left(\frac{\partial}{\partial r'}\right)^j a(r,r',z)\,|_{r'=r} = \left(\frac{r'}{r}-1\right)^j b_j(r,z)$. Choose a cut-off function $\psi(\beta)$ as in Theorem 2. Set $\phi_j(\beta) = (\beta^{-1}-1)^j\psi(\beta)\ln^{-j}\beta \in C_0^\infty(\mathbb{R}_+)$. We then define a $d_j(r,z) \in C^\infty(\overline{\mathbb{R}}_+, L^{m-j}(X;\Gamma_0))$ by

$$
\begin{aligned}
k(d_j)(r,\beta) &= (\ln^j \beta)\phi_j(\beta)k(b_j)(r,\beta) \\
&= (\ln^j \beta)k(b_{j,\phi_j})(r,\beta) \\
&= k(D_\varrho^j b_{j,\phi_j})(r,\beta).
\end{aligned}
$$

Then we have

$$
op_M^{\frac{1}{2}}(d_j) - op_M^{\frac{1}{2}}(f_j) \in M_{\frac{n+1}{2}} L^{-\infty}(\overline{\mathbb{R}}_+ \times X).
$$

It follows from (7.205) that

$$op_M^{\frac{1}{2}}(a) = op_M^{\frac{1}{2}}\left(\sum_{j=0}^{N} d_j\right) + op_M^{\frac{1}{2}}(r_N) \bmod \ M_{\frac{n+1}{2}} L^{-\infty}(\overline{\mathbb{R}}_+ \times X)$$

and hence, we may set $\underline{a}(r,z) \sim \sum_{j=0}^{\infty} d_j(r,z)$, with the asymptotic sum being carried out in $C^{\infty}(\overline{\mathbb{R}}_+, L^m(X;\Gamma_0))$. $\qquad\qquad\square$

Exercise 11. Show that the symbol $\underline{a}(r,z)$ of Theorem 10 allows the asymptotic expansion

$$\underline{a}(r,z) \sim \sum_{k=0}^{\infty} \frac{1}{k!}\left(-r'\frac{\partial}{\partial r'}\right)^k \partial_z^k a(r,r',z)\,|_{r'=r} \ . \tag{7.206}$$

Theorem 12. *$A \in M_{\frac{n+1}{2}} L^m(\overline{\mathbb{R}}_+ \times X)$ implies $A^* \in M_{\frac{n+1}{2}} L^m(\overline{\mathbb{R}}_+ \times X)$ with A^* as the formal adjoint of A. If $A = op_M^{\frac{1}{2}}(a) + C$ with $a(r,z) \in C^{\infty}(\overline{\mathbb{R}}_+, L^m(X;\Gamma_0))$, $C \in M_{\frac{n+1}{2}} L^{-\infty}(\overline{\mathbb{R}}_+ \times X)$, then there is an $h(r,z) \in C^{\infty}(\overline{\mathbb{R}}_+, L^m(X;\Gamma_0))$ with $A^* = op_M^{\frac{1}{2}}(h) + C_1$ with some $C_1 \in M_{\frac{n+1}{2}} L^{-\infty}(\overline{\mathbb{R}}_+ \times X)$. In an analogous manner it follows that ${}^tA \in M_{\frac{n+1}{2}} L^m(\overline{\mathbb{R}}_+ \times X)$, and ${}^tA = op_M^{\frac{1}{2}}(f) + C_2$ for some $f(r,z) \in C^{\infty}(\overline{\mathbb{R}}_+, L^m(X;\Gamma_0))$ and some $C_2 \in M_{\frac{n+1}{2}} L^{-\infty}(\overline{\mathbb{R}}_+ \times X)$. (Here ${}^tA$ is the transposed of A with respect to the bilinear pairing.)*

Proof. We have for $A = op_M^{\frac{1}{2}}(a)$

$$
\begin{aligned}
(Au,v) &= \int\left(\iint \left(\frac{r}{r'}\right)^{-i\varrho} a(r,i\varrho)u(r')\frac{dr'}{r'}d\varrho, v(r)\right)_{L^2(X)} \frac{dr}{r} \\
&= \int\left(u(r'), \iint \left(\frac{r'}{r}\right)^{-i\varrho} a(r,i\varrho)^{(*)}v(r)\frac{dr}{r}d\varrho\right)_{L^2(X)} \frac{dr'}{r'} \tag{7.207} \\
&= (u, A^*v)
\end{aligned}
$$

with $u,v \in C_0^{\infty}(\mathbb{R}_+ \times X)$. Here $a(r,i\varrho)^{(*)}$ denotes the pointwise formal adjoint with respect to operators along X. Thus

$$A^*v(r) = \iint \left(\frac{r}{r'}\right)^{-i\varrho} a(r',i\varrho)^{(*)}v(r')\frac{dr'}{r'}.$$

Using Theorem 10 we can easily derive an asymptotic expansion of a complete symbol $h(r,z)$ of A^* in terms of a. The second statement follows in an analogous manner. $\qquad\qquad\square$

Corollary 13. *For every* $A \in M_{\frac{n+1}{2}} L^m(\overline{\mathbb{R}}_+ \times X)$ *there is a* $b(r',z) \in C^\infty(\overline{\mathbb{R}}_+, L^m(X; \Gamma_0))$ *with* $A = op_M^{\frac{1}{2}}(b)$ *mod* $M_{\frac{n+1}{2}} L^{-\infty}(\overline{\mathbb{R}}_+ \times X)$, *and*

$$b(r', z) \sim \sum_{k=0}^{\infty} \frac{1}{k!} \partial_z^k \left(-r' \frac{\partial}{\partial r'} \right)^k a(r', z). \tag{7.208}$$

In fact, it suffices to observe that $(A^*)^* = A$ and to express A^* by $h(r, z)$ as in the preceding theorem. Then $(A^*)^*$ follows by applying (7.207) to $op_M^{\frac{1}{2}}(h)$, which yields $b(r', z) = h(r', z)$.

Theorem 14. $A \in M_{\frac{n+1}{2}} L^m(\overline{\mathbb{R}}_+ \times X)$, $B \in M_{\frac{n+1}{2}} L^{m'}(\overline{\mathbb{R}}_+ \times X)$ *implies* $A\omega B \in M_{\frac{n+1}{2}} L^{m+m'}(\overline{\mathbb{R}}_+ \times X)$ *for every cut-off function* $\omega(r)$. *If* $A = op_M^{\frac{1}{2}}(a)$ *mod* $M_{\frac{n+1}{2}} L^{-\infty}(\overline{\mathbb{R}}_+ \times X)$, $B = op_M^{\frac{1}{2}}(b)$ *mod* $M_{\frac{n+1}{2}} L^{-\infty}(\overline{\mathbb{R}}_+ \times X)$, *with symbols* $a(r, z) \in C^\infty(\overline{\mathbb{R}}_+, L^m(X; \Gamma_0))$, $b(r,z) \in C^\infty(\overline{\mathbb{R}}_+, L^{m'}(X; \Gamma_0))$ *then* $c(r,z) \in C^\infty(\overline{\mathbb{R}}_+, L^{m+m'}(X; \Gamma_0))$, *obtained by*

$$c(r, z) \sim \sum_{k=0}^{\infty} \frac{1}{k!} \partial_z^k a(r, z) \left(-r \frac{\partial}{\partial r} \right)^k b(r, z) \tag{7.209}$$

satisfies $A\omega B\omega_0 = op_M^{\frac{1}{2}}(c)\omega_0$ *mod* $M_{\frac{n+1}{2}} L^{-\infty}(\overline{\mathbb{R}}_+ \times X)$ *for every cut-off function* ω_0 *with* $\omega\omega_0 = \omega_0$.

Proof. According to Corollary 13, the operator ωB can be written as $\omega B = op_M^{\frac{1}{2}}(h)$ mod $M_{\frac{n+1}{2}} L^{-\infty}(\overline{\mathbb{R}}_+ \times X)$ with some $h(r', z) \in C^\infty(\overline{\mathbb{R}}_+, L^{m'}(X; \Gamma_0))$. Here we can assume that $h(r', z)$ has bounded support in r'. Then $A\omega B = op_M^{\frac{1}{2}}(d)$ mod $M_{\frac{n+1}{2}} L^{-\infty}(\overline{\mathbb{R}}_+ \times X)$ with

$$d(r, r', z) = a(r, z)h(r', z) \in C^\infty(\overline{\mathbb{R}}_+ \times \overline{\mathbb{R}}_+, L^{m+m'}(X; \Gamma_0)).$$

This yields $A\omega B \in M_{\frac{n+1}{2}} L^{m+m'}(\overline{\mathbb{R}}_+ \times X)$. The asymptotic expansion (7.209) follows immediately from Exercise 11 and Corollary 13 (details are left to the reader as an exercise!). $\square$

7.2.5 Meromorphic Fredholm families

In the description of the symbolic structure of operators on manifolds with conical singularities there arise meromorphic operator-valued functions in the complex plane that are pointwise Fredholm between Hilbert spaces. We will need meromorphic Fredholm families again when we discuss the symbolic structures of edge pseudo-differential operators in Chapter 9. Let us begin with some remarks about

Fredholm families parametrized by the points of a compact topological space X. We assume that the occurring Hilbert spaces are separable ones.

Proposition 1. *Let H be a Hilbert space and X be a compact topological space,*

$$a : \ X \longrightarrow \mathcal{L}(H) \tag{7.210}$$

a continuous operator function, $\mathcal{L}(H)$ being equipped with the norm topology. Assume that

$$a(x) : \ H \longrightarrow H$$

is a Fredholm operator for every $x \in X$. Then there is an $N \in \mathbb{N}$ and a linear continuous operator $k : \ \mathbb{C}^N \longrightarrow H$ such that

$$(a(x), k) : \ \begin{array}{c} H \\ \oplus \\ \mathbb{C}^N \end{array} \longrightarrow H \tag{7.211}$$

is surjective for every $x \in X$.

Proof. Fix $x_0 \in X$ and choose a finite-dimensional subspace $V(x_0) \subset H$ such that

$$V(x_0) \oplus \operatorname{im} a(x_0) = H.$$

The existence of those $V(x_0)$ is a consequence of the Fredholm property of $a(x_0)$. Let $N(x_0) = \dim V(x_0)$ and

$$k(x_0) : \ \mathbb{C}^{N(x_0)} \longrightarrow V(x_0)$$

be an arbitrary isomorphism. Then the row matrix of operators

$$(a(x), k(x_0)) : \ \begin{array}{c} H \\ \oplus \\ \mathbb{C}^{N(x_0)} \end{array} \longrightarrow H \tag{7.212}$$

is surjective for $x = x_0$. Since the surjective operators form an open subset of $\mathcal{L}(H \oplus \mathbb{C}^{N(x_0)}, H)$, the operator family (7.212) is surjective for all x in a certain open neighbourhood $\Omega(x_0)$ of x_0. Now if x_0 runs through X we obtain an open covering of X by those $\Omega(x_0)$. Since X is compact, there is a finite subcovering $\{\Omega(x_0), \Omega(x_1), \ldots, \Omega(x_m)\}$ belonging to certain $x_0, \ldots, x_m \in X$. Let us form for every x_j the spaces $V(x_j)$ and isomorphisms $k(x_j) : \ \mathbb{C}^{N(x_j)} \longrightarrow V(x_j)$. Then the row matrix of operators

$$k = (k(x_0), \ldots, k(x_m)) : \ \oplus_{j=0}^{m} \mathbb{C}^{N(x_j)} \longrightarrow \sum_{j=0}^{m} V(x_j)$$

obviously has the asserted properties, with $N = \sum_{j=0}^{m} N(x_j)$. $\qquad \square$

Since the operator k in (7.211) is of finite dimension, the operator family (7.211) is Fredholm, again, between $H \oplus \mathbb{C}^N$ and H. Let us assume henceforth for simplicity that X is arc-wise connected. Then

$$\operatorname{ind}(a(x), k) = \dim \ker(a(x), k)$$

is independent of x, since (7.211) is surjective. Our next observation is that the vector spaces

$$\tilde{J}_x = \{(u, v) \in H \oplus \mathbb{C}^N : (u, v) \in \ker(a(x), k)\}$$

form a finite-dimensional vector sub-bundle $\tilde{J}$ of the trivial bundle $X \times (H \oplus \mathbb{C}^N)$. Let us construct a system of trivializations. Let

$$\pi(x) : H \oplus \mathbb{C}^N \longrightarrow \tilde{J}_x$$

be the orthogonal projection to $\tilde{J}_x$. Denote by

$$e : H \longrightarrow H \oplus \mathbb{C}^N, \qquad e' : \mathbb{C}^N \longrightarrow H \oplus \mathbb{C}^N$$

the corresponding canonical embeddings, such that $H \oplus \mathbb{C}^N = eH \oplus e'\mathbb{C}^N$. Let $M = \dim \tilde{J}_x$ and

$$j(x) : \tilde{J}_x \longrightarrow \mathbb{C}^M$$

be an arbitrary isomorphism. Set

$$b(x) = j(x)\pi(x)e, \qquad r(x) = j(x)\pi(x)e'. \tag{7.213}$$

Then

$$\begin{pmatrix} a(x) & k \\ b(x) & r(x) \end{pmatrix} : \begin{matrix} H \\ \oplus \\ \mathbb{C}^N \end{matrix} \longrightarrow \begin{matrix} H \\ \oplus \\ \mathbb{C}^M \end{matrix}$$

is an isomorphism for every fixed $x \in X$. Since the isomorphisms in $\mathcal{L}(H \oplus \mathbb{C}^N, H \oplus \mathbb{C}^M)$ form an open subset in the norm topology,

$$\begin{pmatrix} a(x) & k \\ b(x_0) & r(x_0) \end{pmatrix} : \begin{matrix} H \\ \oplus \\ \mathbb{C}^N \end{matrix} \longrightarrow \begin{matrix} H \\ \oplus \\ \mathbb{C}^M \end{matrix} \tag{7.214}$$

is an isomorphism for all x in a certain open neighbourhood $\Omega(x_0)$ of x_0. A trivial algebraic consideration tells us that a matrix like (7.214) is an isomorphism if (7.211) is surjective and

$$(b(x_0), r(x_0)) : \ker(a(x), k) \longrightarrow \mathbb{C}^M$$

is an isomorphism. In other words

$$(b(x_0), r(x_0)) : \tilde{J}_x \longrightarrow \mathbb{C}^M$$

is an isomorphism for all $x \in \Omega(x_0)$. Thus we get a trivialization

$$\tilde{J}\,|_{\Omega(x_0)} \cong \Omega(x_0) \times \mathbb{C}^M. \tag{7.215}$$

In view of the compactness of X there is a finite subset $\{x_0, \ldots, x_m\} \subset X$ such that $\{\Omega(x_0), \ldots, \Omega(x_m)\}$ is an open covering of X. From the corresponding trivialization $\tilde{J}\,|_{\Omega(x_l)} \cong \Omega(x_l) \times \mathbb{C}^M$, $l = 0, \ldots, m$, we can easily construct the transition homeomorphisms

$$(\Omega(x_l) \cap \Omega(x_i)) \times \mathbb{C}^M \longrightarrow (\Omega(x_i) \cap \Omega(x_l)) \times \mathbb{C}^M \tag{7.216}$$

for the corresponding (continuous) vector bundle J over X, $J \cong \tilde{J}$.

Remark 2. There is an evident extension of the preceding discussion including Proposition 1 to the case of Fredholm families

$$a: \ X \longrightarrow \mathcal{L}(H_1, H_2)$$

with different Hilbert spaces H_1, H_2. It suffices to choose isomorphisms $d_i : H_i \to H$, $i = 1, 2$ and to pass to the Fredholm family $\tilde{a} = d_2 a d_1^{-1}$.

If $U \subseteq \mathbb{C}$ is an open set and E a Banach or Fréchet space, we denote the space of all holomorphic E-valued functions in U by $\mathcal{A}(U, E)$ as usual.

Remark 3. Let $X = \overline{U}$ with an open subset $U \subseteq \mathbb{C}$ and assume that the restriction of (7.210) to U is holomorphic. Then, under the conditions of Proposition 1, the above vector bundle J over X is holomorphic over U (i.e. the transition functions in (7.216) are holomorphic $M \times M$ matrix functions over U).

Theorem 4. *Let $U \subseteq \mathbb{C}$ be open and*

$$a(z) \in \mathcal{A}(U, \mathcal{L}(H_1, H_2))$$

be an operator function between Hilbert spaces H_1, H_2. Assume that

$$a(z): \ H_1 \longrightarrow H_2 \tag{7.217}$$

is a Fredholm operator for every $z \in U$ and let (7.217) be an isomorphism for at least one $z = \tilde{z}$ in every connected component of U. Then there is a countable subset $D = \bigcup_{j \in \mathbb{Z}_+} \{d_j\} \subset U$ with $D \cap K$ finite for every compact subset $K \subset\subset U$, such that (7.217) is an isomorphism for all $z \in U \setminus D$. The inverse $a^{-1}(z) \in \mathcal{A}(U \setminus D, \mathcal{L}(H_2, H_1))$ has an extension to a meromorphic operator function on U with poles at $d_j \in D$ of multiplicities $m_j + 1$, $j \in \mathbb{Z}_+$, and the Laurent coefficients at $(z - d_j)^{-(k+1)}$, $0 \le k \le m_j$, are finite-dimensional operators.

Proof. Without loss of generality we may assume $H = H_1 = H_2$ (cf. Remark 2). Furthermore it obviously suffices to consider a connected open set U with compact closure. The operator family $a(z)$ can be regarded as a homomorphism of Hilbert bundles

$$a: \ X \times H \longrightarrow X \times H,$$

$X = \overline{U}$. From the assumptions it follows that ind $a(z) = 0$ for all $z \in X$. To simplify notation we shall say that a function $h(z)$ is holomorphic on a compact subset K of U if it can be extended to a holomorphic function in an open neighbourhood of K. Applying the constructions of the proof of Proposition 1, we find for every compact subset K of U a finite-dimensional subspace $V \subset H$, such that

$$(1-p)a(z): \ K \times H \longrightarrow K \times V^{\perp}$$

is surjective for all $z \in K$. Here $p: \ H \to V$ is the orthogonal projection, and $V^{\perp} = \ker p$. Let $W_z = \ker(1-p)a(z)$ and $q(z): \ H \to W_z$ be the orthogonal projection, $W_z^{\perp} = \ker q(z)$. The families of subspaces W_z and $W_z^{\perp}$ constitute holomorphic vector bundles W and $W^{\perp}$, respectively, over K, where $K \times H = W^{\perp} \oplus W$. The operator function $a(z)$ can be identified with a morphism

$$A = \begin{pmatrix} E & C \\ B & R \end{pmatrix} : \quad \begin{matrix} W^{\perp} \\ \oplus \\ W \end{matrix} \quad \longrightarrow \quad \begin{matrix} K \times V^{\perp} \\ \oplus \\ K \times V \end{matrix} \quad .$$

The fibre dimension of W coincides with $M = \dim V$.

$$E: \ W^{\perp} \longrightarrow K \times V^{\perp}$$

is an isomorphism and E^{-1} is holomorphic. Now

$$L_1 \ = \ \begin{pmatrix} 1 & 0 \\ -BE^{-1} & 1 \end{pmatrix} : \quad \begin{matrix} K \times V^{\perp} \\ \oplus \\ K \times V \end{matrix} \quad \longrightarrow \quad \begin{matrix} K \times V^{\perp} \\ \oplus \\ K \times V \end{matrix} \quad ,$$

$$L_2 \ = \ \begin{pmatrix} 1 & -E^{-1}C \\ 0 & 1 \end{pmatrix} : \quad \begin{matrix} W^{\perp} \\ \oplus \\ W \end{matrix} \quad \longrightarrow \quad \begin{matrix} W^{\perp} \\ \oplus \\ W \end{matrix}$$

are holomorphic isomorphisms, and we have

$$F = \begin{pmatrix} E & 0 \\ 0 & R - BE^{-1}C \end{pmatrix} = L_1 A L_2. \tag{7.218}$$

Thus A is an isomorphism at precisely those points $z \in K$ where the finite-dimensional morphism

$$R - BE^{-1}C: \ W \longrightarrow K \times V$$

induces an isomorphism $W_z \to V$. Our consideration is local in nature. So we may think of small disks when we talk about K. The bundle W is then trivial over K. Choose a holomorphic trivialization $T : W \to K \times \mathbb{C}^M$ and an isomorphism $S : K \times V \to K \times \mathbb{C}^M$, the latter being induced by $V \cong \mathbb{C}^M$. Then

$$f(z) := S(R - BE^{-1}C)T^{-1} : \; K \times \mathbb{C}^M \longrightarrow K \times \mathbb{C}^M$$

can be regarded as a holomorphic $M \times M$ matrix function. It follows altogether that $a(z)$ is an isomorphism if $\det f(z) \neq 0$. By assumption there is at least one $\tilde{z}$ where $\det f(\tilde{z}) \neq 0$. Since $\det f(z)$ is holomorphic, it vanishes at most over a discrete subset D with the asserted properties. Thus the first part of Theorem 4 is proved.

The inverse $a^{-1}(z)$ can be obtained by $A^{-1} = L_2 F^{-1} L_1$ (cf. (7.218)), i.e.

$$A^{-1} = L_2 \begin{pmatrix} 1 & 0 \\ 0 & T^{-1} \end{pmatrix} \begin{pmatrix} E^{-1} & 0 \\ 0 & f^{-1} \end{pmatrix} \begin{pmatrix} 1 & 0 \\ 0 & S \end{pmatrix} L_1. \qquad (7.219)$$

The $M \times M$-matrix $f^{-1}(z)$ on $U \setminus D$ obviously has an extension to a meromorphic matrix function on U. Let $d \in D$ be a fixed pole of multiplicity $m+1$. Then there is an open neighbourhood Ω of d with $f^{-1}(z) = h(z) + l(z)$, $h(z) \in \mathcal{A}(\Omega) \otimes \mathbb{C}^M \otimes \mathbb{C}^M$,

$$l(z) = \sum_{k=0}^{m} \gamma_k (z - d)^{-(k+1)}, \quad \gamma_k \in \mathbb{C}^M \otimes \mathbb{C}^M.$$

Thus $a^{-1}(z)$ has an extension to a meromorphic operator function, and

$$L_2 \begin{pmatrix} 1 & 0 \\ 0 & T^{-1} \end{pmatrix} \begin{pmatrix} 0 & 0 \\ 0 & \gamma_k \end{pmatrix} \begin{pmatrix} 1 & 0 \\ 0 & S \end{pmatrix} L_1$$

is the coefficient at $(z - d)^{-(k+1)}$ of the Laurent expansion of $a^{-1}(z)$ at d. It is of finite dimension, since γ_k is an $M \times M$-matrix. $\qquad \square$

We will also need an extension of Theorem (4) to the case of meromorphic operator functions.

Lemma 5. *Let $U = \{z \in \mathbb{C} : |z| < c\}$ for some $c > 0$, X a Banach space, and let $a_1, \ldots, a_N \in \mathcal{L}(X)$ be operators with $\dim \operatorname{im} a_k < \infty$ for all $k = 1, \ldots, N$. Let $K_0 \subset X$ be a vector subspace with $\dim X/K_0 < \infty$ and assume that for a given operator function $h(z) \in \mathcal{A}(U, \mathcal{L}(X))$ we have $h(z)u = 0$ for all $u \in K_0$. Moreover, let $f(z) := 1 + h(z) + \sum_{k=1}^{N} a_k z^{-k}$ be invertible for some $\tilde{z} \in U, \tilde{z} \neq 0$. Then there exists a $\delta > 0$ such that $f(z)$ is invertible for all $0 < |z| < \delta$.*

Proof. $K_1 := \bigcap_k \ker a_k$ is a closed subspace of X and we have $\dim(X/K_1) \leq \sum_{k=1}^{N} \dim(X/\ker a_k) < \infty$. If we set $K = K_0 \cap K_1$, we obtain $\dim X/K < \infty$. Thus there is a finite-dimensional subspace $L \subset X$ with $X = K \oplus L, K \cap L = \{0\}$.

Let $p : X \to L$ be a projection to L along K. Then $f(z)$ can be written as a block matrix

$$f(z) = \begin{pmatrix} D & F_1 \\ E & F_2 \end{pmatrix} : \begin{matrix} K \\ \oplus \\ L \end{matrix} \to \begin{matrix} K \\ \oplus \\ L \end{matrix}$$

for $D = (1-p)f(1-p), E = pf(1-p)$. For $u \in K$ we have $f(z)u = u$, i.e.

$$f(z) = \begin{pmatrix} I & F_1 \\ 0 & F_2 \end{pmatrix} : X \to X.$$

Hence $f(z)$ is invertible if and only if $F_2 : L \to L$ is invertible.

$$F_2(z) = I_L + ph(z)p + \sum_{k=1}^{N} pa_k pz^{-k}$$

can be regarded as a meromorphic $\dim L \times \dim L$-matrix. There exists a $\delta > 0$ with $\det(z^N F_2(z)) \neq 0$ for all $0 < |z| < \delta$ (otherwise we find a sequence $\{z_n\}$ with $z_n \to 0$ and $\det(z_n^N F_2(z_n)) = 0$; by virtue of the holomorphy this would imply $\det(z^N F_2(z)) = 0$ for all z, i.e. also $\det(F_2(z)) = 0$ for $z \neq 0$, which contradicts the invertibility of $f(\tilde{z})$). Thus $\det F_2(z) \neq 0$ for all $0 < |z| < \delta$. $\square$

Theorem 6. *Let $U \subseteq \mathbb{C}$ be an open connected set, and $D_0 \subset U$ be a discrete subset (i.e. countable, with $D_0 \cap M$ finite for every compact subset $M \subset U$). Let H be a Hilbert space and $h(z) \in \mathcal{A}(U \setminus D_0, \mathcal{L}(H))$ be an operator function that extends to a meromorphic $h(z)$ in U with finite-dimensional Laurent coefficients at all $z \in D_0$. Assume that $h(z)$ takes values in the Fredholm operators for all $z \in U \setminus D_0$ and that $h(\tilde{z})$ is an isomorphism for a $\tilde{z} \in U \setminus D_0$. Then there is a discrete subset $D \subset U \setminus D_0$ such that $h(z)$ is an isomorphism for all $z \in U \setminus (D_0 \cup D)$ and $D_0 \cup D$ is discrete, again.*

Proof. In view of Theorem 4 the only point to prove is that $D_0 \cup D$ is discrete. Let $w \in D_0$, i.e. a pole of $h(z)$. Then there exist an $M \in \mathbb{N}$ and finite-dimensional operators F_j such that $h(z) = h_0(z) + \sum_{j=1}^{M} F_j(z-w)^{-j}$ for some $h_0(z)$ which is holomorphic near w. Moreover, there is a $\tilde{z}$ near w such that $h(\tilde{z})$ is an isomorphism. Then, since ind $h(\tilde{z}) = $ ind $h_0(\tilde{z}) = 0$, there is a finite-dimensional operator F_0 such that $h_0(\tilde{z}) - F_0$ is an isomorphism. Set $h_1(z) = h_0(z) - F_0$. Then $h_1(z)$ is holomorphic in a neighbourhood of w and invertible at $\tilde{z}$, and we find a $\delta_0 > 0$ such that $h_1^{-1}(z)$ exists on $V_{\delta_0} := \{z : |w - z| < \delta_0\} \setminus \{w\}$. Set $r(z) = \sum_{j=0}^{M} F_j(z-w)^{-j}$. Then $h(z) = h_1(z)(I + h_1^{-1}(z)r(z))$ on V_{δ_0}. Moreover, there is an operator function $l(z)$ which is holomorphic near w such that $g(z) = h_1^{-1}(z)r(z) = l(z) + \sum_{k=1}^{N} a_k(z-w)^{-k}$ holds with finite-dimensional operators $a_1, \ldots, a_N$. Then $g(z)u = 0$ for all $u \in N_0 := \bigcap_{j=0}^{M} \ker F_j$. Put $N_1 = \bigcap_{k=1}^{N} \ker a_k$. Then $l(z)u = 0$ for all $u \in N_0 \cap N_1$. In view of Lemma 5, the operator $1 + g(z)$ is invertible in $0 < |z - w| < \delta_1$ for some $\delta_1 > 0$, and hence $h(z)$ is invertible on V_δ for $\delta = \min\{\delta_0, \delta_1\}$. Thus D has no accumulation point.

$\square$

Let us return once again to the discussions in the proof of Proposition 1. First it can easily be verified that when X is a closed compact C^∞ manifold and

$$a \in C^\infty(X, \mathcal{L}(H_1, H_2)) \tag{7.220}$$

with Hilbert spaces H_1, H_2, then

$$J \in \mathrm{Vect}(X).$$

Here $\mathrm{Vect}(X)$ is as usual the set (of isomorphy classes) of finite-dimensional complex C^∞ vector bundles over X.

For brevity we will adopt the notation $\mathrm{Vect}(X)$ also in the case when X is only a compact topological space. Then $\mathrm{Vect}(X)$ denotes the set of (isomorphy classes of) finite-dimensional complex C^0-vector bundles over X. For a closed compact C^∞ manifold X the notation coincides. Denote by $K(X)$ the K-group over X, i.e. the set of all equivalence classes of pairs $(G, J) \in \mathrm{Vect}(X) \times \mathrm{Vect}(X)$, where

$$(G, J) \sim (\widetilde{G}, \tilde{J}) \quad \Longleftrightarrow \quad \text{there are } F, \widetilde{F} \in \mathrm{Vect}(X) \text{ with}$$
$$G \oplus F \cong \widetilde{G} \oplus \widetilde{F}, J \oplus F \cong \tilde{J} \oplus \widetilde{F}$$

(cf. also [At], [P]). Denote by $[G]$–$[J]$ the equivalence class represented by the pair (G, J).

The above construction has provided a pair (G, J) for every Fredholm function (7.219), with J being the trivial bundle $J = X \times \mathbb{C}^N$ and $G = \tilde{J}$. We set

$$\mathrm{ind}_X a = [G]\text{–}[J] \in K(X), \tag{7.221}$$

called the index element of (7.219).

Chapter 8

Pseudo-differential operators on manifolds with conical singularities

8.1 The cone algebra with asymptotics

8.1.1 Weighted Sobolev spaces with asymptotics and Green operators

We now return to the weighted Sobolev spaces

$$\mathcal{H}^{s,\gamma}(\mathbb{B}), \qquad \mathcal{K}^{s,\gamma}(X^\wedge)$$

of 7.1.2 Definition 10 and (7.72). We shall introduce Fréchet subspaces with conormal asymptotics. First, by definition, there is a diffeomorphism κ of a collar neighbourhood V of $\partial\mathbb{B}$ to $[0,c) \times X$, $X \cong \partial\mathbb{B}$, $c > 0$, that is kept fixed. For every cut-off function $\omega(r)$ supported by $[0,c)$ we get a cut-off function $\tilde{\omega} = \kappa^*\omega$ on $\mathbb{B}$. Then, by construction,

$$\tilde{\omega}\mathcal{H}^{s,\gamma}(\mathbb{B}) \cong \omega\mathcal{K}^{s,\gamma}(X^\wedge). \tag{8.1}$$

Since the asymptotics refer to a neighbourhood of the (stretched) conical singularity, locally being reached by $r \to 0$, we shall mainly look at the case $X^\wedge$. Then, via (8.1), we can easily pass to $\mathbb{B}$.

The asymptotics will be of the form

$$u(r,x) \sim \sum_j \sum_{k=0}^{m_j} c_{jk}(x) r^{-p_j} \ln^k r \tag{8.2}$$

for $r \to 0$, with coefficients $c_{jk} \in C^\infty(X)$ and

$$p_j \in \mathbb{C}, \qquad \mathrm{Re}\, p_j < \frac{n+1}{2} - \gamma \tag{8.3}$$

where every strip

$$\{\frac{n+1}{2} - \gamma + \vartheta < \mathrm{Re}\, z < \frac{n+1}{2} - \gamma\}, \quad \vartheta < 0, \tag{8.4}$$

only contains finitely many points p_j. We shall see that the solutions of elliptic equations on manifolds with conical singularities have such asymptotic expansions. In that case the first sum in (8.2) runs over $j \in \mathbb{Z}_+$.

For convenience we shall also allow finite asymptotics, where (8.4) is fixed and the sum over j in (8.2) is finite. First we set $\Theta = (\vartheta, 0]$ for $\vartheta < 0$, and define

$$\mathcal{K}_{\Theta}^{s,\gamma}(X^{\wedge}) = \bigcap_{\varepsilon > 0} \mathcal{K}^{s,\gamma-\vartheta-\varepsilon}(X^{\wedge}), \tag{8.5}$$

endowed with the Fréchet structure of the projective limit. The elements of (8.5) may be regarded as being flat of order $-\vartheta - 0$, relative to the weight γ.

We will say that there is given an *asymptotic type*

$$P = \{(p_j, m_j, L_j)\}_{j=0,\dots,N}, \quad N = N(P), \tag{8.6}$$

associated with the *weight data* (γ, Θ), $\Theta = (\vartheta, 0]$, if all p_j belong to the *weight strip* (8.4), $m_j \in \mathbb{Z}_+$, and if $L_j \subset C^{\infty}(X)$ is a sequence of finite-dimensional subspaces, $j = 0, \dots, N$. In an analogous manner we define an asymptotic type

$$P = \{(p_j, m_j, L_j)\}_{j \in \mathbb{Z}_+} \tag{8.7}$$

to $(\gamma, (-\infty, 0])$ with given p_j satisfying (8.3), Re $p_j \to -\infty$ as $j \to \infty$, and finite-dimensional subspaces $L_j \subset C^{\infty}(X)$, $j \in \mathbb{Z}_+$. Set

$$\pi_{\mathbb{C}} P = \{p_j\}_{j=0,\dots,N}$$

with $N = N(P)$ for (8.6) and $N = \infty$ for (8.7).

Definition 1. Let P be an asymptotic type associated with the weight data (γ, Θ), $\Theta = (\vartheta, 0]$, $-\infty < \vartheta < 0$. Then we define

$$\mathcal{K}_P^{s,\gamma}(X^{\wedge}) := \{u \in \mathcal{K}^{s,\gamma}(X^{\wedge}) :$$
$$\text{there are } c_{jk} \in L_j, \quad 0 \le k \le m_j, \quad j = 0, \dots, N,$$
$$\text{with} \ \ u - \omega \sum_{j=0}^{N} \sum_{k=0}^{m_j} c_{jk} r^{-p_j} \ln^k r \in \mathcal{K}_{\Theta}^{s,\gamma}(X^{\wedge})\}. \tag{8.8}$$

Here ω is an arbitrary fixed cut-off function.

This is a correct definition because of

$$\phi(r) \sum_{j=0}^{N} \sum_{k=0}^{m_j} c_{jk} r^{-p_j} \ln^k r \in \mathcal{K}^{\infty,\infty}(X^{\wedge})$$

for arbitrary $\phi \in C_0^{\infty}(\mathbb{R}_+)$, and $\mathcal{K}^{\infty,\infty}(X^{\wedge}) \subset \mathcal{K}_{\Theta}^{s,\gamma}(X^{\wedge})$ for all $s, \gamma \in \mathbb{R}$ and all ϑ. Incidentally we shall also write

$$\mathcal{K}_{\Theta}^{s,\gamma}(X^{\wedge}) = \mathcal{K}^{s,\infty}(X^{\wedge}) \ \text{ for } \ \Theta = (-\infty, 0].$$

Note that $u(r,x) = c(x) r^{-p} \ln^k r \, \omega(r)$ for $c(x) \in C^\infty(X)$ belongs to $\mathcal{K}^{s,\varrho}(X^\wedge)$ for every $p \in \mathbb{C}$ with $\mathrm{Re}\, p < \frac{n+1}{2} - \varrho$ and all $s \in \mathbb{R}$. However, $u(r,x) \notin \mathcal{K}^{s,\varrho}(X^\wedge)$ once $\mathrm{Re}\, p = \frac{n+1}{2} - \varrho$.

Remark 2. Set

$$\mathcal{E}_P(X^\wedge) = \omega \left\{ \sum_{j=0}^{N} \sum_{k=0}^{m_j} c_{jk} r^{-p_j} \ln^k r \; : \; c_{jk} \in L_j, \quad 0 \le k \le m_j, \quad j = 0, \ldots, N \right\} \tag{8.9}$$

with an arbitrary fixed cut-off function $\omega(r)$. Then

$$\mathcal{K}_P^{s,\gamma}(X^\wedge) = \mathcal{K}_\Theta^{s,\gamma}(X^\wedge) + \mathcal{E}_P(X^\wedge) \tag{8.10}$$

is a direct decomposition.

Since $L := \dim \mathcal{E}_P(X^\wedge)$ is finite, there is an isomorphism $\mathcal{E}_P(X^\wedge) \cong \mathbb{C}^L$. This gives us a natural topology in the space (8.9). Hence $\mathcal{K}_P^{s,\gamma}(X^\wedge)$ can be endowed with the topology of the direct sum which is Fréchet and independent of the choice of ω.

Let us also introduce a Fréchet structure in $\mathcal{K}_P^{s,\gamma}(X^\wedge)$ for (8.7). To this end we define for every $l \in \mathbb{N}$

$$P_l = \left\{ (p,m,L) \in P : \; \frac{n+1}{2} - \gamma - l < \mathrm{Re}\; p < \frac{n+1}{2} - \gamma \right\}$$

which is an asymptotic type associated with (γ, Θ_l), $\Theta_l = (-l, 0]$. It is clear that then $P_l \subseteq P_{l+1}$ and there are continuous embeddings

$$\mathcal{K}_{P_{l+1}}^{s,\gamma}(X^\wedge) \hookrightarrow \mathcal{K}_{P_l}^{s,\gamma}(X^\wedge)$$

for all l. We then define

$$\mathcal{K}_P^{s,\gamma}(X^\wedge) = \bigcap_{l \in \mathbb{N}} \mathcal{K}_{P_l}^{s,\gamma}(X^\wedge), \tag{8.11}$$

with the Fréchet topology of the projective limit.

Exercise 3. $\mathcal{K}_\Theta^{s,\gamma}(X^\wedge)$ is the subspace of all $u \in \mathcal{K}^{s,\gamma}(X^\wedge)$ such that for an arbitrary cut-off function $\omega(r)$ the weighted Mellin transform (cf. (7.42))

$$h(z) = (\mathcal{M}_{\gamma', r \to z} \omega u)(z), \qquad \gamma' := \gamma - \frac{n}{2}$$

(interpreted as a transform on vector-valued distributions) has the following properties:

(i) $h(z)$, first given on $\Gamma_{\frac{n+1}{2} - \gamma}$, has an extension to an element in $\mathcal{A}(G(\gamma, \vartheta), H^s(X))$ for

$$G(\gamma, \vartheta) = \left\{ z : \; \frac{n+1}{2} - \gamma + \vartheta < \mathrm{Re}\; z \right\},$$

(ii) if $h_\beta(\varrho) := h(\beta + i\varrho)$, then we have for every coordinate neighbourhood U on X with local coordinates $x \in \mathbb{R}^n$ and every $\phi(x) \in C_0^\infty(U)$,

$$(\mathcal{F}_{\varrho \to t}^{-1} h_\beta)(t)\phi(x) \in H^s(\mathbb{R}_{t,x}^{n+1}) \tag{8.12}$$

uniformly in $c \le \beta \le c'$ for every $\frac{n+1}{2} - \gamma + \vartheta < c < c' < \infty$. Here $\mathcal{F} = \mathcal{F}_{t \to \varrho}$ is the one-dimensional Fourier transform.

Hint: Employ the higher-dimensional analogues of relations like (7.49), generalized to arbitrary weights.

Exercise 4. Let P be an asymptotic type to (γ, Θ) with finite $\Theta = (\vartheta, 0]$. Then

$$M_{\gamma', r \to z} \mathcal{E}_P(X^\wedge) \subset \mathcal{A}(G(\gamma, \vartheta) \setminus \pi_\mathbb{C} P, C^\infty(X)) \tag{8.13}$$

for $\gamma' = \gamma - \frac{n}{2}$. The elements $h(z)$ on the left of (8.13) are meromorphic in $G(\gamma, \vartheta)$ with poles at all $p_j \in \pi_\mathbb{C} P$ of multiplicities $m_j + 1$ and Laurent coefficients at $(z - p_j)^{-(k+1)}$ in L_j, $0 \le k \le m_j$. Further

$$\chi(z) h(z) \mid_{\Gamma_\beta} \in \mathcal{S}(\Gamma_\beta, C^\infty(X)) \tag{8.14}$$

for every $\pi_\mathbb{C} P$-excision function χ (i.e. $\chi(z) \in C^\infty(\mathbb{C})$, $\chi(z) = 0$ for dist $(z, \pi_\mathbb{C} P) < \varepsilon_0$, $\chi(z) = 1$ for dist $(z, \pi_\mathbb{C} P) > \varepsilon_1$ with certain $0 < \varepsilon_0 < \varepsilon_1 < \infty$), and (8.14) holds uniformly in $c \le \beta \le c'$ for every $\frac{n+1}{2} - \gamma - \vartheta < c < c' < \infty$.

From (8.8) and (8.11), respectively, we can pass to further variants of spaces with asymptotics. We set

$$\mathcal{S}_P^\gamma(X^\wedge) = \{u \in \mathcal{K}_P^{\infty,\gamma}(X^\wedge) : (1 - \omega)u \in \mathcal{S}(\mathbb{R}_+, C^\infty(X))\} \tag{8.15}$$

for any fixed cut-off function $\omega(r)$ and $\mathcal{S}(\mathbb{R}_+, C^\infty(X)) := \mathcal{S}(\mathbb{R}, C^\infty(X))|_{\mathbb{R}_+}$. The space (8.15) is Fréchet in a natural way (the topology is nuclear). Further, we introduce

$$\mathcal{H}_P^{s,\gamma}(\mathbb{B}) \tag{8.16}$$

as the subspace of all $u \in \mathcal{H}^{s,\gamma}(\mathbb{B})$ for which $\tilde{\omega} u$ corresponds via (8.1) to an element in $\mathcal{K}_P^{s,\gamma}(X^\wedge)$. The space (8.16) is also Fréchet in a natural way.

Exercise 5. Let $\chi : X^\wedge \to X^\wedge$ be a diffeomorphism which is the restriction of a diffeomorphism $\tilde{\chi} : \mathbb{R} \times X \to \mathbb{R} \times X$ to $X^\wedge$ and let $\chi(r, x) = (r, x)$ for all $r \ge c$ with some constant $c > 0$. Then the pull-back $\chi^* : C^\infty(X^\wedge) \to C^\infty(X^\wedge)$ has an extension to an isomorphism

$$\chi^* : \mathcal{K}^{s,\gamma}(X^\wedge) \longrightarrow \mathcal{K}^{s,\gamma}(X^\wedge) \tag{8.17}$$

for all $s, \gamma \in \mathbb{R}$. Furthermore for every asymptotic type P associated with (γ, Θ) for any $\Theta = (\vartheta, 0]$, $-\infty \le \vartheta < 0$, there is an asymptotic type Q associated with (γ, Θ), such that (8.17) induces by restriction an isomorphism

$$\chi^* : \mathcal{K}_P^{s,\gamma}(X^\wedge) \longrightarrow \mathcal{K}_Q^{s,\gamma}(X^\wedge).$$

An analogous result holds for diffeomorphisms $\chi : \mathbb{B} \to \mathbb{B}$ with respect to the $\mathcal{H}^{s,\gamma}(\mathbb{B})$-spaces.

Remark 6. The role of the spaces $\mathcal{H}_P^{\infty,\gamma}(\mathbb{B})$ is to express an adequate notion of smoothness on (stretched) manifolds $\mathbb{B}$ with conical singularities. Adequate means that the kernels of elliptic operators A of Fuchs type on $\mathbb{B}$ belong to such spaces. General asymptotics P are necessary even in the simplest cases, when $\mathbb{B}$ is the interval $[0,1]$ with 0 and 1 as conical singularities. Below we shall see that the spaces $\mathcal{H}_P^{s,\gamma}(\mathbb{B})$ are involved in the elliptic regularity on $\mathbb{B}$.

The algebras of pseudo-differential operators on a manifold with conical singularities will contain corresponding smoothing operators. According to Remark 6 the smoothing property has to reflect the asymptotics of smooth functions occurring as the C^∞ solutions of elliptic equations. The corresponding operators will be called Green ones. We shall fix a scalar product in the spaces $\mathcal{H}^0(\mathbb{B}) = \mathcal{H}^{0,0}(\mathbb{B})$ and $\mathcal{K}^0(X^\wedge) = \mathcal{K}^{0,0}(X^\wedge)$, respectively.

First consider $\mathcal{K}^0(X^\wedge)$ with a fixed identification

$$\mathcal{K}^0(X^\wedge) = r^{-\frac{n}{2}} L_2(\mathbb{R}_+ \times X)$$

where the space $L_2(\mathbb{R}_+ \times X)$ refers to the scalar product induced by $dr\,dx$ with a density dx on X based on a fixed Riemannian metric. This gives us a scalar product in $\mathcal{K}^0(X^\wedge)$. The nondegenerate sesquilinear pairing

$$(\cdot,\cdot)_{\mathcal{K}^0(X^\wedge)} : \ C_0^\infty(X^\wedge) \times C_0^\infty(X^\wedge) \longrightarrow \mathbb{C}$$

(cf. 7.1.2 Exercise 20) has an extension to a nondegenerate sesquilinear pairing

$$(\cdot,\cdot) : \ \mathcal{K}^{s,\gamma}(X^\wedge) \times \mathcal{K}^{-s,-\gamma}(X^\wedge) \longrightarrow \mathbb{C} \tag{8.18}$$

for all $s,\gamma \in \mathbb{R}$. Then, if

$$A : \ \mathcal{K}^{s,\gamma}(X^\wedge) \longrightarrow \mathcal{K}^{s-m,\delta}(X^\wedge)$$

is continuous for all $s \in \mathbb{R}$ with given $m,\gamma,\delta \in \mathbb{R}$, we can pass to the formal adjoint A^* with respect to (8.18) which induces continuous operators

$$A^* : \ \mathcal{K}^{s,-\delta}(X^\wedge) \longrightarrow \mathcal{K}^{s-m,-\gamma}(X^\wedge)$$

for all $s \in \mathbb{R}$.

Definition 7. Let P and Q be asymptotic types associated with the weight data (δ,Θ) and $(-\gamma,\Theta)$, respectively, for fixed $\gamma,\delta \in \mathbb{R}$ and a weight strip $\Theta = (\vartheta,0]$, $-\infty \leq \vartheta < 0$. Then an operator

$$G \in \bigcap_{s \in \mathbb{R}} \mathcal{L}(\mathcal{K}^{s,\gamma}(X^\wedge), \mathcal{K}^{\infty,\delta}(X^\wedge))$$

is called Green (on $X^\wedge$) with the asymptotic types P,Q if G induces continuous operators

$$G : \ \mathcal{K}^{s,\gamma}(X^\wedge) \longrightarrow \mathcal{S}_P^\delta(X^\wedge), \tag{8.19}$$
$$G^* : \ \mathcal{K}^{s,-\delta}(X^\wedge) \longrightarrow \mathcal{S}_Q^{-\gamma}(X^\wedge) \tag{8.20}$$

for all $s \in \mathbb{R}$. Analogously an operator

$$G \in \bigcap_{s \in \mathbb{R}} \mathcal{L}(\mathcal{H}^{s,\gamma}(\mathbb{B}), \mathcal{H}^{\infty,\delta}(\mathbb{B}))$$

is called Green (on $\mathbb{B}$) with the asymptotic types P, Q if G induces continuous operators

$$G : \mathcal{H}^{s,\gamma}(\mathbb{B}) \longrightarrow \mathcal{H}_P^{\infty,\delta}(\mathbb{B}), \tag{8.21}$$

$$G^* : \mathcal{H}^{s,-\delta}(\mathbb{B}) \longrightarrow \mathcal{H}_Q^{\infty,-\gamma}(\mathbb{B}) \tag{8.22}$$

for all $s \in \mathbb{R}$. Here the formal adjoint refers to the pairing mentioned above in Section 7.1.2.

The notation "Green" is borrowed from Boutet de Monvel's algebra, where analogous objects on the level of boundary symbols on $\mathbb{R}_+$ with Taylor asymptotics occur as parts of the structure of Green functions of elliptic boundary value problems.

$$C_G(X^\wedge, \underline{g}) \quad \text{and} \quad C_G(\mathbb{B}, \underline{g}) \tag{8.23}$$

will denote the spaces of all Green operators on $X^\wedge$ and $\mathbb{B}$, respectively, for arbitrary asymptotic types P, Q. Here

$$\underline{g} = (\gamma, \delta, \Theta)$$

is the tuple of involved weight data.

Proposition 8. *Every $G \in C_G(X^\wedge, \underline{g})$ induces compact operators*

$$G : \mathcal{K}^{s,\gamma}(X^\wedge) \longrightarrow \mathcal{K}^{s',\delta}(X^\wedge),$$

and every $G \in C_G(\mathbb{B}, \underline{g})$ induces compact operators

$$G : \mathcal{H}^{s,\gamma}(\mathbb{B}) \longrightarrow \mathcal{H}^{s',\delta}(\mathbb{B})$$

for all $s, s' \in \mathbb{R}$.

Proof. There are compact embeddings

$$(1 + r^2)^{-\frac{\nu}{2}} \mathcal{K}^{s+\varepsilon,\delta+\varepsilon}(X^\wedge) \longrightarrow \mathcal{K}^{s,\delta}(X^\wedge) \tag{8.24}$$

for every $\varepsilon, \nu > 0$ and all $s, \delta \in \mathbb{R}$. Clearly the space on the left of (8.24) consists of all products $(1 + r^2)^{-\frac{\nu}{2}} u(r, x)$ for $u(r, x) \in \mathcal{K}^{s+\varepsilon,\delta+\varepsilon}(X^\wedge)$, endowed with a canonical norm. For every $G \in C_G(X^\wedge)$ there is an $\varepsilon > 0$ such that

$$G : \mathcal{K}^{s,\gamma}(X^\wedge) \longrightarrow (1 + r^2)^{-\frac{\varepsilon}{2}} \mathcal{K}^{s+\varepsilon,\delta+\varepsilon}(X^\wedge)$$

is continuous. This shows the first assertion. The second one follows in an analogous manner. $\square$

Proposition 9. *Let $G \in C_G(X^\wedge, \underline{g})$ for $\underline{g} = (\gamma, \gamma, \Theta)$, and assume that*

$$1 + G: \ \mathcal{K}^{s,\gamma}(X^\wedge) \longrightarrow \mathcal{K}^{s,\gamma}(X^\wedge) \tag{8.25}$$

is invertible for an $s = s_0$. Then (8.25) is invertible for all $s \in \mathbb{R}$, and there is a $G_1 \in C_G(X^\wedge, \underline{g})$ with

$$(1 + G)^{-1} = 1 + G_1.$$

An analogous result holds for Green operators over $\mathbb{B}$.

Proof. Let us look, for instance, at $X^\wedge$. The arguments for $\mathbb{B}$ are analogous. Let $k^\gamma(r)$ be an arbitrary strictly positive C^∞ function on $\mathbb{R}_+$ with

$$k^\gamma(r) = \begin{cases} r^\gamma & \text{for} \quad 0 < r < c_0, \\ 1 & \text{for} \quad c_1 < r \le \infty, \end{cases}$$

with constants $0 < c_0 < c_1$. Then

$$k^{-\gamma} C_G(X^\wedge, \underline{g}) k^\gamma = C_G(X^\wedge, \underline{g}_0)$$

with $\underline{g}_0 = (0, 0, \Theta)$. Thus, it suffices to consider the case $\gamma = 0$. If $G \in C_G(X^\wedge, \underline{g}_0)$ is a Green operator, then the kernel and cokernel of $1+G$ in $\mathcal{K}^s(X^\wedge)$ will not depend on s. Thus $1 + G$ is invertible at the same time for all $s \in \mathbb{R}$ once it is invertible for $s = s_0$. Furthermore, we have $(1+G)^{-1} = 1+G_1$ with some $G_1 \in \mathcal{L}(\mathcal{K}^s(X^\wedge))$. It follows that $(1 + G)(1 + G_1) = 1 + G + G_1 + GG_1 = 1$, i.e. $G_1 = -G - GG_1$. Thus G_1 maps $\mathcal{K}^s(X^\wedge)$ to $\mathcal{S}_P^0(X^\wedge)$ for some P. The adjoint can be characterized analogously by looking at the adjoint of $(1 + G_1)(1 + G) = 1$. This shows that G_1 is of Green type. $\qquad\square$

Denote by

$$C_G(X^\wedge, \underline{g})_{P,Q} \tag{8.26}$$

the subspace of Green operators with fixed asymptotic types P, Q. Then $C_G(X^\wedge, \underline{g})_{P,Q}$ is a Fréchet space in a natural way if we use the canonical Fréchet topologies of

$$\bigcap_s \mathcal{L}(\mathcal{K}^{s,\gamma}(X^\wedge), \mathcal{S}_P^\delta(X^\wedge)), \quad \bigcap_s \mathcal{L}(\mathcal{K}^{s,-\delta}(X^\wedge), \mathcal{S}_Q^{-\gamma}(X^\wedge)).$$

The latter spaces just by definition characterize the Green operators on $X^\wedge$ of the given P, Q.

8.1.2 Smoothing Mellin operators

The full algebra of cone pseudo-differential operators, cf. Section 8.1.4 below, will contain a further class of smoothing operators, quite different from the Green operators. These are the smoothing Mellin operators. As we shall see, together with the Green operators they form an algebra with a symbolic structure.

We shall first consider the corresponding operators on $X^\wedge$. As in Section 8.1.1 it will then be easy to pass to $\mathbb{B}$. These operators on $X^\wedge$ are, in particular, elements of

$$\bigcap_{s \in \mathbb{R}} \mathcal{L}(\mathcal{K}^{s,\gamma}(X^\wedge), \mathcal{K}^{\infty,\gamma-m}(X^\wedge)) \tag{8.27}$$

for prescribed $\gamma, m \in \mathbb{R}$. The description of asymptotic properties will refer to a fixed weight strip Θ, which is assumed first to be finite, namely

$$\Theta = (-k, 0] \quad \text{with some} \ \ k \in \mathbb{N}.$$

In other words, our subclass of (8.27) will be associated with the weight data

$$g = (\gamma, \gamma - m, \Theta). \tag{8.28}$$

Definition 1. An *asymptotic type R of Mellin symbols* is a sequence

$$R = \{(q_j, m_j, N_j)\}_{j \in \mathbb{Z}} \tag{8.29}$$

with $q_j \in \mathbb{C}$, $m_j \in \mathbb{Z}_+$, and finite-dimensional subspaces N_j of $L^{-\infty}(X)$ of finite-dimensional operators, $j \in \mathbb{Z}$, and it is assumed that $\pi_\mathbb{C} R := \{q_j\}_{j \in \mathbb{Z}}$ intersects $\{c < \mathrm{Re}\ z < c'\}$ only in a finite set for every $c < c'$.

Now

$$M_R^{-\infty}(X) \tag{8.30}$$

will denote the space of all

$$h(z) \in \mathcal{A}(\mathbb{C} \setminus \pi_\mathbb{C} R, L^{-\infty}(X)) \tag{8.31}$$

which are meromorphic with poles at q_j of multiplicities $m_j + 1$ and Laurent coefficients at $(z - q_j)^{-(k+1)}$ belonging to N_j for $0 \le k \le m_j$, furthermore

$$(\chi h)(\beta + i\varrho) \in \mathcal{S}(\mathbb{R}_\varrho, L^{-\infty}(X)) \tag{8.32}$$

uniformly in $c \le \beta \le c'$ for every $c < c'$ and every $\pi_\mathbb{C} R$-excision function $\chi(z)$ (i.e. $\chi(z) \in C^\infty(\mathbb{C})$, $\chi(z) = 0$ for $\mathrm{dist}(z, \pi_\mathbb{C} R) < \varepsilon_0$, $\chi(z) = 1$ for $\mathrm{dist}(z, \pi_\mathbb{C} R) > \varepsilon_1$ for certain $0 < \varepsilon_0 < \varepsilon_1$). We will write

$$sg(h) = \pi_\mathbb{C} R \quad \text{once} \ \ h \in M_R^{-\infty}(X). \tag{8.33}$$

We set

$$(T^\beta h)(z) = h(z + \beta)$$

and

$$op_M^\gamma(h)u(r) = r^\gamma op_M(T^{-\gamma} h) r^{-\gamma} u. \tag{8.34}$$

Here $op_M(\cdot)$ is the Mellin action along $\mathbb{R}_+$ (cf. (7.52)) whereas the values of $h(z)$ act as operators along X with respect to the dependence of u on x. As usual we write $u = u(r)$ when the action along r is separately indicated, whereas the action in x is involved in the interpretation of h as an operator-valued Mellin symbol. In (8.34) it is always assumed that

$$sg(h) \cap \Gamma_{\frac{1}{2}-\gamma} = \emptyset.$$

Our smoothing Mellin operators will be defined as finite linear combinations of operators of the form

$$A_j = r^{-m+j} \omega op_M^{\alpha_j}(h_j)\tilde{\omega}, \tag{8.35}$$

$j \in \mathbb{Z}_+$, with arbitrary cut-off functions $\omega(r), \tilde{\omega}(r)$ and

$$h_j \in M_{R_j}^{-\infty}(X), \qquad \pi_{\mathbb{C}} R_j \cap \Gamma_{\frac{1}{2}-\alpha_j} = \emptyset$$

for certain asymptotic types R_j of Mellin symbols. In order to get continuous operators

$$A_j : \ \mathcal{K}^{s,\gamma}(X^\wedge) \longrightarrow \mathcal{K}^{\infty,\gamma-m}(X^\wedge), \tag{8.36}$$

first for $s = 0$, the weights α_j have to be chosen in such a way that

$$\tilde{\omega} r^{-\alpha_j} r^{\gamma-\frac{n}{2}} L_2(\mathbb{R}_+ \times X) \subset L_2(\mathbb{R}_+ \times X)$$

and

$$\omega r^j r^{\alpha_j} L_2(\mathbb{R}_+ \times X) \subset r^{\gamma-\frac{n}{2}} L_2(\mathbb{R}_+ \times X).$$

This requires the conditions $\gamma - \frac{n}{2} - \alpha_j \geq 0$, $j + \alpha_j \geq \gamma - \frac{n}{2}$, i.e.

$$\gamma - \frac{n}{2} - j \leq \alpha_j \leq \gamma - \frac{n}{2}. \tag{8.37}$$

Definition 2. We denote by

$$C_{M+G}(X^\wedge, \underline{g}) \ \ \text{for} \ \ \underline{g} = (\gamma, \gamma - m, \Theta) \tag{8.38}$$

with $\gamma, m \in \mathbb{R}$, $\Theta = (-k, 0]$, $k \in \mathbb{N}$, the space of all operators

$$A = M + G \ \ \text{with} \ \ M = \sum_{j=0}^{k-1} A_j$$

with arbitrary A_j of the form (8.35), the α_j satisfying (8.37), and Green operators $G \in C_G(X^\wedge, \underline{g})$. In an analogous manner we define

$$C_{M+G}(\mathbb{B}, \underline{g})$$

by applying a push-forward to M under a diffeomorphism $[0, c) \times X$ to a collar neighbourhood V of $\partial \mathbb{B} \cong X$ (cf. the notations from the beginning of 8.1.1) and taking $G \in C_G(\mathbb{B}, \underline{g})$. We set

$$\sigma_M^{m-j}(A)(z) = h_j(z), \quad j = 0, \dots, k-1, \tag{8.39}$$

called the conormal symbol of A of conormal order $m - j$.

It is clear that the C_{M+G}-classes for the infinite weight interval $\Theta = (-\infty, 0]$ may be defined by taking intersections over those for $\Theta_k = (-k, 0]$, $k \in \mathbb{N}$. For simplicity we will mainly discuss the case of finite weight intervals, though the basic results also hold for infinite Θ.

Theorem 3. *Every $A \in C_{M+G}(X^{\wedge}, \underline{g})$ induces continuous operators (8.36) for all $s \in \mathbb{R}$. Furthermore, for every asymptotic type P to (γ, Θ) there is an asymptotic type Q to $(\gamma - m, \Theta)$ such that*

$$A : \ \mathcal{K}_P^{s,\gamma}(X^{\wedge}) \longrightarrow \mathcal{K}_Q^{\infty,\gamma-m}(X^{\wedge})$$

is continuous for all $s \in \mathbb{R}$. An analogous result holds for $A \in C_{M+G}(\mathbb{B}, \underline{g})$ with the spaces $\mathcal{H}^{s,\gamma}(\mathbb{B})$ and $\mathcal{H}_P^{s,\gamma}(\mathbb{B})$, respectively.

Proof. The first assertion is obvious. It remains to consider the action of A on $\mathcal{K}_P^{s,\gamma}(X^{\wedge})$. Let us set $D_{\gamma} = \left\{ z \in \mathbb{C} : \ \frac{n+1}{2} - \gamma - k < \mathrm{Re}\ z < \frac{n+1}{2} - \gamma \right\}$ and denote by

$$\mathcal{A}_P^{s,\gamma}(X) \tag{8.40}$$

for an asymptotic type $P = \{(p_j, m_j, L_j)\}$ to (γ, Θ), $\Theta = (-k, 0]$, the space of all

$$f(z) \in \mathcal{A}(D_{\gamma} \setminus \pi_{\mathbb{C}} P, H^s(X))$$

which are meromorphic with poles at p_j of multiplicities $m_j + 1$ and Laurent coefficients at the powers $(z - p_j)^{-(k+1)}$ in L_j, $0 \le k \le m_j$, for all j, and such that for every $\pi_{\mathbb{C}} P$-excision function $\chi(z)$,

$$b^s(\varrho)\chi(\beta + i\varrho)f(\beta + i\varrho) \in L_2(\mathbb{R}_{\varrho} \times X)$$

for every real $\beta \in D_{\gamma}$, uniformly in $\frac{n+1}{2} - \gamma - k + \varepsilon \le \beta < \frac{n+1}{2} - \gamma$ for every $\varepsilon > 0$, with corresponding L_2-limit for $\beta \to \frac{n+1}{2} - \gamma$. Remember that $b^s(\varrho)$ was introduced in connection with 7.2.3 Definition 3. The spaces $\mathcal{A}_P^{s,\gamma}(X)$ are Fréchet in a natural way. Moreover, it is an immediate consequence of 8.1.1 Definition 1 that

$$\mathcal{M}\tilde{\omega} : \ \mathcal{K}_P^{s,\gamma}(X^{\wedge}) \longrightarrow \mathcal{A}_P^{s,\gamma}(X)$$

is continuous. Here $\mathcal{M}$ is the weighted Mellin transform for the line $\Gamma_{\frac{n+1}{2}-\gamma}$. The multiplication by any $h(z) \in M_R^{-\infty}(X)$ induces a continuous operator

$$\mathcal{A}_P^{s,\gamma}(X) \longrightarrow \mathcal{A}_Q^{\infty,\gamma}(X) \tag{8.41}$$

with a resulting asymptotic type Q. Now, again by 8.1.1 Definition 1, by applying the inverse Mellin transform to the image under (8.41) and the translation of weights, according to the power $-m + j$ in (8.35), we obtain continuous operators $A_j : \ \mathcal{K}_P^{s,\gamma}(X^{\wedge}) \longrightarrow \mathcal{K}_Q^{\infty,\gamma-m}(X^{\wedge})$. This yields corresponding continuous operators for A itself, with another resulting Q, since the Green operators have the asserted mapping properties anyway. The consideration for $\mathbb{B}$ is analogous. $\qquad\square$

Exercise 4. (8.36) induce compact operators

$$A_j : \mathcal{K}^{s,\gamma}(X^\wedge) \longrightarrow \mathcal{K}^{s,\gamma-m}(X^\wedge)$$

for all $j \geq 1$, $s \in \mathbb{R}$.

Theorem 5. *For every $k \in \mathbb{Z}$, $\gamma \in \mathbb{R}$, there exists an $f(z) \in M_T^{-\infty}(X)$ with some asymptotic type T and $\pi_\mathbb{C} T \cap \Gamma_{\frac{n+1}{2}-\gamma} = \emptyset$, such that for*

$$M(f) := \omega(r) op_M^{\gamma-\frac{n}{2}}(f)\omega(r)$$

with an arbitrary cut-off function ω, the operator

$$1 + M(f) : \mathcal{K}^{s,\gamma}(X^\wedge) \longrightarrow \mathcal{K}^{s,\gamma}(X^\wedge)$$

is Fredholm with $\mathrm{ind}(1 + M(f)) = k$ for all $s \in \mathbb{R}$.

A proof of this theorem for the case $X^\wedge = \mathbb{R}_+$ and $s = 0$, $\gamma = 0$ may be found in [Es]. An explicit construction of $f(z)$ is also given in [Sc5]. There it is also shown how to pass to arbitrary $X^\wedge$. Since kernels and cokernels are independent of s, cf. Theorem 3, we get the same for all s. A simple weight shift then gives the result for all γ.

Theorem 6. *Let $h(z) \in M_R^{-\infty}(X)$ and assume $sg(h) \cap \Gamma_{\frac{1}{2}-\gamma} = sg(h) \cap \Gamma_{\frac{1}{2}-\beta} = \emptyset$ for certain $\beta, \gamma \in \mathbb{R}$. Then, for arbitrary cut-off functions $\omega(r), \tilde{\omega}(r)$ we have*

$$\omega(r) op_M^\gamma(h)\tilde{\omega}(r) - \omega(r) op_M^\beta(h)\tilde{\omega}(r) \in C_G(X^\wedge, \underline{g}) \tag{8.42}$$

for $\underline{g} = \left(\max\left(\beta + \frac{n}{2}, \gamma + \frac{n}{2}\right), \min\left(\beta + \frac{n}{2}, \gamma + \frac{n}{2}\right), (-\infty, 0]\right)$. The operators in (8.42) are finite-dimensional.

Proof. For convenience, let us consider only $n = 0$. The general case is completely analogous and is left to the reader. Assume, for instance, $\gamma < \beta$. Then, for $u \in \mathcal{K}^{s,\beta}(\mathbb{R}_+)$ and $v := \tilde{\omega}u$ we obtain, if $G_{\gamma,\beta}$ denotes the operator on the left of (8.42),

$$G_{\gamma,\beta}u(r) = \omega\frac{1}{2\pi i}\{\int_{\Gamma_{\frac{1}{2}-\gamma}} r^{-z}h(z)(Mv)(z)dz - \int_{\Gamma_{\frac{1}{2}-\beta}} r^{-z}h(z)(Mv)(z)dz\}$$

$$= \omega\frac{1}{2\pi i}\int_{\Delta_{\gamma,\beta}} r^{-z}h(z)(Mv)(z)dz \tag{8.43}$$

with $\Delta_{\gamma,\beta} = \Gamma_{\frac{1}{2}-\gamma} \cup \Gamma_{\frac{1}{2}-\beta}$ in corresponding orientation. The function $Mv(z)$ is holomorphic in the strip $\{\frac{1}{2} - \beta < \mathrm{Re}\, z < \frac{1}{2} - \gamma\}$, and because of the Schwartz space behaviour of $h(z)$ for $|\mathrm{Im}\, z| \to \infty$ the functions under the integrals also decrease like Schwartz functions for $|\operatorname{Im} z| \to \infty$. By Cauchy's integral formula we may replace $\Delta_{\gamma,\beta}$ by a finite curve $C_{\gamma,\beta}$ in the strip referred to, surrounding the poles of the meromorphic function $h(z)$. If q_j are those poles of multiplicities n_j+1, for $j = 0, \ldots, N$, there exist coefficients $d_{jk} = d_{jk}(u)$, $0 \le k \le n_j$, $j = 0, \ldots, N$, such that for a cut-off function $\omega_0(r)$,

$$h(z)(Mv)(z) - \sum_{j=0}^{N} \sum_{k=0}^{n_j} d_{jk} M(\omega_0(r) r^{-q_j} \ln^k r)(z) \tag{8.44}$$

is holomorphic in the strip. The reason is that the Mellin transform of $\omega_0(r) \cdot$ $\cdot r^{-p} \ln^k r$ is meromorphic with pole at $z = p$ of multiplicity $k+1$. The coefficients d_{jk} are linear forms in u. Since (8.44) is holomorphic, we can replace $h(z)(Mv)(z)$ in (8.43) by the corresponding sum in (8.44). This shows that

$$G_{\gamma,\beta} u(r) = \omega_1 \sum_{j=0}^{N} \sum_{k=0}^{n_j} d_{jk} r^{-q_j} \ln^k r,$$

with $\omega_1 = \omega \omega_0$. Hence $G_{\gamma,\beta}$ is of finite dimension and has the mapping properties which are required in 8.1.1 Definition 7. Analogous arguments can be applied for the adjoints. $\square$

Exercise 7. If $j \ge k$ the operator (8.35) with the properties mentioned is Green with respect to the weight data (8.38), $\Theta = (-k, 0]$.

Remark 8. Let $f(z) \in M_O^m(X)$ and $\beta, \gamma \in \mathbb{R}$ be arbitrary. Then

$$\omega(r) \operatorname{op}_M^\gamma(f) \tilde{\omega}(r) - \omega(r) \operatorname{op}_M^\beta(f) \tilde{\omega}(r) = 0$$

on the space $\mathcal{K}^{s,\beta+\frac{n}{2}}(X^\wedge)$ (for $\gamma < \beta$).

In fact, we can first insert $u \in C_0^\infty(X^\wedge)$. Then the analogues of the integrals (8.43) converge, since $Mu(z)$ is strongly decreasing for $|\operatorname{Im} z| \to \infty$. The density of $C_0^\infty(X^\wedge)$ in the cone Sobolev spaces then shows that the operator vanishes on $\mathcal{K}^{s,\beta+\frac{n}{2}}(X^\wedge)$ identically.

If R is an asymptotic type for Mellin symbols we will set

$$M_P^m(X) = M_O^m(X) + M_P^{-\infty}(X), \tag{8.45}$$

the notation with obvious meaning. The space (8.45) is Fréchet in a natural way, cf. (7.140), and we have

$$C^\infty(\overline{\mathbb{R}}_+, M_P^m(X)) = C^\infty(\overline{\mathbb{R}}_+, M_O^m(X)) + C^\infty(\overline{\mathbb{R}}_+, M_P^{-\infty}(X))$$

(exercise!). An analogous identity holds with $\overline{\mathbb{R}}_+ \times \overline{\mathbb{R}}_+$ instead of $\overline{\mathbb{R}}_+$.

Exercise 9. Let $h(r, r', z) \in C^\infty(\overline{\mathbb{R}}_+ \times \overline{\mathbb{R}}_+, M_R^m(X))$ and let $\pi_{\mathbb{C}} R \cap \Gamma_{\frac{1}{2}-\gamma} = \pi_{\mathbb{C}} R \cap \Gamma_{\frac{1}{2}-\beta} = \emptyset$. Then (8.42) holds with the weight data $\underline{g}$ mentioned above.

(Hint: Employ Taylor expansions of h in r and r' near 0 and show first that $\omega op_M^\delta(h)\tilde{\omega}$ is Green with respect to any finite Θ once h vanishes in r or r' at 0 of sufficiently high order.)

Corollary 10. *Let $h(r, r', z) \in C^\infty(\overline{\mathbb{R}}_+ \times \overline{\mathbb{R}}_+, M_R^{-\infty}(X))$ and $\delta \in \mathbb{R}$, $\pi_\mathbb{C} R \cap \Gamma_{\frac{1}{2}-\delta} = \emptyset$. Then, if $\omega, \tilde{\omega}, \omega_0, \tilde{\omega}_0$ are arbitrary cut-off functions, we have*

$$\omega op_M^\delta(h)\tilde{\omega} - \omega_0 op_M^\delta(h)\tilde{\omega}_0 \in C_G(X^\wedge, \underline{g}) \quad \text{with} \quad \underline{g} = \left(\delta + \frac{n}{2}, \delta + \frac{n}{2}, (-\infty, 0]\right).$$

Another consequence of Remark 8 and Corollary 10 is the following:

Proposition 11. *The conormal symbols $\sigma_M^{m-j}(A), j = 0, \ldots, k - 1$, of an $A \in C_{M+G}(X^\wedge, \underline{g})$ for $\Theta = (-k, 0]$ are uniquely determined by A.*

Remark 12. The results of Theorem 6, Remark 8 and of the subsequent observations may be regarded as assertions on commutators of Mellin operators and powers of r. It follows, in particular, that

$$\omega(r)r^{-\beta} op_M^\gamma(T^\beta h)\tilde{\omega}(r) - \omega(r) op_M^\gamma(h) r^{-\beta}\tilde{\omega}(r) \tag{8.46}$$

is of Green type of the corresponding weights where $h(r,r',z) \in C^\infty(\overline{\mathbb{R}}_+ \times \overline{\mathbb{R}}_+, M_R^m(X))$ satisfies $sg(h) \cap \Gamma_{\frac{1}{2}-\gamma} = sg(T^\beta h) \cap \Gamma_{\frac{1}{2}-\gamma} = \emptyset$.

This gives us the means to characterize formal adjoints and compositions of operators in (8.38) (when the weight data fit together).

Theorem 13. *$A \in C_{M+G}(X^\wedge, \underline{g})$ for $\underline{g} = (\gamma, \gamma - m, \Theta)$, $\Theta = (-k, 0]$, implies $A^* \in C_{M+G}(X^\wedge, \underline{h})$ for $\underline{h} = (-\gamma + m, -\gamma, \Theta)$ and*

$$\sigma_M^{m-j}(A^*)(z) = \sigma_M^{m-j}(A)^{(*)}(z + m - j),$$

$j = 0, \ldots, k-1$, with $f^{(*)}(z) = f(n+1-\overline{z})^*$, where $*$ on the right is the pointwise formal adjoint in $L_2(X)$.

Theorem 14. *Let $A \in C_{M+G}(X^\wedge, \underline{g})$, $\underline{g} = (\gamma, \gamma - m, \Theta)$, $\tilde{A} \in C_{M+G}(X^\wedge, \tilde{\underline{g}})$, $\tilde{\underline{g}} = (\gamma - m, \gamma - (\tilde{m} + m), \Theta)$ with $\Theta = (-k, 0]$. Then $\tilde{A}A \in C_{M+G}(X^\wedge, \underline{h})$ with $\underline{h} = (\gamma, \gamma - (\tilde{m} + m), \Theta)$, and we have*

$$\sigma_M^{\tilde{m}+m-j}(\tilde{A}A)(z) = \sum_{p+q=j} (T^{m-q}\sigma_M^{\tilde{m}-p}(\tilde{A}))(z)\sigma_M^{m-q}(A)(z) \tag{8.47}$$

for $j = 0, \ldots, k-1$. An analogous result holds for the operators in $C_{M+G}(\mathbb{B}, \ldots)$.

The proofs of Theorems 13 and 14 are straightforward after the commutator arguments which show how to shift powers of r to the left of a Mellin operator. In the proof of Theorem 14 the fact that operators of the form

$$\omega r^l op_M^\alpha(\tilde{h})(1 - \omega_1)r^j op_M^\beta(h)\omega_2 \tag{8.48}$$

with cut-off functions $\omega, \omega_1, \omega_2$ and corresponding weights α, β are always of Green type also plays a role. The latter assertion as well as the complete details of the proofs may be regarded as exercises.

Proposition 15. *Let $A \in C_{M+G}(X^\wedge, \underline{g})$ with $\underline{g} = (\gamma, \gamma, \Theta)$, $\Theta = (-k, 0]$, be arbitrary, and assume that*

$$1 + A : \ \mathcal{K}^{s,\gamma}(X^\wedge) \longrightarrow \mathcal{K}^{s,\gamma}(X^\wedge) \tag{8.49}$$

is invertible for one $s = s_0 \in \mathbb{R}$. Then (8.49) is invertible for all $s \in \mathbb{R}$, and there is an $A_1 \in C_{M+G}(X^\wedge, \underline{g})$ with

$$(1 + A)^{-1} = 1 + A_1. \tag{8.50}$$

Proof. From Theorem 3 it follows that $\ker(1 + A)$ is independent of s. The same is true of the cokernel if we argue via the formal adjoint $(1 + A)^*$ combined with a weight shift which reduces the assertion to $\gamma = 0$. The invertibility means, in particular, that $1 + A$ is Fredholm. According to the general concept of ellipticity below, cf. 8.1.5 Theorem 7, this implies that

$$1 + \sigma_M^0(A)(z) : \ H^s(X) \longrightarrow H^s(X)$$

is an isomorphism for all $z \in \Gamma_{\frac{n+1}{2}-\gamma}$. Now there is an $f_0(z) \in M_R^{-\infty}(X)$ for an asymptotic type R such that

$$(1 + \sigma_M^0(A)(z))^{-1} = 1 + f_0(z),$$

where $\pi_\mathbb{C} R \cap \Gamma_{\frac{n+1}{2}-\gamma} = \emptyset$. According to (8.47) we have

$$(1 + A)(1 + \omega op_M^{\gamma-\frac{n}{2}}(f_0)\omega) = 1 + A_0$$

with $\sigma_M^0(A_0)(z) = 0$. Thus, $A_0 : \ \mathcal{K}^{s,\gamma}(X^\wedge) \longrightarrow \mathcal{K}^{s,\gamma}(X^\wedge)$ is a compact operator, and hence,

$$\mathrm{ind}(1 + \omega op_M^{\gamma-\frac{n}{2}}(f_0)\omega) = 0.$$

We have

$$(1 + A_0)\left(\sum_{j=0}^{N}(-1)^j A_0^j\right) = 1 + G_1$$

with some $G_1 \in C_G(X^\wedge, \underline{g})$, when $N \geq k$. Thus, if we set

$$A_2 = (1 + \omega op_M^{\gamma - \frac{n}{2}}(f_0)\omega) \left(\sum_{j=0}^{N} (-1)^j A_0^j \right) - 1,$$

which belongs to $C_{M+G}(X^\wedge, \underline{g})$, we get

$$(1 + A)(1 + A_2) = 1 + G_1.$$

Also, $(1 + A_2)$ is of index 0. Since $C_0^\infty(X^\wedge)$ is dense in $\mathcal{K}^{s,\gamma}(X^\wedge)$, there exists a finite-dimensional operator G_2 with kernel in $C_0^\infty(X^\wedge \times X^\wedge)$ (which belongs to $C_G(X^\wedge, \underline{g})$) such that

$$1 + A_2 + G_2 : \ \mathcal{K}^{s,\gamma}(X^\wedge) \longrightarrow \mathcal{K}^{s,\gamma}(X^\wedge)$$

is an isomorphism. Then

$$(1 + A)(1 + A_2 + G_2) = 1 + G_3$$

is also an isomorphism. Now, for some $G_4 \in C_G(X^\wedge, \underline{g})$ we have $(1 + G_3)^{-1} = 1 + G_4$, cf. 8.1.1 Proposition 9. It follows that

$$(1 + A)^{-1} = (1 + A_2 + G_2)(1 + G_4) = 1 + A_1$$

with $A_1 \in C_{M+G}(X^\wedge, \underline{g})$, according to Theorem 14. $\qquad\square$

Remark 16. Proposition 15 allows an easy variant for the situation of operators

$$1 + \begin{pmatrix} A & G_{12} \\ G_{21} & G_{22} \end{pmatrix} : \ \begin{matrix} \mathcal{K}^{s,\gamma}(X^\wedge) \\ \oplus \\ \mathbb{C}^N \end{matrix} \ \longrightarrow \ \begin{matrix} K^{s,\gamma}(X^\wedge) \\ \oplus \\ \mathbb{C}^N \end{matrix} \tag{8.51}$$

for $A \in C_{M+G}(X^\wedge, \underline{g})$ and finite-dimensional entries

$$G_{12}c \ = \ \sum_{j=1}^{N} g_{12,j} c_j, \quad c = (c_1, \ldots, c_N) \in \mathbb{C}^N,$$

$$G_{21}u \ = \ \left(\int\!\!\int g_{21,j}(r, x) u(r, x) dr dx \right)_{j=1,\ldots,N},$$

with functions $g_{12,j}(r, x) \in \mathcal{S}_P^\gamma(X^\wedge)$, $g_{21,j}(r, x) \in \mathcal{S}_Q^\gamma(X^\wedge)$ for certain asymptotic types P, Q. If (8.51) is invertible, then the inverse is of analogous nature. A similar result holds when the operators in the block matrix (8.51) continuously (smoothly) depend on parameters varying over a compact parameter space (or a compact C^∞ manifold). Then the entries of the inverse are also continuous (smooth) in the parameter variables.

8.1.3 A Mellin operator convention

We now return once again to the discussion of 7.2.3 Remark 2 in connection with the choice of interior symbol classes for the cone operator algebra to be established below. The quality of interior symbols will be motivated by the problem of expressing parametrices of elliptic differential operators of Fuchs type. Since the typical degeneracy concerns $(r, x) \in X^\wedge$ for $r \to 0$, it suffices for the moment to look at a neighbourhood of $r = 0$.

The differential operators A of Fuchs type were defined in terms of operator-valued functions $r^{-m}a(r, z)$ with

$$a(r, z) = \sum_{j=0}^{m} a_j(r) z^j, \quad a_j(r) \in C^\infty(\overline{\mathbb{R}}_+, Diff^{m-j}(X)),$$

cf. 7.1.2(7.30). We have to insert $-r\frac{\partial}{\partial r}$ for z in order to obtain the operator

$$A: \ \mathcal{K}^{s,\gamma}(X^\wedge) \longrightarrow \mathcal{K}^{s-m,\gamma-m}(X^\wedge) \tag{8.52}$$

(after the corresponding assumptions for $r \to \infty$, cf. Section 8.2.4 below). Let us fix a covering

$$\{U_1, \ldots, U_N\} \tag{8.53}$$

of X by coordinate neighbourhoods and charts $\kappa_k : \ U_k \to \mathbb{R}^n$, $k = 1, \ldots, N$. Let $\{\phi_1, \ldots, \phi_N\}$ be a subordinate partition of unity and $\{\psi_1, \ldots, \psi_N\}$ be another system of functions $\psi_k \in C_0^\infty(U_k)$ with $\phi_k \psi_k = \phi_k$ for all k. Then $a_j(r)$ can be written as

$$a_j(r) = \sum_{k=1}^{N} \phi_k(\kappa_k^*\{\mathcal{F}_{\xi \to x}^{-1} b_{jk}(r, x, \xi)\mathcal{F}_{x' \to \xi}\})\psi_k \tag{8.54}$$

with symbols $b_{jk}(r, x, \xi) \in S_{cl}^m(\overline{\mathbb{R}}_+ \times \mathbb{R}_x^n \times \mathbb{R}_\xi^n)$, and κ_k^* being the operator pull-back. The operator $a\left(r, -r\frac{\partial}{\partial r}\right)$ can be expressed locally in coordinates of U_k by symbols

$$a_k(r, x, \tilde{\varrho}, \xi)\,|_{\tilde{\varrho}=r\varrho}, \quad k = 1, \ldots, N,$$

with $a_k(r, x, \tilde{\varrho}, \xi) \in S_{cl}^m(\overline{\mathbb{R}}_+ \times \mathbb{R}^n \times \mathbb{R}_{\tilde{\varrho},\xi}^{1+n})$. Let $a_{k,(m)}$ indicate the homogeneous principal part of order m. Then the conditions of ellipticity of A contain

$$a_{k,(m)}(r, x, \tilde{\varrho}, \xi) \neq 0 \ \text{ for all } \ (r, x) \in \overline{\mathbb{R}}_+ \times \mathbb{R}^n, \quad (\tilde{\varrho}, \xi) \neq 0.$$

Clearly A itself is then also elliptic on $X^\wedge$ in the standard sense, though *degenerate* for $r \to 0$. The parametrix construction requires to calculate symbols

$$p_k(r, x, \tilde{\varrho}, \xi) \in S_{cl}^{-m}(\overline{\mathbb{R}}_+ \times \mathbb{R}^n \times \mathbb{R}_{\tilde{\varrho},\xi}^{1+n}) \tag{8.55}$$

such that

$$a_k(r, x, r\varrho, \xi) \ \# \ p_k(r, x, r\varrho, \xi) \sim 1 \tag{8.56}$$

for $k = 1, \ldots, N$. Here $\#$ is the Leibniz product with respect to the r, x-variables.

Exercise 1. Let $a(r, x, \tilde{\varrho}, \xi) \in S_{cl}^m(\overline{\mathbb{R}}_+ \times \mathbb{R}^n \times \mathbb{R}_{\tilde{\varrho}, \xi}^{1+n})$ be elliptic in the above sense. Then, there is a $p(r, x, \tilde{\varrho}, \xi) \in S_{cl}^{-m}(\overline{\mathbb{R}}_+ \times \mathbb{R}^n \times \mathbb{R}_{\tilde{\varrho}, \xi}^{1+n})$ such that

$$1 \sim a \mid_{\tilde{\varrho}=r\varrho} \ \# \ p \mid_{\tilde{\varrho}=r\varrho} = \sum_\alpha \frac{1}{\alpha!} \partial_{\varrho, \xi}^\alpha a(r, x, r\varrho, \xi) D_{r, x}^\alpha p(r, x, r\varrho, \xi).$$

Note that the differentiation with respect to r applied to p also contains the ϱ-derivatives.

It is now a basic problem of the calculus near $r = 0$ to pass from the system of local Leibniz inverses $\{p_1, \ldots, p_N\}$ (times r^m) to an operator

$$P: \ \mathcal{K}^{s-m, \gamma-m}(X^\wedge) \longrightarrow \mathcal{K}^{s, \gamma}(X^\wedge) \tag{8.57}$$

which is a parametrix of A. An arbitrary choice of a representative of P in $L_{cl}^{-m}(X^\wedge)$ mod $L^{-\infty}(X^\wedge)$ will not transform the space with weights from $\gamma - m$ to γ in the desired way. It is actually by no means obvious that a parametrix P with (8.57) exists. But this follows from the Mellin operator convention, namely that for

$$P_0 = \sum_{k=1}^N \phi_k \kappa_k^* Op_{(r,x)}(p_k) \psi_k \in L^{-m}(X^\wedge) \tag{8.58}$$

there exists a P_1 with $P_0 - P_1 \in L^{-\infty}(X^\wedge)$ such that $P = P_1 r^m$ is a map like (8.57). Here $Op_{(r,x)}(p) = \mathcal{F}_{(\varrho, \xi) \to (r, x)}^{-1} p \mathcal{F}_{(r', x') \to (\varrho, \xi)}$.

Theorem 2. *Let $m \in \mathbb{R}$ and $a_k(r, x, \tilde{\varrho}, \xi) \in S_{cl}^m(\overline{\mathbb{R}}_+ \times \mathbb{R}^n \times \mathbb{R}_{\tilde{\varrho}, \xi}^{1+n})$ be arbitrary. Set*

$$A_0 = \sum_{k=1}^N \phi_k \kappa_k^* Op_{(r,x)}(a_k \mid_{\tilde{\varrho}=r\varrho}) \psi_k.$$

Then, there is an $h(r, z) \in C^\infty(\overline{\mathbb{R}}_+, M_O^m(X))$, cf. 7.2.4 Definition 5, such that

$$A_0 - op_M^\delta(h) \in L^{-\infty}(X^\wedge) \tag{8.59}$$

for every $\delta \in \mathbb{R}$.

Proof. We shall first show the following result. For every $a(r, x, \tilde{\varrho}, \xi) \in S_{cl}^m(\overline{\mathbb{R}}_+ \times \mathbb{R}^n \times \mathbb{R}_{\tilde{\varrho}, \xi}^{1+n})$ there is an $f(r, x, z, \xi) \in S_{cl}^m(\overline{\mathbb{R}}_+ \times \mathbb{R}^n \times \Gamma_0 \times \mathbb{R}_\xi^n)$ with

$$Op_{(r,x)}(a \mid_{\tilde{\varrho}=r\varrho}) = op_{M,r}^{\frac{1}{2}} Op_x(f) \mod L^{-\infty}(\mathbb{R}_+ \times \mathbb{R}^n). \tag{8.60}$$

Set $f_0(r, x, i\varrho, \xi) = a(r, x, -\varrho, \xi)$ and for brevity write

$$op_M^{\frac{1}{2}} Op_x(f_0) u(r) =: op_M^{\frac{1}{2}}(f_0) u(r) \tag{8.61}$$

for $u(r) \in C_0^\infty(\mathbb{R}_+, C_0^\infty(\mathbb{R}^n))$. In this notation the Mellin symbol $f_0 = f_0(r, i\varrho)$ on the right of (8.61) is interpreted as an $L_{cl}^m(\mathbb{R}^n)$-valued one, where the action along x-variables is carried out tacitly. In other words we treat the Mellin actions in an analogous manner as to that in the one-dimensional theory, here with symbols in $C^\infty(\mathbb{R}_+, L_{cl}^m(\mathbb{R}^n; \Gamma_0))$. Then

$$op_M^{\frac{1}{2}}(f_0)u(r) \;=\; \int\limits_{-\infty}^{\infty}\int\limits_{0}^{\infty} \left(\frac{r}{r'}\right)^{i\varrho} f_0(r, i\varrho)u(r')\frac{dr'}{r'}\,d\varrho$$

$$\;=\; (\kappa^*)^{-1}Op_t(b_0)v(t)$$

with $\kappa(t) := r = e^{-t}$, $v = \kappa^* u$, $Op_t(b_0) = \mathcal{F}_{\varrho \to t}^{-1} b_0(t, \varrho)\mathcal{F}_{t' \to \varrho}$, and

$$b_0(t, \varrho) = f_0(e^{-t}, i\varrho),$$

cf. (7.169). Thus,

$$op_M^{\frac{1}{2}}(f_0) = \kappa_* Op_t(b_0),$$

where κ_* is the push-forward of pseudo-differential operators. We now apply the symbolic rule for expressing push-forwards of pseudo-differential operators under $\kappa : \mathbb{R} \to \mathbb{R}_+$, $t \to r$. Let (t, τ) and (r, ϱ) denote the corresponding variables together with the covariables. Then we find a symbol $c(r, \varrho)$ with

$$c(r, \varrho)\,|_{r=\kappa(t)} \sim \sum \frac{1}{k!} (\partial_\tau^k b_0)(t, {}^t d\kappa(t)\varrho)\Phi_k(t, \varrho)$$

where here ${}^t d\kappa(t)$ is the multiplication by $-r$, whereas

$$\Phi_k(t, \varrho) = D_s^k e^{i\delta(s,t)\varrho}\,|_{s=t}, \quad k \in \mathbb{Z}_+,$$

with $\delta(s, t) = \kappa(s) - \kappa(t) - d\kappa(t)(s - t)$. It follows easily that $c(r, \varrho)$ may actually be chosen in the form $c_0(r, \tilde{\varrho})\,|_{\tilde{\varrho}=r\varrho}$, in other words $\kappa_* Op_t(b_0) = Op_r(c_0\,|_{\tilde{\varrho}=r\varrho})$ mod $L^{-\infty}(X^\wedge)$ with a

$$c_0(r, \tilde{\varrho}) \in C^\infty(\overline{\mathbb{R}}_+, L_{cl}^m(\mathbb{R}^n; \mathbb{R}_{\tilde{\varrho}}))$$

and $a(r, \tilde{\varrho}) - c_0(r, \tilde{\varrho}) =: a_1(r, \tilde{\varrho}) \in C^\infty(\overline{\mathbb{R}}_+, L_{cl}^{m-1}(\mathbb{R}^n; \mathbb{R}_{\tilde{\varrho}}))$. Thus

$$Op_r(a\,|_{\tilde{\varrho}=r\varrho}) = op_M^{\frac{1}{2}}(f_0) + Op_r(a_1\,|_{\tilde{\varrho}=r\varrho}) \mod L^{-\infty}(X^\wedge).$$

By iterating this procedure it follows for $f_1(r, i\varrho) := a_1(r, -\varrho)$ that

$$Op_r(a\,|_{\tilde{\varrho}=r\varrho}) = op_M^{\frac{1}{2}}(f_0) + op_M^{\frac{1}{2}}(f_1) + Op_r(a_2\,|_{\tilde{\varrho}=r\varrho}) \mod L^{-\infty}(X^\wedge),$$

and then inductively

$$Op_r(a\,|_{\tilde{\varrho}=r\varrho}) = \sum_{j=0}^{M} op_M^{\frac{1}{2}}(f_j) + Op_r(a_{M+1}\,|_{\tilde{\varrho}=r\varrho}) \tag{8.62}$$

with certain

$$f_j(r, i\varrho) \in C^\infty(\overline{\mathbb{R}}_+, L^{m-j}(\mathbb{R}^n; \Gamma_0)), a_{M+1}(r, \tilde{\varrho}) \in C^\infty(\overline{\mathbb{R}}_+, L^{m-(M+1)}(\mathbb{R}^n; \mathbb{R}_{\tilde{\varrho}})),$$

for every M. Let $f(r, z) \sim \sum_{j=0}^\infty f_j(r, z)$ be the asymptotic sum of the f_j, carried out in $C^\infty(\overline{\mathbb{R}}_+, L^m_{cl}(\mathbb{R}^n; \Gamma_0))$. This could be expressed as well by an $f(r, x, z, \xi)$ as indicated in the beginning. Then (8.60) follows immediately from (8.62). This construction can be carried out for every chart on X. Denote the corresponding local Mellin symbols by $f_{(k)}$, $k = 1, \ldots, N$, for simplicity interpreted this in the operator-valued form along x-variables, we get by

$$\tilde{f}(r, z) = \sum_{k=1}^N \phi_k f_{(k)}(r, z) \psi_k$$

an $\tilde{f}(r, z) \in C^\infty(\overline{\mathbb{R}}_+, L^m_{cl}(\mathbb{R}^n; \Gamma_0))$ with

$$A_0 = op_M^{\frac{1}{2}}(\tilde{f}) \mod L^{-\infty}(X^\wedge).$$

Applying 7.2.4 Exercise 6, we obtain $h(r, z) = H(\psi)\tilde{f}(r, z)$ with the asserted properties, where we use

$$op_M^\delta(h) = op_M^\gamma(h)$$

for all $\delta, \gamma \in \mathbb{R}$ which is a consequence of the holomorphy of h in z, cf. 8.1.2 Remark 12. $\qquad\square$

8.1.4 The cone algebra

Let $\mathbb{B}$ be the stretched manifold of a compact manifold B with conical singularities. As usual we fix an identification map

$$\chi : V \longrightarrow \overline{\mathbb{R}}_+ \times X$$

for a collar neighbourhood V of $\partial\mathbb{B} \cong X$, $n = \dim X$. (Remember that we always allow X to have several connected components; thus we may assume without loss of generality that B has exactly one conical point.)

Definition 1. Let $\underline{g} = (\gamma, \gamma - m, \Theta)$ with $m, \gamma \in \mathbb{R}$ and $\Theta = (-k, 0]$, $k \in \mathbb{N}$, or $\Theta = (-\infty, 0]$. Then

$$C^m(\mathbb{B}, \underline{g}) \tag{8.63}$$

is the subspace of all $A \in L^m_{cl}(\mathrm{int}\,\mathbb{B})$ of the form

$$A = \chi^*(\omega r^{-m} op_M^{\gamma - \frac{n}{2}}(h)\omega_0) + (1 - \tilde{\omega})\tilde{A}(1 - \tilde{\omega}_1) + W + G \tag{8.64}$$

with arbitrary $h(r, z) \in C^\infty(\overline{\mathbb{R}}_+, M^m_O(X))$, elements $\tilde{A} \in L^m_{cl}(\mathrm{int}\,\mathbb{B})$, $W + G \in C_{M+G}(\mathbb{B}, \underline{g})$, and cut-off functions $\omega, \omega_0, \omega_1$, where $\tilde{\omega} = \chi^*\omega$, $\tilde{\omega}_1 = \chi^*\omega_1$.

Observe that every differential operator of order m of Fuchs type on $\mathbb{B}$ belongs to $C^m(\mathbb{B}, \underline{g})$, cf. 7.1.2 Definition 2. Let us introduce the principal symbolic levels

$$\sigma_\psi^m(A), \quad \sigma_M^m(A),$$

called the interior and the conormal symbol, respectively, of the operator A. Analgously to (7.61) we can also introduce a symbol $\sigma_{\psi,b}^m(A)$ which is in the coordinates (r, x) in a neighbourhood of $r = 0$ defined by

$$\sigma_{\psi,b}^m(A)(r, x, \varrho, \xi) = \sigma_\psi^m(A)(r, x, r^{-1}\varrho, \xi).$$

Since $\sigma_{\psi,b}^m(A)$ in a neighbourhood of $\partial\mathbb{B}$ is uniquely determined by $\sigma_\psi^m(A)$, we shall mainly talk about σ_ψ^m.

The conormal symbol of A is defined as the operator family

$$\sigma_M^m(A)(z) = h(0, z) + h_0(z) : \ H^s(X) \longrightarrow H^{s-m}(X) \tag{8.65}$$

with $h_0(z) = \sigma_M^m(W)(z)$, cf. (8.39), and z varies along $\Gamma_{\frac{n+1}{2}-\gamma}$. Below $\sigma_M^m(A)(z)$ will also be interpreted as a meromorphic operator family in the complex plane. The poles are determined by the asymptotic type R of $h_0(z)$, more precisely by $\pi_{\mathbb{C}}R$. We also have the conormal symbols of lower orders

$$\sigma_M^{m-j}(A)(z) = \frac{1}{j!} \left(\frac{\partial^j}{\partial r^j} h \right) (0, z) + \sigma_M^{m-j}(W)(z) \tag{8.66}$$

for $j = 0, \ldots, k-1$, $\Theta = (-k, 0]$, cf. (8.39).

Remark 2. We have

$$C^m(\mathbb{B}, \underline{g}) \cap L^{-\infty}(\text{int } \mathbb{B}) = C_{M+G}(\mathbb{B}, \underline{g}).$$

Theorem 3. *Every* $A \in C^m(\mathbb{B}, \underline{g})$, $\underline{g} = (\gamma, \gamma - m, \Theta)$, *induces continuous operators*

$$A : \ \mathcal{H}^{s,\gamma}(\mathbb{B}) \longrightarrow \mathcal{H}^{s-m,\gamma-m}(\mathbb{B}) \tag{8.67}$$

for all $s \in \mathbb{R}$. *Furthermore, for every asymptotic type* P *to the weight data* (γ, Θ) *there is an asymptotic type* Q *to* $(\gamma - m, \Theta)$ *such that* (8.67) *induces continuous operators*

$$A : \ \mathcal{H}_P^{s,\gamma}(\mathbb{B}) \longrightarrow \mathcal{H}_Q^{s-m,\gamma-m}(\mathbb{B}) \tag{8.68}$$

for all $s \in \mathbb{R}$.

Proof. In view of 8.1.2 Theorem 3 it suffices to consider the case

$$A = \chi^*(\omega r^{-m} op_M^{\gamma - \frac{n}{2}}(h)\omega_0) + (1 - \tilde{\omega})\tilde{A}(1 - \tilde{\omega}_1). \tag{8.69}$$

It is obvious that the second item on the right of (8.69) has the asserted continuity property. Thus, it suffices to verify that

$$\omega r^{-m} op_M^{\gamma - \frac{n}{2}}(h)\omega_0 : \ \mathcal{K}^{s,\gamma}(X^\wedge) \longrightarrow \mathcal{K}^{s-m,\gamma-m}(X^\wedge) \tag{8.70}$$

is continuous as well as the corresponding restriction to the subspace with asymptotics. (8.70) was obtained in 7.2.3 Theorem 9. The arguments for subspaces with asymptotics in the case of r-independent $h(z)$ are completely analogous to those in the proof of 8.1.2 Theorem 3. The general $h(r, z)$ can be reduced to this case by applying a finite Taylor expansion in r, namely

$$h(r, z) = \sum_{j=0}^{N} r^j h_j(z) + h_{(N+1)}(r, z), \tag{8.71}$$

where $h_{(N+1)}$ is of flatness order $N + 1$ in r in a neighbourhood of $r = 0$. This gives us

$$\omega r^{-m} op_M^{\gamma - \frac{n}{2}}(h_{(N+1)})\omega_0 : \ \mathcal{K}_P^{s,\gamma}(X^\wedge) \longrightarrow \mathcal{K}_\Theta^{s-m,\gamma-m}(X^\wedge)$$

for $\Theta = (-k, 0]$, $N \geq k$. The items of the finite sum on the right of (8.71) can be reduced to the case of constant coefficients, up to the multiplications by r^j which cause only corresponding changed exponents in the asymptotics. $\qquad \square$

Remark 4. Let $A \in C^m(\mathbb{B}, \underline{g})$, $\underline{g} = (\gamma, \gamma - m, \Theta)$, and let $\sigma_\psi^m(A) = 0$, $\sigma_M^m(A) = 0$. Then

$$A : \ \mathcal{H}^{s,\gamma}(\mathbb{B}) \longrightarrow \mathcal{H}^{s-m,\gamma-m}(\mathbb{B})$$

is a compact operator for all $s \in \mathbb{R}$.

Let us define $C^{m-1}(\mathbb{B}, \underline{g})$ as the subspace of all $A \in C^m(\mathbb{B}, \underline{g})$ for which $\sigma_\psi^m(A) = 0, \sigma_M^m(A) = 0$ holds.
We then get a map $A : \ \mathcal{H}^{s,\gamma}(\mathbb{B}) \longrightarrow \mathcal{H}^{s-m+1,\gamma-m+\varepsilon}(\mathbb{B})$ for an $\varepsilon > 0$.
But the embedding $\mathcal{H}^{s',\gamma'}(\mathbb{B}) \longrightarrow \mathcal{H}^{s,\gamma}(\mathbb{B})$ for $s' > s, \gamma' > \gamma$ is compact.

Theorem 5. $A \in C^m(\mathbb{B}, \underline{g})$ *for* $\underline{g} = (\gamma, \gamma - m, \Theta)$ *implies* $A^* \in C^m(\mathbb{B}, \underline{g}^*)$ *for* $\underline{g}^* = (-\gamma + m, -\gamma, \Theta)$, *and we have*

$$\sigma_\psi^m(A^*) = \overline{\sigma_\psi^m(A)}, \quad \sigma_M^{m-j}(A^*)(z) = \sigma_M^{m-j}(A)^{(*)}(z + m - j), \tag{8.72}$$

$j = 0, \ldots, k - 1$, *with* $f^{(*)}(z) = f^*(n + 1 - \bar{z})$, *where* $*$ *on the right is the formal adjoint in* $L_2(X)$, $n = \dim X$.

This result is left to the reader as an exercise.

Theorem 6. *Let $A \in C^m(\mathbb{B}, \underline{g})$, $\underline{g} = (\gamma, \gamma - m, \Theta)$, $\tilde{A} \in C^{\tilde{m}}(\mathbb{B}, \underline{\tilde{g}})$, $\underline{\tilde{g}} = (\gamma - m, \gamma - (\tilde{m} + m), \Theta)$. Then $\tilde{A}A \in C^{\tilde{m}+m}(\mathbb{B}, \underline{h})$, with $\underline{h} = (\gamma, \gamma - (\tilde{m} + m), \Theta)$, and*

$$\sigma_\psi^{\tilde{m}+m}(\tilde{A}A) = \sigma_\psi^{\tilde{m}}(\tilde{A})\sigma_\psi^m(A), \quad \sigma_M^{\tilde{m}+m}(\tilde{A}A) = \{T^m \sigma_M^{\tilde{m}}(\tilde{A})\}\sigma_M^m(A). \tag{8.73}$$

Proof. As usual, it suffices to consider the case of finite Θ. Then the result for $\Theta = (-\infty, 0]$ is an obvious consequence. Let us write

$$A = \omega A_0 \omega_0 + \chi A_1 \chi_1 + W + G \tag{8.74}$$

with $\chi = 1 - \omega$, $\chi_1 = 1 - \omega_1$, and

$$A_0 = r^{-m} op_M^{\gamma - \frac{n}{2}}(h) \tag{8.75}$$

with a certain $h(r, z) \in C^\infty(\overline{\mathbb{R}}_+, M_O^m(X))$. In other words, to simplify notation we drop the obvious pull-backs of operators from the collar neighbourhood V of $\partial\mathbb{B}$ to $[0, 1) \times X$. The meaning of the other notation in (8.74) follows immediately from (8.64). Analogously we write

$$\tilde{A} = \tilde{\omega}\tilde{A}_0\tilde{\omega}_0 + \tilde{\chi}\tilde{A}_1\tilde{\chi}_1 + \tilde{W} + \tilde{G}$$

with

$$\tilde{A}_0 = r^{-\tilde{m}} op_M^{\gamma - m - \frac{n}{2}}(\tilde{h}),$$

$\tilde{h}(r, z) \in C^\infty(\overline{\mathbb{R}}_+, M_O^{\tilde{m}}(X))$. Then

$$\tilde{A}A = B_0 + B_1 + D_0 + D_1 + C_0 + C_1 + E$$

with

$$
\begin{array}{rclcrcl}
B_0 & = & \tilde{\omega}\tilde{A}_0\tilde{\omega}_0\omega A_0\omega_0 & , & B_1 & = & \tilde{\chi}\tilde{A}_1\tilde{\chi}_1\chi A_1\chi_1 \qquad , \\
D_0 & = & \tilde{\omega}\tilde{A}_0\tilde{\omega}_0\chi A_1\chi_1 & , & D_1 & = & \tilde{\chi}\tilde{A}_1\tilde{\chi}_1\omega A_0\omega_0 \qquad , \\
C_0 & = & (\tilde{\omega}\tilde{A}_0\tilde{\omega}_0 + \tilde{\chi}\tilde{A}_1\tilde{\chi}_1)(W + G) & , & C_1 & = & (\tilde{W} + \tilde{G})(\omega A_0\omega_0 + \chi A_1\chi_1) \quad , \\
E & = & (\tilde{W} + \tilde{G})(W + G). & & & &
\end{array}
$$

Then, with $\delta := \gamma - \frac{n}{2}$, we have

$$
\begin{aligned}
B_0 & = & \tilde{\omega}r^{-\tilde{m}} op_M^{\delta - m}(\tilde{h})\tilde{\omega}_0\omega r^{-m} op_M^\delta(h)\omega_0 \\
& = & \tilde{\omega}r^{-\tilde{m}-m} op_M^\delta(T^m\tilde{h})\omega_2 op_M^\delta(h)\omega_0,
\end{aligned}
$$

$\omega_2 := \tilde{\omega}_0\omega$. Here, 8.1.2 Remark 12 was used, where Green remainders vanish because of the holomorphy of $\tilde{h}$ in z. Let us set $\tilde{f} = T^m\tilde{h}$, $f = \omega_2 h$. Then

$$B_0 = \tilde{\omega}r^{-\tilde{m}-m} op_M^\delta(\tilde{f}) op_M^\delta(f)\omega_0.$$

Applying the Taylor expansion of h around $r = 0$

$$h(r, z) = \sum_{j=0}^{N} r^j h_j(z) + r^{N+1} h_{(N)}(r, z)$$

with $h_{(N)}(r, z) \in C^\infty(\overline{\mathbb{R}}_+, M_O^m(X))$, we obtain

$$\tilde{\omega} r^{-\tilde{m}-m} op_M^\delta(\tilde{f}) op_M^\delta(f) \omega_0 = \tilde{\omega} r^{-\tilde{m}-m} op_M^\delta(\tilde{f}) \omega_2 \left\{ \sum_{j=0}^{N} r^j op_M^\delta(h_j) \right\} \omega_0 + R_N$$

with

$$R_N = \tilde{\omega} r^{-\tilde{m}-m} op_M^\delta(\tilde{f}) \omega_2 r^{N+1} op_M^\delta(h_{(N)}) \omega_0.$$

In the sequel we will systematically use the commutation argument of 8.1.2 Remark 12. Then, in particular,

$$B_0 = \tilde{\omega} r^{-\tilde{m}-m} \sum_{j=0}^{N} r^j op_M^\delta((T^{-j}\tilde{f}) h_j) \omega_0 - S_N + R_N \tag{8.76}$$

with

$$S_N = \tilde{\omega} r^{-\tilde{m}-m} \sum_{j=0}^{N} r^j op_M^\delta(T^{-j}\tilde{f})(1 - \omega_2) op_M^\delta(h_j) \omega_0.$$

Writing $F_j = r^{-\tilde{m}-m} r^j op_M^\delta(T^{-j}\tilde{f})$, $H_j = op_M^\delta(h_j)$, we obtain for $\tilde{\omega} = \omega' + \phi$, $\omega_0 = \omega'' + \psi$ with cut off functions ω', ω'' and $\phi, \psi \in C_0^\infty(\mathbb{R}_+)$

$$S_N = \sum_{j=0}^{N} S_{N,j}$$

for

$$\begin{aligned}
S_{N,j} &= \tilde{\omega} F_j (1 - \omega_2) H_j \omega_0 \\
&= \omega' F_j (1 - \omega_2) H_j \omega'' + \phi F_j (1 - \omega_2) H_j \omega'' \\
&\quad + \omega' F_j (1 - \omega_2) H_j \psi + \phi F_j (1 - \omega_2) H_j \psi.
\end{aligned}$$

We choose ω', ω'' in such a way that $\omega'(1 - \omega_2) = \omega''(1 - \omega_2) = 0$. Then 7.2.4 Proposition 9 together with Theorem 3 easily gives us

$$\omega' F_j (1 - \omega_2) H_j \omega'', \quad \phi F_j (1 - \omega_2) H_j \omega'', \quad \omega' F_j (1 - \omega_2) H_j \psi \ \in \ C_G(\mathbb{B}, \underline{h}). \tag{8.77}$$

Moreover, 7.2.3 Proposition 1 shows $\phi F_j (1 - \omega_2) H_j \psi \in \phi L_{cl}^{\tilde{m}+m}(X^\wedge) \psi$ (henceforth we identify a collar neighbourhood of $\partial\mathbb{B}$ with a corresponding piece of $\mathbb{R}_+ \times X$). In other words, we have obtained $S_N \in C^{\tilde{m}+m}(\mathbb{B}, \underline{h})$. To characterize R_N we assume $N = 2k - 1$. Then

$$R_N = \tilde{\omega} r^{-\tilde{m}-m+k} op_M^{\delta}(T^{-k}\tilde{f})\omega_2 op_M^{\delta}(T^k h_{(N)})r^k \omega_0.$$

Putting $a(r,z) = T^{-\delta+\frac{1}{2}}T^{-k}\tilde{f}(r,z)$, $b(r,z) = \omega_2(r)T^{-\delta+\frac{1}{2}}T^k h_{(N)}(r,z)$, the operator R_N can be rewritten as

$$R_N = \tilde{\omega} r^{-\tilde{m}-m+k} r^{\delta-\frac{1}{2}} op_M^{\frac{1}{2}}(a) op_M^{\frac{1}{2}}(b) r^{-\delta+\frac{1}{2}} r^k \omega_0.$$

In the middle we can use 7.2.4 Theorem 14. This yields a $c(r,z) \in C^{\infty}(\overline{\mathbb{R}}_+, L_{cl}^{\tilde{m}+m}(X; \Gamma_0))$ such that

$$R_N = \tilde{\omega} r^{-\tilde{m}-m+k+\delta-\frac{1}{2}}\{op_M^{\frac{1}{2}}(c) + K\}r^{-\delta+\frac{1}{2}+k}\omega_0,$$

where $K \in M_{\frac{1}{2}}L^{-\infty}(\overline{\mathbb{R}}_+ \times X)$. Using the result of 7.2.4 Exercise 6, we can write $c = c_0 + d$ with $c_0(r,z) \in C^{\infty}(\overline{\mathbb{R}}_+, M_O^{\tilde{m}+m}(X))$, $d(r,z) \in C^{\infty}(\overline{\mathbb{R}}_+, L^{-\infty}(\overline{\mathbb{R}}_+ \times X; \Gamma_0))$. It is then obvious that

$$\tilde{\omega} r^{-\tilde{m}-m+k+\delta-\frac{1}{2}}\{op_M^{\frac{1}{2}}(d) + K\}r^{-\delta+\frac{1}{2}+k}\omega_0 \in C_G(\mathbb{B}, \underline{h}).$$

Thus,

$$R_N = \tilde{\omega} r^{-\tilde{m}-m+2k} op_M^{\delta}(T^{-k+\delta-\frac{1}{2}}c_0)\omega_0 \mod C_G(\mathbb{B}, \underline{h}).$$

It follows that

$$B_0 = \tilde{\omega} r^{-\tilde{m}-m} op_M^{\delta}(l)\tilde{\omega}_0 \mod \{\phi L_{cl}^{\tilde{m}+m}(X^{\wedge})\psi + C_G(\mathbb{B}, \underline{h})\}$$

with

$$l(r,z) = \sum_{j=0}^{N} r^j (T^{-j+m}\tilde{h}(r,z))h_j(z) + r^{2k}T^{-k+\delta-\frac{1}{2}}c_0(r,z) \in C^{\infty}(\overline{\mathbb{R}}_+, M_O^{\tilde{m}+m}(X)).$$

Hence $B_0 \in C^{\tilde{m}+m}(\mathbb{B}, \underline{h})$. To characterize D_0 we set $\phi = \tilde{\omega}_0\chi$ which is in $C_0^{\infty}(\mathbb{R}_+)$, and write $\tilde{\omega} = \omega' + \psi$ with a cut-off function ω' such that $\omega'\phi = 0$. Then

$$D_0 = \omega'\tilde{A}_0\phi A_1\chi_1 + \psi\tilde{A}_0\phi A_1\chi_1.$$

$\psi\tilde{A}_0\phi A_1\chi_1 \in C^{\tilde{m}+m}(\mathbb{B}, \underline{h})$ is clear and further $\omega'\tilde{A}_0\phi A_1\chi_1 \in C_G(\mathbb{B}, \underline{h})$, by arguments analogous to those in (8.77). Thus $D_0 \in C^{\tilde{m}+m}(\mathbb{B}, \underline{h})$. For D_1 we can proceed in an analogous manner.

From Theorem 3 and 8.1.1 Definition 7 we see that the compositions in C_0 with G are of Green type, again. Further, the cut-off factors in W may be chosen conveniently, since the operators with different cut-off functions are equal modulo Green remainders, cf. 8.1.2 Corollary 10. Therefore we may assume $\tilde{\chi}W = 0$. For C_0 it remains to consider $\tilde{\omega}\tilde{A}_0\tilde{\omega}_0W$. However from the material of Section 8.1.2 it is an easy task to verify that this belongs to $C_{M+G}(\mathbb{B}, \underline{h})$. The operator C_1 can be treated in an analogous manner. The symbolic rules (8.73) are obvious. $\square$

Exercise 7. The conormal symbols of $\tilde{A}A$ can be calculated by the Mellin translation product, i.e.

$$\sigma_M^{m+\tilde{m}-j}(\tilde{A}A)(z) = \sum_{p+q=j} (T^{m-q}\sigma_M^{\tilde{m}-p}(\tilde{A}))(z)\sigma_M^{m-q}(A)(z) \qquad (8.78)$$

for $j = 0, \ldots, k-1$.

Remark 8. Let the operator A in Theorem 6 belong to the corresponding C_{M+G} (C_G) subclass. Then the same is true of $A\tilde{A}$. An analogous statement holds when $\tilde{A}$ is in C_{M+G} (C_G).

Let $k^\alpha \in C^\infty(\text{int } \mathbb{B})$ for $\alpha \in \mathbb{R}$ be an arbitrary strictly positive function with

$$k^\alpha(r) = r^\alpha \qquad (8.79)$$

in a collar neighbourhood of $\partial\mathbb{B}$ in the coordinates (r, x).

Remark 9. For arbitrary $\alpha \in \mathbb{R}$ we have

$$k^\alpha C^m(\mathbb{B}, \underline{g})k^{-\alpha} = C^m(\mathbb{B}, \underline{g}_\alpha)$$

for $\underline{g} = (\gamma, \gamma - m, \Theta)$ with $\underline{g}_\alpha = (\gamma + \alpha, \gamma + \alpha - m, \Theta)$.

It will sometimes be convenient to generalize Definition 1 and to introduce the class

$$C^l(\mathbb{B}, \underline{g}) \qquad (8.80)$$

with weight data $\underline{g} = (\gamma, \gamma - m, \Theta)$ as above and an arbitrary l with $m - l \in \mathbb{Z}_+$. The operators $A \in C^l(\mathbb{B}, \underline{g})$ consist of all

$$A = \chi^*(\omega r^{-l}op_M^{\gamma - \frac{n}{2}}(h)\omega_0) + (1 - \tilde{\omega})\tilde{A}(1 - \tilde{\omega}_1) + W + G$$

with $h(r, z) \in C^\infty(\overline{\mathbb{R}}_+, M_O^l(X))$, $\tilde{A} \in L_{cl}^l(\text{int}\mathbb{B})$, $G \in C_G(\mathbb{B}, \underline{g})$, and W being a finite linear combination of operators like

$$r^{-l+j}\omega op_M^{\alpha_j}(h_j)\tilde{\omega}$$

with arbitrary cut-off functions $\omega, \tilde{\omega}$, $j \in \mathbb{Z}_+$ and $h_j \in M_{R_j}^{-\infty}(X)$, $\pi_{\mathbb{C}}R_j \cap \Gamma_{\frac{1}{2}-\alpha_j} = \emptyset$, and

$$\gamma - \frac{n}{2} - (m-l) - j \leq \alpha_j \leq \gamma - \frac{n}{2}$$

for all j. Then, in particular,

$$A \in C^m(\mathbb{B}, \underline{g}), \quad \sigma_\psi^m(A) = 0, \quad \sigma_M^m(A) = 0$$

will imply $A \in C^{m-1}(\mathbb{B}, \underline{g})$, cf. Remark 4.

Note that Theorem 6 can be generalized to $A \in C^l(\mathbb{B}, g)$, $\tilde{A} \in C^{\tilde{l}}(\mathbb{B}, \tilde{g})$ with the indicated g, $\tilde{g}$. Then $\tilde{A}A \in C^{\tilde{l}+l}(\mathbb{B}, \underline{h})$, and we get an obvious extension of the symbol rule (8.73).

The proof follows by arguments analogous to those used above for Theorem 6. Let us now draw some conclusion from this composition result which gives another insight into the structure of cone pseudo-differential operators. The results of Section 8.1.3 show that every $A \in C^l(\mathbb{B}, g)$ with $g = (\gamma, \gamma - m, \Theta)$ belongs to $L^l_{cl}(\mathrm{int}\mathbb{B})$ where, in a neighbourhood of $\partial\mathbb{B}$ in local coordinates (r, x, ϱ, ξ), the operator has a complete symbol

$$r^{-l}a(r, x, \tilde{\varrho}, \xi)\,|_{\tilde{\varrho}=r\varrho} \quad \text{with} \quad a(r, x, \tilde{\varrho}, \xi) \in S^l_{cl}(\overline{\mathbb{R}}_+ \times \mathbb{R}^n \times \mathbb{R}^{1+n}_{\tilde{\varrho}, \xi}).$$

Similarly to Theorem 2 of Section 8.1.3 we start with a system

$$\{r^{-l}a_k(r, x, \tilde{\varrho}, \xi)\,|_{\tilde{\varrho}=r\varrho}\}_{k=1,\dots,N}$$

of symbols in the local coordinates from an open covering of X by coordinate neighbourhoods. Then we associate $h(r, z) \in C^\infty(\overline{\mathbb{R}}_+, M^l_O(X))$ to it in such a way that

$$A - r^{-l}\mathrm{op}_M^{\gamma - \frac{n}{2}}(h)\,|_{(0,\varepsilon)\times X} \in L^{-\infty}((0, \varepsilon) \times X)$$

for an $\varepsilon > 0$. Now if $\tilde{A} \in C^{\tilde{l}}(\mathbb{B}, \underline{g}^l)$ with $\tilde{g} = (\gamma - m, \gamma - (\tilde{m} + m), \Theta)$ is another operator, with a system of complete symbols $\{r^{-\tilde{l}}\tilde{a}_k(r, x, \tilde{\varrho}, \xi)\,|_{\tilde{\varrho}=r\varrho}\}_{k=1,\dots,N}$, here of order $\tilde{l}$, and an associated $\tilde{h}(r, x) \in C^\infty(\overline{\mathbb{R}}_+, M^{\tilde{l}}_O(X))$, then we may investigate $\tilde{A}A$ in terms of the system of complete symbols

$$\{r^{-\tilde{l}}\tilde{a}_k(r, x, r\varrho, \xi)\#r^{-l}a_k(r, x, r\varrho, \xi)\}_{k=1,\dots,N}.$$

As before, $\#$ indicates the Leibniz product with respect to r, x. It can easily be verified that it has the form

$$\{r^{-(\tilde{l}+l)}b_k(r, x, r\varrho, \xi)\}_{k=1,\dots,N}$$

with $b_k(r, x, \tilde{\varrho}, \xi) \in S^{l+\tilde{l}}_{cl}(\overline{\mathbb{R}}_+ \times \mathbb{R}^n \times \tilde{\mathbb{R}}^{1+n}_{\tilde{\varrho}, \xi})$. Using Theorem 2 we find an $f(r, z) \in C^\infty(\overline{\mathbb{R}}_+, M^{l+\tilde{l}}_O(X))$ such that $\mathrm{op}_M^\delta(f)$ has those complete symbols, too. Now we form

$$B := \omega r^{-(\tilde{l}+l)}\mathrm{op}_M^{\gamma - \frac{n}{2}}(f)\omega_0 + (1 - \omega)\tilde{A}A(1 - \omega_1) \tag{8.81}$$

with $\tilde{A}A$ being interpreted as a composition in $L^{l+\tilde{l}}_{cl}(\mathrm{int}\mathbb{B})$, one factor supposed without loss of generality to be properly supported. By definition the operator B belongs to $C^{l+\tilde{l}}(\mathbb{B}, \underline{h})$, $\underline{h} = (\gamma, \gamma - (m + \tilde{m}), \Theta)$. It is clear that $B - \tilde{A}A$ is also an operator in $C^{l+\tilde{l}}(\mathbb{B}, \underline{h})$ and we even have

$$B - \tilde{A}A \in C_{M+G}(\mathbb{B}, \underline{h}).$$

The latter relation is a consequence of the fact that $f(r,z)$ coincides with the holomorphic Mellin symbol from the proof of Theorem 5 (for the orders l and $\tilde{l}$, respectively), modulo some $m(t,z) \in C^\infty(\overline{\mathbb{R}}_+, M_O^{-\infty}(X))$. Then, it suffices to observe

$$r^{-(l+\tilde{l})}\omega op_M^{\gamma-\frac{n}{2}}(m)\tilde{\omega} \in C_{M+G}(\mathbb{B}, \underline{h}).$$

Summing up we have obtained the following:

Remark 10. Let $A \in C^l(\mathbb{B}, \underline{g})$, $\tilde{A} \in C^{\tilde{l}}(\mathbb{B}, \underline{\tilde{g}})$ with the above notation. Then $\tilde{A}A$ may be obtained in the form (8.81) mod $C_{M+G}(\mathbb{B}, \underline{h})$ by applying the Mellin operator convention from 8.1.3 Theorem 2 to the Leibniz products between the complete symbols of $\tilde{A}$ and A, respectively.

8.1.5 Ellipticity and regularity with asymptotics

We now extend the notion of ellipticity from 7.1.2 Definition 14 to the operator class $C^m(\mathbb{B}, \underline{g})$.

Definition 1. An operator $A \in C^m(\mathbb{B}, \underline{g})$ for $\underline{g} = (\gamma, \gamma - m, \Theta)$ is called *elliptic* if

(i) $\sigma_\psi^m(A) \neq 0$ on $T^*(\text{int }\mathbb{B}\backslash 0)$ and if in a neighbourhood of $\partial\mathbb{B}$ in the coordinates $(r,x) \in \mathbb{R}_+ \times X$

$$r^m \sigma_\psi^m(A)(r,x,r^{-1}\varrho,\xi) \neq 0 \quad \text{for all} \quad (r,x) \in \overline{\mathbb{R}}_+ \times X, \quad (\varrho,\xi) \neq 0,$$

(ii) the operators

$$\sigma_M^m(A)(z):\ H^s(X) \longrightarrow H^{s-m}(X) \tag{8.82}$$

are isomorphisms for an $s = s_0 \in \mathbb{R}$ and all $z \in \Gamma_{\frac{n+1}{2}-\gamma}$.

Remark 2. Let $\sigma_M^m(A)(z) = h(0,z) + h_0(z)$ (cf. the notations in (8.65)) and let R be the asymptotic type of $h_0(z)$. Then the condition (i) of Definition 1 implies that $\sigma_M^m(A)(z)$ for all $z \notin \pi_\mathbb{C}R$ is an elliptic element of $L_{cl}^m(X)$. Furthermore, $\sigma_M^m(A)(\beta + i\varrho) \in L_{cl}^m(X;\mathbb{R}_\varrho)$ is parameter-dependent elliptic for all $\beta \in \mathbb{R}$ with $\Gamma_\beta \cap \pi_\mathbb{C}R = \emptyset$.

Remark 3. Remark 2 together with 7.2.2 Theorem 17 shows that (8.82) is an isomorphism for all sufficiently large $|\varrho|$. Thus, by virtue of 7.2.5 Theorem 4, there exists a discrete set $D \subset \mathbb{C}$ with countable intersections with all compact subsets of $\mathbb{C}$ such that (8.82) are isomorphisms for all $z \in \mathbb{C} \setminus D$. Because of the ellipticity this holds then for all $s \in \mathbb{R}$ at the same time.

Exercise 4. Let $A \in C^m(\mathbb{B}, \underline{g})$ be elliptic. Then the Laurent coefficients of $\sigma_M^m(A)^{-1}(z)$ at the poles $d_j \in D$ at $(d_j - z)^{-(k+1)}$, $0 \leq k \leq m_j + 1$ (where m_j is the multiplicity of d_j) are finite-dimensional operators in $L^{-\infty}(X \times X)$.

Definition 5. Let $A \in C^m(\mathbb{B}, \underline{g})$ for $\underline{g} = (\gamma, \gamma - m, \Theta)$ be given. An operator $P \in C^{-m}(\mathbb{B}, \underline{h})$ for $\underline{h} = (\gamma - m, \gamma, \Theta)$ is called a *parametrix* of A if

$$AP - I \in C_G(\mathbb{B}, \underline{g}_r), \quad PA - I \in C_G(\mathbb{B}, \underline{g}_l) \tag{8.83}$$

with $\underline{g}_r = (\gamma - m, \gamma - m, \Theta)$, $\underline{g}_l = (\gamma, \gamma, \Theta)$.

Remark 6. Note that the condition (i) of Definition 1 implies the ellipticity of A in the sense of the class $L^m_{cl}(\mathrm{int}\ \mathbb{B})$. Thus, according to the standard pseudo-differential calculus, there is then a properly supported parametrix $P_0 \in L^{-m}_{cl}(\mathrm{int}\ \mathbb{B})$ of A. As we know, it satisfies $AP_0 - I, P_0 A - I \in L^{-\infty}(\mathrm{int}\ \mathbb{B})$. We could also relax the condition (8.83) by requiring the existence of an operator P_1 such that

$$AP_1 - I \in C_{M+G}(\mathbb{B}, \underline{g}_r), \quad P_1 A - I \in C_{M+G}(\mathbb{B}, \underline{g}_l), \tag{8.84}$$

because $C_{M+G}(\mathbb{B}, \dots) \subset L^{-\infty}(\mathrm{int}\ \mathbb{B})$. In other words the notion of a parametrix is not canonically defined.

Theorem 7. *Let $A \in C^m(\mathbb{B}, \underline{g})$, $\underline{g} = (\gamma, \gamma - m, \Theta)$, be given. Then the following conditions are equivalent:*

(i) *A is elliptic in the sense of Definition 1,*

(ii) *the operator*

$$A : \ \mathcal{H}^{s,\gamma}(\mathbb{B}) \longrightarrow \mathcal{H}^{s-m,\gamma-m}(\mathbb{B}) \tag{8.85}$$

is Fredholm for an $s = s_0 \in \mathbb{R}$.

If A is elliptic then (8.85) is Fredholm for all $s \in \mathbb{R}$. There exists a parametrix P of A in the sense of Definition 5. Furthermore

$$Au = f \in \mathcal{H}^{s-m,\gamma-m}(\mathbb{B}),$$

for any $s \in \mathbb{R}$ and

$$u \in \mathcal{H}^{-\infty,\gamma}(\mathbb{B}) \tag{8.86}$$

implies $u \in \mathcal{H}^{s,\gamma}(\mathbb{B})$. Finally

$$Au = f \in \mathcal{H}^{s-m,\gamma-m}_Q(\mathbb{B}) \tag{8.87}$$

for some asymptotic type Q to $(\gamma - m, \Theta)$ and (8.86) implies $u \in \mathcal{H}^{s,\gamma}_R(\mathbb{B})$ with an asymptotic type R to (γ, Θ). This holds for all $s \in \mathbb{R}$.

Proof. We shall only prove here $(i) \Rightarrow (ii)$. For the converse direction we refer to [Sc1][2.2.1. Theorem 14]. The Fredholm property of (8.85) will follow from the construction of a parametrix. First, the ellipticity of A implies the ellipticity of A interpreted as an element in $L_{cl}^m(\text{int } \mathbb{B})$. Thus, there is a parametrix $P_{int} \in L_{cl}^m(\text{int } \mathbb{B})$. We can construct P_{int} in a neighbourhood of $\partial\mathbb{B}$ in a rather precise manner. We use the coordinates $(r, x) \in X^\wedge$ in a collar neighbourhood of $\partial\mathbb{B}$. Then A has a complete symbol of the form $r^{-m}a(r, x, r\varrho, \xi)$ with $a(r, x, \tilde{\varrho}, \xi) \in S_{cl}^m(\overline{\mathbb{R}}_+ \times \mathbb{R}^n \times \mathbb{R}_{\tilde{\varrho},\xi}^{1+n})$ (this refers to a chart $U \to \mathbb{R}^n$ of a coordinate neighbourhood U on X). According to 8.1.3 Exercise 1 we choose a Leibniz inverse $p(r, x, r\varrho, \xi)$ of $a(r, x, r\varrho, \xi)$. Then we form the operator (8.58) and apply 8.1.3 Theorem 2. An $f(r, z) \in C^\infty(\overline{\mathbb{R}}_+, M_O^{-m}(X))$ then follows, with

$$P_0 - op_M^\delta(f) \in L^{-\infty}(X^\wedge)$$

for every $\delta \in \mathbb{R}$. Moreover it is easy to see that

$$\sigma_M^m(A)(z)^{-1} - f(0, z) =: \tilde{f}(z) \in M_R^{-\infty}(X)$$

for some asymptotic type R (exercise!). Let us set

$$\tilde{P} = \omega r^m op_M^{\gamma - m - \frac{n}{2}}(T^{-m}f)\omega_0 + (1 - \omega)P_0(1 - \omega_1) + W \qquad (8.88)$$

with $W := \tilde{\omega} r^m op_M^{\gamma - m - \frac{n}{2}}(T^{-m}\tilde{f})\tilde{\omega}$, where $\tilde{\omega}, \omega, \omega_0, \omega_1$ are cut-off functions with $\omega\omega_0 = \omega$, $\omega\omega_1 = \omega$. Then, we obtain

$$\tilde{P}A = 1 + W_l, \qquad A\tilde{P} = 1 + W_r,$$

with $W_l \in C_{M+G}(\mathbb{B}, \underline{g}_l)$, $W_r \in C_{M+G}(\mathbb{B}, \underline{g}_r)$. This is an immediate consequence of $\tilde{P} \in C^{-m}(\mathbb{B}, \underline{h})$ and of $\tilde{P}A - 1, A\tilde{P} - 1 \in L^{-\infty}(\text{int } \mathbb{B})$, where in addition we use 8.1.4 Remark 2. From the second formula of (8.73) we see that

$$\{T^m \sigma_M^{-m}(P_1)\}\sigma_M^m(A) = 1, \quad \{T^{-m}\sigma_M^m(A)\}\sigma_M^{-m}(P_1) = 1,$$

which is a consequence of the particular choice of $\tilde{f}(z)$. Thus,

$$\sigma_M^0(W_l)(z) = \sigma_M^0(W_r)(z) = 0.$$

From (8.78) we now obtain that for $N \geq k$ (where $\Theta = (-k, 0]$)

$$G := -1 + \sum_{j=0}^{N}(-1)^j W_l^j(1 + W_l) \in C_G(\mathbb{B}, \underline{g}_l). \qquad (8.89)$$

In fact, $\sigma_M^{-j}(G)(z) = 0$ for $j = 0, \ldots, k - 1$ implies (8.89), cf. 8.1.2 Proposition 11. Then we get a left parametrix P of A by putting $P = \{\sum_{j=0}^{N}(-1)^j W_l^j\}\tilde{P}$. In an analogous manner we obtain a right parametrix. Then, P is a parametrix of A in the sense of Definition 5.

The second part of Theorem 7 follows by using P as a left parametrix. Let us consider, for instance, the part of the assertion that refers to the asymptotics. The case without asymptotics is simpler and left to the reader. From (8.87) we get $PAu = Pf$. Applying 8.1.4 Theorem 3 we get $Pf \in \mathcal{H}_{Q_1}^{s,\gamma}(\mathbb{B})$ with some new asymptotic type Q_1. $PAu = u + Gu$ with some $G \in C_G(\mathbb{B}, \underline{g}_l)$ follows from (8.83). Now (8.86) and (8.21) show $Gu \in \mathcal{H}_{Q_2}^{\infty;\gamma}(\mathbb{B})$ with some asymptotic type Q_2. This yields $u \in \mathcal{H}_{Q_2}^{\infty,\gamma}(\mathbb{B}) + \mathcal{H}_{Q_1}^{s,\gamma}(\mathbb{B})$, but the latter space is contained in $\mathcal{H}_R^{s,\gamma}(\mathbb{B})$ for some R. $\qquad\square$

Remark 8. The condition (i) of Definition 1 implies the existence of a parametrix P_1 in the sense of (8.84). In fact, with the notation of formula (8.88) it suffices to set

$$P_1 = \omega r^m op_M^{\gamma - m - \frac{n}{2}}(T^{-m}f)\omega_0 + (1 - \omega)P_0(1 - \omega_1)$$

with cut-off functions $\omega, \omega_0, \omega_1$, supported by a collar neighbourhood of $\partial\mathbb{B}$, with $\omega\omega_0 = \omega$, $\omega\omega_1 = \omega$.

8.2   The algebra on the infinite cone

8.2.1   Symbols in $\mathbb{R}^n$ with exit behaviour

The goal of the remaining part of this chapter is to extend the analysis on manifolds with conical singularities to the case of the infinite (open streched) cone $X^\wedge = \mathbb{R}_+ \times X$. The main technical points are the specific effects for $r \to \infty$ and the ellipticity under which the corresponding operators are Fredholm between $\mathcal{K}^{s,\gamma}(X^\wedge)$-spaces. Since the properties for finite r and for $r \to \infty$ can be discussed separately, we will first neglect the conical singularity and talk about the contribution from $r \to \infty$. The global results will follow from local considerations by a partition of unity argument and from the material of Section 8.1. So we shall begin with pseudo-differential operators globally in $\mathbb{R}^n$, cf. also [Co]. The dimension n is arbitrary and has nothing to do with the above dimension of X.

The present theory may be regarded as an aspect of the general important question of analyzing pseudo-differential operators on a non-compact manifold, here with emphasis on the *exits* $r \to \infty$. The cases we are interested in are motivated by the edge symbolic calculus below which deals with operator families on $X^\wedge$, parametrized by points of the cotangent bundle of the edge. In this sense the material will play an auxiliary role. For this reason we will often look at classical operators which occur in our applications.

Let $m, \mu \in \mathbb{R}$ and denote by

$$S^{m,\mu}(\mathbb{R}^n \times \mathbb{R}^n) \tag{8.90}$$

the space of all $a(x,\xi) \in C^\infty(\mathbb{R}^n \times \mathbb{R}^n)$ such that

$$|D_x^\alpha D_\xi^\beta a(x,\xi)| \leq c(1+|\xi|)^{m-|\beta|}(1+|x|)^{\mu-|\alpha|} \tag{8.91}$$

with a constant $c = c(\alpha,\beta) > 0$ for all $(x,\xi) \in \mathbb{R}^n \times \mathbb{R}^n$ and all multi-indices $\alpha, \beta \in \mathbb{Z}_+^n$. Note that in contrast to the "usual" symbol spaces $S^m(\mathbb{R}^n \times \mathbb{R}^n)$ we have here the additional order μ in x. This controls a corresponding behaviour for $|x| \to \infty$. We intend to associate with the symbols $a(x,\xi) \in S^{m,\mu}(\mathbb{R}^n \times \mathbb{R}^n)$ pseudo-differential operators in $\mathbb{R}^n$, acting between the spaces

$$H^{s,\delta}(\mathbb{R}^n) := (1+|x|^2)^{-\frac{\delta}{2}} H^s(\mathbb{R}^n).$$

More precisely, we will set

$$L^{m,\mu}(\mathbb{R}^n) = \{Op(a) + C : a(x,\xi) \in S^{m,\mu}(\mathbb{R}^n \times \mathbb{R}^n), \;\; C \in L^{-\infty,-\infty}(\mathbb{R}^n)\} \tag{8.92}$$

where $L^{-\infty,-\infty}(\mathbb{R}^n)$ is defined as the space of all integral operators

$$Cu(x) = \int\limits_{\mathbb{R}^n} c(x,x')u(x')dx'$$

with kernels $c(x,x') \in \mathcal{S}(\mathbb{R}^n \times \mathbb{R}^n)$. (We shall see below that the summands C in (8.92) are unnecessary in this definition, since they will also have the form $Op(a_\infty)$ for an $a_\infty(x,\xi) \in S^{-\infty,-\infty}(\mathbb{R}^n \times \mathbb{R}^n)$.) Many formal properties of the spaces (8.90) and (8.92) are rather similar to the corresponding ones of the local theory on an open subset in $\mathbb{R}^n$. The present section will consider the symbol spaces.

The best constants $c(\alpha,\beta)$ in the estimates (8.91) form a semi-norm system on $S^{m,\mu}(\mathbb{R}^n \times \mathbb{R}^n)$. Then $S^{m,\mu}(\mathbb{R}^n \times \mathbb{R}^n)$ becomes a Fréchet space.

Remark 1. For every $m \leq m'$, $\mu \leq \mu'$ there are continuous embeddings

$$S^{m,\mu}(\mathbb{R}^n \times \mathbb{R}^n) \hookrightarrow S^{m',\mu'}(\mathbb{R}^n \times \mathbb{R}^n).$$

Moreover, we have for arbitrary $m, \tilde{m}, \mu, \tilde{\mu}$,

$$S^{m,\mu}(\mathbb{R}^n \times \mathbb{R}^n) \cdot S^{\tilde{m},\tilde{\mu}}(\mathbb{R}^n \times \mathbb{R}^n) \subseteq S^{m+\tilde{m},\mu+\tilde{\mu}}(\mathbb{R}^n \times \mathbb{R}^n).$$

Note that

$$a^{m,\mu}(x,\xi) := (1+|x|^2)^{\frac{\mu}{2}}(1+|\xi|)^{\frac{m}{2}} \in S^{m,\mu}(\mathbb{R}^n \times \mathbb{R}^n). \tag{8.93}$$

The multiplication by $a^{m,\mu}(x,\xi)$ induces isomorphisms

$$S^{\tilde{m},\tilde{\mu}}(\mathbb{R}^n \times \mathbb{R}^n) \to S^{m+\tilde{m},\mu+\tilde{\mu}}(\mathbb{R}^n \times \mathbb{R}^n)$$

for all $\tilde{m}, \tilde{\mu}$.

We set

$$S^{-\infty,-\infty}(\mathbb{R}^n \times \mathbb{R}^n) := \bigcap_{m,\mu\in\mathbb{R}} S^{m,\mu}(\mathbb{R}^n \times \mathbb{R}^n).$$

Analogously we can form the space $S^{-\infty,\mu}(\mathbb{R}^n \times \mathbb{R}^n)$ and $S^{m,-\infty}(\mathbb{R}^n \times \mathbb{R}^n)$. Note that

$$\begin{aligned}
S^{-\infty,-\infty}(\mathbb{R}^n \times \mathbb{R}^n) &\cong \mathcal{S}(\mathbb{R}^n \times \mathbb{R}^n), \\
S^{-\infty,\mu}(\mathbb{R}^n \times \mathbb{R}^n) &\cong \mathcal{S}(\mathbb{R}^n_\xi, S^\mu(\mathbb{R}^n_x)), \\
S^{m,-\infty}(\mathbb{R}^n \times \mathbb{R}^n) &\cong \mathcal{S}(\mathbb{R}^n_x, S^m(\mathbb{R}^n_\xi))
\end{aligned}$$

(remember that $S^m(\mathbb{R}^n_\xi)$ is the space of all symbols in ξ of order m that are independent of x). The following two theorems may be regarded as exercises for the reader.

Theorem 2. *Let $a_j(x,\xi) \in S^{m_j,\mu_j}(\mathbb{R}^n \times \mathbb{R}^n)$, $j \in \mathbb{Z}_+$, be an arbitrary sequence with $m_j \to -\infty$, $\mu_j \to -\infty$ as $j \to \infty$. Then there is an $a(x,\xi) \in S^{m,\mu}(\mathbb{R}^n \times \mathbb{R}^n)$ for $m = \max\{m_j\}$, $\mu = \max\{\mu_j\}$ such that for every $M \in \mathbb{R}$ there is an $N = N(M) \in \mathbb{Z}_+$ with*

$$a(x,\xi) - \sum_{j=0}^{N} a_j(x,\xi) \in S^{m-M,\mu-M}(\mathbb{R}^n \times \mathbb{R}^n),$$

and $a(x,\xi)$ is unique mod $S^{-\infty,-\infty}(\mathbb{R}^n \times \mathbb{R}^n)$.

If $\chi(x,\xi)$ is an excision function in the variables $(x,\xi) \in \mathbb{R}^{2n}$ (i.e., $\chi \in C^\infty(\mathbb{R}^{2n})$, $\chi(x,\xi) = 0$ in a neighbourhood of $(x,\xi) = 0$, $\chi(x,\xi) = 1$ for $|x|^2 + |\xi|^2 > c$ with a $c > 0$) then $a(x,\xi)$ may be obtained from a convergent series in $S^{m,\mu}(\mathbb{R}^n \times \mathbb{R}^n)$

$$a(x,\xi) = \sum_{j=0}^{\infty} \chi(\frac{x}{c_j}, \frac{\xi}{c_j}) a_j(x,\xi)$$

with suitable $c_j > 0$, $c_j \to \infty$ sufficiently fast as $j \to \infty$.

We call a the *asymptotic sum* of a_j and also write $a \sim \sum a_j$. Analogous notation will be used in connection with the following:

Theorem 3. *(i) Let $a_j(x,\xi) \in S^{m_j,\mu}(\mathbb{R}^n \times \mathbb{R}^n)$, $j \in \mathbb{Z}_+$, be an arbitrary sequence with $m_j \to -\infty$ as $j \to \infty$. Then there is an $a(x,\xi) \in S^{m,\mu}(\mathbb{R}^n \times \mathbb{R}^n)$ with $m = \max\{m_j\}$, such that for every $M \in \mathbb{R}$ there is an $N = N(M) \in \mathbb{Z}_+$ with*

$$a(x,\xi) - \sum_{j=0}^{N} a_j(x,\xi) \in S^{m-M,\mu}(\mathbb{R}^n \times \mathbb{R}^m),$$

and $a(x,\xi)$ is unique mod $S^{-\infty,\mu}(\mathbb{R}^n \times \mathbb{R}^n)$.

(ii) Let $b_k(x,\xi) \in S^{m,\mu_k}(\mathbb{R}^n \times \mathbb{R}^n)$, $k \in \mathbb{Z}_+$, be arbitrary, $\mu_k \to -\infty$ as $k \to \infty$. Then there is a $b(x,\xi) \in S^{m,\mu}(\mathbb{R}^n \times \mathbb{R}^n)$ with $\mu = \max\{\mu_k\}$, such that for every $M \in \mathbb{R}$ there is an $N = N(M) \in \mathbb{Z}_+$ with

$$b(x,\xi) - \sum_{k=0}^{N} b_k(x,\xi) \in S^{m,\mu-M}(\mathbb{R}^n \times \mathbb{R}^n),$$

and $b(x,\xi)$ is unique mod $S^{m,-\infty}(\mathbb{R}^n \times \mathbb{R}^n)$.

The assertion (ii) follows from (i) by interchanging the roles of x and ξ. If $\chi(\xi)$ is an excision function in $\xi \in \mathbb{R}^n$, then $a(x,\xi)$ can be obtained as a convergent series

$$a(x,\xi) = \sum_{j=0}^{\infty} \chi(\frac{\xi}{c_j}) a_j(x,\xi)$$

for suitable $c_j > 0$, $c_j \to \infty$ for $j \to \infty$.

A symbol $a(x,\xi) \in S^{m,\mu}(\mathbb{R}^n \times \mathbb{R}^n)$ is called *elliptic* if there are constants $c, R > 0$ such that

$$|a(x,\xi)| \geq c[\xi]^m [x]^\mu \quad \text{for all} \quad |x,\xi| \geq R. \tag{8.94}$$

Here $\xi \to [\xi]$ is a strictly positive C^∞ function in $\mathbb{R}^n$ with $[\xi] = |\xi|$ for $|\xi| > $ const with a constant > 0. If $\chi(x,\xi)$ is an excision function in $\mathbb{R}^{2n}$ that vanishes for $|x,\xi| \leq R$ then (8.94) implies

$$\chi(x,\xi) a^{-1}(x,\xi) \in S^{-m,-\mu}(\mathbb{R}^n \times \mathbb{R}^n). \tag{8.95}$$

8.2.2 Classical symbols

Our next objective is subclasses of $S^{m,\mu}(\mathbb{R}^n \times \mathbb{R}^n)$ which are classical in (x,ξ)-variables or only in x or ξ alone. To this end we first formulate the space of homogeneous functions in the corresponding variables. Let us set

$$S_\xi^{(m)} = \{a(x,\xi) \in C^\infty(\mathbb{R}^n \times (\mathbb{R}^n \setminus \{0\})) : a(x,\lambda\xi) = \lambda^m a(x,\xi)$$
$$\text{for all } \lambda > 0 \text{ and all } (x,\xi) \in \mathbb{R}^n \times (\mathbb{R}^n \setminus \{0\})\},$$

$m \in \mathbb{R}$. By interchanging the roles of x and ξ we get the spaces

$$S_x^{(\mu)}, \quad \mu \in \mathbb{R}$$

analogously. Furthermore, we set

$$S_{\xi,x}^{(m,\mu)} = S_\xi^{(m)} \cap S_x^{(\mu)}.$$

The latter space consists of all $a(x,\xi) \in C^\infty((\mathbb{R}^n \setminus \{0\}) \times (\mathbb{R}^n \setminus \{0\}))$ such that

$$a(\delta x, \lambda \xi) = \delta^\mu \lambda^m a(x,\xi) \quad \text{for all} \quad \lambda, \delta > 0, \; x \neq 0, \; \xi \neq 0.$$

It will also be useful to define the spaces

$$S_\xi^{[m]}$$
$$= \{a(x,\xi) \in C^\infty(\mathbb{R}^n \times \mathbb{R}^n) : a(x,\lambda\xi) = \lambda^m a(x,\xi) \text{ for all } \lambda \geq 1,\ x \in \mathbb{R}^n,\ |\xi| \geq c\}$$

with constants $c > 0$ dependent on a. Analogously we get $S_x^{[\mu]}$, and furthermore

$$S_{\xi,x}^{[m,\mu]} = \{a(x,\xi) \in C^\infty(\mathbb{R}^n \times \mathbb{R}^n) : a(\delta x, \lambda\xi) = \delta^\mu \lambda^m a(x,\xi)$$
$$\text{for all } \delta \geq 1,\ \lambda \geq 1,\ |x| \geq c,\ |\xi| \geq c\}$$

with a-dependent constants $c > 0$. Note that

$$S_{\xi,x}^{[m],[\mu]} := S_\xi^{[m]} \cap S_x^{[\mu]} \subseteq S_{\xi,x}^{[m,\mu]}.$$

Let us also introduce

$$S^{m,[\mu]} := S_x^{[\mu]} \cap S^{m,\mu}, \quad S^{[m],\mu} := S_\xi^{[m]} \cap S^{m,\mu}$$

with the symbol class $S^{m,\mu} = S^{m,\mu}(\mathbb{R}^n \times \mathbb{R}^n)$, cf. (8.90). We can also define the subclasses

$$S_{cl(\xi)}^{m,[\mu]} \subset S^{m,[\mu]}, \quad S_{cl(\xi)}^{m,\mu} \subset S^{m,\mu} \tag{8.96}$$

of elements that are classical in the ξ-variables and analogously

$$S_{cl(x)}^{[m],\mu} \subset S^{[m],\mu}, \quad S_{cl(x)}^{m,\mu} \subset S^{m,\mu} \tag{8.97}$$

the subspaces, classical in the x-variables. Since (8.97) is defined by interchanging the roles of x and ξ, μ and m, it suffices to explain (8.96).

The space $S_{cl(\xi)}^{m,[\mu]}$ is defined as the set of all $a(x,\xi) \in S^{m,[\mu]}$ such that there are $a_k \in S^{[m-k],[\mu]}$, $k \in \mathbb{Z}_+$, with

$$a - \sum_{k=0}^{N} a_k \in S^{m-(N+1),\mu} \quad \text{for all } N \in \mathbb{Z}_+.$$

The space $S_{cl(\xi)}^{m,\mu}$ consists of all $a(x,\xi) \in S^{m,\mu}$ such that there are $a_k \in S^{[m-k],\mu}$, $k \in \mathbb{Z}_+$, with

$$a - \sum_{k=0}^{N} a_k \in S^{m-(N+1),\mu} \quad \text{for all } N \in \mathbb{Z}_+.$$

Definition 1. The space of *classical symbols with exit condition* for $|x| \to \infty$ is defined as the set

$$S_{cl(\xi,x)}^{m,\mu} := S_{cl(\xi,x)}^{m,\mu}(\mathbb{R}^n \times \mathbb{R}^n)$$

of all $a(x,\xi) \in S^{m,\mu}$ for which

(i) there are $a_k(x,\xi) \in S_{cl(x)}^{[m-k],\mu}$, $k \in \mathbb{Z}_+$, such that

$$a(x,\xi) - \sum_{k=0}^{N} a_k(x,\xi) \in S_{cl(x)}^{m-(N+1),\mu} \quad \text{for all} \ \ N \in \mathbb{Z}_+, \tag{8.98}$$

(ii) there are $b_j(x,\xi) \in S_{cl(\xi)}^{m,[\mu-j]}$, $j \in \mathbb{Z}_+$, such that

$$a(x,\xi) - \sum_{j=0}^{N} b_j(x,\xi) \in S_{cl(\xi)}^{m,\mu-(N+1)} \quad \text{for all} \ \ N \in \mathbb{Z}_+. \tag{8.99}$$

Remark 2. For every $m, \mu \in \mathbb{R}$ we have

$$S_{cl(x)}^{[m],\mu} \subset S_{cl(\xi,x)}^{m,\mu}, \quad S_{cl(\xi)}^{m,[\mu]} \subset S_{cl(\xi,x)}^{m,\mu}.$$

The conditions (i), (ii) of Definition 1 allow us to define canonical mappings

$$\sigma_\psi^{m-k} \ : \ S_{cl(\xi,x)}^{m,\mu} \to S_\xi^{(m-k)}, \tag{8.100}$$
$$\sigma_e^{\mu-j} \ : \ S_{cl(\xi,x)}^{m,\mu} \to S_x^{(\mu-j)} \tag{8.101}$$

for all $k, j \in \mathbb{Z}_+$, satisfying

$$\sigma_\psi^{m-k}(a) = a_k(x,\xi) \quad \text{for} \ \ |\xi| \geq c,$$
$$\sigma_e^{\mu-j}(a) = b_j(x,\xi) \quad \text{for} \ \ |x| \geq c$$

for sufficiently large constants $c > 0$. The image under (8.100) consists of classical symbols with respect to x of order μ. Therefore we find for $\sigma_\psi^{m-k}(a)$ the unique sequence of homogeneous components in $x \neq 0$ of order $\mu - j$. If we denote its $(\mu - j)$th component by $\sigma_e^{\mu-j}\sigma_\psi^{m-k}(a)$, we get a sequence of maps

$$\sigma_e^{\mu-j}\sigma_\psi^{m-k} : S_{cl(\xi,x)}^{m,\mu} \to S_{\xi,x}^{(m-k,\mu-j)},$$

$j, k \in \mathbb{Z}_+$. In an analogous manner we obtain

$$\upsilon_\psi^{m-k}\sigma_e^{\mu-j} : S_{cl(\xi,x)}^{m,\mu} \to S_{\xi,x}^{(m-k,\mu-j)}$$

for all $j, k \in \mathbb{Z}_+$.

Exercise 3. We have for every $a \in S_{cl(\xi,x)}^{m,\mu}$,

$$\sigma_e^{\mu-j}\sigma_\psi^{m-k}(a) = \sigma_\psi^{m-k}\sigma_e^{\mu-j}(a)$$

for all $j, k \in \mathbb{Z}_+$.

For brevity, set

$$\sigma_\psi^{m-k}\sigma_e^{\mu-j}(a) =: \sigma_{\psi,e}^{m-k,\mu-j}(a).$$

For ellipticity we will be interested mainly in $k = j = 0$. The pair

$$\sigma_e^\mu(a), \quad \sigma_{\psi,e}^{m,\mu}(a) \tag{8.102}$$

will be called the *principal exit symbols* of $a(x,\xi) \in S_{cl(\xi,x)}^{m,\mu}$.

Exercise 4. $a \in S_{cl(\xi,x)}^{m,\mu}$, $b \in S_{cl(\xi,x)}^{r,\gamma}$ implies $ab \in S_{cl(\xi,x)}^{m+r,\mu+\gamma}$ and

$$\begin{aligned}
\sigma_\psi^{m+r}(ab) &= \sigma_\psi^m(a)\sigma_\psi^r(b), \\
\sigma_e^{\mu+\gamma}(ab) &= \sigma_e^\mu(a)\sigma_e^\gamma(b), \\
\sigma_{\psi,e}^{m+r,\mu+\gamma}(ab) &= \sigma_{\psi,e}^{m,\mu}(a)\sigma_{\psi,e}^{r,\gamma}(b).
\end{aligned}$$

Note that the operator of multiplication by $[\xi]^r[x]^\gamma$ induces isomorphisms

$$[\xi]^r[x]^\gamma : S_{cl(\xi,x)}^{m,\mu} \to S_{cl(\xi,x)}^{m+r,\mu+\gamma}$$

for all $r, \gamma, m, \mu \in \mathbb{R}$.

Remark 5. If $a_j(x,\xi) \in S_{cl(\xi,x)}^{m-j,\mu-j}$, $j \in \mathbb{Z}_+$, is an arbitrary sequence, then $a(x,\xi)\sim \sum_j a_j(x,\xi)$ in the sense of 8.2.1 Theorem 2 belongs to $S_{cl(\xi,x)}^{m,\mu}$.

Exercise 6. Let $a(x,\xi) \in S_{cl(\xi,x)}^{m,\mu}$. Then $\sigma_\psi^m(a) = 0$, $\sigma_e^\mu(a) = 0$ implies $a(x,\xi) \in S_{cl(\xi,x)}^{m-1,\mu-1}$.

Let us set

$$S_{cl(x)}^{(m),\mu} = \sigma_\psi^m S_{cl(\xi,x)}^{m,\mu}, \quad S_{cl(\xi)}^{m,(\mu)} = \sigma_e^\mu S_{cl(\xi,x)}^{m,\mu}.$$

Then, for every

$$p(x,\xi) \in S_{cl(x)}^{(m),\mu}, \quad q(x,\xi) \in S_{cl(\xi)}^{m,(\mu)}$$

with $\sigma_e^\mu(p) = \sigma_\psi^m(q) =: h \in S^{(m,\mu)}$ there exists an $a(x,\xi) \in S_{cl(\xi,x)}^{m,\mu}$ with

$$\sigma_\psi^m(a) = p, \quad \sigma_e^\mu(a) = q, \quad \sigma_{\psi,e}^{m,\mu}(a) = h.$$

In fact, if χ is an excision function in $\mathbb{R}^n$, it suffices to set

$$a(x,\xi) = \chi(\xi)p(x,\xi) + \chi(x)(q(x,\xi) - \chi(\xi)h(x,\xi)). \tag{8.103}$$

Definition 7. A symbol $a(x,\xi) \in S_{cl(\xi,x)}^{m,\mu}(\mathbb{R}^n \times \mathbb{R}^n)$ is called *elliptic* if

$$\begin{aligned}
\sigma_\psi^m(a)(x,\xi) \neq 0 \quad &\text{for all} \quad (x,\xi) \in \mathbb{R}^n \times (\mathbb{R}^n \setminus \{0\}), \\
\sigma_e^\mu(a)(x,\xi) \neq 0 \quad &\text{for all} \quad (x,\xi) \in (\mathbb{R}^n \setminus \{0\}) \times \mathbb{R}^n, \\
\sigma_{\psi,e}^{m,\mu}(a)(x,\xi) \neq 0 \quad &\text{for all} \quad (x,\xi) \in (\mathbb{R}^n \setminus \{0\}) \times (\mathbb{R}^n \setminus \{0\}).
\end{aligned}$$

Exercise 8. $a(x,\xi) \in S^{m,\mu}_{cl(\xi,x)}(\mathbb{R}^n \times \mathbb{R}^n)$ is elliptic iff it is elliptic in the sense of the class $S^{m,\mu}(\mathbb{R}^n \times \mathbb{R}^n)$, cf. (8.94).

Remark 9. Let $a(x,\xi) \in S^{m,\mu}_{cl(\xi,x)}$ be elliptic. Then

$$p(x,\xi) = \sigma^m_\psi(a)(x,\xi)^{-1}, \quad q(x,\xi) = \sigma^\mu_e(a)(x,\xi)^{-1}, \quad h(x,\xi) = \sigma^{m,\mu}_{\psi,e}(a)(x,\xi)^{-1}$$

satisfy $\sigma^{-\mu}_e(p) = \sigma^{-m}_\psi(q) = h$. Hence, according to (8.103), we find by

$$b(x,\xi) = \chi(\xi)\sigma^m_\psi(a)(x,\xi)^{-1} + \chi(x)\{\sigma^\mu_e(a)(x,\xi)^{-1} - \chi(\xi)\sigma^{m,\mu}_{\psi,e}(a)(x,\xi)^{-1}\}$$

a $b(x,\xi) \in S^{-m,-\mu}_{cl(\xi,x)}$ that is elliptic of order $(-m,-\mu)$ and satisfies

$$\begin{aligned}
\sigma^{-m}_\psi(b)(x,\xi) &= \sigma^m_\psi(a)(x,\xi)^{-1}, \\
\sigma^{-\mu}_e(b)(x,\xi) &= \sigma^\mu_e(a)(x,\xi)^{-1}, \\
\sigma^{-m,-\mu}_{\psi,e}(b)(x,\xi) &= \sigma^{m,\mu}_{\psi,e}(a)(x,\xi)^{-1}.
\end{aligned}$$

Let us finally remark that there is an alternative easy description of the symbol classes $S^{m,\mu}_{cl(\xi,x)}(\mathbb{R}^n \times \mathbb{R}^n)$. Set $B^n := \{y \in \mathbb{R}^n : |y| \leq 1\}$ and choose a diffeomorphism $\chi : \mathrm{int}B^n \to \mathbb{R}^n$ which satisfies $\chi(y) = y\{|y|(1 - |y|)\}^{-1}$ for $2/3 < |y| < 1$. Moreover, let $a(x,\xi) \in C^\infty(\mathbb{R}^n \times \mathbb{R}^n)$, and set

$$b(y,\eta) = (1 - [y])^\mu (1 - [\eta])^m (\chi^*_2 a)(y,\eta)$$

with the pull-back χ^*_2 under $\chi_2 := \chi \times \chi$, and $[y]$ being chosen in such a way that $[y] = |y|$ for $2/3 < |y| < 1$ and $1 - [y] \neq 0$ for all $|y| < 1$. Then $a(x,\xi) \to b(y,\eta)$ induces an isomorphism

$$D^{m,\mu} : S^{m,\mu}_{cl(\xi,x)}(\mathbb{R}^n \times \mathbb{R}^n) \to C^\infty(B^n \times B^n).$$

This allows us to introduce the Fréchet topology in $S^{m,\mu}_{cl(\xi,x)}(\mathbb{R}^n \times \mathbb{R}^n)$ by this bijection from $C^\infty(B^n \times B^n)$. Moreover, we have

$$\sigma^m_\psi(a) = (\chi^*_2)^{-1}\left(b|_{B^n \times S^{n-1}}\right), \tag{8.104}$$

$$\sigma^m_e(a) = (\chi^*_2)^{-1}\left(b|_{S^{n-1} \times B^n}\right), \tag{8.105}$$

$$\sigma^{m,\mu}_{\psi,e}(a) = (\chi^*_2)^{-1}\left(b|_{S^{n-1} \times S^{n-1}}\right). \tag{8.106}$$

In particular, the ellipticity of $a(x,\xi)$ in equivalent to the non-vanishing of $D^{m,\mu}a$ on

$$\partial(B^n \times B^n) = B^n \times S^{n-1} \cup S^{n-1} \times B^n = B^n \times S^{n-1} \sqcup_d S^{n-1} \times B^n \sqcup_d S^{n-1} \times S^{n-1},$$

where $\sqcup_d$ means the disjoint union. More generally, if we set

$$p(r,x,\varrho,\xi) = (1 - r)^\mu (1 - \varrho)^m a\left(\frac{x}{1-r}, \frac{\xi}{1-\varrho}\right)$$

for $0 < r, \varrho < 1$ and $|x| = |\xi| = 1$, then

$$\sigma^{m-k}_\psi \sigma^{\mu-j}_e a(x,\xi) = \frac{(-1)^{k+j}}{k!j!}\left(\frac{\partial}{\partial\varrho}\right)^k \left(\frac{\partial}{\partial r}\right)^j p(r,x,\varrho,\xi)|_{r=\varrho=1}.$$

8.2.3 Pseudo-differential operators in $\mathbb{R}^n$ with exit behaviour

We now study the elements of the calculus of pseudo-differential operators of the classes (8.92) and

$$L_{cl}^{m,\mu}(\mathbb{R}^n) = \{Op(a)+C : a(x,\xi) \in S_{cl(\xi,x)}^{m,\mu}(\mathbb{R}^n\times\mathbb{R}^n),\ C \in L^{-\infty,-\infty}(\mathbb{R}^n)\}. \tag{8.107}$$

As in the standard theory, it will be reasonable to consider more general symbol classes

$$S^{m,m',\mu,\mu'}(\mathbb{R}^{2n} \times \mathbb{R}^{2n}) \tag{8.108}$$

consisting of all $a(x,x',\xi,\xi') \in C^\infty(\mathbb{R}_{x,x'}^{2n} \times \mathbb{R}_{\xi,\xi'}^{2n})$ satisfying the estimates

$$|D_x^\alpha D_{x'}^{\alpha'} D_\xi^\beta D_{\xi'}^{\beta'} a(x,x',\xi,\xi')|$$
$$\leq\ c(1 + |\xi|)^{m-|\beta|}(1 + |\xi'|)^{m'-|\beta'|}(1 + |x|)^{\mu-|\alpha|}(1 + |x'|)^{\mu'-|\alpha'|}$$

with constants $c = c(\alpha,\alpha',\beta,\beta') > 0$, for all $(x,x',\xi,\xi') \in \mathbb{R}^{2n} \times \mathbb{R}^{2n}$ and all multi-indices $\alpha,\alpha',\beta,\beta' \in \mathbb{Z}_+^n$.

The assertions of Section 8.2.1 concerning the spaces $S^{m,\mu}(\mathbb{R}^n \times \mathbb{R}^n)$ have natural generalizations to (8.108). In particular, we can form asymptotic sums in (8.108) with respect to the various groups of variables, e.g. in (x,x',ξ,ξ') or ξ,ξ',x,x' alone.

Since $L^{m,\mu}(\mathbb{R}^n) \subset L^m(\mathbb{R}^n)$ with the ordinary class $L^m(\mathbb{R}^n)$ of pseudo-differential operators, every $A \in L^{m,\mu}(\mathbb{R}^n)$ induces a continuous operator

$$A : C_0^\infty(\mathbb{R}^n) \to C^\infty(\mathbb{R}^n). \tag{8.109}$$

Let $C_b^\infty(\mathbb{R}^n)$ be the space of all $u \in C^\infty(\mathbb{R}^n)$ such that

$$\sup_{x\in\mathbb{R}^n} |D_x^\alpha u(x)| < \infty \quad \text{for all}\ \ \alpha \in \mathbb{Z}_+^n.$$

Then $C_b^\infty(\mathbb{R}^n)$ is Fréchet in a natural way. The action

$$A : u \to Op(a)u(x) = \int\int e^{i(x-x')\xi} a(x,\xi)u(x')dx'\,d\xi$$

is interpreted as an oscillatory integral. As noted above in Section 8.2.2, we can always write

$$Au = A_N(1 - \Delta)^N u \tag{8.110}$$

with $\Delta = \sum_{i=1}^n \frac{\partial^2}{\partial x_i^2}$ being the Laplacian and $A_N = Op(a_N)$ with $a_N(x,\xi) = a(x,\xi)(1 + |\xi|^2)^{-N}$. Choosing N large enough, the kernel

$$K_N(x,x') = \int e^{i(x-x')\xi} a_N(x,\xi)d\xi \tag{8.111}$$

exists as a continuous function in $(x,x') \in \mathbb{R}^{2n}$.

Exercise 1. Every $A \in L^{m,0}(\mathbb{R}^n)$ induces a continuous operator

$$A : C_b^\infty(\mathbb{R}^n) \to C_b^\infty(\mathbb{R}^n),$$

and every $A \in L^{m,\mu}(\mathbb{R}^n)$ a continuous operator

$$A : \mathcal{S}(\mathbb{R}^n) \to \mathcal{S}(\mathbb{R}^n).$$

Hint: For every fixed semi-norm in $C_b^\infty(\mathbb{R}^n)$ ($\mathcal{S}(\mathbb{R}^n)$), use the representation (8.110) with sufficiently large N and employ the properties of the kernel function $K_N(x, x')$.

Remark 2. Every $A \in L^{-\infty,-\infty}(\mathbb{R}^n)$ can be written as $A = Op(a)$ with some $a(x, \xi) \in S^{-\infty,-\infty}(\mathbb{R}^n \times \mathbb{R}^n)$.

Exercise 3. Let $a(x, \xi) \in S^{m,0}(\mathbb{R}^n \times \mathbb{R}^n)$, $A = Op(a)$. Then, for $e_\xi(x) = e^{ix\xi}$ we have

$$a(x, \xi) = e_{-\xi}(x)(Ae_\xi)(x).$$

Hint: Use $e_\xi(x) \in C_b^\infty(\mathbb{R}^n)$.

Exercise 4. The mapping

$$Op : S^{m,\mu}(\mathbb{R}^n \times \mathbb{R}^n) \to L^{m,\mu}(\mathbb{R}^n)$$

is bijective.

The symbols $a(x, x', \xi, \xi') \in S^{m,m',\mu,\mu'}(\mathbb{R}^{2n} \times \mathbb{R}^{2n})$ also give rise to operators if we set

$$Op(a)u(x) = \int\int\int\int e^{-i(y\eta+y'\eta')} a(x, x+y, \eta, \eta') u(x+y+y') dy dy' đ\eta đ\eta' \tag{8.112}$$

for $u \in C_b^\infty(\mathbb{R}^n)$, $x \in \mathbb{R}^n$. (8.112) is to be interpreted as an oscillatory integral.

Let $a(x, x', \xi, \xi') \in S^{m,m',\mu,\mu'}(\mathbb{R}^{2n} \times \mathbb{R}^{2n})$ and form

$$b(x, \xi) := \int\int e^{-iy\eta} a(x, x+y, \xi+\eta, \xi) dy đ\eta \tag{8.113}$$

in the sense of oscillatory integrals.

Theorem 5. $a(x, x', \xi, \xi') \in S^{m,m',\mu,\mu'}(\mathbb{R}^{2n} \times \mathbb{R}^{2n})$ *implies* $b(x, \xi) \in S^{m+m',\mu+\mu'}(\mathbb{R}^n \times \mathbb{R}^n)$ *and*

$$Op(a) = Op(b). \tag{8.114}$$

The symbol b allows the asymptotic expansion

$$b(x, \xi) \sim \sum_\alpha \frac{1}{\alpha!} (\partial_\xi^\alpha D_{x'}^\alpha a)(x, x', \xi, \xi')|_{x'=x, \xi'=\xi}. \tag{8.115}$$

Proof. Let us first form

$$b_\theta(x,\xi) = \int\int e^{-iy\eta} a(x, x+y, \xi+\theta\eta, \xi)\,dy\,d\eta,$$

$\theta \in \mathbb{R}$. We shall show that $\{b_\theta\}_{|\theta|\leq 1}$ is a bounded set in $S^{m+m',\mu+\mu'}(\mathbb{R}^n \times \mathbb{R}^n)$. That means, in particular,

$$|b_\theta(x,\xi)| \leq c\langle\xi\rangle^{m+m'}\langle x\rangle^{\mu+\mu'} \tag{8.116}$$

for $\langle\xi\rangle = (1+|\xi|^2)^{\frac{1}{2}}$. In order to obtain this we choose numbers $p, q \in \mathbb{Z}_+$ such that $2p > n + |m|$, $2q > n + |\mu'|$ and set

$$r_\theta(x, y, \xi, \eta) = \langle\eta\rangle^{-2p}\langle D_y\rangle^{2p}\langle y\rangle^{-2q}\langle D_\eta\rangle^{2q} a(x, x+y, \xi+\theta\eta, \xi).$$

Then

$$b_\theta(x,\xi) = \int\int e^{-iy\eta} r_\theta(x, y, \xi, \eta)\,dy\,d\eta.$$

We have

$$\begin{aligned}
|r_\theta(x, y, \xi, \eta)| \;\leq\;& c\langle\eta\rangle^{-2p}\langle y\rangle^{-2q}\max_{|\gamma|\leq 2p}|D_y^\gamma\langle D_\eta\rangle^{2q} a(x, x+y, \xi+\theta\eta, \xi)| \\
\leq\;& c\langle\eta\rangle^{-2p}\langle y\rangle^{-2q}\langle\xi+\theta\eta\rangle^m\langle\xi\rangle^{m'}\langle x\rangle^\mu\langle x+y\rangle^{\mu'}.
\end{aligned}$$

Thus

$$|b_\theta(x,\xi)| \leq c\left\{\int \langle x+y\rangle^{\mu'}\langle y\rangle^{-2q}\langle x\rangle^\mu dy\right\}\left\{\int \langle\xi+\theta\eta\rangle^m\langle\eta\rangle^{-2p}\langle\xi\rangle^{m'} d\eta\right\}. \tag{8.117}$$

Set $D_1 = \{y : |y| \leq \frac{\langle x\rangle}{2}\}$, $D_2 = \{y : |y| > \frac{\langle x\rangle}{2}\}$ and write

$$\int \ldots dy = \int_{D_1} \ldots dy + \int_{D_2} \ldots dy.$$

Let us first look at the integral over D_1. We have

$$\frac{d}{dt}\langle x+ty\rangle = y\frac{x+ty}{\langle x+ty\rangle}$$

and

$$|\langle x+y\rangle - \langle x\rangle| = \left|\int_0^1 y\frac{x+ty}{\langle x+ty\rangle}dt\right| \leq |y|.$$

$y \in D_1$ then implies $|\langle x+y\rangle - \langle x\rangle| \leq \frac{\langle x\rangle}{2}$, and hence

$$\frac{\langle x\rangle}{2} \leq \langle x+y\rangle \leq \frac{3\langle x\rangle}{2}.$$

It follows that

$$
\int_{D_1} \langle x+y\rangle^{\mu'} \langle y\rangle^{-2q} \langle x\rangle^{\mu} dy \;=\; \langle x\rangle^{\mu+\mu'} \int_{D_1} \frac{\langle x+y\rangle^{\mu'}}{\langle x\rangle^{\mu'}} \langle y\rangle^{-2q} dy
$$

$$
\leq\; \langle x\rangle^{\mu+\mu'} c \int_{D_1} \langle y\rangle^{-2q} dy \leq c\langle x\rangle^{\mu+\mu'}.
$$

Here $c > 0$ are different suitable constants. For the integral over D_2 we write $\langle x+y\rangle^2 = \langle x\rangle^2 + |y|^2$. It follows that $\langle x+y\rangle \leq \langle x\rangle + |y| \leq 3|y|$ for $y \in D_2$. Thus

$$
\int_{D_2} \langle x+y\rangle^{\mu'} \langle y\rangle^{-2q} \langle x\rangle^{\mu} dy \;\leq\; \langle x\rangle^{\mu} \int_{D_2} (3|y|)^{\mu'} \langle y\rangle^{-2q} dy
$$

$$
=\; c\langle x\rangle^{\mu} \int_{\frac{\langle x\rangle}{2}}^{\infty} \frac{t^{\mu'+n-1}}{(1+t^2)^q} dt
$$

$$
\leq\; c\langle x\rangle^{\mu} \int_{\frac{\langle x\rangle}{2}}^{\infty} t^{\mu'+n-1-2q} dt
$$

$$
=\; c\langle x\rangle^{\mu} \langle x\rangle^{\mu'+n-2q} \leq c\langle x\rangle^{\mu+\mu'},
$$

with different constants $c > 0$, again. The last inequality is true, since q was chosen sufficiently large. It follows that

$$
\int \langle x+y\rangle^{\mu'} \langle y\rangle^{-2q} \langle x\rangle^{\mu} dy \leq c\langle x\rangle^{\mu+\mu'}.
$$

In an analogous manner we get

$$
\int \langle \xi+\theta\eta\rangle^{m} \langle \eta\rangle^{-2p} \langle \xi\rangle^{m'} d\eta \leq c\langle \xi\rangle^{m+m'}.
$$

From (8.117) we then obtain (8.116). The next step is to prove

$$
|D_x^\alpha D_\xi^\beta b_\theta(x,\xi)| \leq c\langle \xi\rangle^{m+m'-|\beta|} \langle x\rangle^{\mu+\mu'-|\alpha|} \tag{8.118}
$$

for all multi-indices $\alpha, \beta \in \mathbb{Z}_+^n$. To this end we observe that

$$
a(x, x+y, \xi+\theta\eta, \xi)
$$

varies over a bounded set in $S^{m,\mu'}(\mathbb{R}_y^n \times \mathbb{R}_\eta^n)$ when $(x,\xi) \in K \subset\subset \mathbb{R}^{2n}$, $|\theta| \leq 1$, for arbitrary K (exercise!). This allows us to write

$$
D_x^\alpha D_\xi^\beta b_\theta(x,\xi) = \int\int e^{-iy\eta} D_x^\alpha D_\xi^\beta a(x, x+y, \xi+\theta\eta, \xi) dy d\eta,
$$

and we may argue as before, this time with the symbol function $D_x^\alpha D_\xi^\beta a(\ldots)$ under the oscillatory integral. The orders of growth as they were used in (8.117) are diminished in the expected way because of the symbol estimates for a, such that (8.118) is true. Now we look at

$$
\begin{aligned}
Op(b)u(x) &= \int\int e^{i(x-x')\xi}b(x,\xi)u(x')dx'\,d\xi \\
&= \int\int e^{i(x-x')\xi}\left\{\int\int e^{-iy\eta}a(x,x+y,\xi+\eta,\xi)dy\,d\eta\right\}u(x')dx'\,d\xi.
\end{aligned}
$$

For $x' := x + y + y'$, i.e. $x - x' = -(y + y')$, and $\eta' := \xi$, $\tilde{\eta} = \eta + \eta'$, we obtain

$$
Op(b)u(x) = \int\int\int\int e^{-i(y\tilde{\eta}+y'\eta')}a(x,x+y,\tilde{\eta},\eta')u(x+y+y')dy\,dy'\,d\tilde{\eta}\,d\eta'.
$$

This shows the formula (8.114).

It remains to show (8.115). To this end we apply the Taylor expansion of $a(x, x + y, \xi + \eta, \xi)$ in η near $\eta = 0$

$$
\begin{aligned}
a(x,x+y,\xi+\eta,\xi) &= \sum_{|\alpha|<N}\frac{\eta^\alpha}{\alpha!}\partial_\zeta^\alpha a(x,x+y,\zeta,\xi)|_{\zeta=\xi} \\
&+ N\sum_{|\beta|=N}\frac{\eta^\beta}{\beta!}\int_0^1(1-\theta)^{N-1}(\partial_\zeta^\beta a)(x,x+y,\zeta,\xi)|_{\zeta=\xi+\theta\eta}d\theta.
\end{aligned}
$$

This gives us, using (8.113),

$$
\begin{aligned}
b(x,\xi) &= \sum_{|\alpha|<N}\int\int e^{-iy\eta}\frac{\eta^\alpha}{\alpha!}\partial_\zeta^\alpha a(x,x+y,\zeta,\xi)|_{\zeta=\xi}dy\,d\eta \qquad (8.119)\\
&+ N\sum_{|\beta|=N}\int_0^1(1-\theta)^{N-1}\int\int e^{-iy\eta}\frac{\eta^\beta}{\beta!}(\partial_\zeta^\beta a)(x,x+y,\zeta,\xi)|_{\zeta=\xi+\theta\eta}dy\,d\eta\,d\theta.
\end{aligned}
$$

The first sum on the right of (8.119) equals

$$
\begin{aligned}
&\sum_{|\alpha|<N}\frac{1}{\alpha!}\int\int e^{-iy\eta}(D_{x'}^\alpha\partial_\zeta^\alpha a)(x,x',\zeta,\xi)|_{x'=x+y,\zeta=\xi}dy\,d\eta \\
&= \sum_{|\alpha|<N}\frac{1}{\alpha!}(\partial_\zeta^\alpha D_{x'}^\alpha a)(x,x',\zeta,\xi)|_{x'=x,\zeta=\xi}.
\end{aligned}
$$

Here we have used $\int\int e^{-iy\eta}f(y)dy\,d\eta = f(0)$. Remember that the integrals are interpreted in the oscillatory sense.

For the second sum on the right we get

$$N \sum_{|\beta|=N} \int_0^1 (1-\theta)^{N-1} \frac{1}{\beta!} \int\int e^{-iy\eta} \partial_\zeta^\beta D_{x'}^\beta a(x,x',\zeta,\xi)_{x'|=x+y,\zeta=\xi+\theta\eta} \, dy \, d\eta \, d\theta,$$

which belongs to $S^{m+m'-N,\mu+\mu'-N}(\mathbb{R}^n \times \mathbb{R}^n)$ by the first part of the proof. Thus we have obtained (8.115). $\qquad\qquad\square$

Exercise 6. $A \in L^{m,\mu}(\mathbb{R}^n)$, $C \in L^{-\infty,-\infty}(\mathbb{R}^n)$ implies $AC, CA \in L^{-\infty,-\infty}(\mathbb{R}^n)$.

Theorem 7. $A \in L^{m,\mu}(\mathbb{R}^n)$, $B \in L^{\tilde{m},\tilde{\mu}}(\mathbb{R}^n)$ *implies* $AB \in L^{m+\tilde{m},\mu+\tilde{\mu}}(\mathbb{R}^n)$. *For* $A = Op(a)$ *with* $a(x,\xi) \in S^{m,\mu}(\mathbb{R}^n \times \mathbb{R}^n)$, $B = Op(b)$ *with* $b(x,\xi) \in S^{\tilde{m},\tilde{\mu}}(\mathbb{R}^n \times \mathbb{R}^n)$ *we obtain* $AB=Op(c)$ *with a symbol* $c(x,\xi) \in S^{m+\tilde{m},\mu+\tilde{\mu}}(\mathbb{R}^n \times \mathbb{R}^n)$ *that allows the asymptotic expansion*

$$c(x,\xi) \sim \sum_\alpha \frac{1}{\alpha!} \partial_\xi^\alpha a(x,\xi) D_x^\alpha b(x,\xi). \tag{8.120}$$

Proof. Applying Exercise 6 it suffices to assume $A = Op(a)$, $B = Op(b)$ with the indicated a,b. Let us set $p(x,x',\xi,\xi') = a(x,\xi)b(x',\xi')$. Then $AB = Op(p)$ in the sense of (8.112). From Theorem 5 it follows that

$$c(x,\xi) \sim \sum_\alpha \frac{1}{\alpha!} \partial_\xi^\alpha D_{x'}^\alpha p(x,x',\xi,\xi')|_{x'=x,\xi'=\xi} \tag{8.121}$$

satisfies $Op(p) = Op(c)$. It is obvious that (8.121) coincides with (8.120). $\qquad\square$

Corollary 8. $A \in L_{cl}^{m,\mu}(\mathbb{R}^n)$, $B \in L_{cl}^{\tilde{m},\tilde{\mu}}(\mathbb{R}^n)$ *implies* $AB \in L_{cl}^{m+\tilde{m},\mu+\tilde{\mu}}(\mathbb{R}^n)$ *and*

$$\sigma_\psi^{m+\tilde{m}}(AB) = \sigma_\psi^m(A)\sigma_\psi^{\tilde{m}}(B), \quad \sigma_e^{\mu+\tilde{\mu}}(AB) = \sigma_e^\mu(A)\sigma_e^{\tilde{\mu}}(B), \tag{8.122}$$
$$\sigma_{\psi,e}^{m+\tilde{m},\mu+\tilde{\mu}}(AB) = \sigma_{\psi,e}^{m,\mu}(A)\sigma_{\psi,e}^{\tilde{m},\tilde{\mu}}(B). \tag{8.123}$$

Exercise 9. $A \in L^{m,\mu}(\mathbb{R}^n)$ implies $A^* \in L^{m,\mu}(\mathbb{R}^n)$ where $*$ indicates the formal adjoint in the sense $(Au,v) = (u,A^*v)$ for all $u,v \in \mathcal{S}(\mathbb{R}^n)$, $(\cdot,\cdot)$ being the $L_2(\mathbb{R}^n)$-scalar product. For $A = Op(a)$ with $a(x,\xi) \in S^{m,\mu}(\mathbb{R}^n \times \mathbb{R}^n)$ it follows that $A^* = Op(p)$ for $p(x',\xi) = \overline{a(x',\xi)}$ and $A = Op(a^*)$ mod $L^{-\infty,-\infty}(\mathbb{R}^n)$ with $a^*(x,\xi) \in S^{m,\mu}(\mathbb{R}^n \times \mathbb{R}^n)$,

$$a^*(x,\xi) \sim \sum_\alpha \frac{1}{\alpha!} \partial_\xi^\alpha D_x^\alpha \overline{a(x,\xi)}.$$

Hint: Employ Theorem 5.

Corollary 10. $A \in L_{cl}^{m,\mu}(\mathbb{R}^n)$ *implies* $A^* \in L_{cl}^{m,\mu}(\mathbb{R}^n)$ *and*

$$\sigma_\psi^m(A^*) \;=\; \overline{\sigma_\psi^m(A)}, \;\; \sigma_e^\mu(A^*) = \overline{\sigma_e^\mu(A)}, \tag{8.124}$$

$$\sigma_{\psi,e}^{m,\mu}(A^*) \;=\; \overline{\sigma_{\psi,e}^{m,\mu}(A)}. \tag{8.125}$$

Theorem 11. *Let* $A \in L^{m,\mu}(\mathbb{R}^n)$, $m,\mu \in \mathbb{R}$. *Then* (8.109) *has extensions to continuous operators*

$$A : H^{s,\delta}(\mathbb{R}^n) \to H^{s-m,\delta-\mu}(\mathbb{R}^n) \tag{8.126}$$

for all $s, \delta \in \mathbb{R}$.

Proof. The result for $m = \mu = 0$ and $s = \delta = 0$ follows from the Calderón-Vaillancourt theorem. From that we immediately get the result for arbitrary m, s, μ, δ. In fact, using the symbols $a^{m,\mu}(x,\xi)$ from 8.2.1(8.93) we get isomorphisms

$$Op(a^{s,\delta}) : H^{s,\delta}(\mathbb{R}^n) \to H^{0,0}(\mathbb{R}^n) = L_2(\mathbb{R}^n) \tag{8.127}$$

and then (8.126) can be written as a composition

$$A = Op(a^{s-m,\delta-\mu})^{-1} A_0 \, Op(a^{s,\delta}) \tag{8.128}$$

with

$$A_0 = Op(a^{s-m,\delta-\mu})A \, Op(a^{s,\mu})^{-1}.$$

In view of Theorem 7 we have $a_0 \in L^{0,0}(\mathbb{R}^m)$ and hence a continuous operator

$$A_0 : L_2(\mathbb{R}^n) \to L_2(\mathbb{R}^n).$$

Then the asserted continuity of A follows immediately from (8.127) for arbitrary s, δ and from (8.128). $\square$

Definition 12. An $A \in L^{m,\mu}(\mathbb{R}^n)$, written in the form $A = Op(a) + C$ with $a(x,\xi) \in S^{m,\mu}(\mathbb{R}^n \times \mathbb{R}^n)$, $C \in L^{-\infty,-\infty}(\mathbb{R}^n)$, is called *elliptic* (of order (m,μ)), if $a(x,\xi)$ is elliptic in the sense (8.94) (cf. also 8.2.2 Definition 7).

This definition is, of course, independent of the concrete decomposition of A into $Op(a) + C$.

Definition 13. Let $A \in L^{m,\mu}(\mathbb{R}^n)$ be given. Then an operator $P \in L^{-m,-\mu}(\mathbb{R}^n)$ is called a *parametrix* of A if

$$AP - I, \; PA - I \in L^{-\infty,-\infty}(\mathbb{R}^n).$$

Theorem 14. *Let $A \in L^{m,\mu}(\mathbb{R}^n)$ be given, $m, \mu \in \mathbb{R}$. Then the following conditions are equivalent:*

(i) A is elliptic (of order (m, μ)),

(ii) (8.126) is a Fredholm operator for certain $s = s_0$, $\delta = \delta_0$.

If A is elliptic, (8.126) is Fredholm for all $s, \delta \in \mathbb{R}$. Furthermore, A has a parametrix $P \in L^{-m,-\mu}(\mathbb{R}^n)$. Finally,

$$Au = f \in H^{t,\tau}(\mathbb{R}^n) \tag{8.129}$$

for certain $t, \tau \in \mathbb{R}$, and

$$u \in H^{-\infty,-\infty}(\mathbb{R}^n)$$

implies

$$u \in H^{t+m,\tau+\mu}(\mathbb{R}^n). \tag{8.130}$$

Proof. The proof of $(ii) \Longrightarrow (i)$ will not be given here. Details have been elaborated in [Hi]. The result in the nonclassical case is due to Schrohe [S2]. Now let A be elliptic. We shall first construct a parametrix.

The ellipticity of A implies that for some excision function $\chi(x, \xi)$ in $\mathbb{R}^n \times \mathbb{R}^n$

$$\chi(x,\xi)a^{-1}(x,\xi) =: p_1(x,\xi) \in S^{-m,-\mu}(\mathbb{R}^n \times \mathbb{R}^n).$$

From Theorem 7 it follows that $P_1 A - I =: C \in L^{-1,-1}(\mathbb{R}^n)$. The operators C^j belong to $L^{-j,-j}(\mathbb{R}^n)$ for all $j \in \mathbb{Z}_+$. Hence they are of the form $Op(c_j)$ mod $L^{-\infty,-\infty}(\mathbb{R}^n)$ with $c_j(x,\xi) \in S^{-j,-j}(\mathbb{R}^n \times \mathbb{R}^n)$, and we can form the asymptotic sum $\sum_{j=1}^{\infty} (-1)^j c_j(x,\xi) =: d(x,\xi) \in S^{-1,-1}(\mathbb{R}^n \times \mathbb{R}^n)$. It follows for $P = (1 + Op(d))P_1$ that $PA - I \in L^{-\infty,-\infty}(\mathbb{R}^n)$. In an analogous manner we find a $\tilde{P} \in L^{-m,-\mu}(\mathbb{R}^n)$ with $A\tilde{P} - I \in L^{-\infty,-\infty}(\mathbb{R}^n)$. A simple algebraic argument shows that $P - \tilde{P} \in L^{-\infty,-\infty}(\mathbb{R}^n)$. Hence P is a parametrix. The Fredholm property of (8.126) follows from the fact that operators in $L^{-\infty,-\infty}(\mathbb{R}^n)$ are compact in $H^{s,\delta}(\mathbb{R}^n)$ and $H^{s-m,\delta-\mu}(\mathbb{R}^n)$. Now let (8.129) be satisfied for an elliptic A. Then $PAu = Pf \in H^{t+m,\tau+\mu}(\mathbb{R}^n)$, cf. Theorem 11. Since $PAu = (I + C)u$ for some $C \in L^{-\infty,-\infty}(\mathbb{R}^n)$ and $CH^{r,\gamma}(\mathbb{R}^n) \subseteq H^{\infty,\infty}(\mathbb{R}^n)$, we obtain (8.130). $\square$

Remark 15. Let $A \in L_{cl}^{m,\mu}(\mathbb{R}^n)$ be elliptic of order (m, μ). Then every parametrix P in the sense of Definition 13 belongs to $L_{cl}^{-m,-\mu}(\mathbb{R}^n)$, and

$$\sigma_{\psi}^{-m}(P) = \sigma_{\psi}^m(A)^{-1}, \quad \sigma_e^{-\mu}(P) = \sigma_e^\mu(A)^{-1}, \quad \sigma_{\psi,e}^{-m,-\mu}(P) = \sigma_{\psi,e}^{m,\mu}(A)^{-1}.$$

8.2.4 The calculus on the infinite cylinder

The calculus of the preceding section will now be generalized to a C^∞ manifold in the form of an infinite cylinder $\mathbb{R} \times X$ with a closed compact C^∞ manifold X of dimension n. This can immediately be modified for the case of manifolds M for which there is a compact submanifold M_0 with compact C^∞ boundary $\partial M_0 \cong X$ such that $M \setminus M_0 = (c, \infty) \times X$ with some constant $c > 0$. A special case of such an M is the space $\mathbb{R}^{n+1}$ when we set $M_0 = \{\tilde{x} \in \mathbb{R}^{n+1} :| \tilde{x} |\leq 1\}$. Then $X = S^n = \{\tilde{x} \in \mathbb{R}^{n+1} :| \tilde{x} |= 1\}$, and $\mathbb{R}^{n+1} \setminus M_0 \cong (1, \infty) \times S^n \ni (r, x)$ with $r =| \tilde{x} |$, $x = \frac{\tilde{x}}{|\tilde{x}|}$.

Let $U \subseteq \mathbb{R}^{n+1}$ be an open set such that there exists a constant $c > 0$ with

$$\tilde{x} \in U \implies \lambda\tilde{x} \in U \ \text{ for all } \ \lambda \geq 1, \ | \tilde{x} |\geq c.$$

We then define the symbol classes

$$S^{m,\mu}(U \times \mathbb{R}^{n+1}) \ni a(\tilde{x}, \tilde{\xi})$$

by the condition that for every open $U_0 \subset \mathbb{R}^{n+1}$ with $\overline{U_0} \subset U$ and $\tilde{x} \in U_0 \Rightarrow \lambda\tilde{x} \in U_0$ for all $\lambda \geq 1$, $| \tilde{x} |\geq c_0$ with some $c_0 > 0$, there is an $a_0(\tilde{x}, \tilde{\xi}) \in S^{m,\mu}(\mathbb{R}^{n+1}_{\tilde{x}} \times \mathbb{R}^{n+1}_{\tilde{\xi}})$ with $a_0 \mid_{U_0 \times \mathbb{R}^{n+1}} = a \mid_{U_0 \times \mathbb{R}^{n+1}}$. In an analogous manner we can introduce the subclass

$$S^{m,\mu}_{cl}(U \times \mathbb{R}^{n+1})$$

of symbols that are classical in $(\tilde{x}, \tilde{\xi})$ (for brevity we will omit here $cl(\tilde{x}, \tilde{\xi})$ in the case of classical symbols in both variables). Furthermore, we can talk about the space

$$\mathcal{S}(U \times U)$$

defined as the set of all $c(\tilde{x}, \tilde{x}') \in C^\infty(U \times U)$ such that for every open U_0 of the above kind there is a $c_0(\tilde{x}, \tilde{x}') \in \mathcal{S}(\mathbb{R}^{n+1} \times \mathbb{R}^{n+1})$ with $c_0 \mid_{U_0 \times U_0} = c \mid_{U_0 \times U_0}$. This yields the operator classes

$$L^{m,\mu}(U) = \{Op(a) + C : a(\tilde{x}, \tilde{\xi}) \in S^{m,\mu}(U \times \mathbb{R}^{n+1}), \ C \in L^{-\infty,-\infty}(U)\},$$

where $L^{-\infty,-\infty}(U)$ is defined as the space of operators with kernels in $\mathcal{S}(U \times U)$. Analogously we introduce $L^{m,\mu}_{cl}(U)$.

Now let $V \subseteq \mathbb{R}^{n+1}$ be another open set of the same kind as U and let

$$\kappa : U \to V \tag{8.131}$$

be a diffeomorphism satisfying

$$\kappa(\lambda\tilde{x}) = \lambda\kappa(\tilde{x}) \ \text{ for all } \ \lambda \geq 1, \ | \tilde{x} |\geq c$$

with a constant $c > 0$. We then have the push-forward of pseudo-differential operators in the standard sense, i.e.

$$(\kappa_* A)v = (\kappa^*)^{-1} A(\kappa^* v), \quad v \in C_0^\infty(V),$$

where $\kappa^* : C_0^\infty(V) \to C_0^\infty(U)$ is the pull-back of functions, i.e. $(\kappa^* v)(\tilde{x}) = v(\kappa(\tilde{x}))$.

Exercise 1. The push-forward of pseudo-differential operators under (8.131) induces bijections

$$\kappa_* \;:\; L^{m,\mu}(U) \to L^{m,\mu}(V),$$
$$\kappa_* \;:\; L^{m,\mu}_{cl}(U) \to L^{m,\mu}_{cl}(V)$$

for all $m, \mu \in \mathbb{R}$.

Remark 2. For $A = Op(a) \mod L^{-\infty,-\infty}(U)$ with an $a(\tilde{x}, \tilde{\xi}) \in S^{m,\mu}(U \times \mathbb{R}^{n+1})$ we have $\kappa_* A = Op(b) \mod L^{-\infty,-\infty}(V)$ with a $b(\tilde{y}, \tilde{\eta}) \in S^{m,\mu}(V \times \mathbb{R}^{n+1})$ which allows the asymptotic expansion

$$b(\tilde{y}, \tilde{\eta})\,|_{\tilde{y}=\kappa(\tilde{x})} \sim \sum_\alpha (\partial_{\tilde{\xi}}^\alpha a)(\tilde{x}, {}^t d\kappa(\tilde{x})\tilde{\eta}) \Phi_\alpha(\tilde{x}, \tilde{\eta}), \qquad (8.132)$$

where

$$\Phi_\alpha(\tilde{x}, \tilde{\eta}) \;=\; D_{\tilde{z}}^\alpha e^{i\delta(\tilde{z},\tilde{x})\tilde{\eta}}\,|_{\tilde{z}=\tilde{x}}, \qquad \alpha \in \mathbb{Z}_+^{n+1},$$
$$\delta(\tilde{z}, \tilde{x}) \;=\; \kappa(\tilde{z}) - \kappa(\tilde{x}) - d\kappa(\tilde{x})(\tilde{z} - \tilde{x}).$$

(8.132) is the transformation of complete symbols by the same rule as in the standard context of pseudo-differential operators. $\Phi_\alpha(\tilde{x}, \tilde{\eta})$ is a polynominal in $\tilde{\eta}$ of degree $\leq \frac{|\alpha|}{2}$ such that the orders in $\tilde{\eta}$ of the items in the sum on the right tend to $-\infty$ as $|\,\alpha\,| \to \infty$. As usual ${}^t d\kappa(\tilde{x})$ is the transposed Jacobian of κ at $\tilde{x}$. Here it satisfies

$${}^t d\kappa(\lambda\tilde{x}) = {}^t d\kappa(\tilde{x}) \;\text{ for all }\; \lambda \geq 1,\; |\,\tilde{x}\,| \geq c.$$

Thus,

$$(\partial_{\tilde{\xi}}^\alpha a)(\tilde{x}, {}^t d\kappa(\tilde{x})\tilde{\eta}) \in S^{m-|\alpha|,\mu}(U_{\tilde{x}} \times \mathbb{R}_{\tilde{\eta}}^{n+1})$$

for all $\alpha \in \mathbb{Z}_+^{n+1}$. Furthermore, we have for $\;\lambda \geq 1,\; |\,\tilde{x}\,| \geq c$

$$\begin{aligned}
\Phi_\alpha(\tilde{x}, \lambda\tilde{\eta}) &= D_{\tilde{z}}^\alpha e^{i\delta(\tilde{z},\tilde{x})\lambda\tilde{\eta}}\,|_{\tilde{z}=\tilde{x}} \\
&= D_{\tilde{z}}^\alpha e^{i\delta(\lambda\tilde{z},\lambda\tilde{x})\tilde{\eta}}\,|_{\tilde{z}=\tilde{x}} \\
&= \lambda^{|\alpha|} D_{z'}^\alpha e^{i\delta(z',\lambda\tilde{x})\tilde{\eta}}\,|_{z'=\lambda\tilde{x}} \\
&= \lambda^{|\alpha|} \Phi_\alpha(\lambda\tilde{x}, \tilde{\eta}).
\end{aligned}$$

Thus,

$$\Phi_\alpha(\lambda\tilde{x}, \tilde{\eta}) = \lambda^{-|\alpha|} \Phi_\alpha(\tilde{x}, \lambda\tilde{\eta})$$

and hence, since $\Phi_\alpha(\tilde{x}, \tilde{\eta})$ is a polynomial in $\tilde{\eta}$ of degree $\leq \frac{|\alpha|}{2}$, we get

$$\mathrm{ord}_\lambda \Phi_\alpha(\lambda\tilde{x}, \tilde{\eta}) \leq -\frac{|\,\alpha\,|}{2}$$

as a symbol of λ for $\lambda \geq 1$. It follows that the items in the sum on the right of (8.132) are of orders $(m - \frac{|\alpha|}{2}, \mu - \frac{|\alpha|}{2})$. Hence they tend to $(-\infty, -\infty)$ as $|\alpha| \to \infty$. Thus the asymptotic sum on the right of (8.132) can be carried out in $S^{m,\mu}$ with remainders in $S^{-\infty,-\infty}$, cf. 8.2.1 Theorem 2. In particular, we have

$$b(\tilde{y}, \tilde{\eta})\,|_{\tilde{y}=\kappa(\tilde{x})} = a(\tilde{x},\,{}^t d\kappa(\tilde{x})\tilde{\eta}) \mod S^{m-1,\mu-1}(U_{\tilde{x}} \times \mathbb{R}^{n+1}_{\tilde{\eta}}).$$

Remark 3. For every $A \in L^{m,\mu}_{cl}(U)$ we have

$$\begin{aligned}
\sigma^m_\psi(\kappa_* A)(\tilde{y}, \tilde{\eta}) &= \sigma^m_\psi(A)(\tilde{x}, \tilde{\xi}), \\
\sigma^\mu_e(\kappa_* A)(\tilde{y}, \tilde{\eta}) &= \sigma^\mu_e(A)(\tilde{x}, \tilde{\xi}), \\
\sigma^{m,\mu}_{\psi,e}(\kappa_* A)(\tilde{y}, \tilde{\eta}) &= \sigma^{m,\mu}_{\psi,e}(A)(\tilde{x}, \tilde{\xi})
\end{aligned}$$

with $(\tilde{y}, \tilde{\eta}) = (\kappa(\tilde{x}),\,{}^t d\kappa(\tilde{x})^{-1}\tilde{\xi})$.

The invariance of our operator classes $L^{m,\mu}(U)$, $L^{m,\mu}_{cl}(U)$ allows us to introduce the analogous classes $L^{m,\mu}(\mathbb{R} \times \Omega)$, $L^{m,\mu}_{cl}(\mathbb{R} \times \Omega)$ for any coordinate neighbourhood Ω on a C^∞ manifold X. We will need this, in particular, when X is a closed compact C^∞ manifold. Define the space $L^{-\infty,-\infty}(\mathbb{R} \times X)$ of smoothing operators in terms of a fixed density dx on X and the Lebesgue measure dr on $\mathbb{R}$ as the set of all integral operators on $\mathbb{R} \times X$ with kernels in $\mathcal{S}(\mathbb{R} \times X \times \mathbb{R} \times X) \cong \mathcal{S}(\mathbb{R} \times \mathbb{R}, C^\infty(X \times X))$.

Now $L^{m,\mu}(\mathbb{R} \times X)$ is the space of all pseudo-differential operators on $\mathbb{R} \times X$ that have the form $A + C$ with $C \in L^{-\infty,-\infty}(\mathbb{R} \times X)$ and $\varphi A \psi \in L^{m,\mu}(\mathbb{R} \times \Omega)$ for an arbitrary coordinate neighbourhood Ω on X and arbitrary $\varphi, \psi \in C^\infty_0(\Omega)$. In an analogous manner we introduce $L^{m,\mu}_{cl}(\mathbb{R} \times X)$. For operators $A \in L^{m,\mu}_{cl}(\mathbb{R} \times X)$ it follows from Remark 3 that

$$\begin{aligned}
\sigma^m_\psi(A) &\in C^\infty(T^*(\mathbb{R} \times X) \setminus 0), & (8.133) \\
\sigma^\mu_e(A) &\in C^\infty(T^*(\mathbb{R} \times X) \setminus \{r = 0\}), & (8.134) \\
\sigma^{m,\mu}_{\psi,e}(A) &\in C^\infty(T^*(\mathbb{R} \times X) \setminus (0 \cup \{r = 0\})), & (8.135)
\end{aligned}$$

where 0 indicates the zero section of $T^*(\mathbb{R} \times X)$. The meaning of $\{r = 0\}$ is evident.

The following theorem is an immediate consequence of the material of the preceding section and of the above definitions.

Theorem 4. *$A \in L^{m,\mu}(\mathbb{R} \times X)$, $B \in L^{\tilde{m},\tilde{\mu}}(\mathbb{R} \times X)$ implies $AB \in L^{m+\tilde{m},\mu+\tilde{\mu}}(\mathbb{R} \times X)$. Furthermore, the formal adjoint A^* of A with respect to the scalar product, based on $drdx$, belongs to $L^{m,\mu}(\mathbb{R} \times X)$. The subclasses of classical operators are closed under compositions and formal adjoints. In the latter case we have the analogues of the formulas (8.122), (8.123), (8.124), and (8.125).*

Now let us return to the Sobolev spaces $H^s(\mathbb{R} \times X)$, $s \in \mathbb{R}$, from Section 7.1.2 (cf. 7.1.2 Exercise 17). They give rise to the weighted spaces

$$H^{s,\delta}(\mathbb{R} \times X) = (1 + |r|^2)^{-\frac{\delta}{2}} H^s(\mathbb{R} \times X) \tag{8.136}$$

such that $H^{s,0}(\mathbb{R} \times X) = H^s(\mathbb{R} \times X)$. Since the multiplication of $H^s(\mathbb{R} \times X)$ by the weight factor $(1 + |r|^2)^{-\frac{\delta}{2}}$ induces an isomorphism $H^s(\mathbb{R} \times X) \to H^{s,\delta}(\mathbb{R} \times X)$, we get a natural norm on (8.136) by using a corresponding one on $H^s(\mathbb{R} \times X)$. Note that there are canonical embeddings

$$H^{s',\delta'}(\mathbb{R} \times X) \to H^{s,\delta}(\mathbb{R} \times X) \tag{8.137}$$

for $s' \geq s$, $\delta' \geq \delta$. They are compact for $s' > s$, $\delta' > \delta$.

Exercise 5. Every $A \in L^{m,\mu}(\mathbb{R} \times X)$ induces continuous operators

$$A : H^{s,\delta}(\mathbb{R} \times X) \to H^{s-m,\delta-\mu}(\mathbb{R} \times X) \tag{8.138}$$

for all $s, \delta \in \mathbb{R}$.

Hint: Use 8.2.3 Theorem 11 and the above Exercise 1.

Definition 6. An $A \in L_{cl}^{m,\mu}(\mathbb{R} \times X)$ is called *elliptic* (of order (m, μ)) if the three symbol components (8.133), (8.134), and (8.135) never vanish (cf. also 8.2.3 Definition 12).

Exercise 7. Formulate an adequate notion of ellipticity for operators in $L^{m,\mu}(\mathbb{R} \times X)$ (cf. (8.94)).

Definition 8. Let $A \in L^{m,\mu}(\mathbb{R} \times X)$ be given. Then an operator $P \in L^{-m,-\mu}(\mathbb{R} \times X)$ is called a *parametrix* of A if

$$AP - I, \ PA - I \in L^{-\infty,-\infty}(\mathbb{R} \times X).$$

Theorem 9. *Let* $A \in L_{cl}^{m,\mu}(\mathbb{R} \times X)$ *be given; then the following conditions are equivalent:*

(i) A is elliptic (of order (m, μ)),

(ii) (8.138) is a Fredholm operator for certain $s = s_0$, $\delta = \delta_0$.

If A is elliptic, (8.138) is Fredholm for all $s, \delta \in \mathbb{R}$. Furthermore, A has a parametrix $P \in L_{cl}^{-m,-\mu}(\mathbb{R} \times X)$. Finally,

$$Au = f \in H^{s,\delta}(\mathbb{R} \times X)$$

for certain $s, \delta \in \mathbb{R}$ and

$$u \in H^{-\infty,-\infty}(\mathbb{R} \times X)$$

imply

$$u \in H^{s+m,\delta+\mu}(\mathbb{R} \times X).$$

The proof of $(i) \Rightarrow (ii)$ as well as of the second statement of Theorem 9 follows by the same scheme as that of 8.2.3 Theorem 14. The proof of the necessity of the ellipticity for the Fredholm property of (8.138) is left to the reader as an exercise (Hint: Use the method in [RSc] for proving 3.1.1.1. Theorem 7 which gives a localizing effect in $\mathbb{R}^d$, $d = 1 + \dim X$, and do the same by interchanging the role of variables and covariables for the exit symbol; cf. also [Hi]).

8.2.5 The cone algebra on $X^\wedge$

This section will combine the operator classes from 8.1.4 Definition 1 and Section 8.2.4 to an operator class on the infinite stretched cone $X^\wedge = \mathbb{R}_+ \times X$ over a closed compact C^∞ manifold X. In 8.1.2 Definition 2 we introduced

$$C_{M+G}(X^\wedge, \underline{g}) \tag{8.139}$$

for the weight data $\underline{g} = (\gamma, \gamma - m, \Theta)$, $\gamma, m \in \mathbb{R}$, $\Theta = (-k, 0]$, $k \in \mathbb{N}$, or $\Theta = (-\infty, 0]$.

Definition 1. The operator class

$$C^m(X^\wedge, \underline{g}) \tag{8.140}$$

is defined as the subspace of all $A \in L^m_{cl}(X^\wedge)$ of the form

$$A = \omega r^{-m} op_M^{\gamma - \frac{n}{2}}(h)\omega_0 + (1 - \tilde{\omega})\tilde{A}(1 - \tilde{\omega}_1) + W + G \tag{8.141}$$

with arbitrary $h(r, z) \in C^\infty(\overline{\mathbb{R}}_+, M_O^m(X))$, arbitrary cut-off functions $\omega, \tilde{\omega}, \omega_0, \tilde{\omega}_1$, and elements $W + G \in C_{M+G}(X^\wedge, \underline{g})$, $\tilde{A} \in L^{m,0}_{cl}(\mathbb{R} \times X)|_{X^\wedge}$.

According to the definition, we have in (8.140) the symbols

$$\begin{aligned}
\sigma_\psi^m(A) &\in C^\infty(T^*X^\wedge \setminus 0), & (8.142) \\
\sigma_M^m(A) &\in L^m_{cl}(X; \Gamma_{\frac{n+1}{2} - \gamma}), & (8.143) \\
\sigma_e^0(A) &\in C^\infty(T^*X^\wedge \setminus 0), & (8.144) \\
\sigma_{\psi,e}^{m,0}(A) &\in C^\infty(T^*X^\wedge \setminus 0). & (8.145)
\end{aligned}$$

Furthermore, we have the lower order conormal symbols

$$\sigma_M^{m-j}(A) \in L^m_{cl}(X; \Gamma_{\frac{n+1}{2} - \gamma}). \tag{8.146}$$

As usual, (8.143) and (8.146) are also regarded as meromorphic $L^m_{cl}(X)$-valued functions in the complex plane. The following considerations are immediate consequences of the above material on cone operators and on operators globally over $\mathbb{R} \times X$. Thus we may omit further explanations.

Theorem 2. Every $A \in C^m(X^\wedge, \underline{g})$, $\underline{g} = (\gamma, \gamma - m, \Theta)$, induces continuous operators

$$A : \mathcal{K}^{s,\gamma}(X^\wedge) \to \mathcal{K}^{s-m, \gamma-m}(X^\wedge) \tag{8.147}$$

for all $s \in \mathbb{R}$. Furthermore, for every asymptotic type P to the weight data (γ, θ) there is an asymptotic type Q to $(\gamma - m, \theta)$ such that (8.147) induces continuous operators

$$A : \mathcal{K}_P^{s,\gamma}(X^\wedge) \to \mathcal{K}_Q^{s-m,\gamma-m}(X^\wedge) \tag{8.148}$$

for all $s \in \mathbb{R}$.

Remark 3. Let $A \in C^m(X^\wedge, \underline{g})$ and assume that the symbols (8.142)-(8.145) all vanish. Then, the operator (8.147) is compact for all $s \in \mathbb{R}$.

Theorem 4. $A \in C^m(X^\wedge, \underline{g})$ *for* $\underline{g} = (\gamma, \gamma - m, \Theta)$ *implies* $A^* \in C^m(X^\wedge, \underline{g}^*)$ *for* $\underline{g}^* = (-\gamma + m, -\gamma, \Theta)$, *where A^* is the formal adjoint with respect to the $\mathcal{K}^0(X^\wedge)$-scalar product. We have the symbol rules as in 8.1.4 Theorem 5 together with the complex conjugates of the exit symbols.*

Theorem 5. *Let* $A \in C^m(X^\wedge, \underline{g})$, $\underline{g} = (\gamma, \gamma - m, \Theta)$, *and* $\tilde{A} \in C^{\tilde{m}}(X^\wedge, \underline{\tilde{g}})$, $\underline{\tilde{g}} = (\gamma - m, \gamma - (\tilde{m} + m), \Theta)$. *Then* $\tilde{A}A \in C^{\tilde{m}+m}(X^\wedge, \underline{h})$ *with* $\underline{h} = (\gamma, \gamma - (\tilde{m} + m), \Theta)$, *and we have the symbol rules of 8.1.4 Theorem 6 together with the usual multiplications of the exit symbols.*

Definition 6. Let $A \in C^m(X^\wedge, \underline{g})$, $\underline{g} = (\gamma, \gamma - m, \Theta)$. Then A is called *elliptic* if

(i) $\sigma_\psi^m(A) \neq 0$ on $T^*X^\wedge \setminus 0$ and if in a neighbourhood of $r = 0$ in the coordinates $(r, x) \in \mathbb{R}_+ \times X$

$$r^m \sigma_\psi^m(A)(r, x, r^{-1}\varrho, \xi) \neq 0 \quad \text{for all} \quad (r, x) \in \overline{\mathbb{R}}_+ \times X, (\varrho, \xi) \neq 0,$$

(ii) the operators (8.82) are isomorphisms for an $s = s_0 \in \mathbb{R}$ and all $z \in \Gamma_{\frac{n+1}{2} - \gamma}$,

(iii) the symbols (8.142), (8.144) and (8.145) are nowhere vanishing.

Definition 7. Let $A \in C^m(X^\wedge, \underline{g})$ for $\underline{g} = (\gamma, \gamma - m, \Theta)$. An operator $P \in C^{-m}(X^\wedge, \underline{h})$ for $\underline{h} = (\gamma - m, \gamma, \Theta)$ is called a *parametrix* of A if

$$AP - I \in C_G(X^\wedge, \underline{g}_r), \quad PA - I \in C_G(X^\wedge, \underline{g}_l) \tag{8.149}$$

with $\underline{g}_r = (\gamma - m, \gamma - m, \Theta)$, $\underline{g}_l = (\gamma, \gamma, \Theta)$.

Theorem 8. *Let* $A \in C^m(X^\wedge, \underline{g})$, $\underline{g} = (\gamma, \gamma - m, \Theta)$, *be given. Then the following conditions are equivalent:*

(i) *A is elliptic in the sense of Definition 6,*

(ii) *the operator*

$$A : \mathcal{K}^{s,\gamma}(X^\wedge) \to \mathcal{K}^{s-m,\gamma-m}(X^\wedge) \tag{8.150}$$

is Fredholm for an $s = s_0 \in \mathbb{R}$.

If A is elliptic, then (8.150) is a Fredholm operator for all $s \in \mathbb{R}$. There exists a parametrix P of A in the sense of Definition 7. Furthermore,

$$Au = f \in \mathcal{K}^{s-m,\gamma-m}(X^\wedge)$$

for some $s \in \mathbb{R}$ and

$$u \in \mathcal{K}^{-\infty,\gamma}(X^\wedge) \tag{8.151}$$

imply $u \in \mathcal{K}^{s,\gamma}(X^\wedge)$. Finally,

$$Au = f \in \mathcal{K}_Q^{s-m,\gamma-m}(X^\wedge)$$

for some asymptotic type Q to $(\gamma - m, \Theta)$ and (8.151) imply $u \in \mathcal{K}_R^{s,\gamma}(X^\wedge)$ with an asymptotic type R to (γ, Θ). This holds for all $s \in \mathbb{R}$.

For reference we formulate below a part of the necessity of the ellipticity in Theorem 8 as an extra assertion, namely

Proposition 9. *The Fredholm property of A implies that*

$$\sigma_M^m(A)(z) : H^s(X) \to H^{s-m}(X)$$

is an isomorphism for all $z \in \Gamma_{\frac{n+1}{2}-\gamma}$ and all $s \in \mathbb{R}$.

Remark 10. Let $A, \tilde{A} \in C^m(X^\wedge, \underline{g})$, $\underline{g} = (\gamma, \gamma - m, \Theta)$, be elliptic, and assume

$$\sigma_\psi^m(A) = \sigma_\psi^m(\tilde{A}) \quad , \quad \sigma_M^m(A) = \sigma_M^m(\tilde{A}),$$
$$\sigma_e^0(A) = \sigma_e^0(\tilde{A}) \quad , \quad \sigma_{\psi,e}^{m,0}(A) = \sigma_{\psi,e}^{m,0}(\tilde{A}).$$

Then
$$\text{ind } A = \text{ind } \tilde{A}.$$

Furthermore, if $B \in C^{-m}(X^\wedge, \underline{h})$ with $\underline{h} = (\gamma - m, \gamma, \Theta)$ satisfies

$$\sigma_\psi^{-m}(B) = \sigma_\psi^m(A)^{-1} \quad , \quad T^m \sigma_M^{-m}(B) = \sigma_M^m(A)^{-1},$$
$$\sigma_e^0(B) = \sigma_e^0(A)^{-1} \quad , \quad \sigma_{\psi,e}^{m,0}(B) = \sigma_{\psi,e}^{m,0}(A)^{-1},$$

then
$$\text{ind } B = -\text{ind } A.$$

According to the construction at the beginning of Section 8.2.4 we can obtain operators in $L^{m,\mu}(\mathbb{R} \times X)$ ($L_{cl}^{m,\mu}(\mathbb{R} \times X)$) by employing the corresponding classes over $\mathbb{R} \times U_j$ for a finite open covering $\{U_1, \ldots, U_N\}$ of X. We can always write

$$P = \sum_{j=1}^N \varphi_j P_j \psi_j$$

for $P_j \in L^{m,\mu}(\mathbb{R} \times U_j)$, and a partition of unity $\{\varphi_1, \ldots, \varphi_N\}$ on X, and $\psi_j \in C_0^\infty(U_j)$ with $\varphi_j \psi_j = \varphi_j$, $j = 1, \ldots, N$.

This can also be done including further covariables $\eta \in \mathbb{R}^q$.

Assume in particular that we are given symbols

$$p_j(r, x, \tilde{\varrho}, \xi, \tilde{\eta}) \in S_{cl}^m(\mathbb{R}_+ \times V_j \times \mathbb{R}_{\tilde{\varrho}, \xi, \tilde{\eta}}^{1+n+q})$$

for open sets $V_j \subset \mathbb{R}^n, j = 1, \ldots, N$, associated with charts $\chi_j : U_j \to V_j$, where the p_j are classical in $(r, \tilde{\varrho}, \xi, \tilde{\eta})$ for $r \to \infty$.

Set

$$P_j(\eta) = r^{-m} Op_{(r,x)}(p_j \mid_{\tilde{\varrho}=r\eta, \ \tilde{\eta}=r\eta}).$$

Then

$$P(\eta) := (1 - \omega(r)) \sum_{j=1}^N \varphi_j P_j(\eta) \psi_j$$

with some cut-off function $\omega(r)$ will belong to $L_{cl}^{m,0}(\mathbb{R} \times X)$.

Exercise 11. Let $p_{j,(m)}(r, x, \tilde{\varrho}, \xi, \tilde{\eta}) \neq 0$ for $r > \text{const} > 0$, all x and $(\tilde{\varrho}, \xi, \tilde{\eta}) \neq 0$, where sub (m) indicates the homogeneous principal part of order m. Then the operator $P(\eta)$ for every fixed $\eta \neq 0$ satisfies the conditions

$$\sigma_e^0(P(\eta)) \neq 0, \quad \sigma_{\psi,e}^{m,0}(P(\eta)) \neq 0$$

everywhere on $T^* X^\wedge \setminus 0$.

As in (8.80), we can look at the operator class

$$C^l(X^\wedge, \underline{g})$$

with $\underline{g} = (\gamma, \gamma - m, \Theta)$, $l \in \mathbb{R}$, $m - l \in \mathbb{Z}_+$. It consists of all $A \in L_{cl}^l(X^\wedge)$ of the form

$$A = \omega r^{-l} op_M^{\gamma - \frac{n}{2}}(h)\omega_0 + (1 - \tilde{\omega})\tilde{A}(1 - \tilde{\omega}_1) + W + G$$

with arbitrary $h(r, z) \in C^\infty(\overline{\mathbb{R}}_+, M_O^l(X))$, $\tilde{A} \in L_{cl}^{l,0}(\mathbb{R} \times X) \mid_{X^\wedge}$, $G \in C_G(X^\wedge, \underline{g})$, and W being of the same form as for (8.80).

There is a natural generalization of Theorem 5, namely, $A \in C^l(X^\wedge, \underline{g})$, $\tilde{A} \in C^{\tilde{l}}(X^\wedge, \underline{\tilde{g}})$ implies $\tilde{A}A \in C^{\tilde{l}+l}(X^\wedge, \underline{h})$, and we get an obvious extension of the corresponding symbol rules.

Remark 12. Analogously to 8.1.4 Remark 10 the composition $\tilde{A}A$ can also be obtained mod $C_{M+G}(X^\wedge, \underline{h})$ by applying the Mellin operator convention from 8.1.3 Theorem 2 to the Leibniz products between the involved complete symbols. In other words we first find the composition in the form of a $P \in L_{cl}^{l+\tilde{l},0}(X^\wedge)$ such that the complete symbols near $r = 0$ have the form $r^{-(l+\tilde{l})} p_k(r, x, r\varrho, \xi)$ with $p_k(r, x, \tilde{\varrho}, \xi) \in S_{cl}^{l+\tilde{l}}(\overline{\mathbb{R}}_+ \times \mathbb{R}^n \times \mathbb{R}^{1+n})$, and then an $f(r, z) \in C^\infty(\overline{\mathbb{R}}_+, M_O^{l+\tilde{l}}(X))$ can be assigned such that $op_M^\delta(f)$ has the same system of interior symbols. Then

$$\tilde{A}A - r^{-(l+\tilde{l})}\{\omega op_M^{\gamma - \frac{n}{2}}(f)\omega_0 + (1 - \omega)P(1 - \omega_1)\} \in C_{M+G}(X^\wedge, \underline{h}) \qquad (8.152)$$

with arbitrary cut-off functions $\omega, \omega_0, \omega_1$ satisfying $\omega\omega_0 = \omega$, $\omega\omega_1 = \omega_1$.

Remark 13. The cone algebra defined in terms of Mellin symbols with discrete asymptotic types of the form (8.29) and Green operators with pairs of discrete asymptotic types (8.7) can be modified in different ways such that the essential results remain true. In particular, we can pass to continuous asymptotic types, cf. [Sc1], [Sc5], or suppress the asymptotic information up to small strips around the weight lines.

For the edge calculus it will be convenient to replace L_j in (8.7) and N_j in (8.29) by $C^\infty(X)$ and $L^{-\infty}(X)$, respectively, though the more precise calculus with the corresponding finite-dimensional subspace is also possible, however much more technical. For future references we call a (finite or infinite) sequence

$$P = \{(p_j, m_j, L_j)\} \quad \text{for} \quad L_j = C^\infty(X)$$

a weakly discrete asymptotic type for weighted Sobolev distributions and Green operators, and

$$R = \{(q_j, m_j, N_j)\}_{j \in \mathbb{Z}} \quad \text{for} \quad N_j = L^{-\infty}(X)$$

a weakly discrete asymptotic type for Mellin symbols. Here we impose the same assumptions on (p_j, m_j) and (q_j, m_j) as in (8.7) and (8.29), respectively. We then write for brevity

$$P = \{(p_j, m_j)\} \quad \text{for} \quad R = \{(q_j, m_j)\}_{j \in \mathbb{Z}}.$$

This gives rise to corresponding spaces $\mathcal{K}_P^{s,\gamma}(X^\wedge)$, $M_R^\mu(X)$ which are obvious generalizations of (8.10) and (8.45), respectively, as well as of (8.15) and 8.1.1 Definition 7.

Chapter 9

Pseudo-differential operators on manifolds with edges

9.1 Pseudo-differential operators with operator-valued symbols

9.1.1 The operator-valued symbol spaces

The pseudo-differential operators on manifolds with edges can be obtained as a calculus along the edges with operator-valued symbols acting along the model cones of corresponding wedges. The "abstract" background can be formulated independently, and it has applications also to more complicated singularities, cf. [Sc3]. Sections 9.1.1–9.1.3 are devoted to the main definitions and results that are needed below in the concrete edge theory. Proofs and further details may be found in [Sc1], [Sc2]. On the other hand, the theory of pseudo-differential operators with operator-valued symbols is a natural extension of the scalar case, and the reader will recognize the basic ideas of the ordinary calculus.

If E is a Banach space we fix a group of isomorphisms $\{\kappa_\lambda\}_{\lambda \in \mathbb{R}_+}$ of E, with

$$\kappa_\lambda \kappa_\varrho = \kappa_{\lambda\varrho}, \quad \kappa_\lambda^{-1} = \kappa_{\lambda^{-1}} \quad \text{for all} \ \ \lambda, \varrho \in \mathbb{R}.$$

$\mathcal{L}_\sigma(E)$ will denote the space of continuous operators $E \to E$, equipped with the strong operator topology. We then assume

$$\{\kappa_\lambda\}_{\lambda \in \mathbb{R}_+} \in C(\mathbb{R}_+, \mathcal{L}_\sigma(E)).$$

Note that for given $\{\kappa_\lambda\}_{\lambda \in \mathbb{R}_+}$ with these properties, $\{\kappa_{\lambda^{-1}}\}_{\lambda \in \mathbb{R}_+}$ is also a permitted choice of this kind.

Example 1. $E = L_2(\mathbb{R}_+)$ and $(\kappa_\lambda u)(r) = \lambda^{\frac{1}{2}} u(\lambda r)$, $\lambda \in \mathbb{R}_+$, satisfy our assumption. A more general example is

$$E = \mathcal{K}^{s,\gamma}(X^\wedge) \tag{9.1}$$

for arbitrary $s, \gamma \in \mathbb{R}$, and

$$(\kappa_\lambda u)(r, x) = \lambda^{\frac{n+1}{2}} u(\lambda r, x). \tag{9.2}$$

Here, as above, $X^\wedge = \mathbb{R}_+ \times X \ni (r, x)$, with a closed compact C^∞ manifold X, and $n = \dim X$.

Proposition 2. *There are constants c and M such that*

$$\|\kappa_\lambda\|_{\mathcal{L}(E)} \leq \begin{cases} c\lambda^M & for \quad \lambda \geq 1, \\ c\lambda^{-M} & for \quad \lambda \leq 1. \end{cases}$$

This is a well-known result. A proof may be found, for instance, in [Sc1]. We choose a strictly positive function $\eta \to [\eta]$ in $C^\infty(\mathbb{R}^q_\eta)$ with

$$[\eta] = |\eta| \quad \text{for} \quad |\eta| > \text{const},$$

and for brevity set

$$\kappa(\eta) = \kappa_{[\eta]}.$$

Let $\tilde{E}$ be another Banach space with a corresponding group $\{\tilde{\kappa}_\lambda\}_{\lambda \in \mathbb{R}_+} \in C(\mathbb{R}_+, \mathcal{L}_\sigma(\tilde{E}))$ and form $\tilde{\kappa}(\eta)$ similarly.

Definition 3. Let $U \subseteq \mathbb{R}^p$ be an open set, and $m \in \mathbb{R}$. Then

$$S^m(U \times \mathbb{R}^q; E, \tilde{E}) \tag{9.3}$$

denotes the set of all $a(y, \eta) \in C^\infty(U \times \mathbb{R}^q, \mathcal{L}(E, \tilde{E}))$, such that

$$\|\tilde{\kappa}^{-1}(\eta)\{D_y^\alpha D_\eta^\beta a(y, \eta)\}\kappa(\eta)\|_{\mathcal{L}(E, \tilde{E})} \leq c[\eta]^{m-|\beta|} \tag{9.4}$$

for all $y \in K \subset\subset U$ for arbitrary K, all $\eta \in \mathbb{R}^q$, $\alpha \in \mathbb{Z}_+^p$, $\beta \in \mathbb{Z}_+^q$, with constants $c = c(\alpha, \beta, K) > 0$.

Remark 4. The space (9.3) depends, of course, on the choice of $\{\kappa_\lambda\}_{\lambda \in \mathbb{R}_+}$ and $\{\tilde{\kappa}_\lambda\}_{\lambda \in \mathbb{R}_+}$, which are fixed for every concrete pair $(E, \tilde{E})$. But (9.3) is independent of the particular choice of the function $[\eta]$.

The following observations are simple excercises.

The best possible constants c in (9.4) for fixed a, denoted for a moment by $c_{\alpha, \beta, K}^m(a)$, form a semi-norm system on $S^m(U \times \mathbb{R}^q; E, \tilde{E})$ under which this space is Fréchet. There are natural continuous embeddings

$$S^m(U \times \mathbb{R}^q; E, \tilde{E}) \hookrightarrow S^{\tilde{m}}(U \times \mathbb{R}^q; E, \tilde{E})$$

for arbitrary $\tilde{m} \geq m$. The space

$$S^m(\mathbb{R}^q; E, \tilde{E}) \tag{9.5}$$

of y-independent elements of (9.3) is a closed subspace, and we have

$$S^m(U \times \mathbb{R}^q; E, \tilde{E}) = C^\infty(U, S^m(\mathbb{R}^q; E, \tilde{E})).$$

Exercise 5. Let $E = \tilde{E} = H^s(\mathbb{R}^n)$, $s \in \mathbb{R}$ fixed, $(\kappa_\lambda u)(x) = \lambda^{\frac{n}{2}} u(\lambda x)$, and $\varphi(x) \in \mathcal{S}(\mathbb{R}^n)$. Then the operator of multiplication $\mathcal{M}_\varphi$ by φ satisfies

$$\mathcal{M}_\varphi \in S^0(\mathbb{R}^q; H^s(\mathbb{R}^n), H^s(\mathbb{R}^n)).$$

If $\varphi \to 0$ in $\mathcal{S}(\mathbb{R}^n)$, then $\mathcal{M}_\varphi \to 0$ in the topology of the symbol space.

Hint: The symbol estimates (9.4) require the analysis of the behaviour of the operator of multiplication by $\varphi(r, \eta) := \varphi(r[\eta]^{-1})$. Use the method of proving 7.2.3 Theorem 7 in the variant for Schwartz functions φ and spaces $H^s(\mathbb{R}^n)$ to show

$$\|\mathcal{M}_{\varphi(r,\eta)}\|_{\mathcal{L}(H^s(\mathbb{R}^n))} < c_\varphi$$

with an η-independent constant c_φ and $c_\varphi \to 0$ as $\varphi \to 0$.

Let

$$S_{(1)}^m(U \times \mathbb{R}^q; E, \tilde{E})$$

denote the version of the space (9.3) with $\kappa_\lambda = id_E$, $\tilde{\kappa}_\lambda = id_{\tilde{E}}$ for all $\lambda \in \mathbb{R}_+$. Then there are continuous embeddings

$$S_{(1)}^m(U \times \mathbb{R}^q; E, \tilde{E}) \;\hookrightarrow\; S^{m+M+\tilde{M}}(U \times \mathbb{R}^q; E, \tilde{E}),$$
$$S^m(U \times \mathbb{R}^q; E, \tilde{E}) \;\hookrightarrow\; S_{(1)}^{m+M+\tilde{M}}(U \times \mathbb{R}^q; E, \tilde{E})$$

for all $m \in \mathbb{R}$. Here M and $\tilde{M}$ are the constants belonging to $\{\kappa_\lambda\}$ and $\{\tilde{\kappa}_\lambda\}$, respectively, in the sense of Proposition 2. This implies, in particular, that

$$S^{-\infty}(U \times \mathbb{R}^q; E, \tilde{E}) = \bigcap_{m \in \mathbb{R}} S^m(U \times \mathbb{R}^q; E, \tilde{E})$$

is independent of the concrete choice of $\{\kappa_\lambda\}$, $\{\tilde{\kappa}_\lambda\}$. Note that

$$S^{-\infty}(U \times \mathbb{R}^q; E, \tilde{E}) = C^\infty(U, \mathcal{S}(\mathbb{R}^q, \mathcal{L}(E, \tilde{E})))$$

with

$$\mathcal{S}(\mathbb{R}^q, \mathcal{L}(E, \tilde{E}))$$

being the Schwartz space of $\mathcal{L}(E, \tilde{E})$-valued functions. Let $\mathcal{F} = \mathcal{F}_{y \to \eta}$ be the Fourier transform in $\mathbb{R}^q$,

$$\mathcal{F}u(\eta) = \int e^{-iy\eta} u(y) dy,$$

with the inverse

$$(\mathcal{F}^{-1}g)(y) = \int e^{iy\eta} g(\eta) \bar{d}\eta, \quad \bar{d}\eta = (2\pi)^{-q} d\eta.$$

The Fourier transform will also be applied to functions and distributions with values in some locally convex vector space. In particular, if E is a Fréchet space, we can talk about the immediate generalizations of the Schwartz space and the space of temperate distributions on $\mathbb{R}^q$, respectively, to the E-valued case

$$\mathcal{S}(\mathbb{R}^q, E), \quad \mathcal{S}'(\mathbb{R}^q, E) = \mathcal{L}(\mathcal{S}(\mathbb{R}^q), E).$$

Then $\mathcal{F}$ induces isomorphisms

$$\mathcal{F} \;:\; \mathcal{S}(\mathbb{R}^q, E) \to \mathcal{S}(\mathbb{R}^q, E),$$
$$\mathcal{F} \;:\; \mathcal{S}'(\mathbb{R}^q, E) \to \mathcal{S}'(\mathbb{R}^q, E),$$

where $\mathcal{F}$ on $\mathcal{S}(\mathbb{R}^q, E)$ is obtained by $\mathcal{F}u := u \circ \mathcal{F}$. In our applications here E will always be the projective limit of a sequence of Hilbert spaces.

Exercise 6. We have

$$S^m(\mathbb{R}^q; E, \tilde{E}) \subset \mathcal{S}'(\mathbb{R}^q, \mathcal{L}(E, \tilde{E}))$$

and for every $\psi(\zeta) \in C_0^\infty(\mathbb{R}_\zeta^q)$ with $\psi(\zeta) = 1$ in an open neighbourhood of $\zeta = 0$

$$(1 - \psi(\zeta))\mathcal{F}_{\eta \to \zeta}^{-1} S^m(\mathbb{R}_\eta^q; E, \tilde{E}) \subset \mathcal{S}(\mathbb{R}_\zeta^q, \mathcal{L}(E, \tilde{E})). \tag{9.6}$$

In a similar manner one may show

$$(1 - \psi(\zeta))\mathcal{F}_{\eta \to \zeta}^{-1} S^m(U \times \mathbb{R}_\eta^q; E, \tilde{E}) \subset C^\infty(U, \mathcal{S}(\mathbb{R}_\zeta^q, \mathcal{L}(E, \tilde{E}))). \tag{9.7}$$

Let us denote by
$$S^{(m)}(U \times (\mathbb{R}^q \setminus \{0\}); E, \tilde{E})$$
the space of all $f(y, \eta) \in C^\infty(U \times (\mathbb{R}^q \setminus \{0\}), \mathcal{L}(E, \tilde{E}))$ satisfying

$$f(y, \lambda\eta) = \lambda^m \tilde{\kappa}_\lambda f(y, \eta) \kappa_\lambda^{-1} \tag{9.8}$$

for all $\lambda \in \mathbb{R}_+,\, y \in U,\, \eta \in \mathbb{R}^q \setminus \{0\}$.

Example 7. Let

$$A(x, y, D_x, D_y) = \sum_{|\alpha|+|\beta| \leq m} a_{\alpha\beta}(x, y) D_x^\alpha D_y^\beta$$

be a differential operator in $\mathbb{R}^n \times \mathbb{R}^q \ni (x, y)$ with $a_{\alpha\beta}(x, y) \in C^\infty(\mathbb{R}^n \times \mathbb{R}^q)$. Look at the operator family

$$f(y, \eta) := \sum_{|\alpha|+|\beta|=m} a_{\alpha\beta}(0, y) D_x^\alpha \eta^\beta \;\;:\;\; H^s(\mathbb{R}^n) \to H^{s-m}(\mathbb{R}^n). \tag{9.9}$$

If we set $E = H^s(\mathbb{R}^n)$, $(\kappa_\lambda u)(x) = \lambda^{\frac{n}{2}} u(\lambda x)$, $\tilde{E} = H^{s-m}(\mathbb{R}^n)$, $(\tilde{\kappa}_\lambda u)(x) = \lambda^{\frac{n}{2}} u(\lambda x)$, then (9.9) satisfies (9.8).

Exercise 8. Let $\chi(\eta)$ be an excision function (i.e. $\chi \in C^\infty(\mathbb{R}^q)$, $\chi(\eta) = 0$ for $|\eta| < c_0$, $\chi(\eta) = 1$ for $|\eta| > c_1$, with constants $0 < c_0 < c_1 < \infty$). Then

$$\chi(\eta) S^{(m)}(U \times (\mathbb{R}^q \setminus \{0\}); E, \tilde{E}) \subset S^m(U \times \mathbb{R}^q; E, \tilde{E}).$$

Theorem 9. *Let* $a_j \in S^{m_j}(U \times \mathbb{R}^q; E, \tilde{E})$, $j \in \mathbb{Z}_+$, *be an arbitrary sequence with* $m_j \to -\infty$ *as* $j \to \infty$. *Then there is an* $a \in S^m(U \times \mathbb{R}^q; E, \tilde{E})$ *with* $m = \max\{m_j\}$, *such that for every* $M \in \mathbb{N}$ *there is an* $N = N(M)$ *with*

$$a(y, \eta) - \sum_{j=0}^{N} a_j(y, \eta) \in S^{m-M}(U \times \mathbb{R}^q; E, \tilde{E}). \tag{9.10}$$

For any other $\tilde{a}(y, \eta) \in S^m(U \times \mathbb{R}^q; E, \tilde{E})$ *with* (9.10) *it follows that* $a - \tilde{a} \in S^{-\infty}(U \times \mathbb{R}^q; E, \tilde{E})$.

We write analogously to the scalar set-up

$$a(y, \eta) \sim \sum_{j=0}^{\infty} a_j(y, \eta), \tag{9.11}$$

called the *asymptotic sum* of the a_j.

Definition 10. Let $m \in \mathbb{R}$. Then

$$S_{cl}^m(U \times \mathbb{R}^q; E, \tilde{E}) \tag{9.12}$$

is the subspace of all $a(y, \eta) \in S^m(U \times \mathbb{R}^q; E, \tilde{E})$ such that (9.11) holds for a sequence of $a_j(y, \eta)$, $j \in \mathbb{Z}_+$, satisfying

$$a_j(y, \lambda\eta) = \lambda^{m-j} \tilde{\kappa}_\lambda a_j(y, \eta) \kappa_\lambda^{-1}$$

for all $\lambda \geq 1$, $y \in K \subset\subset U$, $|\eta| \geq c = c(K) > 0$, for all compact subsets K. We denote by $S_{cl}^m(\mathbb{R}^q; E, \tilde{E})$ the subspace of (9.12) of y-independent elements.

Exercise 11. For every $a(y, \eta) \in S_{cl}^m(U \times \mathbb{R}^q; E, \tilde{E})$ there is a unique sequence

$$a_{(m-j)}(y, \eta) \in S^{(m-j)}(U \times (\mathbb{R}^q \setminus \{0\}); E, \tilde{E}),$$

such that

$$a(y, \eta) \sim \sum_{j=0}^{\infty} \chi(\eta) a_{(m-j)}(y, \eta)$$

for any excision function $\chi(\eta)$.

$a_{(m)}(y, \eta)$ is also called the *homogeneous principal symbol* of $a(y, \eta) \in S_{cl}^m(U \times \mathbb{R}^q; E, \tilde{E})$ of order m. The map $a_{(m)}(y, \eta) \to a_{(m)}(y, \frac{\eta}{|\eta|})$ induces a bijection

$$S^{(m)}(U \times (\mathbb{R}^q \setminus \{0\}); E, \tilde{E}) \to C^\infty(U \times S^{q-1}, \mathcal{L}(E, \tilde{E})).$$

This yields in $S^{(m)}(U \times (\mathbb{R}^q \setminus \{0\}); E, \tilde{E})$ a natural Fréchet topology. Let us introduce a Fréchet topology also in the space $S_{cl}^m(U \times \mathbb{R}^q; E, \tilde{E})$. The unique mappings $a(y, \eta) \to a_{(m-j)}(y, \eta)$, $j \in \mathbb{Z}_+$, yield linear operators

$$h_j : S_{cl}^m(U \times \mathbb{R}^q; E, \tilde{E}) \to S^{(m-j)}(U \times (\mathbb{R}^q \setminus \{0\}); E, \tilde{E}) \qquad (9.13)$$

for all $j \in \mathbb{Z}_+$. Moreover, for any fixed excision function $\chi(\eta)$ we get by $a(y, \eta) \to a(y, \eta) - \chi(\eta) \sum_{j=0}^{k-1} h_j(a)(y, \eta)$ linear operators

$$r_k : S_{cl}^m(U \times \mathbb{R}^q; E, \tilde{E}) \to S^{m-k}(U \times \mathbb{R}^q; E, \tilde{E}), \qquad (9.14)$$

$k \in \mathbb{Z}_+$. Now $S_{cl}^m(U \times \mathbb{R}^q, E, \tilde{E})$ will be endowed with the Fréchet topology of the projective limit under the mappings h_j, r_k for all $j, k \in \mathbb{Z}_+$. It can easily be checked that this does not depend on the concrete choice of χ. Furthermore, $S_{cl}^m(\mathbb{R}^q; E, \tilde{E})$ is a closed subspace of $S_{cl}^m(U \times \mathbb{R}^q, E, \tilde{E})$ and we have

$$S_{cl}^m(U \times \mathbb{R}^q; E, \tilde{E}) = C^\infty(U, S_{cl}^m(\mathbb{R}^q; E, \tilde{E})).$$

The various notions and results of this section have immediate generalizations to the case when $\tilde{E}$ is a Fréchet space, written as a projective limit

$$\tilde{E} = \varprojlim_{j \in \mathbb{Z}_+} \tilde{E}^j \qquad (9.15)$$

of Banach spaces $\tilde{E}^j$, with continuous embeddings $\tilde{E}^{j+1} \hookrightarrow \tilde{E}^j$ for all $j \in \mathbb{Z}_+$, and such that $\{\tilde{\kappa}_\lambda\}$, first given on $\tilde{E}^0$, induces a corresponding action on every $\tilde{E}^j$. Clearly the constants c, M in Proposition 2 may depend on j.

An example of this situation is

$$\tilde{E} = \mathcal{S}_P^\gamma(X^\wedge),$$

which can always be written as such a projective limit with Hilbert spaces $\tilde{E}^j$, continuously embedded in $\mathcal{K}^{0,\gamma}(X^\wedge)$, where $\tilde{\kappa}_\lambda$ is induced by (9.2).

For (9.15) we then simply set

$$S^m(U \times \mathbb{R}^q; E, \tilde{E}) = \varprojlim_{j \in \mathbb{Z}_+} S^m(U \times \mathbb{R}^q; E, \tilde{E}^j)$$

and

$$S_{cl}^m(U \times \mathbb{R}^q; E, \tilde{E}) = \varprojlim_{j \in \mathbb{Z}_+} S_{cl}^m(U \times \mathbb{R}^q; E, \tilde{E}^j).$$

In the applications, the case $E = \varprojlim_{k} E^k$ will also occur with a sequence E^k, $E^{k+1} \hookrightarrow E^k$, $k \in \mathbb{Z}_+$, and analogous assumptions on the actions of $\{\kappa_\lambda\}$. Then, to every map $b : j \to b(j)$, we can form

$$S^m(U \times \mathbb{R}^q; E, \tilde{E})_b := \varprojlim_{j \in \mathbb{Z}_+} S^m(U \times \mathbb{R}^q; E^{b(j)}, \tilde{E}^j)$$

which is also Fréchet, and define

$$S^m(U \times \mathbb{R}^q; E, \tilde{E}) = \bigcup_b S^m(U \times \mathbb{R}^q; E, \tilde{E})_b \tag{9.16}$$

where the union is taken over all $b : \mathbb{Z}_+ \to \mathbb{Z}_+$. In an analogous manner we proceed with classical symbol spaces. It can be proved that these symbol spaces are closed under the usual algebraic operations when the involved spaces $E, \tilde{E}, \ldots$ fit together.

Let us finish this section by mentioning the special cases of symbol spaces

$$S^m(U \times \mathbb{R}^q; E, \mathbb{C}^{N_+}), \tag{9.17}$$
$$S^m(U \times \mathbb{R}^q; \mathbb{C}^{N_-}, \tilde{E}) \tag{9.18}$$

with certain $N_+, N_- \in \mathbb{N}$. The $\tilde{\kappa}_\lambda(\kappa_\lambda)$-actions on the finite-dimensional spaces $\mathbb{C}^{N_+}(\mathbb{C}^{N_-})$ are always assumed to be the identities for all λ. In the edge theory the elements of (9.17) will have the meaning of edge trace symbols, and those of (9.18) will have the meaning of edge potential symbols.

9.1.2 Pseudo-differential operators

As noted in the preceding section the Fourier transform in $\mathbb{R}^q$ can be applied to temperate distributions with values in E, say for a Banach space E. We have

$$C_0^\infty(\Omega, E) \subset \mathcal{S}(\mathbb{R}^q, E)$$

for any open $\Omega \subseteq \mathbb{R}^q$, where the embedding is defined by extension by zero to the complement of Ω. For every $a(y, y', \eta) \in S^m(\Omega \times \Omega \times \mathbb{R}^q; E, \tilde{E})$ and $u \in C_0^\infty(\Omega, E)$ we can form

$$Op(a)u(y) = F_{\eta \to y}^{-1}\left[F_{y' \to \eta}\{a(y, y', \eta)u(y')\} \right] \tag{9.19}$$

$$= \int \int e^{i(y - y')\eta} a(y, y', \eta) u(y') dy' d\eta. \tag{9.20}$$

The integral (9.20) is interpreted as an oscillatory integral like in the scalar case. This defines a continuous map

$$Op(a) : C_0^\infty(\Omega, E) \to C^\infty(\Omega, \tilde{E}). \tag{9.21}$$

Definition 1. We set

$$L^m(\Omega; E, \tilde{E}) = \left\{ Op(a) : a(y, y', \eta) \in S^m(\Omega \times \Omega \times \mathbb{R}^q; E, \tilde{E}) \right\}. \tag{9.22}$$

In an analogous manner we define

$$L^m_{cl}(\Omega; E, \tilde{E}) \tag{9.23}$$

in terms of $S^m_{cl}(\Omega \times \Omega \times \mathbb{R}^q; E, \tilde{E})$. The operators in $L^m(\Omega; E, \tilde{E})$ are called pseudo-differential operators with operator-valued symbols, the operators of the space $L^m_{cl}(\Omega; E, \tilde{E})$ classical ones.

If $A \in L^m_{cl}(\Omega; E, \tilde{E})$, written as $Op(a)$ with $a(y, y', \eta) \in S^m_{cl}(\Omega \times \Omega \times \mathbb{R}^q; E, \tilde{E})$, we set

$$\sigma^m_\wedge(A)(y, \eta) := a_{(m)}(y, y, \eta) \tag{9.24}$$

where $a_{(m)}$ is the homogeneous principal part of a. We call $\sigma^m_\wedge(A)$ the homogeneous principal symbol of A of order m. Let us set

$$\mathcal{D}'(\Omega, E) = \mathcal{L}(C_0^\infty(\Omega), E)$$

and

$$\mathcal{E}'(\Omega, E) = \{ f \in \mathcal{D}'(\Omega, E) : \text{supp } f \text{ compact} \}.$$

Here supp f is the complement of the largest open $\tilde{\Omega} \subset \Omega$ with $f \mid_{C_0^\infty(\tilde{\Omega})} = 0$.

Similarly to the scalar theory every $A \in L^m(\Omega; E, \tilde{E})$ has a distributional kernel

$$K_A(y, y', y - y') \in \mathcal{D}'(\Omega \times \Omega, \mathcal{L}(E, \tilde{E})), \tag{9.25}$$

defined by

$$\langle K_A(y, y', y - y'), \varphi(y, y') \rangle = \int \int \int e^{i(y - y')\eta} a(y, y', \eta) \varphi(y, y') dy dy' d\eta, \tag{9.26}$$

$\varphi(y, y') \in C_0^\infty(\Omega \times \Omega)$. The formula (9.26) can be regarded as a continuous map

$$K_A : C_0^\infty(\Omega \times \Omega) \to \mathcal{L}(E, \tilde{E})$$

which justifies the interpretation (9.25). Note that for $a(\eta) \in S^m(\mathbb{R}^q; E, \tilde{E})$ the first (y, y')-dependence of K_A disappears, and that then

$$K_A(y - y') = K_A(\zeta) \mid_{\zeta = y - y'},$$

with

$$K_A(\zeta) = \mathcal{F}^{-1}_{\eta \to \zeta} a(\eta) \in \mathcal{S}'(\mathbb{R}^q_\zeta, \mathcal{L}(E, \tilde{E})),$$

cf. 9.1.1 Exercise 6. From (9.7) it follows that

$$\text{sing supp } K_A \subset \text{diag } \Omega \times \Omega := \{ (y, y') \in \Omega \times \Omega : y = y' \}. \tag{9.27}$$

We call an $A \in L^m(\Omega; E, \tilde{E})$ *properly supported* if K_A has a proper support, which has a meaning analogous to that in the scalar theory, cf. Section 7.2.1. For every $A \in L^m(\Omega; E, \tilde{E})$, $A = Op(a)$ we can write

$$A = Op(\omega a) + Op((1 - \omega)a)$$

for any $\omega(y, y^{'}) \in C^\infty(\Omega \times \Omega)$ with proper support and $\omega \equiv 1$ in an open neighbourhood of diag $\Omega \times \Omega$. From Theorem 6 below it will follow that

$$Op((1 - \omega)a) \in L^{-\infty}(\Omega; E, \tilde{E}).$$

Here $L^{-\infty}(\Omega; E, \tilde{E})$ is the space of all pseudo-differential operators with kernel in $C^\infty(\Omega \times \Omega, \mathcal{L}(E, \tilde{E}))$. In this way we obtain the following:

Proposition 2. *Every $A \in L^m(\Omega; E, \tilde{E})$ has the form*

$$A = A_0 + C \tag{9.28}$$

where $A_0 \in L^m(\Omega; E, \tilde{E})$ is properly supported and $C \in L^{-\infty}(\Omega; E, \tilde{E})$.

A properly supported $A \in L^m(\Omega; E, \tilde{E})$ induces continuous operators

$$A : C_0^\infty(\Omega, E) \to C_0^\infty(\Omega, \tilde{E}), \quad A : C^\infty(\Omega, E) \to C^\infty(\Omega, \tilde{E}). \tag{9.29}$$

Remark 3. Every $A \in L^{-\infty}(\Omega; E, \tilde{E})$ has a kernel $K_A \in C^\infty(\Omega \times \Omega, \mathcal{L}(E, \tilde{E}))$. Conversely, for every $c(y, y^{'}) \in C^\infty(\Omega \times \Omega, \mathcal{L}(E, \tilde{E}))$ there exists an $a(y, y^{'}, \eta) \in S^{-\infty}(\Omega \times \Omega \times \mathbb{R}^q; E, \tilde{E})$ such that for $A = Op(a)$

$$K_A(y, y^{'}, y - y^{'}) = c(y, y^{'}).$$

The first statement is obvious. Moreover, if $c(y, y^{'})$ is given as mentioned, then $a(y, y^{'}, \eta) = e^{i(y^{'} - y)\eta} c(y, y^{'}) \psi(\eta)$ for any $\psi(\eta) \in C_0^\infty(\mathbb{R}^q)$ with $\int \psi(\eta) d\eta = 1$ is a symbol in $S^{-\infty}(\Omega \times \Omega \times \mathbb{R}^q; E, \tilde{E})$, and $c(y, y^{'})$ is the kernel of $A = Op(a)$.

Exercise 4. The map $C_0^\infty(\Omega, E) \to C^\infty(\Omega, \mathcal{S}(\mathbb{R}_\eta^q, \tilde{E}))$ defined by

$$u(y) \to \int e^{-iy^{'}\eta} a(y, y^{'}, \eta) u(y^{'}) dy^{'} \tag{9.30}$$

for an $a(y, y^{'}, \eta) \in S^m(\Omega \times \Omega \times \mathbb{R}^q; E, \tilde{E})$ has an extension to an operator

$$\mathcal{E}^{'}(\Omega, E) \to C^\infty(\Omega, \mathcal{S}^{'}(\mathbb{R}_\eta^q, \tilde{E})).$$

Remark 5. If we denote by $b(y, \eta)$ the latter element in $C^\infty(\Omega_y, \mathcal{S}'(\mathbb{R}^q_\eta, \tilde{E}))$ that is determined by $u(y)$ and $a(y, y', \eta)$, then $u(y) \to \int e^{iy\eta} b(y, \eta) d\eta$, interpreted in the distributional sense gives rise to an extension of (9.21) to a continuous operator

$$A = Op(a) : \mathcal{E}'(\Omega, E) \to \mathcal{D}'(\Omega, \tilde{E}).$$

It is pseudo-local by virtue of (9.27). If A is properly supported we obtain

$$A : \mathcal{E}'(\Omega, E) \to \mathcal{E}'(\Omega, \tilde{E}), \quad A : \mathcal{D}'(\Omega, E) \to \mathcal{D}'(\Omega, \tilde{E}),$$

cf. also (9.29).

Theorem 6. *Every $A \in L^m(\Omega; E, \tilde{E})$ can be written as $A = Op(\underline{a}) + C$ with certain $\underline{a}(y, \eta) \in S^m(\Omega \times \mathbb{R}^q; E, \tilde{E})$, $C \in L^{-\infty}(\Omega; E, \tilde{E})$. If A has the form $Op(a)$ with $a(y, y', \eta) \in S^m(\Omega \times \Omega \times \mathbb{R}^q; E, \tilde{E})$, then $\underline{a}(y, \eta)$ allows the asymptotic expansion*

$$\underline{a}(y, \eta) \sim \sum_\alpha \frac{1}{\alpha!} D_\eta^\alpha \partial_{y'}^\alpha a(y, y', \eta) \mid_{y'=y}.$$

Theorem 7. *Let $A \in L^m(\Omega; E, E_0)$, $\tilde{A} \in L^{\tilde{m}}(\Omega; E_0, \tilde{E})$ and A or $\tilde{A}$ be properly supported. Then $\tilde{A}A \in L^{\tilde{m}+m}(\Omega; E, \tilde{E})$. If $A = Op(a) \mod L^{-\infty}$ with $a(y, \eta) \in S^m(\Omega \times \mathbb{R}^q; E, E_0)$, $\tilde{A} = Op(\tilde{a}) \mod L^{-\infty}$ with $\tilde{a}(y, \eta) \in S^{\tilde{m}}(\Omega \times \mathbb{R}^q; E_0, \tilde{E})$, then the composition can be written as $\tilde{A}A = Op(b) \mod L^{-\infty}$ with $b(y, \eta) \in S^{\tilde{m}+m}(\Omega \times \mathbb{R}^q; E, \tilde{E})$, which allows the asymptotic expansion*

$$b(y, \eta) \sim \sum_\alpha \frac{1}{\alpha!} D_\eta^\alpha \tilde{a}(y, \eta) \partial_y^\alpha a(y, \eta). \tag{9.31}$$

Finally, $A \in L^m_{cl}$, $\tilde{A} \in L^{\tilde{m}}_{cl}$ implies $\tilde{A}A \in L^{\tilde{m}+m}_{cl}$ and we then have

$$\sigma_\Lambda^{\tilde{m}+m}(\tilde{A}A) = \sigma_\Lambda^{\tilde{m}}(\tilde{A}) \sigma_\Lambda^m(A). \tag{9.32}$$

Remark 8. The definitions and results of this section have obvious extensions to the case of Fréchet spaces $E, \tilde{E}$, cf. (9.16).

9.1.3　Abstract wedge Sobolev spaces

We now look at a vector-valued analogue of the standard Sobolev spaces of smoothness $s \in \mathbb{R}$, here called abstract spaces, since the values are in a locally convex vector space E. This class of spaces was first introduced in [Sc1], and we shall only summarize some material here. As in Section 9.1.1 we start with the data

$$E, \quad \{\kappa_\lambda\}_{\lambda \in \mathbb{R}_+} \in C(\mathbb{R}_+, \mathcal{L}_\sigma(E)),$$

where $\{\kappa_\lambda\}$ is a group of isomorphisms on the Banach space E.

Definition 1. We denote by

$$\mathcal{W}^s(\mathbb{R}^q, E), \quad s \in \mathbb{R} \tag{9.33}$$

the completion of $\mathcal{S}(\mathbb{R}^q, E)$ with respect to the norm

$$u \to \left\{ \int [\eta]^{2s} \|\kappa^{-1}(\eta)\, (\mathcal{F}_{y \to \eta} u)\, (\eta)\|_E^2 d\eta \right\}^{\frac{1}{2}}. \tag{9.34}$$

Remark 2. For

$$\mathcal{S}'(\mathbb{R}^q, E) = \mathcal{L}(\mathcal{S}(\mathbb{R}^q), E) \tag{9.35}$$

the space $\mathcal{W}^s(\mathbb{R}^q, E)$ can equivalently be defined as the subspace of all $u \in \mathcal{S}'(\mathbb{R}^q, E)$ for which $\|\kappa^{-1}(eta)\, (\mathcal{F}u)(\eta)\|_E^2$ is locally integrable in $\mathbb{R}^q$ and the norm (9.34) is finite.

Many properties of the standard Sobolev spaces have extensions to the vector-valued case. We shall not recall all details here. The choice of the group $\{\kappa_\lambda\}_{\lambda \in \mathbb{R}_+}$ is fixed in every concrete case so it is not indicated in the general notation. An exception is $\kappa_\lambda = id_E$ for all $\lambda \in \mathbb{R}_+$. In that case we write

$$H^s(\mathbb{R}^q, E) \tag{9.36}$$

instead of $\mathcal{W}^s(\mathbb{R}^q, E)$.

We will also have in mind $E = \mathbb{C}^N$, $\kappa_\lambda = id_E$ for all $\lambda \in \mathbb{R}_+$. The identical action on a finite-dimensional space is the only one we take into account. In the applications we will often have

$$E \;=\; \mathcal{K}^{s,\gamma}(X^\wedge) \oplus \mathbb{C}^N, \tag{9.37}$$

$$\kappa_\lambda(u,v) \;:=\; \left(\lambda^{\frac{n+1}{2}} u(\lambda r, x), v\right), \quad \lambda \in \mathbb{R}_+ \tag{9.38}$$

for $(u,v) \in E$.

Let $E = E_1 \oplus E_2$ be a direct sum and suppose that $\{\kappa_\lambda\}_{\lambda \in \mathbb{R}_+}$ induces corresponding actions on E_i, $i = 1, 2$. Then

$$\mathcal{W}^s(\mathbb{R}^q, E_1 \oplus E_2) = \mathcal{W}^s(\mathbb{R}^q, E_1) \oplus \mathcal{W}^s(\mathbb{R}^q, E_2). \tag{9.39}$$

In particular, for (9.37), (9.38) it follows that

$$\mathcal{W}^s(\mathbb{R}^q, \mathcal{K}^{s,\gamma}(X^\wedge) \oplus \mathbb{C}^N) = \mathcal{W}^s(\mathbb{R}^q, \mathcal{K}^{s,\gamma}(X^\wedge)) \oplus H^s(\mathbb{R}^q, \mathbb{C}^N).$$

Note that an immediate consequence of the definition is that

$$\mathcal{F}_{y \to \eta}(\mathcal{W}^s(\mathbb{R}^q, \mathcal{K}^{s,\gamma}(X^\wedge))) \tag{9.40}$$

consists of all $[\eta]^{\frac{n+1}{2}} u(t[\eta], \eta)$ for which $u(t, \eta) \in \mathcal{F}_{y \to \eta} H^s(\mathbb{R}^q, \mathcal{K}^{s,\gamma}(X^\wedge))$. Here, as above, we have suppressed the x-variables in the notation.

Remark 3. In the applications below it will also be reasonable to allow that E_1 or E_2 in the direct decomposition $E = E_1 \oplus E_2$ is not invariant under the action of $\{\kappa_\lambda\}_{\lambda \in \mathbb{R}_+}$. Nevertheless, we may talk about the spaces $\mathcal{W}^s(\mathbb{R}^q, E_i)$, $i = 1, 2$, as the completions of $\mathcal{S}(\mathbb{R}^q, E_i)$ in the norm (9.34). Then we will also obtain (9.39).

An intuitive reason for the definition of $\mathcal{W}^s(\mathbb{R}^q, E)$ is that in the case

$$E = H^s(\mathbb{R}^m_x), \quad (\kappa_\lambda u)(x) = \lambda^{\frac{m}{2}} u(\lambda x) \ \text{ for } \lambda \in \mathbb{R}_+$$

we have

$$\mathcal{W}^s(\mathbb{R}^q, H^s(\mathbb{R}^m)) = H^s(\mathbb{R}^q \times \mathbb{R}^m).$$

In other words $\mathcal{W}^s(\mathbb{R}^q, E)$ gives an anisotropic reformulation of the Sobolev space $H^s(\mathbb{R}^{q+m})$ with respect to a "fictitious edge" $\mathbb{R}^q$. Another example in this sense is

$$E = H^s(\mathbb{R}_+), \quad (\kappa_\lambda u)(r) = \lambda^{\frac{1}{2}} u(\lambda r) \ \text{ for } \lambda \in \mathbb{R}_+,$$

$r \in \mathbb{R}_+$, where $H^s(\mathbb{R}_+) = H^s(\mathbb{R})|_{\mathbb{R}_+}$, and we have

$$\mathcal{W}^s(\mathbb{R}^q, H^s(\mathbb{R}_+)) = H^s(\mathbb{R}^q \times \mathbb{R}_+) = H^s(\mathbb{R}^q \times \mathbb{R})|_{\mathbb{R}^q \times \mathbb{R}_+}.$$

We shall frequently use the following:

Theorem 4. *The operator $\mathcal{M}_\varphi$ of multiplication by a $\varphi \in C_0^\infty(\mathbb{R}^q)$ induces a continuous operator $\mathcal{W}^s(\mathbb{R}^q, E) \to \mathcal{W}^s(\mathbb{R}^q, E)$ and the correspondence $\varphi \to \mathcal{M}_\varphi$ represents a continuous map*

$$C_0^\infty(\mathbb{R}^q) \to \mathcal{L}(\mathcal{W}^s(\mathbb{R}^q, E), \mathcal{W}^s(\mathbb{R}^q, E))$$

for all $s \in \mathbb{R}$.

If $\Omega \subseteq \mathbb{R}^q$ is an open set we introduce

$$\mathcal{W}^s_{\text{comp}}(\Omega, E), \quad \mathcal{W}^s_{\text{loc}}(\Omega, E) \tag{9.41}$$

in the following way. $\mathcal{W}^s_{\text{comp}}(\Omega, E)$ is the space of all $u \in \mathcal{W}^s(\mathbb{R}^q, E)$ with compact $\operatorname{supp} u \subset \Omega$. This is canonically identified with a subspace of $\mathcal{D}'(\Omega, E)$. Moreover $\mathcal{W}^s_{\text{loc}}(\Omega, E)$ is the subspace of all $u \in \mathcal{D}'(\Omega, E)$ with $\varphi u \in \mathcal{W}^s_{\text{comp}}(\Omega, E)$ for all $\varphi \in C_0^\infty(\Omega)$.

Theorem 5. *Every $A \in L^m(\Omega; E, \tilde{E})$ induces continuous operators*

$$A : \mathcal{W}^s_{\text{comp}}(\Omega, E) \to \mathcal{W}^{s-m}_{\text{loc}}(\Omega, \tilde{E}),$$

for all $s \in \mathbb{R}$. If A is properly supported then we can to write comp *or* loc *on both sides.*

Remark 6. Let E be a Fréchet space, written as a projective limit of Banach spaces E^k with $E^{k+1} \hookrightarrow E^k$ for all $k \in \mathbb{Z}_+$. Consider an action κ_λ, $\lambda \in \mathbb{R}_+$, on E that induces strongly continuous actions on every E^k. Then we can form the space

$$\mathcal{W}^s(\mathbb{R}^q, E), \quad s \in \mathbb{R},$$

as the projective limit of the $\mathcal{W}^s(\mathbb{R}^q, E^k)$, $k \in \mathbb{Z}_+$, which is Fréchet. Also the comp- and loc-versions of these spaces over Ω will be used. The results of this section have generalizations to this case in an obvious way.

Example 7. Let $\Theta = (-k, 0]$ and P be an asymptotic type belonging to the weight data (γ, Θ). Then we have

$$\mathcal{K}_P^{s,\gamma}(X^\wedge) = \mathcal{E}_P(X^\wedge) + \mathcal{K}_\Theta^{s,\gamma}(X^\wedge),$$

cf. (8.9), and we can apply Remark 3. Similarly to (9.40) we have

$$\mathcal{F}_{y \to \eta} \mathcal{W}^s(\mathbb{R}^q, \mathcal{E}_P(X^\wedge)) = \left\{ [\eta]^{\frac{n+1}{2}} u(r[\eta], \eta) : \ u(r, \eta) \in \mathcal{F}_{y \to \eta} H^s(\mathbb{R}^q, \mathcal{E}_P(X^\wedge)) \right\}.$$

Now every $u(r, \eta) \in \mathcal{F}_{y \to \eta} H^s(\mathbb{R}^q, \mathcal{E}_P(X^\wedge))$ has the form

$$u(r, \eta) = \omega(r) \sum_{j=0}^N \sum_{k=0}^{m_j} c_{jk}(x, \eta) r^{-p_j} \ln^k r$$

with $c_{jk}(x, \eta) \in \mathcal{F}_{y \to \eta} H^s(\mathbb{R}^q, L_j)$ for all j, k, cf. also the notation of (8.9). In this way we just recover the singular terms of the edge asymptotics, cf. [Sc1].

9.2 The edge symbolic calculus

9.2.1 Green symbols

The calculus of pseudo-differential operators on a manifold with edges Y will be organized in a neighbourhood of the edges in local coordinates $y \in \Omega \subseteq \mathbb{R}^q$ as one along Ω with operator-valued symbols, acting along the model cone $X^\wedge = \mathbb{R}_+ \times X$ of the wedge. Concerning notation, cf. Section 7.1.3.

Outside the edges we have the C^∞ structure and accordingly the standard calculus of pseudo-differential operators. It will then be glued together with the part near the edge by a corresponding partition of unity. Therefore, the specific aspect in the edge calculus concerns a neighbourhood of Y. The operator-valued symbols will be families

$$a(y, \eta) \in S^l(\Omega \times \mathbb{R}^q; \mathcal{K}^{s,\gamma}(X^\wedge), \mathcal{K}^{s-l,\gamma-m}(X^\wedge))$$

(cf. 9.1.1 Definition 3) which are pointwise elements of the cone algebra of Section 8.2.5. The cone operators are defined as sums like

$$a = a_M + a_\psi + w + g \tag{9.42}$$

with a Mellin operator a_M near $r = 0$, a pseudo-differential operator a_ψ away from $r = 0$, a smoothing Mellin operator w and a Green operator g. Now those items will depend on $(y, \eta) \in T^*\Omega \cong \Omega \times \mathbb{R}^q$, or more generally on $(y, y', \eta) \in \Omega \times \Omega \times \mathbb{R}^q$. The novelty in the discussion here is a precise description of that parameter-dependence. This is a priori not canonically defined, since we have even in the case of the cone a rather subtle dependence on asymptotic data. They will now be (y, y')-dependent in general.

In order to avoid too complicated machinery throughout the exposition we shall impose asymptotic data that are constant with respect to (y, y'). A dependence on η would in any case be unnecessary.

Now according to (9.42) we begin with the simplest objects, namely, the parameter-dependent Green operators, here called Green edge symbols. This notation is motivated by a representation of the Green function of an elliptic boundary value problem in terms of an appropriate analogue of our Green operators, modulo a fundamental solution of the elliptic operator (cf. [BdM2]).

Definition 1. A Green edge symbol of order $l \in \mathbb{R}$, associated with the weight data $\underline{g} = (\gamma, \delta, \Theta)$ for $\Theta = (\vartheta, 0]$, $-\infty \leq \vartheta < 0$, $\gamma, \delta \in \mathbb{R}$, is a function

$$g(y, y', \eta) \in \bigcap_{s \in \mathbb{R}} C^\infty(\Omega \times \Omega \times \mathbb{R}^q, \mathcal{L}(\mathcal{K}^{s,\gamma}(X^\wedge), \mathcal{K}^{\infty,\delta}(X^\wedge)))$$

with the property

$$g(y, y', \eta) \in \bigcap_{s \in \mathbb{R}} S_{cl}^l(\Omega \times \Omega \times \mathbb{R}^q; \mathcal{K}^{s,\gamma}(X^\wedge), \mathcal{S}_P^\delta(X^\wedge)),$$

$$g^*(y, y', \eta) \in \bigcap_{s \in \mathbb{R}} S_{cl}^l(\Omega \times \Omega \times \mathbb{R}^q; \mathcal{K}^{s,-\delta}(X^\wedge), \mathcal{S}_Q^{-\gamma}(X^\wedge)).$$

Here P and Q are (weakly discrete) asymptotic types associated with the weight data (δ, Θ) and $(-\gamma, \Theta)$, respectively, dependent on g independent of $y, y' \in \Omega$ however, (cf. also the notation of 8.2.5 Remark 13). g^* is the pointwise formal adjoint in the sense $(gu, v)_{\mathcal{K}^0(X^\wedge)} = (u, g^*v)_{\mathcal{K}^0(X^\wedge)}$ for all $u, v \in C_0^\infty(X^\wedge)$. Symbols are understood in the sense $(\kappa_\lambda u)(t, x) = \lambda^{\frac{n}{2}} u(\lambda t, x)$, $\lambda \in \mathbb{R}$.

Remark 2. As noted in the beginning, the assumption that P and Q are constant with respect to $y, y' \in \Omega$ was made here only for convenience. That condition may be dropped completely but then the calculus needs many more technical precautions (cf. [Sc2][Part XII, XIII).

Remark 3. From the definition of Green edge symbols, a natural locally convex topology follows immediately in the space

$$R_G^l(\Omega \times \Omega \times \mathbb{R}^q, \underline{g})_{P,Q}, \tag{9.43}$$

$\underline{g} = (\gamma, \delta, \Theta)$, of all Green edge symbols with fixed asymptotic types P, Q. The space (9.43) is then Fréchet, since it suffices to take the intersections over $s \in \mathbb{Z}$.

Occasionally we will also use the notation

$$R_G^l(\Omega \times \Omega \times \mathbb{R}^q, \underline{g}) = \bigcup_{P,Q} R_G^l(\Omega \times \Omega \times \mathbb{R}^q, \underline{g})_{P,Q} \tag{9.44}$$

and we write one factor Ω in (9.43) or (9.44) for the set of all corresponding elements that are y'-independent.

Definition 4. A Green edge symbol of order $l \in \mathbb{R}$ with trace and potential part, associated with the weight data $\underline{g} = (\gamma, \delta, \Theta)$ of Definition 1 is a function

$$g(y, y', \eta) \in \bigcap_{s \in \mathbb{R}} C^\infty(\Omega \times \Omega \times \mathbb{R}^q, \mathcal{L}(\mathcal{K}^{s,\gamma}(X^\wedge) \oplus \mathbb{C}^{N_-}, \mathcal{K}^{\infty,\delta}(X^\wedge) \oplus \mathbb{C}^{N_+}))$$

with certain $N_-, N_+ \in \mathbb{Z}_+$ satisfying

$$g(y, y', \eta) \in \bigcap_{s \in \mathbb{R}} S_{cl}^l(\Omega \times \Omega \times \mathbb{R}^q, \mathcal{K}^{s,\gamma}(X^\wedge) \oplus \mathbb{C}^{N_-}, \mathcal{S}_P^\delta(X^\wedge) \oplus \mathbb{C}^{N_+})),$$

$$g^*(y, y', \eta) \in \bigcap_{s \in \mathbb{R}} S_{cl}^l(\Omega \times \Omega \times \mathbb{R}^q, \mathcal{K}^{s,-\delta}(X^\wedge) \oplus \mathbb{C}^{N_+}, \mathcal{S}_Q^{-\gamma}(X^\wedge) \oplus \mathbb{C}^{N_-}))$$

with (weakly discrete) asymptotic types P, Q as in Definition 1.

The $*$ means the pointwise formal adjoint in the sense

$$(gu, v)_{\mathcal{K}^0(X^\wedge) \oplus \mathbb{C}^{N_+}} = (u, g^*v)_{\mathcal{K}^0(X^\wedge) \oplus \mathbb{C}^{N_-}} \tag{9.45}$$

for all $u \in C_0^\infty(X^\wedge) \oplus \mathbb{C}^{N_-}$, $v \in C_0^\infty(X^\wedge) \oplus \mathbb{C}^{N_+}$. Similarly, we can talk about the pointwise transposed operators, indicated by left upper t, with respect to the bilinear pairings

$$\langle f, \tilde{f} \rangle_{\mathcal{K}^0(X^\wedge) \oplus \mathbb{C}^N} = \int r^{-n} f_0(r, x) \tilde{f}_0(r, x) dx dr + \sum_{k=1}^N f_k \tilde{f}_k$$

for $f = (f_0; f_1, \ldots, f_N)$, $\tilde{f} = (\tilde{f}_0; \tilde{f}_1, \ldots, \tilde{f}_N) \in \mathcal{K}^0(X^\wedge) \oplus \mathbb{C}^N$,

$$\langle gu, v \rangle_{\mathcal{K}^0(X^\wedge) \oplus \mathbb{C}^{N_+}} = \langle u, {}^t gv \rangle_{\mathcal{K}^0(X^\wedge) \oplus \mathbb{C}^{N_-}}. \tag{9.46}$$

Remember that the definition of the symbol spaces refers to the group action (9.38). Similarly to (9.43) we can introduce the space

$$\mathcal{R}_G^l(\Omega \times \Omega \times \mathbb{R}^q, \underline{g}; N_-, N_+)_{P,Q} \tag{9.47}$$

of all Green edge symbols in the sense of Definition 4 with fixed asymptotic types P, Q. The elements in (9.47) can be written as matrices of operator functions

$$g = \begin{pmatrix} g_{11} & g_{12} \\ g_{21} & g_{22} \end{pmatrix} : \begin{array}{c} \mathcal{K}^{s,\gamma}(X^\wedge) \\ \oplus \\ \mathbb{C}^{N_-} \end{array} \longrightarrow \begin{array}{c} \mathcal{K}^{\infty,\delta}(X^\wedge) \\ \oplus \\ \mathbb{C}^{N_+} \end{array} \tag{9.48}$$

dependent on y, y', η.

We will also call $g_{21}(y, y', \eta)$ an *edge trace* and $g_{12}(y, y', \eta)$ an *edge potential* symbol of order l, with the corresponding weight data and asymptotic types. Note that $g_{22}(y, y', \eta) \in S_{cl}^l(\Omega \times \Omega \times \mathbb{R}^q) \otimes \mathbb{C}^{N_+} \otimes \mathbb{C}^{N_-}$ with $S_{cl}^l(\Omega \times \Omega \times \mathbb{R}^q)$ being the space of scalar classical symbols of order l.

Similarly to (9.44) we also write

$$\mathcal{R}_G^l(\Omega \times \Omega \times \mathbb{R}^q, \underline{g}; N_-, N_+) \tag{9.49}$$

for the union over all P, Q, and we write Ω instead of $\Omega \times \Omega$ for the corresponding subspaces of y'-independent elements.

An element $g(y, y', \eta) \in$ (9.49) as a classical operator-valued symbol has a principal homogeneous component $g_{(l)}(y, y', \eta)$ that satisfies

$$g_{(l)}(y, y', \lambda\eta) = \lambda^l \begin{pmatrix} \kappa_\lambda & 0 \\ 0 & 1 \end{pmatrix} g_{(l)}(y, y', \eta) \begin{pmatrix} \kappa_\lambda^{-1} & 0 \\ 0 & 1 \end{pmatrix}$$

for all $y, y' \in \Omega$, $\eta \in \mathbb{R}^q \setminus \{0\}$, $\lambda \in \mathbb{R}_+$. We set

$$\sigma_\wedge^l(g)(y, \eta) = g_{(l)}(y, y, \eta), \tag{9.50}$$

and call this the *homogeneous principal symbol* of g of order l.

Remark 5. The statements for the edge Green symbol space (9.44) (such as Remark 3 or the following Theorem 6) hold in an analogous form also for (9.47). They will be used tacitly below.

Theorem 6. *Let $g_j(y, y', \eta) \in R_G^{l-j}(\Omega \times \Omega \times \mathbb{R}^q, \underline{g})_{P,Q}$, $j \in \mathbb{Z}_+$, be an arbitrary sequence (with P, Q being independent of j). Then there exists a $g(y, y', \eta) \in R_G^l(\Omega \times \Omega \times \mathbb{R}^q, \underline{g})_{P,Q}$ such that*

$$g - \sum_{j=0}^N g_j \in R_G^{l-(N+1)}(\Omega \times \Omega \times \mathbb{R}^q, \underline{g})_{P,Q}$$

for every $N \in \mathbb{Z}_+$. The symbol $g(y, y', \eta)$ is unique mod $R_G^{-\infty}(\Omega \times \Omega \times \mathbb{R}^q, \underline{g})_{P,Q}$ (with $R_G^{-\infty}$ being the intersection over all R_G^l, $l \in \mathbb{R}$).

We shall write $g \sim \sum_{j=0}^{\infty} g_j$, called the asymptotic sum of the g_j (i.e. any representative mod $R_G^{-\infty}$ is by definition an asymptotic sum).

From the point of view of standard technique of handling symbol spaces, the proof of Theorem 6 is an elementary though voluminous exercise, that is recommended for the reader. Let us only sketch the idea. We choose an excision function $\chi(\eta)$ and generate g as a convergent sum

$$g(y, y^{'}, \eta) = \sum_{j=0}^{\infty} \chi(\frac{\eta}{c_j}) g_j(y, y^{'}, \eta) \tag{9.51}$$

with respect to the Fréchet topology in the space R_G^l, cf. Remark 3. To this end we fix a semi-norm p on $\mathcal{R}_G^l(\ldots)_{P,G}$ and show

$$p(\chi_c g_j) \to 0 \quad \text{as} \quad c \to \infty \tag{9.52}$$

for $\chi_c(\eta) := \chi(\frac{\eta}{c})$ once $j \geq j(p)$. Then, by putting c_j so large such that $p(\chi_{c_j} g_j) \leq 2^{-j}$ for $j \geq j(p)$, our sum will converge with respect to the semi-norm p. If our p is of number k in the semi-norm system $\{p_k\}_{k \in \mathbb{Z}_+}$ from the Fréchet structure on $\mathcal{R}_G^l(\ldots)_{P,Q}$ we have found constants $c_j(k)$. For $c_j := c_j(j)$ we then obtain convergence of (9.51) with respect to arbitrary k. An analogous diagonal argument gives us the convergence of

$$\sum_{j=N+1}^{\infty} \chi(\frac{\eta}{c_j}) g_j(y, y^{'}, \eta)$$

in $R_G^{l-(N+1)}(\ldots)_{P,Q}$ for every $N \in \mathbb{Z}_+$, provided the constants c_j are chosen in the appropriate way. In other words, the remaining step is (9.52). But here we are actually in a standard situation for establishing asymptotic sums of symbols in the operator-valued set-up, cf. Definition 1 and 9.1.1 Theorem 9. It suffices then to observe that things work when the range spaces $\tilde{E}$ of 9.1.1 Theorem 9 are Fréchet spaces and that in Definition 1 it suffices to take the intersections over $s \in \mathbb{Z}$. Then, to choose the constants, we can apply a diagonal argument.

Remark 7. If $g(y, y^{'}, \eta) \in R_G^l(\Omega \times \Omega \times \mathbb{R}^q, \underline{g})_{P,Q}$ it follows that, for every multi-index $\alpha = (\alpha_1, \ldots, \alpha_q)$,

$$\partial_{y^{'}}^{\alpha} D_{\eta}^{\alpha} g(y, y^{'}, \eta) \,|_{y^{'}=y} \in R_G^{l-|\alpha|}(\Omega \times \mathbb{R}^q, \underline{g})_{P,Q}.$$

Thus, in particular,

$$\sum_{\alpha} \frac{1}{\alpha!} \partial_{y^{'}}^{\alpha} D_{\eta}^{\alpha} g(y, y^{'}, \eta) \,|_{y^{'}=y} \in R_G^l(\Omega \times \mathbb{R}^q, \underline{g})_{P,Q}$$

can be carried out.

9.2.2 Smoothing Mellin symbols

We now turn to the smoothing Mellin symbols $w(y, y', \eta)$ belonging to the edge symbolic calculus. They will be used in the explicit form only for a finite weight interval $\Theta = (-k, 0]$ with $k \in \mathbb{N}$. The smoothing Mellin symbols will be operator-valued, acting from $\mathcal{K}^{s,\gamma}(X^\wedge)$ to $\mathcal{K}^{\infty,\gamma-m}(X^\wedge)$ for prescribed $\gamma, m \in \mathbb{R}$. In the calculations, operators with images in spaces of better weights also occur. Therefore, we start with the expressions

$$w(y, y', \eta) = \omega(r[\eta]) r^{-l} \sum_{j=0}^{k-1} w_j(y, y', \eta) \tilde{\omega}(r[\eta]) \tag{9.53}$$

for $l \in \mathbb{R}$ with $m - l \in \mathbb{Z}_+$ and fixed cut-off functions $\omega(r), \tilde{\omega}(r)$. Here

$$w_j(y, y', \eta) = r^j \sum_{|\alpha| \leq j} op_M^{\gamma_{j\alpha} - \frac{n}{2}} (h_{j\alpha})(y, y') \eta^\alpha$$

with $n = \dim\ X$,

$$h_{j\alpha}(y, y', z) \in C^\infty(\Omega \times \Omega, M_{P_{j\alpha}}^{-\infty}(X)),$$

where $P_{j\alpha}$ are weakly discrete asymptotic types of operator-valued Mellin symbols in the sense of the notation at the end of Chapter 8, which are constant with respect to $y, y' \in \mathbb{R}$, and

$$0 \leq \gamma - \gamma_{j\alpha} \leq j + m - l \quad \text{for all}\ \ j, \alpha.$$

Furthermore, the weights $\gamma_{j\alpha}$ are chosen in such a way that $\Gamma_{\frac{n+1}{2} - \gamma_{j\alpha}}$ does not intersect any pole of $h_{j\alpha}(y, y', z)$ in the complex plane, for all $y, y' \in \Omega$. It is clear from the cone theory that under these conditions

$$w(y, y', \eta) : \mathcal{K}^{s,\gamma}(X^\wedge) \to \mathcal{K}^{\infty,\gamma-m}(X^\wedge) \tag{9.54}$$

is continuous for every $s \in \mathbb{R}$ and every fixed $(y, y', \eta) \in \Omega \times \Omega \times \mathbb{R}^q$. Moreover, for every (weakly discrete) asymptotic type P associated with the weight data (γ, Θ) (in the sense of (8.6), (8.7)), there is another (weakly discrete) asymptotic type Q, associated with $(\gamma - m, \Theta)$ such that

$$w(y, y', \eta) : \mathcal{K}_P^{s,\gamma}(X^\wedge) \to \mathcal{K}_Q^{\infty,\gamma-m}(X^\wedge) \tag{9.55}$$

is continuous for all $s \in \mathbb{R}$ and every fixed $(y, y', \eta) \in \Omega \times \Omega \times \mathbb{R}^q$, cf. 8.1.2 Theorem 3.

Definition 1. The elements $w(y, y', \eta)$ defined above are called smoothing Mellin edge symbols of order l, associated with the weight data $\underline{g} = (\gamma, \gamma - m, \Theta)$, $\Theta = (-k, 0]$, and with constant (weakly discrete) asymptotics. The operator function

$$\sigma_\wedge^l(w)(y, \eta) := \omega(r|\eta|) r^{-l} \sum_{j=0}^{k-1} r^j \sum_{|\alpha|=j} op_M^{\gamma_{j\alpha} - \frac{n}{2}} (h_{j\alpha})(y, y) \eta^\alpha \tilde{\omega}(r|\eta|) \tag{9.56}$$

is called the homogeneous principal symbol of order l of $w(y, y', \eta)$.

Exercise 2. Show that

$$\sigma_\wedge^l(w)(y, \lambda\eta) = \lambda^l \kappa_\lambda \sigma_\wedge^l(w)(y, \eta) \kappa_\lambda^{-1} \tag{9.57}$$

for all $\lambda > 0$, $y \in \Omega$, $\eta \in \mathbb{R}^q \setminus \{0\}$.

Let us denote by

$$R_{M+G}^l(\Omega \times \Omega \times \mathbb{R}^q, \underline{g}) \quad \text{for} \quad \underline{g} = (\gamma, \gamma - m, \Theta) \tag{9.58}$$

the space of all $w(y, y', \eta) + g(y, y', \eta)$, with w being of the form (9.53) and $g \in R_G^l(\Omega \times \Omega \times \mathbb{R}^q, \underline{g})$. Analogously we write $R_{M+G}^l(\Omega \times \mathbb{R}^q, \underline{g})$ for the subspace of (9.58) of y'-independent elements.

Note that

$$R_{M+G}^l(\Omega \times \Omega \times \mathbb{R}^q, \underline{g}) \subseteq R_G^l(\Omega \times \Omega \times \mathbb{R}^q, \underline{g})$$

once $k \leq m - l$; here, as above, $\Theta = (-k, 0]$.

Exercise 3. We have

$$D_{y,y'}^\alpha D_\eta^\beta R_{M+G}^l(\Omega \times \Omega \times \mathbb{R}^q, \underline{g}) \subseteq R_{M+G}^{l-|\beta|}(\Omega \times \Omega \times \mathbb{R}^q, \underline{g})$$

for all multi-indices $\alpha \in \mathbb{Z}_+^{2q}$, $\beta \in \mathbb{Z}_+^q$, $\underline{g} = (\gamma, \gamma - m, \Theta)$.

Proposition 4. *We have for* $\underline{g} = (\gamma, \gamma - m, \Theta)$, $\Theta = (-k, 0]$, $k \in \mathbb{N}$

$$R_{M+G}^l(\Omega \times \Omega \times \mathbb{R}^q, \underline{g}) \subset \bigcap_{s \in \mathbb{R}} S_{cl}^l(\Omega \times \Omega \times \mathbb{R}^q; \mathcal{K}^{s,\gamma}(X^\wedge), \mathcal{K}^{\infty,\gamma-m}(X^\wedge)), \tag{9.59}$$

for all $l \in \mathbb{Z}_+$. *Furthermore, for every* $w(y, y', \eta) \in R_{M+G}^l(\Omega \times \Omega \times \mathbb{R}^q, \underline{g})$ *and every asymptotic type* P *associated with the weight data* (γ, Θ), *there is an asymptotic type* Q *associated with* $(\gamma - m, \Theta)$, *such that*

$$w(y, y', \eta) \in \bigcap_{s \in \mathbb{R}} S_{cl}^l(\Omega \times \Omega \times \mathbb{R}^q; \mathcal{K}_P^{s,\gamma}(X^\wedge), \mathcal{K}_Q^{\infty,\gamma-m}(X^\wedge)). \tag{9.60}$$

Proof. The Green ingredients are classical by definition. The symbols (9.53) have the form of a sum of items like

$$a(y, y', \eta) = \omega(r[\eta]) r^{-l+j} op_M^\delta(h)(y, y') \eta^\alpha \tilde\omega(r[\eta])$$

with certain Mellin symbols $h(y, y', z)$, weights δ, and $|\alpha| \leq j$. They satisfy

$$a(y, y', \lambda\eta) = \lambda^{l-j+|\alpha|} \kappa_\lambda a(y, y', \eta) \kappa_\lambda^{-1}$$

for all $\lambda \geq 1$, $\mid \eta \mid \geq c$ with some $c > 0$, and all $(y, y') \in \Omega \times \Omega$. Those symbols are classical, cf. 9.1.1 Definition 10 and (9.16) in the version with subscript cl. Here we use representations as countable projective limits

$$\mathcal{K}_P^{s,\gamma}(X^\wedge) = \varprojlim_k E^k, \quad \mathcal{K}_Q^{\infty,\gamma-m}(X^\wedge) = \varprojlim_j \tilde{E}^j,$$

and to every concrete $w(y, y', \eta)$ for which the map

$$w(y, y', \eta) : \mathcal{K}_P^{s,\gamma}(X^\wedge) \longrightarrow \mathcal{K}_Q^{\infty,\gamma-m}(X^\wedge)$$

is continuous for every fixed y, y', η we find a function $b : \mathbb{Z}_+ \to \mathbb{Z}_+$ such that

$$w(y, y', \eta) \in \bigcap_j S_{cl}^l(\Omega \times \Omega \times \mathbb{R}^q, E^{b(j)}, \tilde{E}^j). \qquad \square$$

9.2.3   Complete edge symbols

The main contribution to the edge symbolic calculus comes from the symbol of the given pseudo-differential operator on the manifold with edges Y. The situation is analogous to the cone theory, where we have introduced a Mellin reformulation of a given pseudo-differential operator in a neighbourhood of the singularities. The singularities now are the edges. First we look at local coordinates $(r, x, y) \in \mathbb{R}_+ \times V \times \Omega$. Here $V \subseteq \mathbb{R}^n$ and $\Omega \subseteq \mathbb{R}^q$ are open sets that correspond via charts to coordinate neighbourhoods on X and Y, respectively. Remember that X is the base of the model cone of the wedge in a neighbourhood of y.

Let

$$S_{cl}^m(\overline{\mathbb{R}}_+ \times V \times \Omega \times \mathbb{R}^{1+n+q}) \tag{9.61}$$

be the space of all classical symbols $a(r, x, y, \varrho, \xi, \eta)$ in $\mathbb{R}_+ \times V \times \Omega$ of order $m \in \mathbb{R}$ with $a \in S_{cl}^m(\mathbb{R} \times V \times \Omega \times \mathbb{R}^{1+n+q}) \mid_{\overline{\mathbb{R}}_+}$. In other words, a has an extension to $\mathbb{R} \times V \times \Omega \times \mathbb{R}^{1+n+q}$ as a classical symbol. From Section 7.1.3 we know that the adequate *interior symbols* for the pseudo-differential calculus near edges are of the form

$$r^{-m} p(r, x, y, r\varrho, \xi, r\eta), \tag{9.62}$$

where $p(r, x, y, \tilde{\varrho}, \xi, \tilde{\eta})$ belongs to $S_{cl}^m(\overline{\mathbb{R}}_+ \times V \times \Omega \times \mathbb{R}^{1+n+q})$.

The space of the symbols has many natural properties with respect to the operations at a symbolic level from the associated pseudo-differential calculus. In particular, we have invariance under the symbolic rule from changes of coordinates in the x-space. In other words, if $\chi : V \to V'$ is a diffeomorphism, then there is a (non-canonical) map

$$p(r, x, y, r\varrho, \xi, r\eta) \to p'(r, x', y, r\varrho, \xi', r\eta)$$

where $p'(r, x', y, \tilde{\varrho}, \xi, \tilde{\eta}) \in S_{cl}^m(\overline{\mathbb{R}}_+ \times V' \times \Omega \times \mathbb{R}^{1+n+q})$ has the property

$$\chi_* Op(p) - Op(p') \in L^{-\infty}(\mathbb{R}_+ \times V' \times \Omega).$$

Here $Op(\cdot)$ is the pseudo-differential operator associated with the corresponding symbol, based on the Fourier transform in all variables. The latter observation may be regarded as an exercise, using the symbolic rule for coordinate changes (Section 2.4.1) and the fact that the spaces of symbols which are smooth up to $r = 0$ in the sense of (9.61) are preserved under asymptotic sums. To indicate the variables to which the pseudo-differential action refers with respect to the Fourier transform $\mathcal{F}$ we also write $Op_{(r,x,y)}$. In particular, $Op_{(x,y)}$, Op_r will also occur. For $p(r,x,y,r\varrho,\xi,r\eta)$, the operator family

$$Op_{(r,x)}(p)(y,\eta)$$

is C^∞ in $y \in \Omega$ and belongs to $L_{cl}^m(\mathbb{R}_+ \times V; \mathbb{R}_\eta^q)$ for every fixed y, in the sense of parameter-dependent pseudo-differential operators, cf. Section 7.2.2. Analogously if a z-dependent symbol $h(r,x,y,z,\xi,r\eta)$ is given on a weight line $\Gamma_{\frac{n+1}{2}-\gamma} \ni z$, then the mixed *Mellin-Fourier action*

$$op_{M,r}^{\gamma-\frac{n}{2}} Op_x(h)(y,\eta)$$

makes sense and defines a C^∞-family in y of parameter-dependent pseudo-differential operators over $\mathbb{R}_+ \times V$ with parameter $\eta \in \mathbb{R}^q$.

Theorem 1. *For every $p(r,x,y,r\varrho,\xi,r\eta)$ with $p(r,x,y,\tilde{\varrho},\xi,\tilde{\eta}) \in S^m(\overline{\mathbb{R}}_+ \times V \times \Omega \times \mathbb{R}^{1+n+q})$ there exists an $h(r,x,y,z,\xi,r\eta)$ with*

(i) $h(r,x,y,z,\xi,\tilde{\eta}) \in \mathcal{A}(\mathbb{C}_z, S_{cl}^m(\overline{\mathbb{R}}_+ \times V \times \Omega \times \mathbb{R}_\xi^n \times \mathbb{R}_{\tilde{\eta}}^q))$,

(ii) $h(r,x,y,\beta+i\varrho,\xi,\tilde{\eta}) \in S_{cl}^m(\overline{\mathbb{R}}_+ \times V \times \Omega \times \mathbb{R}_\varrho \times \mathbb{R}_\xi^n \times \mathbb{R}_{\tilde{\eta}}^q)$
for every $\beta \in \mathbb{R}$, uniformly in $c \le \beta \le c'$ for every finite $c < c'$

such that for every $\delta \in \mathbb{R}$

$$\begin{aligned} Op_{(r,x,y)}(p) - op_{M,r}^\delta Op_{(x,y)}(h) &\in L^{-\infty}(\mathbb{R}_+ \times V \times \Omega), & (9.63) \\ Op_{(r,x)}(p)(y,\eta) - op_{M,r}^\delta Op_x(h)(y,\eta) &\in C^\infty(\Omega, L^{-\infty}(\mathbb{R}_+ \times V; \mathbb{R}_\eta^q)). & (9.64) \end{aligned}$$

Proof. First remember that when $h(r,x,y,z,\xi,\tilde{\eta})$ is a symbol with the asserted properties, then $op_M^\delta(h) = op_M^{\delta'}(h)$ for every $\delta, \delta' \in \mathbb{R}$, since h is holomorphic in z. Thus it suffices to show the assertion for a convenient weight δ, where we take $\delta = \frac{1}{2}$. We shall prove that for the symbol $p(r,x,y,r\varrho,\xi,\tilde{\eta})$ there is an

$$h_0(r,x,y,z,\xi,\tilde{\eta}) \in S_{cl}^m(\overline{\mathbb{R}}_+ \times V \times \Omega \times \Gamma_0 \times \mathbb{R}_\xi^n \times \mathbb{R}_{\tilde{\eta}}^q)$$

such that

$$Op_{(r,x)}(p)(y,\tilde{\eta}) - op_{M,r}^{\frac{1}{2}} Op_x(h_0)(y,\tilde{\eta}) = Op_{(r,x)}(p_1)(y,\tilde{\eta}) \qquad (9.65)$$

$\bmod C^\infty(\Omega, L^{-\infty}(\mathbb{R}_+ \times V; \mathbb{R}_{\tilde{\eta}}))$, where $p_1(r,x,y,r\varrho,\xi,\tilde{\eta})$ has the property $p_1(r,x,y,\tilde{\varrho},\xi,\tilde{\eta}) \in S_{cl}^{m-1}(\overline{\mathbb{R}}_+ \times V \times \Omega \times \mathbb{R}^{1+n+q})$. Since $m \in \mathbb{R}$ is arbitrary, this implies

$$Op_{(r,x)}(p_k)(y,\tilde{\eta}) - op_{M,r}^{\frac{1}{2}} Op_x(h_k)(y,\tilde{\eta}) = Op_{(r,x)}(p_{k+1})(y,\tilde{\eta}) \tag{9.66}$$

mod $C^\infty(\Omega, L^{-\infty}(\mathbb{R}_+ \times V; \mathbb{R}_{\tilde{\eta}}^q))$ for every $p_k(r,x,y,r\varrho,\xi,\tilde{\eta})$ with $p_k(r,x,y,\tilde{\varrho},\xi,\tilde{\eta}) \in S_{cl}^{m-k}(\overline{\mathbb{R}}_+ \times V \times \Omega \times \mathbb{R}^{1+n+q})$, a resulting $h_k(r,x,y,z,\xi,\tilde{\eta}) \in S_{cl}^{m-k}(\overline{\mathbb{R}}_+ \times V \times \Omega \times \Gamma_0 \times \mathbb{R}_{\xi,\tilde{\eta}}^{n+q})$, and $p_{k+1}(r,x,y,r\varrho,\xi,\tilde{\eta})$ with $p_{k+1}(r,x,y,\tilde{\varrho},\xi,\tilde{\eta}) \in S_{cl}^{m-k-1}(\overline{\mathbb{R}}_+ \times V \times \Omega \times \mathbb{R}^{1+n+q})$. For $p_0 := p$ we thus obtain the sequence of p_k and h_k, $k \in \mathbb{Z}_+$ inductively. Set

$$f(r,x,y,z,\xi,\tilde{\eta}) = \sum_{k=0}^{\infty} h_k(r,x,y,z,\xi,\tilde{\eta}),$$

with the asymptotic sum being taken in $S_{cl}^m(\overline{\mathbb{R}}_+ \times V \times \Omega \times \Gamma_0 \times \mathbb{R}^{n+q})$. By a kernel cut-off construction we will then find an $h(r,x,y,z,\xi,\tilde{\eta})$ satisfying (i) and

$$f(r,x,y,i\varrho,\xi,\tilde{\eta}) - h(r,x,y,i\varrho,\xi,\tilde{\eta}) \in S^{-\infty}(\overline{\mathbb{R}}_+ \times V \times \Omega \times \Gamma_0 \times \mathbb{R}^{n+q}).$$

From the system of relations (9.66) it is clear that we have for every $N \in \mathbb{Z}_+$

$$Op_{(r,x)}(p)(y,\tilde{\eta}) = \sum_{k=0}^{N} op_{M,r}^{\frac{1}{2}} Op_x(h_k)(y,\tilde{\eta}) + Op_{(r,x)}(p_{N+1})(y,\tilde{\eta})$$

mod $C^\infty(\Omega, L^{-\infty}(\mathbb{R}_+ \times V; \mathbb{R}_{\tilde{\eta}}^q))$. Thus

$$Op_{(r,x)}(p)(y,\tilde{\eta}) = op_{M,r}^{\frac{1}{2}} Op_x(f)(y,\tilde{\eta}) = op_{M,r}^{\frac{1}{2}} Op_x(h)(y,\tilde{\eta})$$

with equality mod $C^\infty(\Omega, L^{-\infty}(\mathbb{R}_+ \times V; \mathbb{R}_{\tilde{\eta}}^q))$. From that, by inserting $\tilde{\eta} = r\eta$ (and interpreting r as an action from the left) the relation (9.64) follows immediately. By applying Op_y to both sides this implies (9.63). It remains to show (9.65). We claim that it suffices to set

$$h_0(r,x,y,i\varrho,\xi,\tilde{\eta}) = p(r,x,y,-\varrho,\xi,\tilde{\eta}).$$

For convenience we shall first look at the case of $n = q = 0$, i.e. where we have the one-dimensional situation $\mathbb{R}_+ \ni r$. If $\chi : \mathbb{R} \to \mathbb{R}_+$ denotes the diffeomorphism $r = \chi(t) = e^{-t}$, we easily obtain for arbitrary $h(r,i\varrho) \in S^m(\overline{\mathbb{R}}_+ \times \Gamma_0)$ the following identity:

$$op_M^{\frac{1}{2}}(h)u(r) = \int_{-\infty}^{\infty} \int_{0}^{\infty} \left(\frac{r}{r'}\right)^{-i\varrho} h(r,i\varrho) u(r') \frac{dr'}{r'} d\varrho$$

$$= (\chi^*)^{-1} \int_{-\infty}^{\infty} \int_{-\infty}^{\infty} e^{i(t-t')\varrho} h(e^{-t},i\varrho) v(t') dt' d\varrho$$

with $v = \chi^* u$, χ^* being the pull-back under χ. In other words,

$$op_M^{\frac{1}{2}}(h) = \chi_* Op_t(p) \quad \text{with} \quad p(t,\varrho) = h(e^{-t}, i\varrho), \tag{9.67}$$

where χ_* is the push-forward of pseudo-differential operators. Now it suffices to apply the asymptotic formula for a symbol $b(r,\varrho)$ such that

$$Op_r(\tilde{b}) - \chi_* Op_t(p) \in L^{-\infty}(\mathbb{R}_+)$$

with $\tilde{b}(r,\varrho) = b(r, r\varrho)$ and to observe

$$b(r,\varrho) - h(r, i\varrho) \in S^{m-1}(\overline{\mathbb{R}}_+ \times \mathbb{R}). \tag{9.68}$$

We obtain from the substitution rule for pseudo-differential operators

$$\tilde{b}(r,\varrho)\,|_{r=\chi(t)} \sim \sum_{k=0}^{\infty} \frac{1}{k!} (\partial_\varrho^k p)(t, \chi^{'}(t)\varrho)\Phi_k(t,\varrho)$$

with $\Phi_k(t,\varrho) = D_s^k e^{i\delta(t,s,\varrho)}\,|_{s=r}$, $\delta(t,s,\varrho) = \{\chi(s) - \chi(t) - \chi^{'}(t)(s-t)\}\varrho$. Next we use $\chi^{'}(t) = -e^{-t} = -r$ which shows that $\tilde{b}(r,\varrho)$ is up to lower terms of the form $p(t, -r\varrho)\,|_{t=-\ln r}$. It is easy to show that $\Phi_k(t,\varrho)\,|_{t=-\ln r}$ is of the form $\varphi_k(r, r\varrho)$ with $\varphi_k(r,\varrho) \in C^\infty(\overline{\mathbb{R}}_+ \times \mathbb{R})$. Thus, $\tilde{b}(r,\varrho)$ exists as a symbol of the class $S^m(\overline{\mathbb{R}}_+ \times \mathbb{R})$. This, together with (9.68), is the asserted result. For arbitrary n, q, i.e. including the (x, y, ξ, η)-variables, the conclusions are practically the same. The latter generalization does of course, employ the fact that differentiation with respect to one covariable diminishes the symbol orders in all covariables. This completes the proof of Theorem 1. $\qquad\square$

Theorem 1 may be regarded as a Mellin operator convention which asserts the Mellin reformulation of a pseudo-differential action in r given by the Fourier transform. The idea is completely analogous to the above cone theory of Section 8.1.3.

Let us fix a finite open covering of X by coordinate neighbourhoods $\{U_1, \ldots, U_n\} =: \mathcal{U}$ and charts $\kappa_j : U_j \to V_j$, $j = 1, \ldots, N$, with open $V_j \subseteq \mathbb{R}^n$. Further, choose a subordinated partition of unity $\{\varphi_1, \ldots, \varphi_N\}$ and a system $\{\psi_1, \ldots, \psi_N\}$ of functions in $C_0^\infty(U_j)$ with $\varphi_j \psi_j = \varphi_j$ for $j = 1, \ldots, N$. Then, for every system of symbols

$$p_j(r, x, y, r\varrho, \xi, r\eta) \quad \text{with} \quad p_j(r, x, y, \tilde{\varrho}, \xi, \tilde{\eta}) \in S_{cl}^m(\overline{\mathbb{R}}_+ \times V_j \times \Omega \times \mathbb{R}^{1+n+q}),$$

$j = 1, \ldots, N$, we can form a (y, η)-dependent pseudo-differential operator $P(y, \eta)$ on $\mathbb{R}_+ \times X = X^\wedge$ by

$$P(y,\eta) = \sum_{j=1}^{N} \varphi_j (1 \times \kappa_j)^* Op_{(r,x)}(p_j)(y,\eta)\psi_j, \tag{9.69}$$

with upper * indicating the operator pull-back. Furthermore, if we form an h_j to every p_j via Theorem 1, then we find an (r, y, z, η)-dependent family of pseudo-differential operators along X which we write as

$$h(r, y, z, r\eta) = \sum_{j=1}^{N} \varphi_j \kappa_j^* Op_x(h_j)(r, y, z, r\eta)\psi_j. \tag{9.70}$$

We now have

$$h(r, y, z, \tilde{\eta}) \in C^\infty(\overline{\mathbb{R}}_+ \times \Omega, M_O^m(X; \mathbb{R}_{\tilde{\eta}}^q)) \tag{9.71}$$

(cf. 7.2.4 Definition 5). As an immediate consequence of Theorem 1 we get

$$P(y, \eta) - op_M^\delta(h)(y, \eta) \in C^\infty(\Omega, L^{-\infty}(X^\wedge; \mathbb{R}^q)) \tag{9.72}$$

(remember that $L^{-\infty}(X^\wedge; \mathbb{R}^q) \cong \mathcal{S}(\mathbb{R}^q, L^{-\infty}(X^\wedge))$). Thus we obtain the following

Theorem 2.   *Let $P(y, \eta) \in C^\infty(\Omega_y, L_{cl}^m(\mathbb{R}_+ \times X; \mathbb{R}_\eta^q))$ be a family of pseudo-differential operators which can be written over $\mathbb{R}_+ \times U$ for every $U \in \mathcal{U}$ in the local coordinates $(r, x) \in \mathbb{R}_+ \times V$ with respect to the chart $\kappa : U \to V$ in the form $P(y, \eta) = Op_{(r,x)}(p)$ modulo $C^\infty(\Omega, L^{-\infty}(\mathbb{R}_+ \times X; \mathbb{R}^q))$ with a $p(r, x, y, r\varrho, \xi, r\eta)$ where $p(r, x, y, \tilde{\varrho}, \xi, \tilde{\eta}) \in (9.61)$. Then there is an $h(r, y, z, r\eta)$ as in (9.71) which satisfies (9.72) for every $\delta \in \mathbb{R}$.*

Remark 3.   Assume that the symbols p_j are elliptic for all j, i.e. $p_{j,(m)}(r,x,y,\tilde{\varrho},\xi,\tilde{\eta}) \neq 0$, where (m) indicates the homogeneous principal part of order m. Then, for every fixed $y_0 \in \Omega$

$$h(0, y_0, z, 0) \mid_{z=\beta+i\varrho} \in L_{cl}^m(X; \mathbb{R}_\varrho)$$

is parameter-dependent elliptic in the sense of Section 7.2.2 for all $\beta \in \mathbb{R} \setminus D(y_0)$, where $D(y_0)$ is a countable set of reals with $D(y_0) \cap [c, c']$ finite for every $c < c'$.

Exercise 4.   Let $\omega(r), \omega_0(r), \omega_1(r)$ be arbitrary cut-off functions satisfying $\omega\omega_0 = \omega$, $\omega\omega_1 = \omega_1$. Then

$$P(y, \eta) - \left\{ \omega(r[\eta]) op_M^\delta(h)(y, \eta)\omega_0(r[\eta]) + (1 - \omega(r[\eta]))P(y, \eta)(1 - \omega_1(r[\eta])) \right\}$$
$$\in \ C^\infty(\Omega, L^{-\infty}(\mathbb{R}_+ \times X; \mathbb{R}^q)).$$

Exercise 5.   Let $P(y, \eta)$ have the form (9.69) and h be as in Theorem 2, and assume that the $p_j(r, x, y, \tilde{p}, \xi, \tilde{\eta})$ involved are classical in r for $r \to \infty$ of order zero in the sense of Section 8.2.2. Moreover, let $\omega, \omega_0, \omega_1, \tilde{\omega}, \tilde{\omega}_0, \tilde{\omega}_0$ be cut-off functions with $\omega\omega_0 = \omega$, $\omega\omega_1 = \omega_1$, $\tilde{\omega}\tilde{\omega}_0 = \tilde{\omega}$, $\tilde{\omega}\tilde{\omega}_1 = \tilde{\omega}_1$. Then

$$r^{-m} \left\{ \omega(r[\eta]) op_M^{\gamma-\frac{n}{2}}(h)(y, \eta)\omega_0(r[\eta]) + (1 - \omega(r[\eta]))P(y, \eta)(1 - \omega_1(r[\eta])) \right\}$$

$$- r^{-m} \left\{ \tilde{\omega}(r[\eta]) op_M^{\gamma-\frac{n}{2}}(h)(y, \eta)\tilde{\omega}_0(r[\eta]) + (1 - \tilde{\omega}(r[\eta]))P(y, \eta)(1 - \breve{\omega}_1(r[\eta])) \right\}$$

$$\in R_G^m(\Omega \times \mathbb{R}^q, (\gamma, \gamma - m, \Theta))$$

for $\Theta = (-k, 0]$ with arbitrary $k \in \mathbb{N}$ (for the notation cf. Section 9.2.1).

Exercise 6. Let $P(y,\eta)$ be an arbitrary operator family of the form (9.69) and $\omega,\tilde{\omega}$ be arbitrary cut-off functions. Then

$$r^{-m}(1-\omega(r[\eta]))P(y,\eta)(1-\tilde{\omega}(r[\eta])) \in S^m(\Omega \times \mathbb{R}^q; \mathcal{K}^{s,\gamma}(X^\wedge), \mathcal{K}^{s-m,\delta}(X^\wedge))$$

for arbitrary $s,\gamma,\delta \in \mathbb{R}$. Moreover, let $\tilde{f}(r,y,z,\tilde{\eta}) \in C^\infty(\overline{\mathbb{R}}_+ \times \Omega, M_O^m(X;\mathbb{R}_{\tilde{\eta}}^q))$ and $N(y,\eta) = op_M^{\gamma-\frac{n}{2}}(f)(y,\eta)$, formed with $f(r,y,z,\eta) := \tilde{f}(r,y,z,r\eta)$; then

$$r^{-m}\omega(r[\eta])N(y,\eta)\tilde{\omega}(r[\eta]) \in S^m(\Omega \times \mathbb{R}^q; \mathcal{K}^{s,\gamma}(X^\wedge), \mathcal{K}^{s-m,\gamma-m}(X^\wedge))$$

for all $s \in \mathbb{R}$.

Exercise 7. Let

$$\tilde{h}(r,y,z,\tilde{\eta}) \in C^\infty(\overline{\mathbb{R}}_+ \times \Omega, M_O^{-\infty}(X;\mathbb{R}_{\tilde{\eta}}^q))$$

and $h(r,y,z,\eta) = \tilde{h}(r,y,z,r\eta)$. Then

$$r^{-l}\omega(r[\eta])op_M^{\gamma-\frac{n}{2}}(h)(y,\eta)\tilde{\omega}(r[\eta]) \in R_{M+G}^l(\Omega \times \mathbb{R}^q, \underline{g})$$

for $\underline{g} = (\gamma, \gamma-m, \Theta)$ with any m such that $m-l \in \mathbb{Z}_+$ and arbitrary $\Theta = (-k,0]$, $k \in \mathbb{N}$. Here $\omega(r),\tilde{\omega}(r)$ are arbitrary cut-off functions.

Analogously to (9.69) we will also form an operator family

$$P_\infty(y,\eta) = \sum_{j=1}^N \varphi_j(1 \times \kappa_j)^* Op_{(r,x)}(p_{j,\infty})(y,\eta)\psi_j$$

with symbols

$$p_{j,\infty}(r,x,y,\varrho,\xi,\eta) \in S_{cl}^m(\mathbb{R}_+ \times V_j \times \Omega \times \mathbb{R}^{1+n+q}),$$

$j = 1,\ldots,N$, that for large r belong to $S_{cl((\varrho,\xi,\eta),r)}^{m,0}$ in the sense of 8.2.2 Definition 1, modified in an obvious manner for $\mathbb{R}_r \times \mathbb{R}_{\varrho,\xi,\eta}^{1+n+q}$ for $r \to \infty$, where the order is 0 for r and m for (ϱ,ξ,η).

Definition 8. Let $l,m \in \mathbb{R}$, $m-l \in \mathbb{Z}_+$, and fix weight data $\underline{g} = (\gamma, \gamma-m, \Theta)$ with $\gamma \in \mathbb{R}$ and $\Theta = (-k,0]$, $k \in \mathbb{N}$. A *complete edge symbol* of the class

$$R^l(\Omega \times \mathbb{R}^q, \underline{g}) \tag{9.73}$$

is an operator family of the form

$$a(y,\eta) = \tilde{\omega}(r)\{a_0(y,\eta) + a_1(y,\eta)\}\tilde{\omega}_0(r) + a_\infty(y,\eta) + w(y,\eta) + g(y,\eta) \tag{9.74}$$

with

$$a_0(y,\eta) = \omega(r[\eta])r^{-l}op_M^{\gamma-\frac{n}{2}}(h)(y,\eta)\omega_0(r[\eta]), \tag{9.75}$$
$$a_1(y,\eta) = (1-\omega(r[\eta]))r^{-l}P(y,\eta)(1-\omega_1(r[\eta])), \tag{9.76}$$
$$a_\infty(y,\eta) = (1-\tilde{\omega}(r))P_\infty(y,\eta)(1-\tilde{\omega}_1(r)),$$

$n = \dim X$, where $h(r,y,z,r\eta)$ satisfies

$$h(r,y,z,\tilde{\eta}) \in C^\infty(\overline{\mathbb{R}}_+ \times \Omega, M_O^l(X;\mathbb{R}_{\tilde{\eta}}^q)), \tag{9.77}$$

cf. (9.71). Furthermore, $P(y,\eta)$ is a family of pseudo-differential operators on $X^{\wedge} = \mathbb{R}_+ \times X$ of the form (9.69), associated with h via (9.7.2), here of order l, and we assume that the symbols $p_j(r,x,y,\tilde{\varrho},\xi,\tilde{\eta})\,|_{\tilde{\varrho}=r\varrho,\ \tilde{\eta}=r\eta}$ in (9.69) are independent for large r; $P_{\infty}(y,\eta)$ is of the above form, now for the order l; $\omega, \omega_0, \omega_1$ are arbitrary cut-off functions with

$$\omega\omega_0 = \omega, \quad \omega\omega_1 = \omega_1, \quad \tilde{\omega}\tilde{\omega}_0 = \tilde{\omega}, \quad \tilde{\omega}\tilde{\omega}_1 = \tilde{\omega}_1, \tag{9.78}$$

and finally, $(w + g)(y,\eta) \in R^l_{M+G}(\Omega \times \mathbb{R}^q, \underline{g})$.

More generally, for arbitrary $N_-, N_+ \in \mathbb{Z}_+$, we will denote by

$$\mathcal{R}^l(\Omega \times \mathbb{R}^q, \underline{g}; N_-, N_+) \tag{9.79}$$

the space of all operator families

$$\underline{a}(y,\eta) := \begin{pmatrix} a & g_{12} \\ g_{21} & g_{22} \end{pmatrix} (y,\eta) \tag{9.80}$$

with $a(y,\eta) \in R^l(\Omega \times \mathbb{R}^q, \underline{g})$ and

$$\begin{pmatrix} 0 & g_{12} \\ g_{21} & g_{22} \end{pmatrix} (y,\eta) \in \mathcal{R}^l_G(\Omega \times \mathbb{R}^q, \underline{g}; N_-, N_+), \tag{9.81}$$

cf. (9.49). Finally,
$$\mathcal{R}^l_{M+G}(\Omega \times \mathbb{R}^q, \underline{g}; N_-, N_+) \tag{9.82}$$

is the set of all elements of (9.79) for which the upper left corner $a(y,\eta)$ belongs to $R^l_{M+G}(\Omega \times \mathbb{R}^q, \underline{g})$.

Up to now we have assumed that the weight interval Θ is finite. For $\Theta = (-\infty, 0]$ we define the operator-valued symbol class (9.79) as the intersection over corresponding ones for $\Theta = (-k, 0]$, $k = 1, 2, \ldots$.

Exercise 9. We have

$$D_y^\alpha D_\eta^\beta \mathcal{R}^l(\Omega \times \mathbb{R}^q, \underline{g}; N_-, N_+) \subseteq \mathcal{R}^{l-|\beta|}(\Omega \times \mathbb{R}^q, \underline{g}; N_-, N_+)$$

for all multi-indices $\alpha, \beta \in \mathbb{Z}_+^q$.

Exercise 10. $a(y,\eta) \in \mathcal{R}^l(\Omega \times \mathbb{R}^q, \underline{g}; N_-, N_+)$ for $\underline{g} = (\gamma, \gamma - m, \Theta)$ implies

$$\begin{aligned} a^*(y,\eta) &\in \mathcal{R}^l(\Omega \times \mathbb{R}, \underline{g}^*; N_+, N_-), \\ {}^t a(y,\eta) &\in \mathcal{R}^l(\Omega \times \mathbb{R}^q, \underline{g}^*; N_+, N_-), \end{aligned}$$

$\underline{g}^* = (-\gamma + m, -\gamma, \Theta)$. Here, * and upper t indicate the pointwise formal adjoints and transposed operators, respectively; cf. also (9.45), (9.46).

Remark 11. The spaces of operator families $\mathcal{R}^l(\Omega \times \mathbb{R}^q, \underline{g}; N_-, N_+)$ can (and will) be considered also in the variant where Ω is replaced by $\Omega \times \Omega \ni (y, y')$, i.e. the objects involved such as the interior symbols (9.62) may depend on (y, y') instead of y. We then use notations such as

$$\mathcal{R}^l_{M+G}(\Omega \times \Omega \times \mathbb{R}^q, \underline{g}; N_-, N_+). \tag{9.83}$$

The assertions on operator families (9.83) are analogous to those with Ω. For simplicity we shall sometimes formulate things only for one of the cases.

Exercise 12. Let $\underline{g} = (\gamma, \gamma - m, \Theta)$, $\Theta = (-k, 0]$, and

$$a(y, y', \eta) \in \mathcal{R}^l(\Omega \times \Omega \times \mathbb{R}^q, \underline{g}; N_-, N_+). \tag{9.84}$$

Then

$$a(y, y', \eta) \in \bigcap_{s \in \mathbb{R}} S^l(\Omega \times \Omega \times \mathbb{R}^q; E^{s,\gamma}; \tilde{E}^{s-l,\gamma-m}) \tag{9.85}$$

for $E^{s,\gamma} = \mathcal{K}^{s,\gamma}(X^\wedge) \oplus \mathbb{C}^{N_-}$, $\tilde{E}^{s-l,\gamma-m} = \mathcal{K}^{s-l,\gamma-m}(X^\wedge) \oplus \mathbb{C}^{N_+}$. Furthermore, if $E^{s,\gamma}_P = \mathcal{K}^{s,\gamma}_P(X^\wedge) \oplus \mathbb{C}^{N_-}$, $\tilde{E}^{s-l,\gamma-m}_Q = \mathcal{K}^{s-l,\gamma-m}_Q(X^\wedge) \oplus \mathbb{C}^{N_+}$, with asymptotic types P and Q connected with the corresponding weight data (γ, Θ) and $(\gamma - m, \Theta)$, respectively, then for every P there is a resulting Q such that

$$a(y, y', \eta) \in \bigcap_{s \in \mathbb{R}} S^l(\Omega \times \Omega \times \mathbb{R}^q; E^{s,\gamma}_P, \tilde{E}^{s-l,\gamma-m}_Q). \tag{9.86}$$

The proof of the assertions of Exercise 12 is lengthly though rather elementary. We shall give some hints and comment which illustrate the results. An observation more elementary than (9.85) is that when

$$p(x, y, \xi, \eta) \in S^l(\mathbb{R}^n \times \Omega \times \mathbb{R}^{n+q}_{\xi, \eta}) \tag{9.87}$$

is a symbol on the open set $\mathbb{R}^n \times \Omega$ in $\mathbb{R}^{n+q}$ with open $\Omega \subseteq \mathbb{R}^q$, of order l, and p being independent of x for $|x| \geq$ const, then

$$Op_x(p)(y, \eta) = \mathcal{F}^{-1}_{\xi \to x} p(x, y, \xi, \eta) \mathcal{F}_{x' \to \xi}$$

is an operator-valued symbol in the space

$$S^l(\Omega \times \mathbb{R}^q; H^s(\mathbb{R}^n), H^{s-l}(\mathbb{R}^n)) \tag{9.88}$$

for every $s \in \mathbb{R}$. The underlying group actions $\{\kappa_\lambda\}$ on the space $E = H^s(\mathbb{R}^n)$, $\tilde{E} = H^{s-l}(\mathbb{R}^n)$ are

$$(\kappa_\lambda u)(x) = \lambda^{\frac{n}{2}} u(\lambda x), \quad \lambda > 0.$$

For p we could have also assumed the exit behaviour of order 0 in x for $|x| \to \infty$, cf. (8.91) with $\mu = 0$. Instead of $\mathbb{R}^n$ we might look at a subset $V \subset \mathbb{R}^n$ that is conical for large $|x|$, cf. the constructions from Section 8.2.4. If in that case we replace $H^s(\mathbb{R}^n)$ by a space of Sobolev distributions in $H^s_{\mathrm{loc}}(V)$ that vanish near the origin and behave like a Sobolev distribution in $H^s(\mathbb{R}^n)|_{V_0}$ for every subset $V_0 \subset V$, conical for large $|x|$, $\overline{V}_0 \subset V$, then we can pass to a globalized version of (9.88) for a cylinder $\mathbb{R} \times X$ with a closed compact C^∞ manifold X, in the same spirit as in Section 8.2.4. This globalization is actually straightforward after the discussion of Section 8.2.4. So we return to the $\mathbb{R}^n$ case. Note that the role of n here is $1 + \dim X$ from (9.85), i.e. we have simplified a little the notations for this discussion. In particular, let

$$p(x, y, \xi, \eta) \in S^l_{cl}(\mathbb{R}^n \times \Omega \times \mathbb{R}^{n+q}_{\xi, \eta}). \tag{9.89}$$

In the case of x-independent symbols we will have

$$Op_x(p)(y, \eta) \in S^l_{cl}(\Omega \times \mathbb{R}^q; H^s(\mathbb{R}^n), H^{s-l}(\mathbb{R}^n)) \tag{9.90}$$

for all $s \in \mathbb{R}$. This follows from the fact that when

$$p_{(m)}(y, \xi, \eta) \in C^\infty(\Omega \times (\mathbb{R}^{n+q} \setminus \{0\}))$$

satisfies $p_{(m)}(y, \lambda\xi, \lambda\eta) = \lambda^m p_{(m)}(y, \xi, \eta)$ for all $\lambda > 0$ and all $(y, \xi, \eta) \in \Omega \times (\mathbb{R}^{n+q} \setminus \{0\})$, then

$$Op_x(p_{(m)})(y, \lambda\eta) = \lambda^m \kappa_\lambda Op_x(p_{(m)})(y, \eta)\kappa_\lambda^{-1}$$

for all $\lambda > 0$, $(y, \eta) \in \Omega \times (\mathbb{R}^q \setminus \{0\})$. If $\{p_k(y, \xi, \eta)\}_{k \in \mathbb{Z}_+}$ is a sequence tending to zero in $S^l_{cl}(\Omega \times \mathbb{R}^{n+q})$ for $k \to \infty$, then $Op_x(p_k)(y, \eta)$ will tend to zero in $S^l_{cl}(\Omega \times \mathbb{R}^q; H^s(\mathbb{R}^n), H^{s-l}(\mathbb{R}^n))$ for $k \to \infty$, for every $s \in \mathbb{R}$. From the x-independent case we can easily pass to the case of symbols with support in $K = \{|\,x\,| \leq c\}$ with respect to x. That symbol space may be identified with

$$C_0^\infty(K, S^l_{cl}(\Omega \times \mathbb{R}^{n+q}))$$

where the notation means functions supported by K in x with values in the Fréchet space of classical x-independent symbols. In view of

$$C_0^\infty(K, S^l_{cl}(\Omega \times \mathbb{R}^{n+q})) = C_0^\infty(K) \otimes_\pi S^l_{cl}(\Omega \times \mathbb{R}^{n+q}),$$

every $p(x, y, \xi, \eta)$, supported by K with respect to x, can be written as a convergent sum

$$p(x, y, \xi, \eta) = \sum_{k=0}^{\infty} \beta_k \varphi_k(x) p_k(y, \xi, \eta)$$

with $\varphi_k \in C_0^\infty(K)$, $p_k \in S^l_{cl}(\Omega \times \mathbb{R}^{n+q})$, both tending to 0 in the corresponding spaces for $k \to \infty$, and $\beta_k \in \mathbb{C}$, $\sum |\beta_k| < \infty$, cf. 7.2.1 Theorem 10. Now a further argument for (9.88) is the assertion of 9.1.1 Exercise 5, namely that the operator of multiplication by some $\varphi(x) \in C_0^\infty(\mathbb{R}^n)$ induces continuous actions

$$\mathcal{M}_\varphi : H^s(\mathbb{R}^n) \to H^s(\mathbb{R}^n),$$

and $\|\mathcal{M}_\varphi\|$ in the norm of $\mathcal{L}(H^s(\mathbb{R}^n))$ tends to zero once φ tends to zero in $C_0^\infty(\mathbb{R}^n)$. This holds for every $s \in \mathbb{R}$. We even have

$$\mathcal{M}_\varphi \in S^0(\Omega \times \mathbb{R}^q; H^s(\mathbb{R}^n), H^s(\mathbb{R}^n))$$

and $\mathcal{M}_\varphi \to 0$ in this symbol space for $\varphi \to 0$ in $C_0^\infty(\mathbb{R}^n)$. Together with (9.90) and the subsequent remarks we then obtain

$$Op_x(p)(y,\eta) = \sum_{k=0}^\infty \beta_k \mathcal{M}_{\varphi_k} Op_x(p_k)(y,\eta)$$

with convergence in $S^l(\Omega \times \mathbb{R}^q; H^s(\mathbb{R}^q, H^{s-l}(\mathbb{R}^q))$. This was the consideration for (9.89). The general case, i.e. (9.88) for (9.87) can be treated in an analogous manner, if we start with the easier relation

$$Op_x(p)(y,\eta) \in S^l(\Omega \times \mathbb{R}^q; H^s(\mathbb{R}^q), H^{s-l}(\mathbb{R}^q))$$

for x-independent p and that $p \to 0$ implies $Op_x(p) \to 0$ for those p. To obtain (9.85) certain considerations are actually more immediate than in the model case (9.88). From the definition of (9.82) and from 9.2.2 Proposition 4 it is clear that the statement holds for symbols in (9.82). Thus it suffices to consider (9.75), (9.76) separately and to ignore N_-, N_+. In other words, it suffices to show as a subexercise that

$$\begin{aligned}
a_0(y,\eta) &\in\ S^l(\Omega \times \mathbb{R}^q; \mathcal{K}^{s,\gamma}(X^\wedge), \mathcal{K}^{s-l,\gamma-m}(X^\wedge)), \\
a_1(y,\eta) &\in\ S^l(\Omega \times \mathbb{R}^q; \mathcal{K}^{s,\gamma}(X^\wedge), \mathcal{K}^{s-l,\gamma-m}(X^\wedge)),
\end{aligned}$$

cf. also Exercise 6.

In the case of r-independent h, i.e. $h(y,z,\tilde\eta) \in C^\infty(\Omega, M_O^l(X; \mathbb{R}_{\tilde\eta}^q))$, we have

$$a_0(y,\lambda\eta) = \lambda^l \kappa_\lambda a_0(y,\eta) \kappa_\lambda^{-1}$$

for all $\lambda \geq 1, |\eta| \geq c$ with a constant $c > 0$. This yields for those h

$$a_0(y,\eta) \in S_{cl}^l(\Omega \times \mathbb{R}^q; \mathcal{K}^{s,\gamma}(X^\wedge), \mathcal{K}^{s-l,\gamma-m}(X^\wedge)).$$

Moreover, analogously to 9.1.1 Exercise 5, we obtain, for every $\varphi(r) \in C_0^\infty(\overline{\mathbb{R}}_+)$,

$$\mathcal{M}_\varphi \in S^0(\Omega \times \mathbb{R}^q; \mathcal{K}^{s,\gamma}(X^\wedge), \mathcal{K}^{s,\gamma}(X^\wedge))$$

and $\mathcal{M}_\varphi \to 0$ in this symbol space for $\varphi \to 0$ in $C_0^\infty(\overline{\mathbb{R}}_+)$. This holds for every $s, \gamma \in \mathbb{R}$. Now a general $h(r,y,z,\tilde\eta)$ may be regarded as an element

$$C_0^\infty([0,c)) \otimes_\pi C^\infty(\Omega, M_O^l(X; \mathbb{R}_{\tilde\eta}^q))$$

for some $c > 0$ (because of the cut-offs in the associated operators). Thus h can be written as a convergent sum

$$h(r, y, z, \tilde{\eta}) = \sum_{k=0}^{\infty} \beta_k \varphi_k(r) h_k(y, z, \tilde{\eta})$$

with $\sum |\beta_k| < \infty$, $\varphi_k \to 0$, $h_k \to 0$ in the corresponding spaces for $k \to \infty$. Putting

$$a_{0,k}(y, \eta) = \omega(r[\eta]) r^{-l} op_M^{\gamma - \frac{n}{2}}(h_k)(y, \eta) \omega_0(r[\eta]),$$

we will obtain for (9.75)

$$a_0(y, \eta) = \sum_{k=0}^{\infty} \beta_k \mathcal{M}_{\varphi_k} a_{0,k}(y, \eta)$$

with convergence in $S^l(\Omega \times \mathbb{R}^q; \mathcal{K}^{s,\gamma}(X^\wedge), \mathcal{K}^{s-l,\gamma-m}(X^\wedge))$. Analogous arguments yield the desired result for $a_1(y, \eta)$. Now (9.86) for asymptotic types P, Q only concerns $a_0(y, \eta)$ and $a_0(y, y', \eta)$, respectively. The corresponding considerations are easy as well. Here we have to use the operator-valued symbols for Fréchet spaces $E, \tilde{E}$ from Section 9.1.1 and the fact that for every s, γ, P the space $\mathcal{K}_P^{s,\gamma}(X^\wedge)$ is a projective limit of Hilbert spaces E^k, $k \in \mathbb{Z}_+$, that are $\{\kappa_\lambda\}$-invariant. Furthermore, we have to employ the mapping properties in the cone algebra from Section 8.2.5 to verify that for each $j \in \mathbb{Z}_+$ there is a $b(j) \in \mathbb{Z}_+$ such that $a_0(y, \eta) : E^{b(j)} \to \tilde{E}^j$ is an operator-valued symbol in the sense of Section 9.1.1.

Exercise 13. For every we have two cut-off functions $\omega(t)$, $\tilde{\omega}(t)$,

$$\omega(t) \left\{ R^l(\Omega \times \mathbb{R}^q, \underline{g}) \cap C^\infty(\Omega; L^{-\infty}(X^\wedge; \mathbb{R}^q)) \right\} \tilde{\omega}(t) \subseteq R_{M+G}^l(\Omega \times \mathbb{R}^q, \underline{h}).$$

Next we will introduce the principal symbolic structure of the operator families in (9.79). The Green entries of the operator matrices (9.80) were considered in Section 9.2.1. The formula (9.50) has defined the principal edge symbols of those Green operator families. Thus it remains to look at the upper left corner, i.e. the elements of (9.73). For (9.74) we set

$$\sigma_\wedge^l(a)(y, \eta) = \sigma_\wedge^l(a_0 + a_1)(y, \eta) + \sigma_\wedge^l(w)(y, \eta) + \sigma_\wedge^l(g)(y, \eta),$$

with $\sigma_\wedge^l(w)$ from 9.2.2 Definition 1 and $\sigma_\wedge^l(a_0 + a_1)$ given by

$$\begin{aligned}
\sigma_\wedge^l(a_0 + a_1)(y, \eta) &= \omega(r|\eta|) r^{-l} op_M^{\gamma - \frac{n}{2}}(h_0)(y, \eta) \omega_0(r|\eta|) \\
&\quad + (1 - \omega(r|\eta|)) r^{-l} P_0(y, \eta)(1 - \omega_1(r|\eta|)).
\end{aligned}$$

Here, $h_0(r, y, z, \eta) := h(0, y, z, r\eta)$ and

$$P_0(y, \eta) = \sum_{j=1}^{N} \varphi_j (1 \times \kappa_j)^* Op_{(r,x)}(p_{j,0})(y, \eta) \psi_j, \tag{9.91}$$

cf. analogously to (9.69), with

$$p_{j,0}(r,x,y,\varrho,\xi,\eta) := p_j(0,x,y,r\varrho,\xi,r\eta), \quad j = 1,\ldots,N.$$

For (9.80) we then have

$$\sigma_\wedge^l(\underline{a})(y,\lambda\eta) = \lambda^l \begin{pmatrix} \kappa_\lambda & 0 \\ 0 & 1 \end{pmatrix} \sigma_\wedge^l(\underline{a})(y,\eta) \begin{pmatrix} \kappa_\lambda & 0 \\ 0 & 1 \end{pmatrix}^{-1} \tag{9.92}$$

for all $y \in \Omega$, $\eta \in \mathbb{R}^q \setminus \{0\}$ and all $\lambda \in \mathbb{R}_+$. We call $\sigma_\wedge^l(\underline{a})(y,\eta)$ the *homogeneous principal edge symbol* of $\underline{a}(y,\eta)$ of order l. It is an operator family

$$\sigma_\wedge^l(\underline{a})(y,\eta) : \mathcal{K}^{s,\gamma}(X^\wedge) \oplus \mathbb{C}^{N_-} \to \mathcal{K}^{s-l,\gamma-m}(X^\wedge) \oplus \mathbb{C}^{N_+}, \tag{9.93}$$

for every $s \in \mathbb{R}$. (9.93) is an analogue of the principal boundary symbol from boundary valued problems. In the case $l = m$ it will take part below in the notion of ellipticity of the pseudo-differential operator to $\underline{a}(y,\eta)$. The ellipticity also requires a condition for the principal interior symbol $\sigma_\psi^l(\cdot)$. In the coordinates (r,x,y,ϱ,ξ,η) of the cotangent bundle of $X^\wedge \times \Omega$ near $r = 0$, we set

$$\sigma_\psi^l(\underline{a})(r,x,y,\varrho,\xi,\eta) = \sigma_\psi^l(a)(r,x,y,\varrho,\xi,\eta),$$

cf. (9.80), which is defined as the (invariant) homogeneous principal symbol of $r^{-l}P(y,\eta)$, now of order l. It is of the form

$$r^{-l} \sum_{j=1}^N \varphi_j p_{j,(l)}(r,x,y,r\varrho,\xi,r\eta), \tag{9.94}$$

where $\text{sub}(l)$ indicates the homogeneous component of p_j of order l in the covariables (the pull-back to X was omitted here for brevity). Outside a neighbourhood of $r = 0$ we define $\sigma_\psi^l(a)$ as the ordinary homogeneous principal symbol of a of order l, with η being interpreted as covariable. There is a further invariant homogeneous object, namely,

$$\sum_{j=1}^N \varphi_j p_{j,(l)}(r,x,y,\varrho,\xi,\eta). \tag{9.95}$$

This will be denoted by

$$\sigma_{\psi,b}^l(a)(r,x,y,\varrho,\xi,\eta) \tag{9.96}$$

and we set $\sigma_{\psi,b}^l(\underline{a}) = \sigma_{\psi,b}^l(a)$. Note that (9.96) is C^∞ in r up to $r = 0$. Clearly, by definition, near $r = 0$

$$\sigma_{\psi,b}^l(a)(r,x,y,\varrho,\xi,\eta) = r^l \sigma_\psi^l(a)(r,x,y,r^{-1}\varrho,\xi,r^{-1}\eta).$$

Theorem 14. *Let $a(y,\eta) \in \mathcal{R}^l(\Omega \times \mathbb{R}^q, (\gamma, \gamma - m, \Theta); N_-, N_0)$, $\tilde{a}(y,\eta) \in \mathcal{R}^{\tilde{l}}(\Omega \times \mathbb{R}^q, (\gamma - m, \gamma - m - \tilde{m}, \Theta); N_0, N_+)$, where we impose the usual assumptions such as $m - l \in \mathbb{Z}_+$, $\tilde{m} - \tilde{l} \in \mathbb{Z}_+$. Then*

$$(\tilde{a}a)(y,\eta) \in \mathcal{R}^{l+\tilde{l}}(\Omega \times \mathbb{R}^q, (\gamma, \gamma - m - \tilde{m}, \Theta); N_-, N_+),$$

and

$$\sigma_\psi^{l+\tilde{l}}(\tilde{a}a) = \sigma_\psi^{\tilde{l}}(\tilde{a})\sigma_\psi^l(a), \quad \sigma_\wedge^{l+\tilde{l}}(\tilde{a}a) = \sigma_\wedge^{\tilde{l}}(\tilde{a})\sigma_\wedge^l(a). \tag{9.97}$$

Proof. Let us first look at the case when N_-, N_0, N_+ are zero. According to Definition 8, we then have $a(y,\eta)$ in the form

$$a(y,\eta) = a_0(y,\eta) + a_1(y,\eta) + a_\infty(y,\eta) + w(y,\eta) + g(y,\eta),$$

and similarly

$$\tilde{a}(y,\eta) = \tilde{a}_0(y,\eta) + \tilde{a}_1(y,\eta) + \tilde{a}_\infty(y,\eta) + \tilde{w}(y,\eta) + \tilde{g}(y,\eta).$$

$a_0(y,\eta)$ is defined by means of an operator-valued function

$$h(r,y,z,\tilde{\eta}) \in C^\infty(\overline{\mathbb{R}}_+ \times \Omega, M_O^l(X; \mathbb{R}_{\tilde{\eta}}^q)), \quad \tilde{\eta} = r\eta,$$

and $a_1(y,\eta)$ is determined by the pseudo-differential family $P(y,\eta)$ as in Definition 8. Analogously, the objects with tilde are associated with $\tilde{h}(r,y,z,\tilde{\eta}) \in C^\infty(\overline{\mathbb{R}}_+ \times \Omega, M_O^l(X; \mathbb{R}_{\tilde{\eta}}^q))$, $\tilde{\eta} = r\eta$, and $\tilde{P}(y,\eta)$, respectively. As we know from 8.2.5 Definition 1, the given $a(y,\eta)$, $\tilde{a}(y,\eta)$ are (y,η)-dependent families in $C^l(X^\wedge, g)$ and $C^{\tilde{l}}(X^\wedge, \tilde{g})$, respectively, with $g = (\gamma, \gamma - m, \Theta)$, $\tilde{g} = (\gamma - m, \gamma - m - \tilde{m}, \Theta)$. To characterize the composition $\tilde{a}(y,\eta)a(y,\eta)$ we now proceed analogously to 8.2.5 Remark 12. The desired form of our composition is

$$\tilde{a}(y,\eta)a(y,\eta) = b_0(y,\eta) + b_1(y,\eta) + b_\infty(y,\eta) + v(y,\eta) + c(y,\eta), \tag{9.98}$$

with $b_0(y,\eta), b_1(y,\eta), b_\infty(y,\eta)$ being of structure analogous to that in Definition 8, here with respect to the orders $\tilde{l} + l$ and the required weight data $h = (\gamma - m, \gamma - m - \tilde{m}, \Theta)$, and $(v + c)(y,\eta) \in R_{M+G}^{l+\tilde{l}}(\Omega \times \mathbb{R}^q, \underline{h})$, where $c(y,\eta)$ stands for the Green part, and $v(y,\eta)$ for the smoothing Mellin part. We form the operator family

$$Q(y,\eta) = \tilde{P}(y,\eta)P(y,\eta)$$

which is on $X^\wedge$ outside any neighbourhood of $r = 0$ in a way that we may employ for the pseudo-differential part $b_1(y,\eta)$, namely,

$$b_1(y,\eta) = (1 - \omega(r[\eta])) \, Q(y,\eta) \, (1 - \omega_1(r[\eta]))$$

with cut-off functions ω, ω_1 that together with another cut-off function ω_0 satisfy the relations

$$\omega\omega_0 = \omega, \quad \omega\omega_1 = \omega_1.$$

Concerning $Q(y, \eta)$ near $r = 0$, we are in the same situation as above in the formula (9.72). This means we can pass from a system of complete symbols for $Q(y, \eta)$ with respect to an open covering of X to an

$$\tilde{f}(r, y, z, \tilde{\eta}) \in C^\infty(\overline{\mathbb{R}}_+ \times \Omega, M_O^{l+\tilde{l}}(X; \mathbb{R}^q_{\tilde{\eta}})),$$

where $f(r, y, z, \eta) = \tilde{f}(r, y, z, r\eta)$ satisfies

$$Q(y, \eta) - op_M^\delta(f)(y, \eta) \in C^\infty(\Omega, L^{-\infty}(X^\wedge; \mathbb{R}^q)) \tag{9.99}$$

for every $\delta \in \mathbb{R}$. Now we set

$$b_0(y, \eta) = r^{-(\tilde{l}+l)} \omega(r[\eta]) op_M^{\gamma - \frac{n}{2}}(f)(y, \eta) \omega_0(r[\eta]).$$

Then, analogously to (8.152) we obtain that

$$\tilde{a}(y, \eta) a(y, \eta) - \{b_0(y, \eta) + b_1(y, \eta) + b_\infty(y, \eta)\} =: v(y, \eta) + c(y, \eta)$$

is a family of elements in $C_{M+G}(X^\wedge, \underline{h})$. The proof of 8.1.4 Theorem 6 in the version with parameters (y, η) and for the orders $\tilde{l}$ and l, respectively, gives rise to some $f_0(r, y, z, \eta) = \tilde{f}_0(r, y, z, r\eta)$ with $\tilde{f}_0(r, y, z, \tilde{\eta}) \in C^\infty(\overline{\mathbb{R}}_+ \times \Omega, M_O^{l+\tilde{l}}(X; \mathbb{R}^q_{\tilde{\eta}}))$, such that for

$$d_0(y, \eta) = \omega(r[\eta]) r^{-(\tilde{l}+l)} op_M^{\gamma - \frac{n}{2}}(f_0)(y, \eta) \omega_0(r[\eta])$$

we also have

$$\tilde{a}(y, \eta) a(y, \eta) - \{d_0(y, \eta) + b_1(y, \eta) + b_\infty(y, \eta)\} \in C_{M+G}(X^\wedge, \underline{h}) \tag{9.100}$$

for every (y, η). Here, since the above $\tilde{f}(r, y, z, \tilde{\eta})$ satisfying (9.99) is unique mod $C^\infty(\overline{\mathbb{R}}_+ \times \Omega, M_O^{-\infty}(X; \mathbb{R}^q_{\tilde{\eta}}))$, we have

$$\tilde{f}(r, y, z, \tilde{\eta}) - \tilde{f}_0(r, y, z, \tilde{\eta}) \in C^\infty(\overline{\mathbb{R}}_+ \times \Omega, M_O^{-\infty}(X; \mathbb{R}^q_{\tilde{\eta}})).$$

In the calculation of (9.100) via the method of proving 8.1.4 Theorem 6, it becomes evident that the smoothing Mellin part on the right may be written like the elements in $R_{M+G}^{\tilde{l}+l}(\Omega \times \mathbb{R}^q, \underline{h})$, i.e. with cut-offs $\omega(r[\eta])$ and $\tilde{\omega}(r[\eta])$ on both sides of the Mellin operators, cf. (9.53), and these Mellin operators in the middle are polynomials in η, combined with the corresponding powers of r. Since, furthermore,

$$\omega(r[\eta]) r^{-(\tilde{l}+l)} op_M^{\gamma - \frac{n}{2}}(f - f_0) \tilde{\omega}(r[\eta]) \in R_{M+G}^{\tilde{l}+l}(\Omega \times \mathbb{R}^q, \underline{h}),$$

cf. Exercise 7, we obtain that for some $v(y, \eta) \in R_{M+G}^{\tilde{l}+l}(\Omega \times \mathbb{R}^q, \underline{h})$

$$\tilde{a}(y, \eta) a(y, \eta) - \{b_0(y, \eta) + b_1(y, \eta) + b_\infty(y, \eta) + v(y, \eta)\} \in C_G(X^\wedge, \underline{h}) \tag{9.101}$$

for every (y, η). The left-hand side of (9.101) is also an operator-valued symbol in

$$S^{\tilde{l}+l}(\Omega \times \mathbb{R}^q; \mathcal{K}^{s,\gamma}(X^\wedge), \mathcal{K}^{\infty,\gamma-(\tilde{m}+m)}(X^\wedge)) \tag{9.102}$$

for every $s \in \mathbb{R}$. This follows from the corresponding properties of the various items, cf. Exercise 12 above. In other words, the left hand side of (9.101) is a $C_G(X^\wedge, \underline{h})$-valued symbol. The only remaining point is to show that it is also classical. This is seen immediately once the r-dependence in the involved complete symbols and in the operator-valued Mellin symbols (like $\tilde{f}(r, y, z, \tilde{\eta})$ etc.) disappears. In the general case this follows from Taylor expansions in all r-variables near $r = 0$ in all those symbols (cf. analogous Taylor expansion arguments in the proof of 8.1.4 Theorem 6). Then, up to remainders of η-orders tending to minus infinity, we will obtain symbols with a sequence of homogeneous components, and hence (9.102) will be a classical Green operator-valued symbol also in the general case. This proves our composition result (9.98).

It remains to look at the case of general N_-, N_+. However, the additional matrix elements belong to $\mathcal{R}_G^l(\dots)$ and $\mathcal{R}_G^{\tilde{l}}(\dots)$, respectively. Their structure is analogous to that of Green symbols of the upper left corner that were already treated. The additional entries are actually simpler, so we will drop the details. The symbolic rule (9.97) is also rather obvious, since for this we may restrict considerations to r-independence of symbols in the first arguments. Then the corresponding operator-valued symbols are classical (in the operator-valued sense), and it is then clear that the homogeneous principal symbols are composed under compositions of the complete symbols. $\qquad\square$

Remark 15. If one factor in Theorem 14 belongs to the class with subscript $M+G$ or G, then so does the composition.

The first observation is a consequence of the fact that, for instance, $\tilde{a}(y, \eta) \in \mathcal{R}_{M+G}^{\tilde{l}}(\dots)$ leads to a smoothing composition in the sense

$$\tilde{a}(y, \eta)a(y, \eta) \in C^\infty(\Omega, L^{-\infty}(X^\wedge; \mathbb{R}^q)).$$

Then it suffices to use Exercise 13.

Exercise 16. Let $a_j(y, \eta) \in \mathcal{R}^{l-j}(\Omega \times \mathbb{R}^q, \underline{g}; N_-, N_+)$, $j \in \mathbb{Z}_+$, be an arbitrary sequence where the summands in $\mathcal{R}_G^{l-j}(\Omega \times \mathbb{R}, \underline{g}; N_-, N_+)$ have asymptotic types that are independent of j. Then there is an $a(y, \eta) \in \mathcal{R}^l(\Omega \times \mathbb{R}^q, \underline{g}; N_-, N_+)$ such that

$$a(y, \eta) - \sum_{j=0}^N a_j(y, \eta) \in \mathcal{R}^{l-(N+1)}(\Omega \times \mathbb{R}^q, \underline{g}; N_-, N_+)$$

for every N, and $a(y, \eta)$ is unique mod $\mathcal{R}^{-\infty}(\Omega \times \mathbb{R}^q, \underline{g}; N_-, N_+)$.

9.3 Edge pseudo-differential operators

9.3.1 Edge Sobolev spaces with discrete asymptotics

We shall now apply the constructions of Section 9.1.3 to the spaces

$$E = \mathcal{K}^{s,\gamma}(X^{\wedge}) \quad \text{for} \quad s, \gamma \in \mathbb{R} \tag{9.103}$$

with

$$(\kappa_{\lambda} u)(t, x) = \lambda^{\frac{n+1}{2}} u(\lambda t, x), \quad n = \dim\ X, \tag{9.104}$$

$\kappa(\eta) := \kappa_{[\eta]}$. The space

$$\mathcal{W}^{s,\gamma}(X^{\wedge} \times \mathbb{R}^{q}) = \mathcal{W}^{s}(\mathbb{R}^{q}, \mathcal{K}^{s,\gamma}(X^{\wedge})) \tag{9.105}$$

is called a weighted wedge Sobolev space of smoothness $s \in \mathbb{R}$ and weight $\gamma \in \mathbb{R}$. For every open $\Omega \subseteq \mathbb{R}^{q}$ we can also form the spaces

$$\begin{aligned}
\mathcal{W}^{s,\gamma}_{\mathrm{comp}(y)}(X^{\wedge} \times \Omega) \quad &:= \quad \mathcal{W}^{s}_{\mathrm{comp}}(\Omega, \mathcal{K}^{s,\gamma}(X^{\wedge})), \\
\mathcal{W}^{s,\gamma}_{\mathrm{loc}(y)}(X^{\wedge} \times \Omega) \quad &:= \quad \mathcal{W}^{s}_{\mathrm{loc}}(\Omega, \mathcal{K}^{s,\gamma}(X^{\wedge})),
\end{aligned}$$

cf. (9.41).

Exercise 1. $C_0^{\infty}(X^{\wedge} \times \mathbb{R}^{q})$ is dense in $\mathcal{W}^{s,\gamma}(X^{\wedge} \times \mathbb{R}^{q})$ for every $s, \gamma \in \mathbb{R}$.

We consider on $\mathcal{W}^{0,0}(X^{\wedge} \times \mathbb{R}^{q})$ the Hilbert space scalar product

$$(u, v)_{\mathcal{W}^{0,0}(X^{\wedge} \times \mathbb{R}^{q})} = \int (\kappa^{-1}(\eta) \mathcal{F} u(\eta), \kappa^{-1}(\eta) \mathcal{F} v(\eta))_{\mathcal{K}^{0,0}(X^{\wedge})} d\eta \tag{9.106}$$

where $(\cdot, \cdot)_{\mathcal{K}^{0,0}(X^{\wedge})}$ is the scalar product on the space $\mathcal{K}^{0,0}(X^{\wedge})$. Note that $\kappa^{-1}(\eta)$ on the right-hand side of (9.106) may be omitted, since κ_{λ} is unitary on $\mathcal{K}^{0,0}(X^{\wedge})$. It can easily be proved that (9.106), first taken for $(u, v) \in C_0^{\infty}(X^{\wedge} \times \mathbb{R}^{q}) \times C_0^{\infty}(X^{\wedge} \times \mathbb{R}^{q})$, has an extension to a nondegenerate sesquilinear pairing

$$(\cdot, \cdot) : \mathcal{W}^{s,\gamma}(X^{\wedge} \times \mathbb{R}^{q}) \times \mathcal{W}^{-s,-\gamma}(X^{\wedge} \times \mathbb{R}^{q}) \to \mathbb{C} \tag{9.107}$$

for all $s, \gamma \in \mathbb{R}$.

We will set

$$\mathcal{W}^{0}(X^{\wedge} \times \Omega) = \mathcal{W}^{0,0}(X^{\wedge} \times \mathbb{R}^{q})\,|_{X^{\wedge} \times \Omega} \tag{9.108}$$

for any open $\Omega \subseteq \mathbb{R}^{q}$. This is also a Hilbert space (it can be regarded as a closed subspace of $\mathcal{W}^{0,0}(X^{\wedge} \times \mathbb{R}^{q})$).

Remark 2. We have for open $\Omega \subseteq \mathbb{R}^{q}$

$$\mathcal{W}^{0,0}(X^{\wedge} \times \Omega^{q}) = r^{-\frac{n}{2}} L_2(X^{\wedge} \times \Omega^{q}),$$

where the L_2 space on the right refers to *drdxdy*, *dx* being associated with a Riemannian metric on X.

Remark 3. Let $k^\gamma(r) \in C^\infty(\mathbb{R}_+)$ be a strictly positive function with

$$k^\gamma(r) = \begin{cases} r^\gamma & \text{for} \quad 0 < r < c_0 \\ 1 & \text{for} \quad c_1 < r < \infty \end{cases},$$

with constants $0 < c_0 < c_1$. Further, let $\omega(r)$ be a cut-off function. Then

$$p^\gamma(\eta) := k^\gamma(r[\eta])\omega(r[\eta]) + (1 - \omega(r[\eta])),$$

interpreted as an η-dependent family of multiplication operators $\mathcal{K}^{s,\mu}(X^\wedge) \to \mathcal{K}^{s,\mu+\gamma}(X^\wedge)$, is a symbol

$$p^\gamma(\eta) \in S^0(\mathbb{R}^q; \mathcal{K}^{s,\mu}(X^\wedge), \mathcal{K}^{s,\gamma+\mu}(X^\wedge))$$

for all $s, \gamma, \mu \in \mathbb{R}$. The associated pseudo-differential operator

$$P^\gamma = \mathcal{F}^{-1}_{\eta \to y} p^\gamma(\eta) \mathcal{F}_{y' \to \eta} : \mathcal{W}^{s,\mu}(X^\wedge \times \mathbb{R}^q) \to \mathcal{W}^{s,\mu+\gamma}(X^\wedge \times \mathbb{R}^q)$$

is an isomorphism for all $s, \gamma, \mu \in \mathbb{R}$. We could take $p^\gamma(\eta) = k^\gamma(r[\eta])$ as well.

Note that the space (9.105) can be written in various equivalent ways. Set $E = \mathcal{K}^{s,\gamma}(X^\wedge)$ and $E_1 = [\omega]\mathcal{H}^{s,\gamma}(X^\wedge)$, $E_2 = [1 - \omega]H^s(X^\wedge)$, with a fixed cut-off function $\omega(r)$; concerning notation, cf. (7.141). Then, $E = E_1 + E_2$ is a non-direct sum in the sense of (7.139). According to 9.1.3 Remark 3 we can form the spaces $\mathcal{W}^s(\mathbb{R}^q, E_i)$, $i = 1, 2$. The assertion of the following exercise extends the identity (9.39) to the case of such spaces.

Exercise 4. $E = E_1 + E_2$ implies

$$\mathcal{W}^s(\mathbb{R}^q, E) = \mathcal{V}^s(\mathbb{R}^q, E_1) + \mathcal{V}^s(\mathbb{R}^q, E_2),$$

where $\mathcal{V}^s(\mathbb{R}^q, E_i)$ is the closure of $\mathcal{S}(\mathbb{R}^q, E_i)$ in $\mathcal{W}^s(\mathbb{R}^q, E)$, $i = 1, 2$.

Proposition 5. *For every $s, \gamma \in \mathbb{R}$ we have*

$$\mathcal{W}^{s,\gamma}(X^\wedge \times \mathbb{R}^q) \subset H^s_{\text{loc}}(X^\wedge \times \mathbb{R}^q).$$

Proof. Let us first look at a special case, namely, $X = S^n = \{\tilde{x} \in \mathbb{R}^{n+1} : |\tilde{x}| = 1\}$. Then

$$\mathcal{K}^{s,\gamma}(X^\wedge) = \omega\mathcal{H}^{s,\gamma}(\mathbb{R}^{n+1} \setminus \{0\}) + (1 - \omega)H^s(\mathbb{R}^{n+1}).$$

We will show the following property. For every $\varepsilon > 0$ there are constants $c_1(\varepsilon)$, $c_2(\varepsilon) > 0$ such that for all $u \in C_0^\infty(\mathbb{R}^{1+n+q})$ with supp $u \subset \{(\tilde{x}, y) : |\tilde{x}| \geq \varepsilon\}$ we have

$$c_1(\varepsilon)\|u\|^2_{H^s(\mathbb{R}^{1+n+q})} \leq \|u\|^2_{\mathcal{W}^s(\mathbb{R}^q, \mathcal{K}^{s,\gamma}(X^\wedge))} \leq c_2(\varepsilon)\|u\|^2_{H^s(\mathbb{R}^{1+n+q})}. \tag{9.109}$$

First we know that for every $\delta > 0$ there are constants $b_1(\delta)$, $b_2(\delta) > 0$ such that

$$b_1(\delta)\|v\|^2_{H^s(\mathbb{R}^{n+1})} \leq \|v\|^2_{\mathcal{K}^{s,\gamma}(X^\wedge)} \leq b_2(\delta)\|v\|^2_{H^s(\mathbb{R}^{n+1})}$$

for all $v \in C_0^\infty(\mathbb{R}^{n+1})$ with supp $v \subseteq \{\tilde{x} : |\tilde{x}| \geq \delta\}$. Let us apply this to the η-dependent family of functions $v(\tilde{x}, \eta) = [\eta]^{-\frac{n+1}{2}} f([\eta]^{-1}\tilde{x}, \eta)$, where $f(\tilde{x}, \eta) = (\mathcal{F}_{y \to \eta} u)(\tilde{x}, \eta)$. Then the above assumption on the support of $u(\tilde{x}, y)$ will imply that those $v(\tilde{x}, \eta)$ are supported by $\{\tilde{x} : |\tilde{x}| \geq \delta\}$ with some $\delta = \delta(\varepsilon) > 0$. This holds for all $\eta \in \mathbb{R}^q$. It follows that

$$\begin{aligned}
b_1(\delta)[\eta]^{2s}\|v(\cdot, \eta)\|^2_{H^s(\mathbb{R}^{n+1})} &\leq [\eta]^{2s}\|v(\cdot, \eta)\|^2_{\mathcal{K}^{s,\gamma}(X^\wedge)} \\
&\leq b_2(\delta)[\eta]^{2s}\|v(\cdot, \eta)\|^2_{H^s(\mathbb{R}^{n+1})}
\end{aligned}$$

for all η. Integrating over $\eta \in \mathbb{R}^q$ immediately yields (9.109) for $c_i(\varepsilon) = b_i(\delta(\varepsilon))$. Now we easily get (9.109) also for all $u \in \mathcal{W}^s(\mathbb{R}^q, \mathcal{K}^{s,\gamma}(X^\wedge))$ with support in $\{(\tilde{x}, y) : |\tilde{x}| \geq \varepsilon\}$, cf. Exercise 1. Thus $\mathcal{W}^{s,\gamma}(X^\wedge \times \mathbb{R}^q) \subset H^s_{\mathrm{loc}}(X^\wedge \times \mathbb{R}^q)$ in the special case $X = S^n$. The case of general $X^\wedge$ can be reduced by restricting first the consideration to elements u that are supported with respect to (r, x) in a set $[\varepsilon, \infty) \times U_0$ with $\overline{U}_0$ compact, $\overline{U}_0 \subset U$ for some coordinate neighbourhood U on X. This actually corresponds to the above situation, since the set of all $u \in H^s(X^\wedge)$ with the support mentioned may be interpreted as a subspace of $H^s(\mathbb{R}^{n+1})$. Finally, an arbitrary $u \in \mathcal{W}^{s,\gamma}(X^\wedge \times \mathbb{R}^q)$, supported by $[\varepsilon, \infty) \times \mathbb{R}^q$, can be written as a finite sum of elements with compact supports with respect to x, in coordinate neighbourhoods on X. Hence the general statement follows. $\square$

Now let $\mathbb{W}$ be the stretched object to a manifold W with edge Y in the sense of the notation of Section 7.1.3. We will assume that W is compact, then $\mathbb{W}$ is a compact C^∞ manifold with compact C^∞ boundary $\partial\mathbb{W}$.

For convenience we shall henceforth assume that there is fixed on $\mathbb{W}$ a Riemannian metric such that a collar neighbourhood C of $\partial\mathbb{W}$ has the product metric of $[0, 1) \times X \times Y$ with chosen Riemannian metrics on X and Y. Let $\beta_j : D_j \to \Omega_j$, $j = 1, \ldots, N$, be an atlas on Y with open $\Omega_j \subseteq \mathbb{R}^q$ and write $\mathbb{W}$ as a union

$$\mathbb{W} = \left\{\bigcup_{j=1}^{N} \mathbb{V}_j\right\} \cup Z$$

with $\mathbb{V}_j = [0, 1) \times X \times D_j$ such that $C = \bigcup \mathbb{V}_j$, and an open subset $Z \subset \mathbb{W} \setminus \partial\mathbb{W}$. Let

$$\alpha_j : (0, 1) \times X \to \mathbb{R}_+ \times X = X^\wedge, \quad j = 1, \ldots, N$$

be a system of diffeomorphisms such that $\alpha_j(r, x) = (r, x)$ for all $0 < r < \frac{1}{2}$. Then C can also be regarded as a Cartesian product $X^\wedge \times Y$, locally described by $X^\wedge \times \Omega_j$, $j = 1, \ldots, N$. In other words we have

$$\chi_j := \alpha_j \times \beta_j : (0, 1) \times X \times D_j \to X^\wedge \times \Omega_j, \quad j = 1, \ldots, N. \tag{9.110}$$

In order to introduce the spaces

$$\mathcal{W}^{s,\gamma}(\mathbb{W}) \tag{9.111}$$

we choose a cut-off function $\omega \in C^\infty(\mathbb{W})$, i.e. an $\omega \in C^\infty(\mathbb{W})$ which is supported by $[0,1) \times X \times Y$, with $\omega = 0$ for $r > 1 - \varepsilon$ for some $0 < \varepsilon < 1$. Further, fix a partition of unity $\{\psi_1, \ldots, \psi_N\}$ on Y belonging to the open covering $\{D_1, \ldots, D_N\}$. For every $u \in C_0^\infty(\mathrm{int}\,\mathbb{W})$ we form $\omega\psi_j u$ which is supported by $(0,1) \times X \times D_j$. The support of $v_j(r,x,y) = \omega\psi_j u \circ \chi_j$ with respect to y is compact. So v_j may be interpreted as being defined on $X^\wedge \times \mathbb{R}^q$, and hence it makes sense to form

$$\|\omega\psi_j u \circ \chi_j^{-1}\|_{\mathcal{W}^{s,\gamma}(X^\wedge \times \mathbb{R}^q)}.$$

Furthermore, $(1 - \omega)u$ can be extended by zero to the double $2\mathbb{W}$ of $\mathbb{W}$ which is defined as the closed compact C^∞ manifold obtained by glueing together two copies of $\mathbb{W}$ along their common boundary $\partial\mathbb{W}$. Hence we may talk about the norm of $(1 - \omega)u$ in the Sobolev space $H^s(2\mathbb{W})$.

Definition 6. The space $\mathcal{W}^{s,\gamma}(\mathbb{W})$ for $s, \gamma \in \mathbb{R}$ is defined as the closure of $C_0^\infty(\mathrm{int}\,\mathbb{W})$ with respect to the norm

$$\left\{ \sum_{j=1}^N \|\omega\psi_j u \circ \chi_j^{-1}\|^2_{\mathcal{W}^{s,\gamma}(X^\wedge \times \mathbb{R}^q)} + \|(1 - \omega)u\|^2_{H^s(2\mathbb{W})} \right\}^{\frac{1}{2}}.$$

Note that $\mathcal{W}^{s,\gamma}(\mathbb{W})$ is independent of the particular choice of ω because of Proposition 5. This implies

$$\mathcal{W}^{s,\gamma}(\mathbb{W}) \subset H^s_{\mathrm{loc}}(\mathrm{int}\,\mathbb{W})$$

for all $s, \gamma \in \mathbb{R}$. It should be verified that $\mathcal{W}^{s,\gamma}(\mathbb{W})$ is independent also of the other data, namely, the charts on Y and the partition of unity $\{\psi_j\}$. The main point is then the invariance of the space $\mathcal{W}^{s,\gamma}_{\mathrm{comp}(y)}(X^\wedge \times \Omega)$ under diffeomorphisms in the y-coordinates.

Exercise 7. Let $\beta : \Omega \to \tilde{\Omega}$ be a diffeomorphism. Then $(1 \times \beta)^* : u \to u \circ (1 \times \beta)$, first defined for $u \in C_0^\infty(X^\wedge \times \tilde{\Omega})$, has an extension to an isomorphism

$$(1 \times \beta)^* : \mathcal{W}^{s,\gamma}_{\mathrm{comp}(\tilde{y})}(X^\wedge \times \tilde{\Omega}) \to \mathcal{W}^{s,\gamma}_{\mathrm{comp}(y)}(X^\wedge \times \Omega)$$

for all $s, \gamma \in \mathbb{R}$.

Hints: Reduce the assertion to the case $\Omega = \tilde{\Omega} = \mathbb{R}^q$ and $\beta(y) = y$ for $|y| > c$ with constant $c > 0$. Consider then first the case $\gamma = 0$. Use the fact that for every $m \in \mathbb{R}$ there exists an isomorphism

$$a^m : \mathcal{K}^{m,0}(X^\wedge) \to \mathcal{K}^{0,0}(X^\wedge)$$

such that the family of isomorphisms $b^m(\lambda) := \kappa_\lambda a^m \kappa_\lambda^{-1}$ is C^∞-dependent on $\lambda \in \mathbb{R}_+$. Form the operator-valued symbol

$$r^m(\eta) := [\eta]^m b^m([\eta]) \in S^m(\mathbb{R}^q; \mathcal{K}^{m,0}(X^\wedge), \mathcal{K}^{0,0}(X^\wedge)).$$

For $s = m$ pass to the associated isomorphism

$$\mathcal{F}_{\eta \to y}^{-1} r^s(\eta) \mathcal{F}_{y \to \eta} = R^s : \mathcal{W}^{s,0}(X^\wedge \times \mathbb{R}^q) \to \mathcal{W}^{0,0}(X^\wedge \times \mathbb{R}^q).$$

Use

$$(1 \times \beta)^* : \mathcal{W}^{0,0}(X^\wedge \times \mathbb{R}_{\tilde{y}}^q) \xrightarrow[\cong]{} \mathcal{W}^{0,0}(X^\wedge \times \mathbb{R}_y^q)$$

according to Remark 2. Form the push-forward of the pseudo-differential operator R^s under $(1 \times \beta)$, namely, $\tilde{R}^s = (1 \times \beta)_* R^s = (1 \times \beta^*)^{-1} R^s (1 \times \beta^*)$ with β^* being the function pull-back. Employ the fact that

$$(1 \times \beta)^* = (R^s)^{-1}(1 \times \beta)^* \tilde{R}^s : \mathcal{W}^{s,0}(X^\wedge \times \mathbb{R}_{\tilde{y}}^q) \to \mathcal{W}^{s,0}(X^\wedge \times \mathbb{R}_{\tilde{y}}^q)$$

is continuous. Pass to arbitrary weights by a weight conjugation

$$\mathcal{F}_{\eta \to y}^{-1} k^\gamma(\eta) \mathcal{F}_{y' \to \eta} : \mathcal{W}^{s,0}(X^\wedge \times \mathbb{R}^q) \xrightarrow[\cong]{} \mathcal{W}^{s,\gamma}(X^\wedge \times \mathbb{R}^q)$$

with $k^\gamma(\eta) = (r[\eta])^\gamma \omega(r[\eta]) + (1 - \omega(r[\eta]))$ for a cut-off function $\omega(r)$, cf. Remark 3. Note that a consequence of 9.1.3 Definition 1 is

$$\mathcal{F}_{y \to \eta} \mathcal{W}^{s,\gamma}(X^\wedge \times \mathbb{R}^q) =$$
$$\{ f(r, x, \eta) \in \mathcal{D}'(X^\wedge \times \mathbb{R}_\eta^q) : \ f(r, x, \eta) = [\eta]^{\frac{n+1}{2}} \tilde{v}(r[\eta], x, \eta)$$
$$\text{for some } v(r, x, y) \in H^s(\mathbb{R}_y^q, \mathcal{K}^{s,\gamma}(X^\wedge)) \}. \tag{9.112}$$

Here $\tilde{v}$ indicates the Fourier transform of v with respect to y.

Now let
$$P = \{(p_j, m_j)\}_{j=0,\ldots,N}, \ N = N(P) \in \mathbb{Z}_+,$$

be a weakly discrete asymptotic type belonging to the weight data (γ, Θ) with $\gamma \in \mathbb{R}$ and a finite weight strip $\Theta = (\vartheta, 0]$, $\vartheta < 0$. Then, by definition,

$$\mathcal{K}_P^{s,\gamma}(X^\wedge) = \mathcal{K}_\Theta^{s,\gamma}(X^\wedge) + \mathcal{E}_P(X^\wedge),$$

cf. Section 8.1.1, where L_j from (8.9) is to be replaced by $C^\infty(X)$ for all j. According to 9.1.3 Remark 3, applied to the corresponding version with Fréchet spaces, we obtain a direct decomposition

$$\mathcal{W}^s(\mathbb{R}^q, \mathcal{K}_P^{s,\gamma}(X^\wedge)) = \mathcal{W}^s(\mathbb{R}^q, \mathcal{K}_\Theta^{s,\gamma}(X^\wedge)) + \mathcal{V}^s(\mathbb{R}^q, \mathcal{E}_P(X^\wedge)),$$

cf. the notation in 9.3.1 Exercise 4. This gives us exactly the notion of edge asymptotics corresponding to the weight data (γ, Θ) and an asymptotic type P. Similarly to (9.112) the space $\mathcal{F}_{y\to\eta}\mathcal{W}^s(\mathbb{R}^q, \mathcal{E}_P(X^\wedge))$ is the linear span of all functions

$$\omega(r[\eta])[\eta]^{\frac{n+1}{2}}(r[\eta])^{-p_j}\ln^k(r[\eta])\tilde{v}_{jk}(x,\eta) \tag{9.113}$$

with arbitrary $v_{jk}(x,y) \in H^s(\mathbb{R}^q_y, C^\infty(X))$, $0 \le k \le m_j$, $j = 1,\ldots,N$.

Remark 8. The elements $\mathcal{F}^{-1}_{\eta\to y}(9.113)$ may be regarded as the singular functions of the edge asymptotics. They are typically non-smooth along the edge variable y, dependent on Re p_j and on s.

In the case of an infinite weakly discrete asymptotic type

$$P = \{(p_j, m_j)\}_{j\in\mathbb{Z}_+} \tag{9.114}$$

for (γ, Θ), $\Theta = (-\infty, 0]$, we can form as in Section 8.1.1

$$P_k = \{(p, m) \in P : \frac{n+1}{2} - \gamma - k < \mathrm{Re}\ p\},\ k \in \mathbb{N}.$$

Then $\mathcal{K}^{s,\gamma}_P(X^\wedge) = \varprojlim_k \mathcal{K}^{s,\gamma}_{P_k}(X^\wedge)$ gives rise to the weighted wedge Sobolev spaces with infinite asymptotics

$$\mathcal{W}^s(\mathbb{R}^q, \mathcal{K}^{s,\gamma}_P(X^\wedge)) = \varprojlim_k \mathcal{W}^s(\mathbb{R}^q, \mathcal{K}^{s,\gamma}_{P_k}(X^\wedge)).$$

We can summarize this as follows.

Remark 9. Every $u(r, x, y) \in \mathcal{W}^s(\mathbb{R}^q, \mathcal{K}^{s,\gamma}_P(X^\wedge))$ for P with $(\gamma, (-\infty, 0])$ has an asymptotic expansion

$$u(r,x,y) \sim \mathcal{F}^{-1}_{\eta\to y}\{\sum_{j,k}\omega(r[\eta])[\eta]^{\frac{n+1}{2}}(r[\eta])^{-p_j}\ln^k(r[\eta])\tilde{v}_{jk}(x,\eta)\}$$

with certain $v_{jk}(x,y) \in H^s(\mathbb{R}^q, C^\infty(X))$ in the sense that for every $l \in \mathbb{Z}_+$ there is an $N = N(l)$ with

$$u(r,x,y) - \mathcal{F}^{-1}_{\eta\to y}\{\sum_{j=0}^{N}\sum_{k=0}^{m_j}\omega(r[\eta])[\eta]^{\frac{n+1}{2}}(r[\eta])^{-p_j}\ln^k(r[\eta])\tilde{v}_{jk}(x,\eta)\}$$
$$\in \mathcal{W}^s(\mathbb{R}^q, \mathcal{K}^{s,\gamma}_{(-l,0]}(X^\wedge)).$$

We will also write

$$\mathcal{W}_P^{s,\gamma}(X^\wedge \times \mathbb{R}^q) := \mathcal{W}^s(\mathbb{R}^q, \mathcal{K}_P^{s,\gamma}(X^\wedge)).$$

According to the general scheme of Section 9.1.3 we can also form the spaces

$$\begin{aligned}
\mathcal{W}_{P,\mathrm{loc}(y)}^{s,\gamma}(X^\wedge \times \Omega) &:= \mathcal{W}_{\mathrm{loc}}^s(\Omega, \mathcal{K}_P^{s,\gamma}(X^\wedge)), \\
\mathcal{W}_{P,\mathrm{comp}(y)}^{s,\gamma}(X^\wedge \times \Omega) &:= \mathcal{W}_{\mathrm{comp}}^s(\Omega, \mathcal{K}_P^{s,\gamma}(X^\wedge))
\end{aligned}$$

for every open $\Omega \subseteq \mathbb{R}^q$.

Exercise 10. Let $g(y, y', \eta) \in \mathcal{R}_G^m(\Omega \times \Omega \times \mathbb{R}^q, (\gamma, \varrho, \Theta); N_-, N_+)$ (cf. (9.44)). Then $\mathcal{G} = Op(g)$ induces continuous operators

$$\mathcal{G} : \mathcal{W}_{\mathrm{comp}(y)}^{s,\gamma}(X^\wedge \times \Omega) \oplus H_{\mathrm{comp}}^s(\Omega, \mathbb{C}^{N_-}) \to \mathcal{W}_{P,\mathrm{loc}(y)}^{s-m,\varrho}(X^\wedge \times \Omega) \oplus H_{\mathrm{loc}}^{s-m}(\Omega, \mathbb{C}^{N_+}),$$

for all $s \in \mathbb{R}$, with some asymptotic type P dependent on $\mathcal{G}$.

Exercise 11. Let $g(y, y', \eta) \in R_G^m(\Omega \times \Omega \times \mathbb{R}^q, (\gamma, \varrho, \Theta))$ and $\alpha, \beta \in \mathbb{R}$. Then

$$(r[\eta])^\alpha g(y, y', \eta)(r[\eta])^{-\beta}$$

belongs to $R_G^m(\Omega \times \Omega \times \mathbb{R}^q, (\gamma + \alpha, \varrho + \beta, \Theta))$.

Note that $g(y, y', \eta) \in R_G^m(\Omega \times \Omega \times \mathbb{R}^q, (\gamma, \varrho, \Theta))$ also implies

$$r^\alpha g(y, y', \eta) r^{-\beta} \in R^{m-\alpha+\beta}(\Omega \times \Omega \times \mathbb{R}^q, (\gamma + \alpha, \varrho + \beta, \Theta)).$$

9.3.2 The algebra of edge pseudo-differential operators

We are now in a position to introduce the algebra of edge pseudo-differential operators. The idea is analogous to the scalar pseudo-differential operators in an open domain $\Omega \subseteq \mathbb{R}^q$, where the operators are $A = Op(a) + C$ with a symbol $a(y, \eta)$, $Op(a) = \mathcal{F}^{-1} a \mathcal{F}$, and a smoothing operator C. We know in the classical theory that C may always be written as $C = Op(c)$ with an amplitude function $c(y, y', \eta)$ of order $-\infty$, cf. the scalar analogue of 9.1.2 Remark 3. Here in the edge calculus it is more convenient to define the smoothing operators by the mapping properties.

If

$$\mathcal{G} : \mathcal{W}_{\mathrm{comp}(y)}^{s,\gamma}(X^\wedge \times \Omega) \oplus H_{\mathrm{comp}}^s(\Omega, \mathbb{C}^{N_-}) \to \mathcal{W}_{\mathrm{loc}(y)}^{r,\varrho}(X^\wedge \times \Omega) \oplus H_{\mathrm{loc}}^r(\Omega, \mathbb{C}^{N_+})$$

$$\tag{9.115}$$

is a continuous operator for every $s \in \mathbb{R}$ and $r = s - m$ with some $m \in \mathbb{R}$, then the formal adjoint $\mathcal{G}^*$, defined by

$$(\mathcal{G}u, v)_{\mathcal{W}^0(X^\wedge \times \Omega) \oplus H^0(\Omega, \mathbb{C}^{N_+})} = (u, \mathcal{G}^* v)_{\mathcal{W}^0(X^\wedge \times \Omega) \oplus H^0(\Omega, \mathbb{C}^{N_-})} \tag{9.116}$$

for all $u \in C_0^\infty(X^\wedge \times \Omega) \oplus C_0^\infty(\Omega, \mathbb{C}^{N_-})$, $v \in C_0^\infty(X^\wedge \times \Omega) \oplus C_0^\infty(\Omega, \mathbb{C}^{N_+})$ induces continuous operators

$$\mathcal{G}^* : \mathcal{W}_{\mathrm{comp}(y)}^{s,-\varrho}(X^\wedge \times \Omega) \oplus H_{\mathrm{comp}}^s(\Omega, \mathbb{C}^{N_+}) \to \mathcal{W}_{\mathrm{loc}(y)}^{r,-\gamma}(X^\wedge \times \Omega) \oplus H_{\mathrm{loc}}^r(\Omega, \mathbb{C}^{N_-})$$

for all $s \in \mathbb{R}$. In (9.116) the scalar product from (9.108) was employed.

Definition 1. Let $\Omega \subseteq \mathbb{R}^q$ be open, $N_-, N_+ \in \mathbb{Z}_+$, and fix the weight data $\underline{g} = (\gamma, \varrho, \Theta)$ with $\gamma, \varrho \in \mathbb{R}$, $\Theta = (\vartheta, 0]$, $-\infty \leq \vartheta < 0$. Then

$$\mathcal{Y}^{-\infty}(X^\wedge \times \Omega, \underline{g}; N_-, N_+) \tag{9.117}$$

is the space of all operators (9.115) that are continuous for all $s, r \in \mathbb{R}$, such that $\mathcal{G}$ and $\mathcal{G}^*$ induce continuous operators

$$\mathcal{G} : \mathcal{W}_{\mathrm{comp}(y)}^{s,\gamma}(X^\wedge \times \Omega) \oplus H_{\mathrm{comp}}^s(\Omega, \mathbb{C}^{N_-}) \to \mathcal{W}_{P,\mathrm{loc}(y)}^{\infty,\varrho}(X^\wedge \times \Omega) \oplus H_{\mathrm{loc}}^\infty(\Omega, \mathbb{C}^{N_+})$$

and

$$\mathcal{G}^* : \mathcal{W}_{\mathrm{comp}(y)}^{s,-\varrho}(X^\wedge \times \Omega) \oplus H_{\mathrm{comp}}^s(\Omega, \mathbb{C}^{N_+}) \to \mathcal{W}_{Q,\mathrm{loc}(y)}^{\infty,-\gamma}(X^\wedge \times \Omega) \oplus H_{\mathrm{loc}}^\infty(\Omega, \mathbb{C}^{N_-}),$$

respectively, for all $s \in \mathbb{R}$, with asymptotic types P and Q, related to (ϱ, Θ) and $(-\gamma, \Theta)$, respectively.

The space of upper left corners of the operator matrices in (9.117) will also be denoted by

$$Y^{-\infty}(X^\wedge \times \Omega, \underline{g}). \tag{9.118}$$

Exercise 2. Let $g(y, y', \eta) \in \mathcal{R}_G^{-\infty}(\Omega \times \Omega \times \mathbb{R}^q, \underline{g}; N_-, N_+)$ (cf. (9.49)). Then $Op(g) \in \mathcal{Y}^{-\infty}(X^\wedge \times \Omega, \underline{g}; N_-, N_+)$.

In order to introduce the edge pseudo-differential operators we simply employ 9.2.3 Definition 8.

Definition 3. Let $\Omega \subseteq \mathbb{R}^q$ be open, $N_-, N_+ \in \mathbb{Z}_+$, $\gamma, l, m \in \mathbb{R}$, $m - l \in \mathbb{Z}_+$, $\Theta = (-k, 0]$, $k \in \mathbb{N}$. Then

$$\mathcal{Y}^l(X^\wedge \times \Omega, \underline{g}; N_-, N_+) \tag{9.119}$$

for the weight data $\underline{g} = (\gamma, \gamma - m, \Theta)$ denotes the space of all operators

$$\mathcal{A} = Op(a) + \mathcal{C} \tag{9.120}$$

with $a(y, \eta) \in \mathcal{R}^l(\Omega \times \mathbb{R}^q, \underline{g}; N_-, N_+)$ and $\mathcal{C} \in \mathcal{Y}^{-\infty}(X^\wedge \times \Omega, \underline{g}; N_-, N_+)$. We shall write

$$\mathcal{Y}_{M+G}^l(X^\wedge \times \Omega, \underline{g}; N_-, N_+) \tag{9.121}$$

and

$$\mathcal{Y}_G^l(X^\wedge \times \Omega, \underline{g}; N_-, N_+) \tag{9.122}$$

for the subclasses of $\mathcal{Y}^l(X^\wedge \times \Omega, \underline{g}; N_-, N_+)$ defined by $a(y, \eta) \in \mathcal{R}_{M+G}^l(\Omega \times \mathbb{R}^q, \underline{g}; N_-, N_+)$ and $a(y, \eta) \in \mathcal{R}_G^l(\Omega \times \mathbb{R}^q, \underline{g}; N_-, N_+)$, respectively.

By construction an edge operator has the form

$$\mathcal{A} = \begin{pmatrix} A + M + G + C & K + K_0 \\ T + T_0 & Q + Q_0 \end{pmatrix} \quad \text{with} \quad \mathcal{C} = \begin{pmatrix} C & K_0 \\ T_0 & Q_0 \end{pmatrix} \in \mathcal{Y}^{-\infty}. \quad (9.123)$$

The main objects of the calculus are the operators in the upper left corners of the block matrices. The spaces of operators in the upper left corners of (9.119), (9.121) and (9.122) will be denoted by

$$Y^l(X^\wedge \times \Omega, \underline{g}), \quad (9.124)$$

$$Y^l_{M+G}(X^\wedge \times \Omega, \underline{g}) \quad \text{and} \quad Y^l_G(X^\wedge \times \Omega, \underline{g}), \quad (9.125)$$

respectively. In particular, the elements of (9.124) have the form $A + M + G + C$ with $C \in Y^{-\infty}(X^\wedge \times \Omega, (\gamma, \gamma - m, \Theta))$. It will be convenient to call

A an interior pseudo-differential operator, $A = Op(a_0 + a_1 + a_\infty)$, cf. (9.74)
M a smoothing Mellin operator,
G a Green operator,
C a smoothing operator (with asymptotics)

of the edge pseudo-differential calculus. The meanings of the other entries of $\mathcal{A}$ are as follows

$T + T_0$ is an edge trace operator,
$K + K_0$ is an edge potential operator,
$Q + Q_0$ is an $N_+ \times N_-$ system of pseudo-differential operators on Ω.

Remark 4. An operator $\mathcal{A} \in \mathcal{Y}^l(X^\wedge \times \Omega, \underline{g}; N_-, N_+)$ is equivalent to an edge problem to the pseudo-differential operator in the upper left corner. This is analogous to the interpretation of pseudo-differential boundary value problems in the framework of [BdM1] or more generally [Sc5], where the model cone of the wedge is $\mathbb{R}_+$ and the edge is the boundary. The trace operators play the role of boundary conditions, the potential operators are in a sense dual to them, and the operators $Q + Q_0$ appear for algebraic reasons in the operator matrices. The latter also play a role in reducing a boundary problem to the boundary. In an analogous manner edge problems can be reduced to the edges.

By virtue of

$$\mathcal{R}^l(\Omega \times \mathbb{R}^q, \underline{g}; N_-, N_+) \subset \bigcap_{s \in \mathbb{R}} S^l(\Omega \times \mathbb{R}^q; \mathcal{K}^{s,\gamma}(X^\wedge) \oplus \mathbb{C}^{N_-}, \mathcal{K}^{s-l,\gamma-m}(X^\wedge) \oplus \mathbb{C}^{N_+})$$

$$(9.126)$$

(cf. (9.85)), for every $\mathcal{R}^l(\Omega \times \mathbb{R}^q, \underline{g}; N_-, N_+)$ we have

$$Op(a) \in \bigcap_{s \in \mathbb{R}} L^l(\Omega; \mathcal{K}^{s,\gamma}(X^\wedge) \oplus \mathbb{C}^{N_-}, \mathcal{K}^{s-l,\gamma-m}(X^\wedge) \oplus \mathbb{C}^{N_+}). \quad (9.127)$$

More precisely, if $a(y, \eta) \in \mathcal{R}^l(\Omega \times \mathbb{R}^q, \underline{g}; N_-, N_+)$, then for every asymptotic type P associated with the weight data (γ, Θ), there is an assymptotic type Q associated with $(\gamma - m, \Theta)$, such that

$$a(y, \eta) \in \bigcap_{s \in \mathbb{R}} S^l(\Omega \times \mathbb{R}^q; \mathcal{K}_P^{s,\gamma}(X^\wedge) \oplus \mathbb{C}^{N_-}, \mathcal{K}_Q^{s-l,\gamma-m}(X^\wedge) \oplus \mathbb{C}^{N_+})$$

and

$$Op(a) \in \bigcap_{s \in \mathbb{R}} L^l(\Omega; \mathcal{K}_P^{s,\gamma}(X^\wedge) \oplus \mathbb{C}^{N_-}, \mathcal{K}_Q^{s-l,\gamma-m}(X^\wedge) \oplus \mathbb{C}^{N_+}), \qquad (9.128)$$

cf. (9.86). The following results are now an immediate consequence of 9.1.3 Theorem 5 and of the mapping properties of Definition 1.

Theorem 5. *Every* $\mathcal{A} \in \mathcal{Y}^l(X^\wedge \times \Omega, \underline{g}; N_-, N_+)$ *induces continuous operators*

$$\mathcal{A} : \mathcal{W}^{s,\gamma}_{\text{comp}(y)}(X^\wedge \times \Omega) \oplus H^s_{\text{comp}}(\Omega, \mathbb{C}^{N_-}) \to \mathcal{W}^{s-l,\gamma-m}_{\text{loc}(y)}(X^\wedge \times \Omega) \oplus H^{s-l}_{\text{loc}}(\Omega, \mathbb{C}^{N_+})$$
$$(9.129)$$

for all $s \in \mathbb{R}$.

Theorem 6. *If* $\mathcal{A} \in \mathcal{Y}^l(X^\wedge \times \Omega, \underline{g}; N_-, N_+)$, *then for every asymptotic type* P *associated with the weight data* (γ, Θ), $\Theta = (-k, 0]$, *there exists an asymptotic type* Q *associated with* $(\gamma - m, \Theta)$, *such that* (9.129) *induces continuous operators*

$$\mathcal{A} : \mathcal{W}^{s,\gamma}_{P,\text{comp}(y)}(X^\wedge \times \Omega) \oplus H^s_{\text{comp}}(\Omega, \mathbb{C}^{N_-}) \to \mathcal{W}^{s-l,\gamma-m}_{Q,\text{loc}(y)}(X^\wedge \times \Omega) \oplus H^{s-l}_{\text{loc}}(\Omega, \mathbb{C}^{N_+})$$
$$(9.130)$$

for all $s \in \mathbb{R}$.

Theorem 7. *Let* $a_1(y, y', \eta) \in \mathcal{R}^l(\Omega \times \Omega \times \mathbb{R}^q, \underline{g}; N_-, N_+)$, $\underline{g} = (\gamma, \gamma - m, \Theta)$. *Then we have* $Op(a_1) \in \mathcal{Y}^l(X^\wedge \times \Omega, \underline{g}; N_-, N_+)$, *and there is a representation* $Op(a_1) = Op(a) + \mathcal{C}$ *with some* $\mathcal{C} \in \mathcal{Y}^{-\infty}$, $a(y, \eta) \in \mathcal{R}^l(\Omega \times \mathbb{R}^q, \underline{g}; N_-, N_+)$ *which allows the asymptotic expansion*

$$a(y, \eta) \sim \sum_\alpha \frac{1}{\alpha!} D_\eta^\alpha \partial_{y'}^\alpha a(y, y', \eta) \big|_{y'=y} .$$

Proposition 8. *To every* $\mathcal{A} \in$ (9.119) *there is an* $\mathcal{A}_0 \in$ (9.119) *which is properly supported in the y-variables such that* $\mathcal{A} - \mathcal{A}_0 \in \mathcal{Y}^{-\infty}(X^\wedge \times \Omega, \underline{g}; N_-, N_+)$. *If* $\mathcal{A}_0 \in$ (9.119) *is properly supported in that sense, then* (9.129) *and* (9.130) *may be replaced by corresponding mappings with* comp *or* loc *on both sides, e.g.*

$$\mathcal{A} : \mathcal{W}^{s,\gamma}_{\text{comp}(y)}(X^\wedge \times \Omega) \oplus H^s_{\text{comp}}(\Omega, \mathbb{C}^{N_-}) \to \mathcal{W}^{s-l,\gamma-m}_{\text{comp}(y)}(X^\wedge \times \Omega) \oplus H^{s-l}_{\text{comp}}(\Omega, \mathbb{C}^{N_+}).$$

Proof. In view of 9.1.3 Theorem 5 it remains to show that $\mathcal{A}$ allows a properly supported representative $\mathcal{A}_0$ in the class mod $\mathcal{Y}^{-\infty}$. This follows analogously to the general argument of 9.1.2 Proposition 2. $\qquad \square$

Exercise 9. Let $\mathcal{A} \in \mathcal{Y}^l(X^\wedge \times \Omega, \underline{g}; N_-, N_+)$ and $\varphi_0, \varphi \in C_0^\infty(\Omega)$ with $\varphi_0 \varphi = \varphi$. Then $(1 - \varphi_0)\mathcal{A}\varphi \in \mathcal{Y}^{-\infty}(X^\wedge \times \Omega, \underline{g}; N_-, N_+)$.

Hint: Use Theorem 7.

Remark 10. We have

$$Y^l(X^\wedge \times \Omega, \underline{g}) \subset L_{cl}^l(X^\wedge \times \Omega). \tag{9.131}$$

Moreover, $\mathcal{Y}_{M+G}^l(X^\wedge \times \Omega, \underline{g}; N_-, N_+)$ is contained in the space of smoothing operators in the sense that the elements induce continuous maps

$$H_{\mathrm{comp}}^s(X^\wedge \times \Omega) \oplus H_{\mathrm{comp}}^s(\Omega, \mathbb{C}^{N_-}) \to C^\infty(X^\wedge \times \Omega) \oplus C^\infty(\Omega, \mathbb{C}^{N_+})$$

for all $s \in \mathbb{R}$ and analogously the formal adjoints.

Next we will define the principal symbolic structure of the operator space $\mathcal{Y}^l(X^\wedge \times \Omega, \underline{g}; N_-, N_+)$. By virtue of (9.131), every $\underline{\mathcal{A}} \in Y^l(X^\wedge \times \Omega, \underline{g})$ has a homogeneous principal symbol $\sigma_\psi^l(\underline{\mathcal{A}})$ of order l which belongs to

$$S^{(l)}(T^*(X^\wedge \times \Omega) \setminus 0),$$

cf. (7.123). We shall also write

$$\sigma_\psi^l(\mathcal{A}) := \sigma_\psi^l(\text{left upper corner of } \mathcal{A}) \tag{9.132}$$

for every $\mathcal{A} \in \mathcal{Y}^l(X^\wedge \times \Omega, \underline{g}; N_-, N_+)$, called the *homogeneous principal interior symbol* of $\mathcal{A}$. Note that the symbols $\sigma_\psi^l(\mathcal{A})$ in the coordinates $(r, x, y, \varrho, \xi, \eta)$ have the property that

$$r^l \sigma_\psi^l(\mathcal{A})(t, x, y, r^{-1}\varrho, \xi, r^{-1}\eta)$$

are C^∞ in r up to $r = 0$. Furthermore, we define

$$\sigma_\wedge^l(\mathcal{A}) = \sigma_\wedge^l(a), \tag{9.133}$$

for $\mathcal{A} = Op(a) + \mathcal{C}$ as in Definition 3 above. $\sigma_\wedge^l(\mathcal{A})$ is called the *homogeneous principal edge symbol* of $\mathcal{A}$. By construction, $\sigma_\wedge^l(\underline{A})$ for the upper left corner $\underline{A}$ of $\mathcal{A}$ is an operator family

$$\sigma_\wedge^l(\underline{A})(y, \eta) \in C^l(X^\wedge, \underline{g}) \tag{9.134}$$

parametrized by $(y, \eta) \in T^*\Omega \setminus 0$. Here $C^l(X^\wedge, \underline{g})$ means the cone operator space with weakly discrete asymptotics. Thus we can apply the principal symbolic maps from the class $C^l(X^\wedge, \underline{g})$ to (9.134) for every fixed y, η. This gives us

$$\sigma_\psi^l \sigma_\wedge^l(\cdot), \quad \sigma_{\psi,e}^{l,0} \sigma_\wedge^m(\cdot), \quad \sigma_e^0 \sigma_\wedge^l(\cdot), \quad \sigma_M^l \sigma_\wedge^l(\cdot). \tag{9.135}$$

In $(\cdot)$ we shall insert $\underline{A}$ or likewise $\mathcal{A}$ which always by definition refers to the upper left corner. Note that the operator family

$$\sigma_M^l \sigma_\wedge^l(\underline{A})(y, z) : H^s(X) \to H^{s-l}(X) \tag{9.136}$$

is independent of η.

Remark 11. Let $\mathcal{A} \in \mathcal{Y}^l(X^\wedge \times \Omega, \underline{g}; N_-, N_+)$ for $\underline{g} = (\gamma, \gamma - m, \Theta)$. Then

$$\sigma_\psi^l(\mathcal{A}) = 0, \quad \sigma_\wedge^l(\mathcal{A}) = 0$$

implies $\mathcal{A} \in \mathcal{Y}^{l-1}(X^\wedge \times \Omega, \underline{g}; N_-, N_+)$.

Exercise 12. Let $\mathcal{A}_j \in \mathcal{Y}^{m-j}(X^\wedge \times \Omega, \underline{g}; N_-, N_+)$, $j \in \mathbb{Z}_+$ be an arbitrary sequence. Then there exists an $\mathcal{A} \in \mathcal{Y}^m(X^\wedge \times \Omega, \underline{g}; N_-, N_+)$ such that for every $N \in \mathbb{Z}_+$

$$\mathcal{A} - \sum_{j=0}^N \mathcal{A}_j \in \mathcal{Y}^{m-(N+1)}(X^\wedge \times \Omega, \underline{g}; N_-, N_+).$$

$\mathcal{A}$ is unique mod $\mathcal{Y}^{-\infty}$.

Theorem 13. *Let $\mathcal{A} \in \mathcal{Y}^l(X^\wedge \times \Omega, \underline{g}; N_-, N_+)$, $\underline{g} = (\gamma, \gamma - m, \Theta)$, and define the formal adjoint $\mathcal{A}^*$ as above. Then $\mathcal{A}^* \in \mathcal{Y}^l(X^\wedge \times \Omega, \underline{h}; N_+, N_-)$, $\underline{h} = (-\gamma + m, -\gamma, \Theta)$. For $\mathcal{A} = Op(a) + \mathcal{C}$ with $a(y, \eta) \in \mathcal{R}^l(\Omega \times \mathbb{R}^q, \underline{g}; N_-, N_+)$, $\mathcal{C} \in \mathcal{Y}^{-\infty}$, we have $\mathcal{A}^* = Op(a^{(*)}) + \mathcal{C}_1$ with $\mathcal{C}_1 \in \mathcal{Y}^{-\infty}$, and $a^{(*)}(y, \eta) \in \mathcal{R}^l(\Omega \times \mathbb{R}^q, \underline{h}; N_+, N_-)$ allows the asymptotic expansion*

$$a^{(*)}(y, \eta) \sim \sum_\alpha \frac{1}{\alpha!} D_\eta^\alpha \partial_y^\alpha a^*(y, \eta)$$

where $*$ indicates the pointwise formal adjoints in the sense of 9.2.3 Exercise 10.

In an analogous manner the formal transposed operator ${}^t\mathcal{A}$ of $\mathcal{A}$ with respect to the corresponding bilinear pairing belongs to $\mathcal{Y}^l(X^\wedge \times \Omega, \underline{h}; N_+, N_-)$, and ${}^t\mathcal{A} = Op({}^{(t)}a) + \mathcal{C}_2$ with $\mathcal{C}_2 \in \mathcal{Y}^{-\infty}$, whereas ${}^{(t)}a(y, \eta)$ allows the asymptotic expansion

$$ {}^{(t)}a(y, \eta) \sim \sum_\alpha \frac{1}{\alpha!} D_\eta^\alpha \partial_y^\alpha \, {}^ta(y, -\eta)$$

with ${}^ta$ being the pointwise transposed operator of a in the sense of 9.2.3 Exercise 10.

Corollary 14. *Every $\mathcal{A} = Op(a) + \mathcal{C} \in \mathcal{Y}^l(X^\wedge \times \Omega, \underline{g}; N_-, N_+)$ can be written in the form $\mathcal{A} = Op(a') + \mathcal{C}'$ with $\mathcal{C}' \in \mathcal{Y}^{-\infty}$ and $a'(y', \eta) \in \mathcal{R}^m(\Omega \times \mathbb{R}^q, \underline{g}; N_-, N_+)$ with the asymptotic expansion*

$$a'(y', \eta) \sim \sum_\alpha \frac{1}{\alpha!}(-D_\eta)^\alpha \partial_y^\alpha a(y, \eta)|_{y=y'}. \tag{9.137}$$

Theorem 15. *Let* $\mathcal{A} \in \mathcal{Y}^l(X^\wedge \times \Omega, \underline{g}; N_-, N_0)$, $\underline{g} = (\gamma, \gamma - m, \Theta)$, *and* $\tilde{\mathcal{A}} \in \mathcal{Y}^{\tilde{l}}(X^\wedge \times \Omega, \tilde{\underline{g}}; N_0, N_+)$, $\tilde{\underline{g}} = (\gamma - m, \gamma - m - \tilde{m}, \Theta)$, *and let* $\mathcal{A}$ *or* $\tilde{\mathcal{A}}$ *be properly supported with respect to the y-variables. Then* $\tilde{\mathcal{A}}\mathcal{A} \in \mathcal{Y}^{l+\tilde{l}}(X^\wedge \times \Omega, \underline{h}; N_-, N_+)$ *for* $\underline{h} = (\gamma, \gamma - m - \tilde{m}, \Theta)$, *and*

$$\sigma_\psi^{l+\tilde{l}}(\tilde{\mathcal{A}}\mathcal{A}) = \sigma_\psi^{\tilde{l}}(\tilde{\mathcal{A}})\sigma_\psi^l(\mathcal{A}), \quad \sigma_\wedge^{l+\tilde{l}}(\tilde{\mathcal{A}}\mathcal{A}) = \sigma_\wedge^{\tilde{l}}(\tilde{\mathcal{A}})\sigma_\wedge^l(\mathcal{A}). \tag{9.138}$$

If $\mathcal{A} \in \mathcal{Y}_{M+G}^l$ *or* $\tilde{\mathcal{A}} \in \mathcal{Y}_{M+G}^{\tilde{l}}$, *then* $\tilde{\mathcal{A}}\mathcal{A} \in \mathcal{Y}_{M+G}^{l+\tilde{l}}$; *similarly,* $\mathcal{A} \in \mathcal{Y}_G^l$ *or* $\tilde{\mathcal{A}} \in \mathcal{Y}_G^{\tilde{l}}$ *implies* $\tilde{\mathcal{A}}\mathcal{A} \in \mathcal{Y}_G^{l+\tilde{l}}$. *If* $\mathcal{A}$ *and* $\tilde{\mathcal{A}}$ *have the form* $\mathcal{A} = Op(a) \bmod \mathcal{Y}^{-\infty}$, $\tilde{\mathcal{A}} = Op(\tilde{a}) \bmod \mathcal{Y}^{-\infty}$ *with corresponding* $a(y, \eta)$ *and* $\tilde{a}(y, \eta)$ *as in Definition 3, then* $\tilde{\mathcal{A}}\mathcal{A} = Op(c) \bmod \mathcal{Y}^{-\infty}$ *with some* $c(y, \eta) \in \mathcal{R}^{l+\tilde{l}}(\Omega \times \mathbb{R}^q, \underline{h}; N_-, N_+)$, *and*

$$c(y, \eta) \sim \sum_\alpha \frac{1}{\alpha!} D_\eta^\alpha \tilde{a}(y, \eta) \partial_y^\alpha a(y, \eta). \tag{9.139}$$

Proof. According to the definition of the operator classes we write $\mathcal{A} = Op(a) + \mathcal{C}$, $\tilde{\mathcal{A}} = Op(\tilde{a}) + \tilde{\mathcal{C}}$ with the corresponding symbols $a(y, \eta)$ and $\tilde{a}(y, \eta)$, respectively. Let us assume, for instance, that $\mathcal{A}$ is properly supported. Then, if we fix a compact set $K \subset\subset \Omega$ and first look at the argument functions $u(y)$ supported in K, then there is a $\varphi \in C_0^\infty(\Omega)$ such that $\varphi \mathcal{A}u = \mathcal{A}u$ for all those u. Since for every $\varphi_0 \in C_0^\infty(\Omega)$ with $\varphi_0 \varphi = \varphi$ we have $(1 - \varphi_0)\tilde{\mathcal{A}}\varphi \in \mathcal{Y}^{-\infty}$, it suffices to characterize $\varphi_0 \tilde{\mathcal{A}}\varphi\mathcal{A}$ for every choice of $\varphi_0, \varphi \in C_0^\infty(\Omega)$ with $\varphi_0 \varphi = \varphi$. The simple arguments for $\varphi_0 \tilde{\mathcal{C}}\varphi\mathcal{A}$, $\varphi_0\tilde{\mathcal{A}}\varphi\mathcal{C}$, $\varphi_0\tilde{\mathcal{C}}\varphi\mathcal{C} \in \mathcal{Y}^{-\infty}$ will be dropped. Thus, $\varphi_0 Op(\tilde{a})\varphi Op(a)$ remain. Since $\varphi_0(y)\tilde{a}(y, \eta)$ and $\varphi(y)a(y, \eta)$ are symbols, we may change the notations and assume from now on that $\tilde{a}(y, \eta)$ and $a(y, \eta)$ are of compact support with respect to y.

According to Corollary 14 we can write $Op(a) = Op(a') + \mathcal{C}'$ with $\mathcal{C}' \in \mathcal{Y}^{-\infty}$ and $a' = a'(y', \eta) \in \mathcal{R}^l$. Since the composition with $\mathcal{C}'$ leads to an operator in $\mathcal{Y}^{-\infty}$, again it suffices to characterize $Op(\tilde{a})Op(a') = Op(b)$ with $b(y, y', \eta) = \tilde{a}(y, \eta)a'(y', \eta)$. From Theorem 7 we get $Op(b) = Op(c) \bmod \mathcal{Y}^{-\infty}$ with

$$c(y, \eta) \sim \sum_\alpha \frac{1}{\alpha!} D_\eta^\alpha \{\tilde{a}(y, \eta) \partial_y^\alpha a'(y, \eta)\}.$$

Using (9.137) it is now a straightforward calculation to show that (9.139) holds. $\square$

9.3.3 Ellipicity and regularity with discrete edge asymptotics

This section will study the concept of ellipticity of operators in

$$\mathcal{Y}^m(X^\wedge \times \Omega, \underline{g}; N_-, N_+), \tag{9.140}$$

$\underline{g} = (\gamma, \gamma - m, \Theta)$, and the nature of the parametrix construction.

Definition 1. An operator $\mathcal{A} \in$ (9.140) is called *elliptic* if the following conditions are satisfied:

(i) $\sigma_{\psi}^{m}(\mathcal{A})(r, x, y, \varrho, \xi, \eta) \neq 0$ for all $(r, x, y, \varrho, \xi, \eta) \in T^{*}(X^{\wedge} \times \Omega) \setminus 0$ and
$r^{m}\sigma_{\psi}^{m}(\mathcal{A})(r, x, y, r^{-1}\varrho, \xi, r^{-1}\eta) \neq 0$ for all
$(r, x, y, \varrho, \xi, \eta) \in T^{*}(\overline{\mathbb{R}}_{+} \times X \times \Omega) \setminus 0,$

(ii) $\qquad \sigma_{\wedge}^{m}(\mathcal{A})(y, \eta) : \mathcal{K}^{s,\gamma}(X^{\wedge}) \oplus \mathbb{C}^{N_-} \to \mathcal{K}^{s-m,\gamma-m}(X^{\wedge}) \oplus \mathbb{C}^{N_+}$ (9.141)
is an isomorphism for all $(y, \eta) \in T^{*}\Omega \setminus 0.$

The condition (i) will also be called the *interior ellipticity*. It only concerns the upper left corner of $\mathcal{A}$. The condition (ii) expresses the ellipticity of the additional edge conditions of trace and potential type. This is a generalization of the Shapiro-Lopatinskij condition for boundary value problems, cf. 9.3.2 Remark 4.

Remark 2. Let $\underline{A} := A + M + G + C$ be the upper left corner of $\mathcal{A}$, cf. (9.123). Then the isomorphy of (9.3.3) implies that

$$\sigma_{\wedge}^{m}(\underline{A})(y, \eta) : \mathcal{K}^{s,\gamma}(X^{\wedge}) \to \mathcal{K}^{s-m,\gamma-m}(X^{\wedge}) \qquad (9.142)$$

is a family of Fredholm operators with index

$$\text{ind } \sigma_{\wedge}^{m}(\underline{A})(y, \eta) = N_+ - N_-. \qquad (9.143)$$

The operators $\sigma_{\wedge}^{m}(G + C)(y, \eta) = \sigma_{\wedge}^{m}(G)(y, \eta)$ are compact. Thus (9.143) is independent of G and C.

We shall study here the ellipticity under a technical restriction which is not essential in principle but considerably simplifies many proofs. First remember that the Fredholm property of (9.142) implies that

$$h(y, z) := (\sigma_{M}^{m}\sigma_{\wedge}^{m}(\underline{A}))(y, z) : H^{s}(X) \to H^{s-m}(X) \qquad (9.144)$$

is an isomorphism for all $z \in \Gamma_{\frac{n+1}{2}-\gamma}$, $y \in \Omega$ and all $s \in \mathbb{R}$, cf. 8.2.5 Proposition 9. Let $S \subset \mathbb{C}$ be the system of poles of $h(y, z)$, and set $B_{\varepsilon} = \{z : \text{dist}(z, S) > \varepsilon\}$ for any $\varepsilon > 0$. It is now clear from 7.2.5 Theorem 4 that for every $y \in \Omega$ there is a discrete subset $D(y) \subset \mathbb{C}$ with $D(y) \cap K$ finite for every compact subset $K \Subset B_{\varepsilon}$ for everey $\varepsilon > 0$, such that (9.144) is an isomorphism for all $z \in \mathbb{C} \setminus D(y)$. In general the set $D(y)$ will depend on $y \in \Omega$. For fixed $y = y_0$ we can calculate the inverse of (9.144) at all $z \in \mathbb{C} \setminus D(y)$. By 7.2.5 Theorem 4 the function $h^{-1}(y_0, z)$ has an extension to a meromorphic family $f(y_0, z)$ of operators. Then $D(y_0) = \{d_j(y_0)\}_{j \in \mathbb{Z}}$ is the set of poles of multiplicities $m_j(y_0) + 1$, and the Laurent coefficients $c_{jk}(y_0)$ of $f(y_0, z)$ at $(z - d_j(y_0))^{-(k+1)}$, $0 \leq k \leq m_j(y_0)$, are finite-dimensional operators in $L^{-\infty}(X)$ (cf. also 8.1.5 Execise 4). Our technical assumption is now that

$$D := D(y), \quad m_j := m_j(y), \quad j \in \mathbb{Z}, \quad 0 \leq k \leq m_j \qquad (9.145)$$

are independent of $y \in \Omega$ and that the elements of S are not accumulation points of D.

Note that the program of this section under arbitrary dependence of $D(y)$, $m_{jk}(y)$, $c_{jk}(y)$ on y can be carried out by the same methods as in [Sc2] [parts VII, VIII].

Definition 3. Let $\mathcal{A} \in \mathcal{Y}^m(X^\wedge \times \Omega, \underline{g}; N_-, N_+)$ and

$$\mathcal{B} \in \mathcal{Y}^{-m}(X^\wedge \times \Omega, \underline{h}; N_+, N_-), \tag{9.146}$$

$\underline{h} = (\gamma - m, \gamma, \Theta)$. Let $\mathcal{A}$ or $\mathcal{B}$ be properly supported with respect to the y-variables. Then $\mathcal{B}$ is called a *parametrix* of $\mathcal{A}$ if

$$\begin{aligned}
\mathcal{BA} - I &\in \mathcal{Y}^{-\infty}(X^\wedge \times \Omega, \underline{a}; N_-, N_-), &\tag{9.147}\\
\mathcal{AB} - I &\in \mathcal{Y}^{-\infty}(X^\wedge \times \Omega, \underline{b}; N_+, N_+), &\tag{9.148}
\end{aligned}$$

$\underline{a} = (\gamma, \gamma, \Theta)$, $\underline{b} = (\gamma - m, \gamma - m, \Theta)$. Here I is the identical operator in the corresponding spaces.

Remark 4. To construct a parametrix $\mathcal{B}$ of $\mathcal{A}$ it suffices to find $\mathcal{B}$ in a small open neighbourhood $\Omega(y_0)$ of every fixed $y_0 \in \Omega$, $\mathrm{mod}\,\mathcal{Y}^{-\infty}(X^\wedge \times \Omega(y_0), \underline{h}; N_+, N_-)$. Then we find a locally finite open covering $\{\Omega(y_j)\}_{j \in \mathbb{Z}_+}$ of Ω with such neighbourhoods, and we may set

$$\mathcal{B} = \sum_j \alpha_j(y)\mathcal{B}_j\beta_j(y)$$

if $\mathcal{B}_j$ denotes the parametrix over $\Omega(y_j)$, $\{\alpha_j\}_{j \in \mathbb{Z}_+}$ a subordinate partition of unity, and $\beta_j \in C_0^\infty(\Omega(y_j))$ are arbitrary functions with $\alpha_j\beta_j = \alpha_j$ for all $j \in \mathbb{Z}_+$.

Remark 5. The specific influence of the edge is located in $\{(r, x, y) \in \mathbb{R}_+ \times X \times \Omega : r < \varepsilon\}$ for every $\varepsilon > 0$. Outside that neighbourhood only the operator $A \in L_{cl}^m(X^\wedge \times \Omega)$ from the upper left corner $\underline{A} = A + M + G + C$ plays a role, since the other items M, G, C are smoothing. For the parametrix construction under ellipticity it therefore suffices to think of edge effects for $r < \varepsilon$. If $\mathcal{B}_\varepsilon$ is a parametrix of $\mathcal{A}$ for $r < \varepsilon$ in the sense that (9.147), (9.148) only hold for ω, ω_0 with $\omega_0 = 0$ for $r \geq \varepsilon$, then we get a parametrix everywhere by putting

$$\mathcal{B} = (\omega \oplus 1)\mathcal{B}_\varepsilon(\omega_0 \oplus 1) + ((1 - \omega) \oplus 0)(A^{(-1)} \oplus 0)((1 - \omega_1) \oplus 0) \tag{9.149}$$

with a parametrix $A^{(-1)}$ of A in $L_{cl}^{-m}(X^\wedge \times \Omega)$ in the standard sense, where ω_1 is a cut-off function with $\omega\omega_1 = \omega_1$.

Theorem 6. *Let* $\mathcal{A} \in \mathcal{Y}^m(X^\wedge \times \Omega, \underline{g}; N_-, N_+)$ *be elliptic and suppose* (9.145). *Then there exists a parametrix* $\mathcal{B} \in \mathcal{Y}^{-m}(X^\wedge \times \Omega, \underline{h}; N_+, N_-)$ *of* $\mathcal{A}$*. Moreover,*

$$\mathcal{A}u = f \in \mathcal{W}_{\mathrm{loc}(y)}^{r, \gamma - m}(X^\wedge \times \Omega) \oplus H_{\mathrm{loc}}^r(\Omega, \mathbb{C}^{N_+}) \tag{9.150}$$

for an $r \in \mathbb{R}$ and

$$u \in \mathcal{W}^{-\infty,\gamma}_{\mathrm{comp}(y)}(X^\wedge \times \Omega) \oplus H^{-\infty}_{\mathrm{comp}}(\Omega, \mathbb{C}^{N_-}), \quad \mathrm{supp}\, u \text{ bounded in } r \qquad (9.151)$$

imply

$$u \in \mathcal{W}^{r+m,\gamma}_{\mathrm{comp}(y)}(X^\wedge \times \Omega) \oplus H^{r+m}_{\mathrm{comp}}(\Omega, \mathbb{C}^{N_-}). \qquad (9.152)$$

Furthermore,

$$\mathcal{A}u = f \in \mathcal{W}^{r,\gamma-m}_{Q,\mathrm{loc}(y)}(X^\wedge \times \Omega) \oplus H^{r}_{\mathrm{loc}}(\Omega, \mathbb{C}^{N_+}) \qquad (9.153)$$

with $r \in \mathbb{R}$ and an asymptotic type Q associated with weight data $(\gamma - m, \Theta)$, $\Theta = (-k, 0]$ with arbitrary $k \in \mathbb{N}$, implies together with (9.151)

$$u \in \mathcal{W}^{r+m,\gamma}_{P,\mathrm{comp}(y)}(X^\wedge \times \Omega) \oplus H^{r+m}_{\mathrm{comp}}(\Omega, \mathbb{C}^{N_-}) \qquad (9.154)$$

with a resulting asymptotic type P associated with the weight data (γ, Θ). (Upper subsrcipt $-\infty$ means the union of all spaces in question over all $s \in \mathbb{R}$.)

We shall prove Theorem 6 in several steps. The main part will be the parametrix construction. Assume for a moment that $\mathcal{B} \in$ (9.146) is already constructed. Then we can easily obtain the elliptic regularity (9.150), (9.151) $\Rightarrow$ (9.152) as well as the elliptic regularity with asymptotics (9.153), (9.151) $\Rightarrow$ (9.154). In fact, first by 9.3.2 Proposition 8 we can always pass to a parametrix which is properly supported with respect to y-variables. Multiplying (9.150) by $\mathcal{B}$ from the left we get

$$\mathcal{B}\mathcal{A}u = \mathcal{B}f \in \mathcal{W}^{r+m,\gamma}_{\mathrm{loc}(y)}(X^\wedge \times \Omega) \oplus H^{r+m}_{\mathrm{loc}}(\Omega, \mathbb{C}^{N_-}). \qquad (9.155)$$

Here, 9.3.2 Theorem 5 was used. Now from (9.147) we obtain

$$\mathcal{B}\mathcal{A}u = u + \mathcal{C}u \quad \text{with} \quad \mathcal{C} \in \mathcal{Y}^{-\infty}(X^\wedge \times \Omega, (\gamma, \gamma, \Theta); N_-, N_-). \qquad (9.156)$$

From 9.3.2 Definition 1 we obtain

$$\mathcal{C}u \in \mathcal{W}^{\infty,\gamma}_{P_1,\mathrm{loc}(y)}(X^\wedge \times \Omega) \oplus H^{\infty}_{\mathrm{loc}}(\Omega, \mathbb{C}^{N_-}) \qquad (9.157)$$

with an asymptotic type P_1. The right hand side of (9.157) is embedded into the space on the right of (9.155). Thus (9.155), (9.156), (9.157) yield (9.152).

Now under the condition (9.153) we obtain similarly to (9.155)

$$\mathcal{B}\mathcal{A}u \in \mathcal{W}^{r+m,\gamma}_{P_2,\mathrm{loc}(y)}(X^\wedge \times \Omega) \oplus H^{r+m}_{\mathrm{loc}}(\Omega, \mathbb{C}^{N_-}), \qquad (9.158)$$

with an asymptotic type P_2, dependent on $\mathcal{A}$ and Q. Here, 9.3.2 Theorem 6 was used. The relation (9.158) remains true if we enlarge P_2 to $P := P_1 \cup P_2$ with P_1 from (9.157). Since the spaces on the right of (9.157) and (9.158) are embedded into the space on the right of (9.154) with that P, (9.154) follows from (9.158), (9.156), (9.157). It remains to construct $\mathcal{B}$.

First we look at the complete interior symbol of the upper left corner of $\mathcal{A}$ in the coordinates $(r, x, y) \in \mathbb{R}_+ \times V \times \Omega$ with open $V \subseteq \mathbb{R}^n$. This has the form $r^{-m} p(r, x, y, r\varrho, \xi, r\eta)$ with

$$p(r, x, y, \tilde{\varrho}, \xi, \tilde{\eta}) \in S^m_{cl}(\overline{\mathbb{R}}_+ \times V \times \Omega \times \mathbb{R}^{1+n+q}), \qquad (9.159)$$

cf. (9.61). Let $p_{(m)}(r, x, y, \tilde{\varrho}, \xi, \tilde{\eta})$ denote the homogeneous principal symbol of (9.159) of order m in $(\tilde{\varrho}, \xi, \tilde{\eta})$. Then condition (i) of Definition 1 implies

$$p_{(m)}(r, x, y, \tilde{\varrho}, \xi, \tilde{\eta}) \neq 0 \quad \text{for all} \quad (r, x, y) \in \overline{\mathbb{R}}_+ \times V \times \Omega \quad \text{and} \quad (\tilde{\varrho}, \xi, \tilde{\eta}) \neq 0.$$

Exercise 7. The symbol $p(r, x, y, r\varrho, \xi, r\eta)$ regarded as an element in $S^m_{cl}(\mathbb{R}_+ \times V \times \Omega \times \mathbb{R}^{1+n+q})$ has a Leibniz inverse $d_1(r, x, y, r\varrho, \xi, r\eta)$ in the sense of the relation

$$\sum_\alpha \frac{1}{\alpha!} \left\{ D^\alpha_{\varrho, \xi, \eta} d_1(r, x, y, r\varrho, \xi, r\eta) \right\} \partial^\alpha_{r, x, y} p(r, x, y, r\varrho, \xi, r\eta) \sim 1$$

and analogously with p in the first, d_1 in the second place in the formula. The symbol d_1 can chosen in such a way that

$$d_1(r, x, y, \tilde{\varrho}, \xi, \tilde{\eta}) \in S^{-m}_{cl}(\overline{\mathbb{R}}_+ \times V \times \Omega \times \mathbb{R}^{1+n+q}).$$

For the parametrix construction we basically need a Leibniz inverse of $r^{-m} p(r, x, y, r\varrho, \xi, r\eta)$. But this follows from d_1 if we carry out the asymptotic sum

$$\sum_j \frac{1}{j!} D^j_\varrho d_1(r, x, y, r\varrho, \xi, r\eta) \partial^j_r r^m$$

which is of the form

$$r^m d(r, x, y, r\varrho, \xi, r\eta)$$

with some $d(r, x, y, \tilde{\varrho}, \xi, \tilde{\eta}) \in S^{-m}_{cl}(\overline{\mathbb{R}}_+ \times V \times \Omega \times \mathbb{R}^{1+n+q})$. Now if $\mathcal{U} = \{U_1, \ldots, U_N\}$ is an open covering of X by coordinate neighbourhoods and if $\kappa_j : U_j \to V_j$ are charts with open $V_j \subseteq \mathbb{R}^n$, $j = 1, \ldots, N$, then we can perform the above procedure with the local representations $r^{-m} p_j(r, x, y, r\varrho, \xi, r\eta)$ of complete symbols of the upper left corner of $\mathcal{A}$. This yields symbols $r^m d_j(r, x, y, r\varrho, \xi, r\eta)$. Let us form the operator family

$$\mathcal{B}(y, \eta) = \sum_{j=1}^N \varphi_j (1 \times \kappa_j)^* Op_{(r, x)}(d_j) \psi_j, \qquad (9.160)$$

cf. similarly (9.69). Then, applying 9.2.3 Theorem 2, we find an

$$f(r, y, z, \tilde{\eta}) \in C^\infty(\overline{\mathbb{R}}_+ \times \Omega, M_O^{-m}(X; \mathbb{R}^q_{\tilde{\eta}})),$$

such that for $f(r, y, z, r\eta)$

$$\mathcal{B}(y, \eta) - op^\delta_M(f)(y, \eta) \in C^\infty(\Omega, L^{-\infty}(X^\wedge; \mathbb{R}^q))$$

holds for every $\delta \in \mathbb{R}$. This will be applied to $\delta = \gamma - m - \frac{n}{2}$. According to 9.2.3 Definition 8 we form

$$
\begin{aligned}
b_0(y,\eta) &= \omega(r[\eta])r^m op_M^{\gamma-m-\frac{n}{2}}(f)(y,\eta)\omega_0(r[\eta]), \\
b_1(y,\eta) &= (1-\omega(r[\eta]))r^m \mathcal{B}(y,\eta)(1-\omega_1(r[\eta])), \\
b_\infty &= (1-\tilde{\omega}(r))Q_\infty(y,\eta)(1-\tilde{\omega}_1(r))
\end{aligned}
$$

where $Q_\infty(y,\eta)$ is obtained from the parametrix construction to $P_\infty(y,\eta)$ in the sense of Section 8.2.1, and

$$
b(y,\eta) = \tilde{\omega}(r)\{b_0(y,\eta) + b_1(y,\eta)\}\tilde{\omega}_0(r) + b_\infty(y,\eta) + v(y,\eta).
$$

We will first set

$$
v(y,\eta) = \omega(r[\eta])r^m op_M^{\gamma-m-\frac{n}{2}}(f_0)(y)\omega(r[\eta]),
$$

with a suitable choice of $f_0(y,z) \in C^\infty(\Omega, M_R^{-\infty}(X))$ with an asymptotic type R of Mellin symbols.

Proposition 8. *If $y_0 \in \Omega$ is fixed there is an open neighbourhood $\Omega(y_0)$ of y_0 and a smoothing Mellin edge symbol $v(y,\eta)$ with $f_0(z) \in M_{T_0}^{-\infty}(X)$ for an asymptotic type T_0 for Mellin symbols such that*

$$
\sigma_\wedge^{-m}(b)(y,\eta) : \mathcal{K}^{s,\gamma-m}(X^\wedge) \to \mathcal{K}^{s+m,\gamma}(X^\wedge) \tag{9.161}
$$

is a family of Fredholm operators for all $y \in \Omega(y_0)$, $\eta \in \mathbb{R}^q \setminus \{0\}$, $s \in \mathbb{R}$, and $f_0(z)$ can be chosen in such a way that

$$
ind\ \sigma_\wedge^{-m}(b)(y,\eta) = N_- - N_+ \tag{9.162}
$$

for all those (y,η).

Proof. Set $b'(y,\eta) = b_0(y,\eta) + b_1(y,\eta)$ and consider the composition

$$
\sigma_\wedge^{-m}(b')(y,\eta)\sigma_\wedge^m(a)(y,\eta),
$$

with $a(y,\eta)$ being the complete edge symbol of the upper left corner of $\mathcal{A}$, cf. (9.74). Then for every fixed $y \in \Omega$ we get

$$
\sigma_M^{-m}\sigma_\wedge^{-m}(b')(y,z+m)\sigma_M^m\sigma_\wedge^m(a)(y,z) = 1
$$

modulo $L^{-\infty}(X;\Gamma_\beta)$ for every $\beta \in \mathbb{R}$. This means

$$
f(0,y,z+m,0)\sigma_M^m\sigma_\wedge^m(a)(y,z) = 1 \ \ mod\ \ L^{-\infty}(X;\Gamma_\beta). \tag{9.163}
$$

From (9.144) we know that

$$
h(0,y,z,0) + h_0(y,z) = \sigma_M^m\sigma_\wedge^m(a)(y,z)
$$

is invertible for all $z \in \Gamma_{\frac{n+1}{2}-\gamma}$. Here the operator family $w(y,\eta)$ in (9.74) is written as

$$w(y,\eta) = \omega(r[\eta] r^{-m} op_M^{\gamma-\frac{n}{2}}(h_0)(y)\omega(r[\eta]) \mod \mathcal{R}_{M+G}^{m-1}(\Omega \times \mathbb{R}^q, (\gamma, \gamma - m, \Theta))$$

with some $h_0(y,z) \in C^\infty(\Omega, M_T^{-\infty}(X))$ for some asymptotic type T for Mellin symbols, such that $\sigma_M^m \sigma_\wedge^m(w)(y,z) = h_0(y,z)$. (9.163) implies that

$$f(0, y_0, z + m, 0) - (h(0, y_0, z, 0) + h_0(y_0, z))^{-1} =: f_0(z) \in M_{T_0}^{-\infty}(X)$$

with an asymptotic type T_0 for Mellin symbols. It follows that for

$$b(y_0, \eta) := b'(y_0, \eta) - \omega(r[\eta]) r^m op_M^{\gamma-m-\frac{n}{2}}(f_0)(y)\omega(r[\eta])$$

the operator family

$$\sigma_\wedge^{-m}(b)(y,\eta) : \mathcal{K}^{s,\gamma-m}(X^\wedge) \to \mathcal{K}^{s+m,\gamma}(X^\wedge) \tag{9.164}$$

consists of Fredholm operators for $y = y_0$. This is then true also for all y in a certain open neighbourhood $\Omega(y_0)$ of y_0, since the operators depend continuously on y, and the Fredholm property is stable under continuous perturbations. By construction we have at y_0

$$\sigma_\psi^{-m}\sigma_\wedge^{-m}(b) = \sigma_\psi^m\sigma_\wedge^m(a)^{-1} \quad , \quad \sigma_{\psi,e}^{-m,0}\sigma_\wedge^{-m}(b) = \sigma_{\psi,e}^{m,0}\sigma_\wedge^m(a)^{-1}$$
$$\sigma_e^0\sigma_\wedge^{-m}(b) = \sigma_e^0\sigma_\wedge^m(a)^{-1} \quad , \quad T^m\sigma_M^{-m}\sigma_\wedge^{-m}(b) = \sigma_M^m\sigma_\wedge^m(a)^{-1}.$$

Therefore, by 8.2.5 Remark 10 it follows that

$$\text{ind } \sigma_\wedge^{-m}(b)(y,\eta) = -\text{ind } \sigma_\wedge^m(a)(y,\eta)$$

at $y = y_0$ and for all $\eta \neq 0$. Since the index is locally constant, we get (9.162) from (9.143). $\qquad\square$

Proposition 9. *Let $b(y,\eta)$ be the operator family of Proposition 8, where we assume that $\overline{\Omega(y_0)} \subset \Omega$ is compact and that (9.161) is Fredholm for all $y \in \overline{\Omega(y_0)}$ (if necessary after shrinking this set a little). Then there is an $N_0 \in \mathbb{Z}_+$ and an element*

$$c(y,\eta) \in \mathcal{R}^{-m}(\Omega(y_0) \times \mathbb{R}^q, \underline{h}; N_+ + N_0, N_- + N_0), \tag{9.165}$$

$\underline{h} = (\gamma - m, \gamma, \Theta)$, *with $b(y,\eta)$ as upper left corner, such that*

$$\sigma_\wedge^{-m}(c)(y,\eta) : \mathcal{K}^{s,\gamma-m}(X^\wedge) \oplus \mathbb{C}^{N_++N_0} \to \mathcal{K}^{s+m,\gamma}(X^\wedge) \oplus \mathbb{C}^{N_-+N_0}$$

is an isomorphism for all $y \in \Omega(y_0)$, $\eta \in \mathbb{R}^q \setminus \{0\}$.

Proof. Let us set $H_1 = \mathcal{K}^{s,\gamma-m}(X^\wedge)$, $H_2 = \mathcal{K}^{s+m,\gamma}(X^\wedge)$ for any fixed $s \in \mathbb{R}$. The family of Fredholm operators $d(y,\eta) := \sigma_\wedge^{-m}(b)(y,\eta)$ satisfies

$$d(y,\lambda\eta) = \lambda^{-m}\kappa_\lambda d(y,\eta)\kappa_\lambda^{-1} \quad \text{for all } \lambda \in \mathbb{R}_+$$

and all $\eta \in \mathbb{R}^q \setminus \{0\}$. We will construct $e(y,\eta) := \sigma_\wedge^{-m}(c)(y,\eta)$ in such a way that

$$e(y,\lambda\eta) = \lambda^{-m} \begin{pmatrix} \kappa_\lambda & 0 \\ 0 & 1 \end{pmatrix} e(y,\eta) \begin{pmatrix} \kappa_\lambda & 0 \\ 0 & 1 \end{pmatrix}^{-1} \quad \text{for all } \lambda \in \mathbb{R}_+ \tag{9.166}$$

and all $\eta \in \mathbb{R}^q \setminus \{0\}$. Here the 1's are the corresponding $(N_- + N_0) \times (N_- + N_0)$- and $(N_+ + N_0) \times (N_+ + N_0)$-unit matrices. Thus it suffices to obtain $e(y,\eta)$ on the η-unit sphere S^{q-1} and then pass to $e(y,\eta)$ itself by extension by <u>homogeneity</u> $-m$, according to (9.166). Throughout this construction y varies over $\overline{\Omega(y_0)}$. Then

$$d(y,\eta) : H_1 \to H_2$$

is a family of Fredholm operators parametrized by the compact space $\overline{\Omega(y_0)} \times S^{q-1}$, and ind $d(y,\eta) = N_- - N_+$. Now we can argue in a manner similar to that in Section 7.2.5 around Proposition 1. The fact that the Hilbert spaces H_1, H_2 may be different does not affect the main arguments. We choose a sufficiently large N and a finite-dimensional operator family $k : \mathbb{C}^N \to H_2$ which is in fact independent of $(y,\eta) \in \overline{\Omega(y_0)} \times S^{q-1}$, such that

$$(d(y,\eta),k) : H_1 \oplus \mathbb{C}^N \to H_2 \tag{9.167}$$

is surjective for all (y,η). $\square$

Exercise 10. k can be chosen in such a way that $k\mathbb{C}^N \subset C_0^\infty(X^\wedge)$.

Hint: $C_0^\infty(X^\wedge)$ is dense in $H_2 = \mathcal{K}^{s+m,\gamma}(X^\wedge)$, hence every element in a fixed direct complement of im $b(y,\eta)$ can be approximated by $C_0^\infty(X^\wedge)$ functions.

Now ker $(d(y,\eta),k)$ is a subvector bundle of $(\overline{\Omega(y_0)} \times S^{q-1}) \times (H_1 \oplus \mathbb{C}^N)$ of finite dimension M, and $M - N = N_- - N_+$. For N sufficiently large that bundle is necessarily trivial (Exercise!). Hence there is an isomorphism

$$\tilde{J}(y,\eta) := \ker (d(y,\eta),k) \to (\Omega(y_0) \times S^{q-1}) \times \mathbb{C}^M. \tag{9.168}$$

In Section 7.2.5 we have clearly seen how to obtain a row of operators

$$(r(y,\eta),g(y,\eta)) : H_1 \oplus \mathbb{C}^N \to \mathbb{C}^M \tag{9.169}$$

that induces on $\tilde{J}(y,\eta)$ the isomorphism (9.168) and vanishes on an orthogonal complement of $\tilde{J}(y,\eta)$ with respect to a fixed scalar product $(\cdot,\cdot)$ in $H_1 \oplus \mathbb{C}^N$. (9.169) completes (9.167) to a family of isomorphisms

$$\begin{pmatrix} d(y,\eta) & k \\ r(y,\eta) & g(y,\eta) \end{pmatrix} : \begin{matrix} H_1 \\ \oplus \\ \mathbb{C}^N \end{matrix} \to \begin{matrix} H_2 \\ \oplus \\ \mathbb{C}^M \end{matrix}. \tag{9.170}$$

There is an easy constructive way of finding candidates for (9.169). It suffices to fix an M-dimensional normal base $l(y,\eta) = (l_1,\ldots,l_M)(y,\eta)$ in $\tilde{J}(y,\eta)$ which is C^∞ in (y,η) and to pass to the complex conjugate $\bar{l}(y,\eta)$. Then, if $u \in H_1 \oplus \mathbb{C}^N$, we obtain by

$$u \to \sum_{j=1}^{M} (u, \bar{l}_j(y,\eta)) l_j(y,\eta)$$

a projection to $\tilde{J}(y,\eta)$ and the map

$$u \to \{(u, \bar{l}_j(y,\eta))\}_{j=1,\ldots,M} \in \mathbb{C}^M$$

is a possible choice for (9.168). Again by an approximation argument we can pass from the elements $\bar{l}_j(y,\eta)$ to new elements $\overline{m}_j(y,\eta)$ such that the corresponding map $u \to \{(u, \overline{m}_j(y,\eta))\}_{j=1,\ldots,M}$ is another permitted choice but where $\overline{m}_j(y,\eta)$ projects under $H_1 \oplus \mathbb{C}^N \to H_1$ to $C_0^\infty(X^\wedge)$ for all j. With that choice, now for convenience called (9.169) again, we form $e(y,\eta) = (9.170)$. The extension by homogeneity according to (9.166) will also be called $e(y,\eta)$, which is now defined for all $\eta \in \mathbb{R}^q \setminus \{0\}$ and all $y \in \overline{\Omega(y_0)}$. This also yields some η dependence for k in the upper right corner. Now, if $\chi(\eta)$ is an η-excision function we get by

$$c(y,\eta) = \left(\begin{array}{cc} b(y,\eta) & \chi(\eta)k(\eta) \\ \chi(\eta)r(y,\eta) & \chi(\eta)g(y,\eta) \end{array} \right)$$

an element (9.165) where $N = N_+ + N_0$, $M = N_- + N_0$. From $c(y,\eta)$ we pass to $\mathcal{C} = Op(c) \mod \mathcal{Y}^{-\infty}(X^\wedge \times \Omega(y_0), \underline{h}; N_+ + N_0, N_- + N_0)$, and we choose $\mathcal{C}$ properly supported in the y-variables. By construction $\mathcal{C}$ is elliptic.

To construct a parametrix $\mathcal{B}$ of $\mathcal{A}$ it suffices to construct a parametrix $\mathcal{B}_0$ of $\mathcal{A}_0 = \mathcal{A} \oplus R$ where R is an $N_0 \times N_0$-diagonal matrix of pseudo-differential operators on Ω with entries having the symbol $[\eta]^m$. Then $\mathcal{B}_0$ will be of the form $\mathcal{B}_0 = \mathcal{B} \oplus R^{(-1)} \mod \mathcal{Y}^{-\infty}(\ldots)$, where $R^{(-1)}$ is a parametrix of R. We choose a representative $\mathcal{A}_0$ that is properly supported in the y-variables. By construction we have

$$\sigma_\psi^0(\mathcal{C}\mathcal{A}_0) = \omega.$$

Furthermore we have

$$\sigma_\wedge^0(\mathcal{C}\mathcal{A}_0)(y,\eta) = 1 + \sigma_\wedge^0(w)(y,\eta)$$

with some $w(y,\eta)$ belonging to

$$\mathcal{R}_{M+G}^0(\Omega(y_0) \times \mathbb{R}^q, \underline{a}; N_- + N_0, N_- + N_0), \tag{9.171}$$

$\underline{a} = (\gamma, \gamma, \Theta)$, with $\sigma_M^0 \sigma_\wedge^0(w)(y,z) = 0$. Moreover, as a consequence of the ellipticity of $\mathcal{C}$ and $\mathcal{A}_0$ with respect to the principal edge symbolic level, it follows from the second relation of (9.138) that

$$1 + \sigma_\wedge^0(w)(y,\eta) : \mathcal{K}^{s,\gamma}(X^\wedge) \oplus \mathbb{C}^{N_- + N_0} \to \mathcal{K}^{s,\gamma}(X^\wedge) \oplus \mathbb{C}^{N_- + N_0}$$

is an isomorphism.

Proposition 11. *There exists a $w_1(y, \eta) \in (9.171)$ with $\sigma_M^0 \sigma_\wedge^0(w_1)(y, z) = 0$ such that*

$$(1 + \sigma_\wedge^0(w)(y, \eta))^{-1} = 1 + \sigma_\wedge^0(w_1)(y, \eta). \tag{9.172}$$

Proof. The assertion for $(y, \eta) \in \overline{\Omega(y_0)} \times S^{q-1}$ corresponds to the parameter-dependent invertibility of 8.1.2 Remark 16. The construction shows that in view of $\sigma_M^0 \sigma_\wedge^0(w)(y, z) = 0$ the resulting smoothing Mellin symbols contained in the upper left corner of $(1 + \sigma_\wedge^0(w)(y, \eta))^{-1}$ have asymptotic types that are independent of y (they are independent of η anyway). The operator (9.172) for all $\eta \neq 0$ follows by extending our family from $|\eta| = 1$ by homogeneity of degree 0, similarly to (9.166) for $m = 0$. $\qquad\square$

Let us now continue the proof of Theorem 6. Set $\mathcal{W}_1 = Op(w_1) \mod \mathcal{Y}^{-\infty}$, and choose $\mathcal{W}_1$ properly supported in the y-variables. Then, if I denotes the corresponding identical operator, $\mathcal{C}_1 := (I + \mathcal{W}_1)\mathcal{C}$ satisfies

$$\sigma_\psi^0(\mathcal{C}_1 \mathcal{A}_0) = 1, \quad \sigma_\wedge^0(\mathcal{C}_1 \mathcal{A}_0) = 1.$$

Applying 9.3.2 Remark 11 it follows that

$$\mathcal{D} := \{\mathcal{C}_1 \mathcal{A}_0 - I\} \in \mathcal{Y}^{-1}(X^\wedge \times \Omega(y_0), \underline{a}; N_- + N_0, N_- + N_0),$$

$\underline{a} = (\gamma, \gamma, \Theta)$. Next we apply a formal Neumann series argument. We form the asymptotic sum

$$\mathcal{D}_1 := \sum_{j=0}^{\infty} (-1)^j \mathcal{D}^j \in \mathcal{Y}^0(X^\wedge \times \Omega(y_0), \underline{a}; N_- + N_0, N_- + N_0),$$

cf. 9.3.2 Exercise 12. Set $\mathcal{B}_\varepsilon = (\tilde{\omega} \oplus 1)\mathcal{D}_1(\tilde{\omega} \oplus 1)\mathcal{C}_1$ with an arbitrary cut-off function $\tilde{\omega}$. Then $\mathcal{B}_\varepsilon$ is a parametrix of $\mathcal{A}_0$ for $r < \varepsilon$ in the sense of Remark 5, when $\varepsilon > 0$ is so small that $\tilde{\omega} = 1$, $\omega = 1$ for $0 < r < \varepsilon$. Analogously to (9.149) we now obtain a parametrix $\mathcal{B}_0$ of $\mathcal{A}_0$, since the ellipticity of $\mathcal{A}_0$ implies the ellipticity of A in the upper left corner. As noted above, by virtue of the definition of $\mathcal{A}_0$ at the same time we get a parametrix $\mathcal{B}$ of $\mathcal{A}$. Thus Theorem 6 is completely proved. $\qquad\square$

As a consequence we obtain a priori estimates for solutions u of $\mathcal{A}u = f$ when we suppose that $\mathcal{A}$ is elliptic in the class $\mathcal{Y}^m(X^\wedge \times \Omega, \underline{g}; N_-, N_+)$, $\underline{g} = (\gamma, \gamma - m, \Theta)$. Assume that u has bounded support in r and compact support in y. By Theorem 6 there is a parametrix $\mathcal{B}$ of $\mathcal{A}$. We can choose $\mathcal{B}$ to be properly supported in y. More precisely if $\psi \in C_0^\infty(\Omega)$ is any function and $\psi_0 \in C_0^\infty(\Omega)$ has the property $\psi\psi_0 = \psi$, then we can choose $\mathcal{B}$ in such a way that $\psi\mathcal{B}(1 - \psi_0) = 0$. We now obtain

$$\mathcal{B}\mathcal{A}u = \mathcal{B}f, \quad \text{i.e.} \quad u = \mathcal{B}f + \mathcal{C}u$$

with a $\mathcal{C} \in \mathcal{Y}^{-\infty}(X^\wedge \times \Omega, \underline{a}; N_-, N_-)$, $\underline{a} = (\gamma, \gamma, \Theta)$. It follows that

$$\psi u = \psi\mathcal{B}\psi_0 f + \psi\mathcal{C}u. \tag{9.173}$$

We get after identifying functions with their extensions in y to $\mathbb{R}^q$ by zero

$$\|\psi \mathcal{B}\psi_0 f\|_{\mathcal{W}^{s,\gamma}(X^\wedge \times \mathbb{R}^q)\oplus H^s(\mathbb{R}^q,\mathbb{C}^{N-})} \leq \|\psi_0 f\|_{\mathcal{W}^{s-m,\gamma-m}(X^\wedge \times \mathbb{R}^q)\oplus H^{s-m}(\mathbb{R}^q,\mathbb{C}^{N+})}.$$

Furthermore

$$\|\psi \mathcal{C} u\|_{\mathcal{W}^{s,\gamma}(X^\wedge \times \mathbb{R}^q)\oplus H^s(\mathbb{R}^q,\mathbb{C}^{N-})} \leq c\|\psi u\|_{\mathcal{W}^{\infty,\gamma}(X^\wedge \times \mathbb{R}^q)\oplus H^\infty(\mathbb{R}^q,\mathbb{C}^{N-})}.$$

From (9.173) it therefore follows that there is a constant $c > 0$ with

$$
\begin{aligned}
\|\psi u\|_{\mathcal{W}^{s,\gamma}(X^\wedge \times \mathbb{R}^q)\oplus H^s(\mathbb{R}^q,\mathbb{C}^{N-})} \;\leq\;\; & c\{\|\psi_0 f\|_{\mathcal{W}^{s-m,\gamma-m}(X^\wedge \times \mathbb{R}^q)\oplus H^{s-m}(\mathbb{R}^q,\mathbb{C}^{N+})} \\
& + \|\psi u\|_{\mathcal{W}^{\infty,\gamma}(X^\wedge \times \mathbb{R}^q)\oplus H^\infty(\mathbb{R}^q,\mathbb{C}^{N-})}\}. \quad (9.174)
\end{aligned}
$$

9.3.4 Global constructions and Fredholm property

We now turn to the global pseudo-differential operators on a compact (stretched) manifold $\mathbb{W}$ that corresponds to a manifold W with edge Y, the latter one being a closed compact C^∞ manifold of dimension q. Recall that we have the weighted Sobolev spaces $\mathcal{W}^{s,\gamma}(\mathbb{W})$ on $\mathbb{W}$ of smoothness $s \in \mathbb{R}$ and weight $\gamma \in \mathbb{R}$, cf. (9.111). For simplicity we have assumed that $\partial \mathbb{W}$ has a collar neighbourhood C with the product metric of $[0,1) \times X \times Y$ and that the transition diffeomorphisms belonging to (9.110) are independent of (r,x) for sufficiently small $r > 0$. We can immediately define the subspaces

$$\mathcal{W}_P^{s,\gamma}(\mathbb{W}) \subset \mathcal{W}^{s,\gamma}(\mathbb{W}) \qquad (9.175)$$

for every asymptotic type P, by requiring that restrictions of elements of $\mathcal{W}_P^{s,\gamma}(\mathbb{W})$ to neighbourhoods $(0,1) \times X \times D_j \cong X^\wedge \times \Omega_j$ near $\partial \mathbb{W}$ belong to $\mathcal{W}_{P,\text{loc}(y)}^{s,\gamma}(X^\wedge \times \Omega_j)$ for all j. The space $\mathcal{W}_P^{s,\gamma}(\mathbb{W})$ has a natural Fréchet topology. It can be obtained as a projective limit with respect to the restrictions mentioned together with the norm induced by (9.175). Similarly to 9.3.1 Remark 2 we have

$$\mathcal{W}^{0,0}(\mathbb{W}) = k^{-\frac{n}{2}} L_2(\mathbb{W})$$

with the L_2 scalar product being defined in terms of the fixed Riemannian metric on $\mathbb{W}$, and a strictly positive function $k^\varrho \in C^\infty(\text{int } \mathbb{W})$ with $k^\varrho = r^\varrho$ near $\partial \mathbb{W}$ in the coordinates (r,x,y). Then $\mathcal{W}^{0,0}(\mathbb{W})$ has also a fixed scalar product. There is a natural nondegenerate sesquilinear pairing

$$\mathcal{W}^{s,\gamma}(\mathbb{W}) \times \mathcal{W}^{-s,\,-\gamma}(\mathbb{W}) \to \mathbb{C}$$

for every $s, \gamma \in \mathbb{R}$ that extends

$$(u,v)_{\mathcal{W}^{0,0}(\mathbb{W})} \quad \text{for} \quad u, v \in C_0^\infty(\text{int } \mathbb{W}).$$

This allows us to define formal adjoints of operators

$$A : \mathcal{W}^{s,\gamma}(\mathbb{W}) \to \mathcal{W}^{s-l,\varrho}(\mathbb{W})$$

that are continuous for all $s \in \mathbb{R}$, namely,

$$A^* : \mathcal{W}^{s,-\varrho}(\mathbb{W}) \to \mathcal{W}^{s-l,-\gamma}(\mathbb{W}),$$

$s \in \mathbb{R}$. This is analogous to formal adjoints of operators between standard Sobolev spaces on Y. They were earlier defined with respect to an L_2 scalar product on Y associated with a given Riemannian metric on Y. This gives us formal adjoints of operators

$$A : \mathcal{W}^{s,\gamma}(\mathbb{W}) \oplus H^s(Y, \mathbb{C}^{N_-}) \to \mathcal{W}^{s-l,\varrho}(\mathbb{W}) \oplus H^{s-l}(Y, \mathbb{C}^{N_+}),$$

namely,

$$A^* : \mathcal{W}^{s,-\varrho}(\mathbb{W}) \oplus H^s(Y, \mathbb{C}^{N_+}) \to \mathcal{W}^{s-l,-\gamma}(\mathbb{W}) \oplus H^{s-l}(Y, \mathbb{C}^{N_-})$$

for all $s \in \mathbb{R}$.

In order to introduce the class

$$\mathcal{Y}^l(\mathbb{W}, \underline{g}; N_-, N_+) \tag{9.176}$$

of pseudo-differential edge problems (analogous to boundary value problems for the case of a manifold with boundary Y, with N_+ trace and N_- potential conditions), we first define the subspace

$$\mathcal{Y}^{-\infty}(\mathbb{W}, \underline{g}; N_-, N_+) \tag{9.177}$$

of smoothing elements. As in (9.140) we have fixed the weight data $\underline{g} = (\gamma, \gamma - m, \Theta)$. By definition, an operator

$$\mathcal{G} : \mathcal{W}^{s,\gamma}(\mathbb{W}) \oplus H^s(Y, \mathbb{C}^{N_-}) \to \mathcal{W}^{\infty,\gamma-m}(\mathbb{W}) \oplus H^\infty(Y, \mathbb{C}^{N_+})$$

belongs to $\mathcal{Y}^{-\infty}(\mathbb{W}, \underline{g}; N_-, N_+)$ if there are asymptotic types P and Q associated with $(\gamma-m, \Theta)$ and $(-\gamma, \Theta)$, respectively, such that $\mathcal{G}$ induces continuous operators

$$\begin{aligned}
\mathcal{G} \quad &: \quad \mathcal{W}^{s,\gamma}(\mathbb{W}) \oplus H^s(Y, \mathbb{C}^{N_-}) \to \mathcal{W}^{\infty,\gamma-m}_P(\mathbb{W}) \oplus H^\infty(Y, \mathbb{C}^{N_+}), \\
\mathcal{G}^* \quad &: \quad \mathcal{W}^{s,-\gamma+m}(\mathbb{W}) \oplus H^s(Y, \mathbb{C}^{N_+}) \to \mathcal{W}^{\infty,-\gamma}_Q(\mathbb{W}) \oplus H^\infty(Y, \mathbb{C}^{N_-})
\end{aligned}$$

for all $s \in \mathbb{R}$.

Definition 1. Let $l, m \in \mathbb{R}$, $m - l \in \mathbb{Z}_+$, $\underline{g} = (\gamma, \gamma - m, \Theta)$ for $\gamma \in \mathbb{R}$. Then

$$\mathcal{Y}^l(\mathbb{W}, \underline{g}; N_-, N_+) \tag{9.178}$$

is the space of all operators

$$\mathcal{A} : \mathcal{W}^{s,\gamma}(\mathbb{W}) \oplus H^s(Y, \mathbb{C}^{N_-}) \to \mathcal{W}^{s-l,\gamma-m}(\mathbb{W}) \oplus H^{s-l}(Y, \mathbb{C}^{N_+}) \tag{9.179}$$

that are continuous for all $s \in \mathbb{R}$, such that

(i) $\mathcal{A} = \mathcal{L} + \mathcal{G}$ with $\mathcal{G} \in \mathcal{Y}^{-\infty}(\mathbb{W}, \underline{g}; N_-, N_+)$, and the restriction $\mathcal{L}_j$ of $\mathcal{L}$ to a neighbourhood $(0,1) \times X \times \overline{D}_j \cong X^\wedge \times \Omega_j$ near $\partial\mathbb{W}$ as indicated above satisfies $\chi_{j*}\mathcal{L}_j \in \mathcal{Y}^l(X^\wedge \times \Omega_j, \underline{g}; N_-, N_+)$, cf. (9.110) and 9.3.2 Definition 3, for all $j = 1, \ldots, N$,

(ii) the upper left corner $\mathcal{A}_{11}$ of $\mathcal{A}$ ($\mathcal{A}$ being regarded as a block matrix $(\mathcal{A}_{ij})_{i,j=1,2}$ in an evident way) belongs to $L_{cl}^l(\mathrm{int}\ \mathbb{W})$,

(iii) $\mathcal{A}_{12}$ is an integral operator with kernel in $C^\infty(\mathrm{int}\ \mathbb{W} \times Y) \otimes \mathbb{C}^{N_-}$, $\mathcal{A}_{21}$ is an integral operator with kernel in $\mathbb{C}^{N_+} \otimes C^\infty(Y \times \mathrm{int}\ \mathbb{W})$.

Remark 2. Note that it follows from the definition that $\mathcal{A}_{22}$ is an $N_+ \times N_-$-matrix of elements in $L_{cl}^l(Y)$ whereas the properties (iii) are compatible with the nature of the potential and trace operators in local form, cf. (9.123). They are in fact operators with corresponding C^∞ kernels in the interior, though being singular near $\partial\mathbb{W}$, such that they are not compact as operators between the Sobolev spaces, cf. 9.3.2 Remark 10. Observe that $\mathcal{Y}^l(\mathbb{W}, \underline{g}; N_-, N_+)$ equals the space of all continuous operators

$$\mathcal{A} : C_0^\infty(\mathrm{int}\mathbb{W}) \oplus C^\infty(Y, \mathbb{C}^{N_-}) \to C^\infty(\mathrm{int}\mathbb{W}) \oplus C^\infty(Y, \mathbb{C}^{N_+})$$

satisfying the conditions (i), (ii), (iii) of Definition 1. The continuous extensions (9.179) are then a consequence.

We can also define the subclass

$$\mathcal{Y}_{M+G}^l(\mathbb{W}, \underline{g}; N_-, N_+) \tag{9.180}$$

of all operators in $\mathcal{Y}^l(\mathbb{W}, \underline{g}; N_-, N_+)$ for which $\chi_{j*}\mathcal{L}_j$ belong to $\mathcal{Y}_{M+G}^l(X^\wedge \times \Omega_j, \underline{g}; N_-, N_+)$ for all j. Analogously we denote by

$$\mathcal{Y}_G^l(\mathbb{W}, \underline{g}; N_-, N_+) \tag{9.181}$$

the subspace of all $\mathcal{G} \in \mathcal{Y}^l(\mathbb{W}, \underline{g}; N_-, N_+)$ with $\chi_{j*}\mathcal{L}_j \in \mathcal{Y}_G^l(X^\wedge \times \Omega_j, \underline{g}; N_-, N_+)$ for all j. Those $\mathcal{G}$ will also be called Green operators whereas elements of (9.180) are also called smoothing Mellin + Green operators. Let us also set

$$Y^l(\mathbb{W}, \underline{g}), \quad Y_{M+G}^l(\mathbb{W}, \underline{g}), \quad Y_G^l(\mathbb{W}, \underline{g}) \tag{9.182}$$

for the corresponding operator classes (9.178), (9.180), (9.181) in the case $N_- = N_+ = 0$.

Remember that the weight interval $\Theta = (-k, 0]$ for $k \in \mathbb{N}$ may also be allowed to be $(-\infty, 0]$. If we set $\underline{g}_k = (\gamma, \gamma - m, (-k, 0])$, first for finite k then

$$\mathcal{Y}^l(\mathbb{W}, \underline{g}_{k+1}; N_-, N_+) \subset \mathcal{Y}^l(\mathbb{W}, \underline{g}_k; N_-, N_+),$$

and we define the operator class for $\Theta = (-\infty, 0]$ as the intersection over those for $\underline{g}_k$, $k \in \mathbb{N}$.

Remark 3. Let $\omega \in C^\infty(\mathbb{W})$ be a cut-off function, i.e. $\omega \equiv 1$ in a collar neighbourhood of $\partial\mathbb{W}$, $\omega \equiv 0$ outside another collar neighbourhood of $\partial\mathbb{W}$. Then, if ω_0, ω_1 are other such cut-off functions with $\omega\omega_0 = \omega$, $\omega\omega_1 = \omega_1$, then $\mathcal{A} \in \mathcal{Y}^l(\mathbb{W}, \underline{g}; N_-, N_+)$ implies

$$\mathcal{A} = \begin{pmatrix} \omega & 0 \\ 0 & 1 \end{pmatrix} \mathcal{A} \begin{pmatrix} \omega_0 & 0 \\ 0 & 1 \end{pmatrix} + \begin{pmatrix} 1-\omega & 0 \\ 0 & 0 \end{pmatrix} \mathcal{A} \begin{pmatrix} 1-\omega_1 & 0 \\ 0 & 0 \end{pmatrix} \qquad (9.183)$$

mod $\mathcal{Y}^{-\infty}(\mathbb{W}, \underline{g}; N_-, N_+)$. Now if $\{\varphi_j\}_{j=1,\ldots,N}$ is a partition of unity belonging to the open covering $\{D_j\}_{j=1,\ldots,N}$ of Y, and $\{\psi_j\}_{j=1,\ldots,N}$ a system of functions $\psi_j \in C_0^\infty(D_j)$ with $\varphi_j\psi_j = \varphi_j$ for all j, then

$$\sum_{j=1}^{N} \begin{pmatrix} \omega\varphi_j & 0 \\ 0 & \varphi_j \end{pmatrix} \mathcal{A} \begin{pmatrix} \omega_0\psi_j & 0 \\ 0 & \psi_j \end{pmatrix} = \begin{pmatrix} \omega & 0 \\ 0 & 1 \end{pmatrix} \mathcal{A} \begin{pmatrix} \omega_0 & 0 \\ 0 & 1 \end{pmatrix} \qquad (9.184)$$

mod $\mathcal{Y}^{-\infty}(\mathbb{W}, \underline{g}; N_-, N_+)$. Thus

$$\mathcal{A} = \sum_{j=1}^{N} \mathcal{A}_j + \mathcal{A}_0 \ \ \text{mod} \ \ \mathcal{Y}^{-\infty}(\mathbb{W}, \underline{g}; N_-, N_+) \qquad (9.185)$$

with

$$\mathcal{A}_0 = \begin{pmatrix} 1-\omega & 0 \\ 0 & 0 \end{pmatrix} \mathcal{A} \begin{pmatrix} 1-\omega_1 & 0 \\ 0 & 0 \end{pmatrix}, \quad \mathcal{A}_j = \begin{pmatrix} \omega\varphi_j & 0 \\ 0 & \varphi_j \end{pmatrix} \mathcal{A} \begin{pmatrix} \omega_0\psi_j & 0 \\ 0 & \psi_j \end{pmatrix}$$

for $j = 1, \ldots, N$. The various statements that lead to these decompositions can be regarded as exercises.

Exercise 4. Let $\mathcal{A}_j \in \mathcal{Y}^{l-j}(\mathbb{W}, \underline{g}; N_-, N_+)$, $j \in \mathbb{Z}_+$, be an arbitrary sequence, where the asymptotic types in the Green operators of $\mathcal{A}_j$ are independent of j. Then there is an $\mathcal{A} \in \mathcal{Y}^l(\mathbb{W}, \underline{g}; N_-, N_+)$ such that

$$\mathcal{A} - \sum_{j=0}^{N} \mathcal{A}_j \in \mathcal{Y}^{l-(N+1)}(\mathbb{W}, \underline{g}; N_-, N_+)$$

for all N, and $\mathcal{A}$ is unique mod $\mathcal{Y}^{-\infty}(\mathbb{W}, \underline{g}; N_-, N_+)$.

An immediate consequence of 9.3.2 Theorem 6 is the following:

Theorem 5. *Every* $\mathcal{A} \in \mathcal{Y}^l(\mathbb{W}, \underline{g}; N_-, N_+)$ *induces continuous operators*

$$\mathcal{A} : \mathcal{W}_P^{s,\gamma}(\mathbb{W}) \oplus H^s(Y, \mathbb{C}^{N_-}) \to \mathcal{W}_Q^{s-l,\gamma-m}(\mathbb{W}) \oplus H^{s-l}(Y, \mathbb{C}^{N_+}) \qquad (9.186)$$

for every $s \in \mathbb{R}$, *where P is an arbitrary asymptotic type associated with* (γ, Θ) *and Q a resulting asymptotic type associated with* $(\gamma - m, \Theta)$, *dependent on P and* $\mathcal{A}$.

Remark 6.   There are various obvious generalizations of the operator spaces $\mathcal{Y}^l(\mathbb{W}, \underline{g}; N_-, N_+)$ to the case of arbitrary complex C^∞ vector bundles J^- and J^+ over Y instead of the trivial ones $Y \times \mathbb{C}^{N_-}$ and $Y \times \mathbb{C}^{N_+}$, respectively, with fixed Hermitean metrics in J^- and J^+, cf. analogously the material in [P]. We shall drop the details here as well as the more general case of operators, where also the operators in the upper left corners act between distributional sections of vector bundles on $\mathbb{W}$. Nevertheless these generalizations are necessary in various applications, especially, when discussing the ellipticity and the Fredholm property. In particular, we obtain in the case of non-trivial J^-, J^+ continuous operators

$$\mathcal{A} : \mathcal{W}^{s,\gamma}(\mathbb{W}) \oplus H^s(Y, J^-) \to \mathcal{W}^{s-l,\gamma-m}(\mathbb{W}) \oplus H^{s-l}(Y, J^+)$$

and

$$\mathcal{A} : \mathcal{W}_P^{s,\gamma}(\mathbb{W}) \oplus H^s(Y, J^-) \to \mathcal{W}_Q^{s-l,\gamma-m}(\mathbb{W}) \oplus H^{s-l}(Y, J^+),$$

instead of (9.179) and (9.186), respectively, where $H^s(Y, J)$ is the space of Sobolev distributional sections of smoothness s of the bundle J.

The local definitions of homogeneous principal interior and edge symbols lead to the corresponding global ones. In other words, every $\mathcal{A} \in \mathcal{Y}^l(\mathbb{W}, \underline{g}; N_-, N_+)$ has a homogeneous principal interior symbol

$$\sigma_\psi^l(\mathcal{A}) \in C^\infty(T^*(\text{int } \mathbb{W}) \setminus 0)$$

defined as the standard one of the upper left corner. Moreover there is a homogeneous principal edge symbol

$$\sigma_\wedge^l(\mathcal{A}) \in C^\infty(T^*Y \setminus 0, \bigcap_{s \in \mathbb{R}} \mathcal{L}(E^s, \tilde{E}^{s-l}))$$

with

$$E^s = \mathcal{K}^{s,\gamma}(X^\wedge) \oplus \mathbb{C}^{N_-}, \quad \tilde{E}^{s-l} = \mathcal{K}^{s-l,\gamma-m}(X^\wedge) \oplus \mathbb{C}^{N_+}.$$

The homogeneity of $\sigma_\psi^l(\mathcal{A})$ is as usual, whereas $\sigma_\wedge^l(\mathcal{A})$ is homogeneous in the sense that

$$\sigma_\wedge^l(\mathcal{A})(y, \lambda\eta) = \lambda^l \begin{pmatrix} \kappa_\lambda & 0 \\ 0 & 1 \end{pmatrix} \sigma_\wedge^l(\mathcal{A})(y, \eta) \begin{pmatrix} \kappa_\lambda & 0 \\ 0 & 1 \end{pmatrix}^{-1} \tag{9.187}$$

for all $\lambda > 0$. For notational convenience we write 1 for the identical operators in $\mathbb{C}^{N_-}$ and $\mathbb{C}^{N_+}$, respectively. Note that the entries of $c(y, \eta) = \sigma_\wedge^l(\mathcal{A})(y, \eta)$ can be characterized by the structure that was described in connection with $\sigma_\wedge^l \mathcal{R}^l(\Omega \times \mathbb{R}^q, \underline{g}; N_-, N_+)$, cf. the notation in Section 9.2.3. We can write

$$c(y, \eta) = \begin{pmatrix} c_{11} & c_{12} \\ c_{21} & c_{22} \end{pmatrix}(y, \eta) \tag{9.188}$$

with $c_{11} = a_{(0)} + a_{(1)} + w + g$, where

$$\begin{pmatrix} g & c_{12} \\ c_{21} & c_{22} \end{pmatrix}(y, \eta) \in \sigma_\wedge^l \mathcal{R}_G^l(\Omega \times \mathbb{R}^q, \underline{g}; N_-, N_+), \tag{9.189}$$

$$(a_{(0)} + a_{(1)})(y,\eta) \;=\; \omega(r|\eta|)r^{-l}op_M^{\gamma-\frac{n}{2}}(h_0)(y,\eta)\omega_0(r|\eta|)$$
$$+\;\; (1 - \omega(r|\eta|))r^{-l}P_0(y,\eta)(1 - \omega_1(r|\eta|)), \qquad (9.190)$$

cf. (9.91), and

$$w(y,\eta) = \omega(r|\eta|)r^{-l}\sum_{j=0}^{k-1} r^j \sum_{|\alpha|=j} op_M^{\gamma_{j\alpha}-\frac{n}{2}}(h_{j\alpha})(y)\eta^\alpha \tilde{\omega}(r|\eta|), \qquad (9.191)$$

cf. 9.2.2 Definition 1. The objects involved may be given independently, i.e. without any reference to an associated element in $\mathcal{R}^l(\Omega \times \mathbb{R}^q, \underline{g}; N_-, N_+)$.

Exercise 7. Call $\mathcal{R}^{(l)}(\Omega \times (\mathbb{R}^q \setminus \{0\}), \underline{g}; N_-, N_+)$ the space of operator families $c(y,\eta)$ of the described form. Then $\sigma_\wedge^l$ induces a surjective map

$$\sigma_\wedge^l : \mathcal{R}^l(\Omega \times \mathbb{R}^q, \underline{g}; N_-, N_+) \to \mathcal{R}^{(l)}(\Omega \times (\mathbb{R}^q \setminus \{0\}), \underline{g}; N_-, N_+).$$

Remark 8. Let $\mathcal{A} \in \mathcal{Y}^m(\mathbb{W}, \underline{g}; N_-, N_+)$ with $\underline{g} = (\gamma, \gamma - m, \Theta)$, $\Theta = (-k, 0]$, $k \in \mathbb{N}$. Then

$$\sigma_\psi^m(\mathcal{A}) = 0, \quad \sigma_\wedge^m(\mathcal{A}) = 0$$

implies $\mathcal{A} \in \mathcal{Y}^{m-1}(\mathbb{W}, \underline{g}; N_-, N_+)$. In this case

$$\mathcal{A} : \mathcal{W}^{s,\gamma}(\mathbb{W}) \oplus H^s(Y, \mathbb{C}^{N_-}) \to \mathcal{W}^{s-m,\gamma-m}(\mathbb{W}) \oplus H^{s-m}(Y, \mathbb{C}^{N_+})$$

is a compact operator for all $s \in \mathbb{R}$.

This follows from 9.3.2 Remark 11 together with the continuity (9.179) and the compactness of embeddings

$$\mathcal{W}^{s',\gamma'}(\mathbb{W}) \oplus H^{s'}(Y, \mathbb{C}^N) \to \mathcal{W}^{s,\gamma}(\mathbb{W}) \oplus H^s(Y, \mathbb{C}^N) \qquad (9.192)$$

for arbitrary $s' > s$, $\gamma' > \gamma$.

Exercise 9. To every $b \in C^\infty(T^*(\text{int } \mathbb{W}) \setminus 0)$ that is homogeneous of order l and for which

$$r^l b(r, x, y, r^{-1}\varrho, \xi, r^{-1}\eta)$$

is C^∞ up to $r = 0$ in the local coordinates (r, x, y) in a collar neighbourhood of $\partial \mathbb{W}$, there exists an operator

$$B \in Y^l(\mathbb{W}, \underline{g})$$

for $\underline{g} = (\gamma, \gamma - m, \Theta)$ with arbitrary $\gamma \in \mathbb{R}$, $m \in \mathbb{R}$ with $m - l \in \mathbb{Z}_+$, $\Theta = (-\infty, 0]$, with $\sigma_\psi^l(B) = b$.

Hint: From symbols in local coordinates with the given homogeneous principal part b, apply 9.2.3 Theorem 2 and form a_0, a_1 according to 9.2.3 Definition 8. Then B may be obtained locally near $\partial \mathbb{W}$ as a pseudo-differential operator $\mathcal{F}^{-1}_{\eta \to y} a(y, \eta) \mathcal{F}_{y' \to \eta}$ in y-direction, with the operator-valued symbol $a(y, \eta) = a_0(y, \eta) + a_1(y, \eta)$. Use the local representatives to obtain B globally in a collar neighbourhood of $\partial \mathbb{W}$ and glue it together with a corresponding pseudo-differential operator in int $\mathbb{W}$.

Remember that under the conditions mentioned b has the form

$$b(r, x, y, \varrho, \xi, \eta) = r^{-l} \tilde{b}(r, x, y, r\varrho, \xi, r\eta),$$

with $\tilde{b}(r, x, y, \tilde{\varrho}, \xi, \tilde{\eta})$ being C^∞ in r up to $r = 0$. Then

$$b_0(r, x, y, \varrho, \xi, \eta) = r^{-l} \tilde{b}(0, x, y, r\varrho, \xi, r\eta)$$

is uniquely determined by $\sigma^l_\wedge(B)$.

Exercise 10. Let (b, c) be a pair, where b is a homogeneous symbol of order l as in Exercise 9, and c a matrix as in (9.188), $c \in C^\infty(T^*Y \setminus 0, \bigcap_s \mathcal{L}(E^s, \tilde{E}^{s-l}))$, locally being in $\mathcal{R}^{(l)}(\Omega \times (\mathbb{R}^q \setminus \{0\}), g; N_-, N_+)$ for every coordinate neighbourhood on Y, with local coordinates $y \in \Omega$. Further let b, c be compatible in the sense that b_0 coincides with the homogeneous principal pseudo-differential symbol of the upper left corner of c (in Section 9.2.3 this was denoted by $r^{-l} p_j(0, x, y, r\varrho, \xi, r\eta)$ in the jth coordinate neighbourhood). Then there is a

$$\mathcal{B} \in \mathcal{Y}^l(\mathbb{W}, \underline{g}; N_-, N_+)$$

with $\sigma^l_\psi(\mathcal{B}) = b$, $\sigma^l_\wedge(\mathcal{B}) = c$.

Theorem 11. $\mathcal{A} \in \mathcal{Y}^l(\mathbb{W}, (\gamma, \gamma - m, \Theta); N_-, N_0)$, $\tilde{\mathcal{A}} \in \mathcal{Y}^{\tilde{l}}(\mathbb{W}, (\gamma - m, \gamma - m - \tilde{m}, \Theta); N_0, N_+)$ *implies*

$$\tilde{\mathcal{A}}\mathcal{A} \in \mathcal{Y}^{l+\tilde{l}}(\mathbb{W}, (\gamma, \gamma - m - \tilde{m}, \Theta); N_-, N_+)$$

and

$$\sigma^{l+\tilde{l}}_\psi(\tilde{\mathcal{A}}\mathcal{A}) = \sigma^{\tilde{l}}_\psi(\tilde{\mathcal{A}})\sigma^l_\psi(\mathcal{A}), \quad \sigma^{l+\tilde{l}}_\wedge(\tilde{\mathcal{A}}\mathcal{A}) = \sigma^{\tilde{l}}_\wedge(\tilde{\mathcal{A}})\sigma^l_\wedge(\mathcal{A}). \tag{9.193}$$

Proof. According to Remark 3 the operator $\mathcal{A}$ can be written as

$$\mathcal{A} = \sum_{j=1}^N \mathcal{A}_j + \mathcal{A}_0 + \mathcal{G}$$

with $\mathcal{G} \in \mathcal{Y}^{-\infty}$, $\chi_{j*}\mathcal{A}_j \in \mathcal{Y}^l(X^\wedge \times \Omega_j; \underline{g}; N_-, N_0)$, and

$$\mathcal{A}_0 = \begin{pmatrix} A_0 & 0 \\ 0 & 0 \end{pmatrix}, \quad A_0 \in L^l_{cl}(\text{int } \mathbb{W}),$$

where A_0 vanishes in a collar neighbourhood of $\partial \mathbb{W}$. An analogous representation holds for $\tilde{\mathcal{A}}$. The open covering $\{D_j\}$ of Y may (and will now) be chosen in such a way that for every j_0 the union $\tilde{D}_{j_0}$ of all D_k with $D_k \cap D_{j_0} \neq \emptyset$ again form a system $\{\tilde{D}_j\}$ of coordinate neighbourhoods of Y. Now

$$\tilde{\mathcal{A}}\mathcal{A} = \{\sum_{j=0}^{N} \tilde{\mathcal{A}}_j\}\{\sum_{k=0}^{N} \mathcal{A}_k\} + \{\sum_{j=0}^{N} \tilde{\mathcal{A}}_j\}\mathcal{G} + \tilde{\mathcal{G}}\{\sum_{k=0}^{N} \mathcal{A}_k\}.$$

From the definition of $\mathcal{Y}^{-\infty}$ and from Theorem 5 together with (9.179) it follows that the last two items on the right belong to $\mathcal{Y}^{-\infty}$, again. Moreover, if $\tilde{\chi}_j :$ $(0,1) \times X \times \tilde{D}_j \to X^{\wedge} \times \tilde{\Omega}_j$ are the charts associated with the covering with tilde, we get $\tilde{\chi}_{j*}(\tilde{\mathcal{A}}_j \mathcal{A}_k) \in \mathcal{Y}^{\tilde{l}+l}(X^{\wedge} \times \tilde{\Omega}_j, \underline{g}; N_-, N_+)$, for all $j \geq 1$, $k \geq 1$ with $D_j \cap D_k \neq \emptyset$. This is an immediate consequence of 9.2.3 Theorem 14. Moreover it is easy to see that $\tilde{\mathcal{A}}_j \mathcal{A}_k \in \mathcal{Y}^{-\infty}$ for $j \geq 1$, $k \geq 1$ and $D_j \cap D_k = \emptyset$ (exercise!). It is then obvious that those compositions are modulo $\mathcal{Y}^{-\infty}$ of the form

$$\begin{pmatrix} B & 0 \\ 0 & 0 \end{pmatrix} \quad \text{with} \quad B \in L_{cl}^{l+\tilde{l}}(\text{int } \mathbb{W}),$$

with B vanishing in a collar neighbourhood of $\partial \mathbb{W}$. Finally (9.193) follows from (9.97). $\qquad \square$

Exercise 12. If one factor in Theorem 11 belongs to the class with subscript $M+G$ or G then so does the composition.

Exercise 13. The formal adjoint $\mathcal{A}^*$ of an $\mathcal{A} \in \mathcal{Y}^l(\mathbb{W}, (\gamma, \gamma-m, \Theta); N_-, N_+)$ belongs to $\mathcal{Y}^l(\mathbb{W}, (-\gamma + m, -\gamma, \Theta); N_+, N_-)$, and $\sigma_\psi^l(\mathcal{A}^*) = \overline{\sigma_\psi^l(\mathcal{A})}$, $\sigma_\wedge^l(\mathcal{A}^*) = \sigma_\wedge^l(\mathcal{A})^*$ where * on the right is the (y, η)-wise formal adjoint in the sense of operators $\mathcal{K}^{s,\gamma}(X^{\wedge}) \oplus \mathbb{C}^{N_-} \to \mathcal{K}^{s-l,\gamma-m}(X^{\wedge}) \oplus \mathbb{C}^{N_+}$.

Exercise 14. Let $\mathcal{G} \in \mathcal{Y}^{-\infty}(\mathbb{W}, \underline{g}; N, N)$ with $\underline{g} = (\gamma, \gamma, \Theta)$ and let

$$\mathcal{I} + \mathcal{G} : \mathcal{W}^{s,\gamma}(\mathbb{W}) \oplus H^s(Y, \mathbb{C}^N) \to \mathcal{W}^{s,\gamma}(\mathbb{W}) \oplus H^s(Y, \mathbb{C}^N)$$

be invertible for an $s \in \mathbb{R}$. Then $\mathcal{I} + \mathcal{G}$ is invertible for all $s \in \mathbb{R}$ and there is a $\mathcal{G}_1 \in \mathcal{Y}^{-\infty}(\mathbb{W}, \underline{g}; N, N)$ with $(\mathcal{I} + \mathcal{G})^{-1} = \mathcal{I} + \mathcal{G}_1$.

Definition 15. An operator $\mathcal{A} \in \mathcal{Y}^m(\mathbb{W}, \underline{g}; N_-, N_+)$ is *elliptic* if the following conditions are satisfied:

(i) $\sigma_\psi^m(\mathcal{A}) \neq 0$ on $T^*(\text{int } \mathbb{W}) \setminus 0$ and in local coordinates (r, x, y) near $\partial \mathbb{W}$

$$r^m \sigma_\psi^m(\mathcal{A})(r, x, y, r^{-1}\varrho, \xi, r^{-1}\eta) \neq 0$$

up to $r = 0$,

(ii)

$$\sigma_\wedge^m(\mathcal{A})(y, \eta) : \mathcal{K}^{s,\gamma}(X^{\wedge}) \oplus \mathbb{C}^{N_-} \to \mathcal{K}^{s-m,\gamma-m}(X^{\wedge}) \oplus \mathbb{C}^{N_+}$$

is an isomorphism for all $(y, \eta) \in T^*\Omega \setminus 0$.

Condition (i) will also be called the interior ellipticity of $\mathcal{A}$. It depends only on the upper left corner of $\mathcal{A}$ and it is independent of the weight γ. The condition (ii) is the analogue of the Shapiro-Lopatinskij condition to the additional trace and potential operators with respect to the upper left corner.

Theorem 16. *Let $\mathcal{A} \in \mathcal{Y}^m(\mathbb{W}, g; N_-, N_+)$ be elliptic, $g = (\gamma, \gamma - m, \Theta)$, and suppose (9.145) together with the subsequent condition. Then there exists a parametrix $\mathcal{B} \in \mathcal{Y}^{-m}(\mathbb{W}, \underline{h}; N_+, N_-)$ for $\underline{h} = (\gamma - m, \gamma, \Theta)$ in the sense*

$$\mathcal{A}\mathcal{B} - \mathcal{I} \; \in \; \mathcal{Y}^{-\infty}(\mathbb{W}, \underline{g}_r; N_+, N_+), \tag{9.194}$$

$$\mathcal{B}\mathcal{A} - \mathcal{I} \; \in \; \mathcal{Y}^{-\infty}(\mathbb{W}, \underline{g}_l; N_-, N_-) \tag{9.195}$$

with $\underline{g}_r = (\gamma - m, \gamma - m, \Theta)$, $\underline{g}_l = (\gamma, \gamma, \Theta)$. Moreover, the operator

$$\mathcal{A} : \mathcal{W}^{s,\gamma}(\mathbb{W}) \oplus H^s(Y, \mathbb{C}^{N_-}) \to \mathcal{W}^{s-m,\gamma-m}(\mathbb{W}) \oplus H^{s-m}(Y, \mathbb{C}^{N_+}) \tag{9.196}$$

is Fredholm for every $s \in \mathbb{R}$, and

$$\mathcal{A}u = f \in \mathcal{W}^{s',\gamma-m}(\mathbb{W}) \oplus H^{s'}(Y, \mathbb{C}^{N_+})$$

for any $s' \in \mathbb{R}$ and $u \in \mathcal{W}^{-\infty,\gamma}(\mathbb{W}) \oplus H^{-\infty}(Y, \mathbb{C}^{N_-})$ imply

$$u \in \mathcal{W}^{s'+m,\gamma}(\mathbb{W}) \oplus H^{s'+m}(Y, \mathbb{C}^{N_+}).$$

Proof. From 9.3.3 Theorem 6 we know that the restriction of $\mathcal{A}$ to every neighbourhood like $(0,1) \times X \times D_j$ has a parametrix, cf. 9.3.3 Definition 3. The upper left corner of $\mathcal{A}$ is elliptic in int $\mathbb{W}$ and also has a parametrix, according to the standard calculus of pseudo-differential operators. If we denote by $\mathcal{P}_j$, $j = 1, \dots, N$, the local parametrix of $\mathcal{A}$ over $(0,1) \times X \times D_j$ and set

$$\mathcal{P}_0 = \begin{pmatrix} P_0 & 0 \\ 0 & 0 \end{pmatrix}$$

where P_0 is a parametrix of the upper left corner of $\mathcal{A}$ in int $\mathbb{W}$, then, with the notation of Remark 3, we set $\tilde{\mathcal{B}} = \sum_{j=0}^{N} \tilde{\mathcal{B}}_j$ with

$$\tilde{\mathcal{B}}_0 \;=\; \begin{pmatrix} 1-\omega & 0 \\ 0 & 0 \end{pmatrix} \mathcal{P}_0 \begin{pmatrix} 1-\omega_1 & 0 \\ 0 & 0 \end{pmatrix},$$

$$\tilde{\mathcal{B}}_j \;=\; \begin{pmatrix} \omega\varphi_j & 0 \\ 0 & \varphi_j \end{pmatrix} \mathcal{P}_j \begin{pmatrix} \omega_0\psi_j & 0 \\ 0 & \psi_j \end{pmatrix}, \quad j = 1, \dots, N.$$

Then $\tilde{\mathcal{B}} \in \mathcal{Y}^{-m}(\mathbb{W}, \underline{h}; N_+, N_-)$ satisfies $\mathcal{A}\tilde{\mathcal{B}} - \mathcal{I} \in \mathcal{Y}^{-1}$, $\tilde{\mathcal{B}}\mathcal{A} - \mathcal{I} \in \mathcal{Y}^{-1}$. Here we have applied Theorem 11. It can be proved that the latter remainders actually belong to $\mathcal{Y}^{-\infty}$. In order to avoid the corresponding lengthly arguments we ignore this information (i.e. that $\tilde{\mathcal{B}}$ is already a parametrix as asserted) and employ a formal Neumann series argument. We set

$$\mathcal{C} = \tilde{\mathcal{B}}\mathcal{A} - \mathcal{I}$$

and form the asymptotic sum $\sum_{j=0}^{\infty}(-1)^j \mathcal{C}^j$, according to Exercise 4. This has the form $\mathcal{I} + \mathcal{D}$ with $\mathcal{D} \in \mathcal{Y}^{-1}(\mathbb{W}, \underline{g}_l; N_-, N_-)$. Then $\mathcal{B}_l = (\mathcal{I} + \mathcal{D})\tilde{\mathcal{B}}$ obviously satisfies (9.195). In an analogous manner we find a $\mathcal{B}_r$ satisfying (9.194). Then an easy algebraic argument (that employs, in particular, that $\mathcal{Y}^{-\infty}$ is an ideal in the algebra of edge pseudo-differential problems) shows $\mathcal{B}_r - \mathcal{B}_l \in \mathcal{Y}^{-\infty}$. Thus we may set $\mathcal{B} = \mathcal{B}_l$ which is a parametrix in the sense of both (9.195) and (9.194). Now $\mathcal{C}_l = \mathcal{B}\mathcal{A} - \mathcal{I} \in \mathcal{Y}^{-\infty}$, $\mathcal{C}_r = \mathcal{A}\mathcal{B} - \mathcal{I} \in \mathcal{Y}^{-\infty}$ together with (9.179) and the compactness of (9.192) show that $\mathcal{C}_l$ and $\mathcal{C}_r$ are compact operators in the spaces $\mathcal{W}^{s,\gamma}(\mathbb{W}) \oplus H^s(Y, \mathbb{C}^{N_-})$ and $\mathcal{W}^{s-m,\gamma-m}(\mathbb{W}) \oplus H^{s-m}(Y, \mathbb{C}^{N_+})$, respectively, for all $s \in \mathbb{R}$. Thus (9.196) is a Fredholm operator. Finally, for the last statement we form $\mathcal{B}\mathcal{A}u = \mathcal{B}f$. From $\mathcal{B}\mathcal{A}u = u + \mathcal{C}_l u$ we get

$$u = \mathcal{B}f - \mathcal{C}_l u \in \mathcal{W}^{s'+m,\gamma}(\mathbb{W}) \oplus H^{s'+m}(Y, \mathbb{C}^{N_-}).$$

Here again the continuity (9.179) was used. $\square$

Theorem 17. *Let $\mathcal{A} \in \mathcal{Y}^m(\mathbb{W}, \underline{g}; N_-, N_+)$ be elliptic, and suppose (9.145). Then if Q is an asymptotic type associated with $(\gamma - m, \Theta)$, there is an asymptotic type P associated with (γ, Θ) such that*

$$\mathcal{A}u = f \in \mathcal{W}_Q^{s',\gamma-m}(\mathbb{W}) \oplus H^{s'}(Y, \mathbb{C}^{N_+})$$

and $u \in \mathcal{W}^{-\infty,\gamma}(\mathbb{W}) \oplus H^{-\infty}(Y, \mathbb{C}^{N_-})$ imply

$$u \in \mathcal{W}_P^{s'+m,\gamma}(\mathbb{W}) \oplus H^{s'+m}(Y, \mathbb{C}^{N_-}).$$

This holds for all $s' \in \mathbb{R}$.

Proof. The arguments are analogous to the last part of the proof of Theorem 16. From Theorem 5 we get $\mathcal{B}f \in \mathcal{W}_R^{s'+m,\gamma}(\mathbb{W}) \oplus H^{s'+m}(Y, \mathbb{C}^{N_-})$ with some asymptotic type R. Moreover, since $\mathcal{C}_l \in \mathcal{Y}^{-\infty}(\mathbb{W}, \underline{g}_l; N_-, N_-)$, we get from the definition of this operator class that $\mathcal{C}_l u \in \mathcal{W}_S^{\infty,\gamma}(\mathbb{W}) \oplus H^{\infty}(Y, \mathbb{C}^{N_-})$ holds with some asymptotic type S. Then $u = \mathcal{B}f - \mathcal{C}_l u$ gives the assertion with the unique minimal asymptotic type P containing R and S. $\square$

Remark 18. It can be proved that the ellipticity of $\mathcal{A} \in \mathcal{Y}^m(\mathbb{W}, \underline{g}; N_-, N_+)$ is necessary for the Fredholm property of the operator (9.196) for one fixed $s \in \mathbb{R}$. The arguments are formally analogous to those for the necessity of the ellipticity of a boundary value problem for the Fredholm property of the associated operator in Sobolev spaces. They may be found, in particular, in [RSc].

Remark 19. The index of the Fredholm operator (9.196) for $\mathcal{A} \in \mathcal{Y}^m(\mathbb{W}, \underline{g}; N_-, N_+)$ is independent of s. As a consequence of Theorem 17 we obtain

$$\ker \ \mathcal{A} \subset \mathcal{W}_P^{\infty,\gamma}(\mathbb{W}) \oplus H^\infty(Y, \mathbb{C}^{N_-})$$

for some asymptotic type P.

Remark 20. The notions and results of this section have immediate generalizations to the case of general vector bundles J^-, J^+ on Y, or more generally to operators where also the upper left corners act between distributional sections of vector bundles on $\mathbb{W}$, cf. Remark 6 above. The details will be dropped here.

Exercise 21. For $\mathcal{A} \in \mathcal{Y}^m(\mathbb{W}, \underline{g}; N_-, N_+)$ with $\underline{g} = (\gamma, \gamma - m, \Theta)$ the invertibility of

$$\mathcal{A} : \mathcal{W}^{s,\gamma}(\mathbb{W}) \oplus H^s(Y, \mathbb{C}^{N_-}) \to \mathcal{W}^{s-m,\gamma-m}(\mathbb{W}) \oplus H^{s-m}(Y, \mathbb{C}^{N_-})$$

for an $s \in \mathbb{R}$ implies the invertibility for all $s \in \mathbb{R}$ and (under (9.145) together with the subsequent condition) $\mathcal{A}^{-1} \in \mathcal{Y}^{-m}(\mathbb{W}, \underline{h}; N_+, N_-)$ for $\underline{h} = (\gamma - m, \gamma, \Theta)$.

Hint: Employ Remark 18 that shows that the operator $\mathcal{A}$ is elliptic. First construct a parametrix $\mathcal{B}$ of $\mathcal{A}$, according to Theorem 16. Use ind $\mathcal{B} = 0$ and choose a $\mathcal{C} \in \mathcal{Y}^{-\infty}(\mathbb{W}, \underline{h}; N_+, N_-)$ such that

$$\mathcal{B} + \mathcal{C} : \mathcal{W}^{s-m,\gamma-m}(\mathbb{W}) \oplus H^{s-m}(Y, \mathbb{C}^{N_+}) \to \mathcal{W}^{s,\gamma}(\mathbb{W}) \oplus H^s(Y, \mathbb{C}^{N_-})$$

is an isomorphism. Then $\mathcal{B} + \mathcal{C}$ is an invertible parametrix of $\mathcal{A}$. Hence $(\mathcal{B} + \mathcal{C})\mathcal{A} = \mathcal{I} + \mathcal{D}$ with $\mathcal{D} \in \mathcal{Y}^{-\infty}(\mathbb{W}, \underline{g}_l; N_-, N_-)$ is invertible. From Exercise 14 it follows that $(\mathcal{I} + \mathcal{D})^{-1} = \mathcal{I} + \mathcal{D}_1$ with some $\mathcal{D}_1 \in \mathcal{Y}^{-\infty}(\mathbb{W}, \underline{g}_l; N_-, N_-)$, and Theorem 11 then implies $(\mathcal{I} + \mathcal{D}_1)(\mathcal{B} + \mathcal{C}) = \mathcal{A}^{-1} \in \mathcal{Y}^{-m}(\mathcal{W}, \underline{h}; N_+, N_-)$.

Theorem 22. *Let $A \in Y^m(\mathbb{W}, \underline{g})$, $\underline{g} = (\gamma, \gamma - m, \Theta)$, be an operator satisfying (i) of Definition 15, (9.145) and the subsequent condition. Then there is a $B \in Y^{-m}(\mathbb{W}, \underline{h})$, $\underline{h} = (\gamma - m, \gamma, \theta)$, with*

$$\begin{aligned}
AB - I &\in Y^0_{M+G}(\mathbb{W}, \underline{g}_r), \\
BA - I &\in Y^0_{M+G}(\mathbb{W}, \underline{g}_l).
\end{aligned}$$

Proof. Consider a neighbourhood on $\mathbb{W}$ near $\partial \mathbb{W}$ with the coordinates (r, x, y) and let

$$\underline{a}(r, x, y, \varrho, \xi, \eta) = r^{-m} a(r, x, y, r\varrho, \xi, r\eta)$$

be a symbol of A. Then we find a Leibniz inverse of $\underline{a}$ of the form

$$\underline{b}(r, x, y, \varrho, \xi, \eta) = r^m b(r, x, y, r\varrho, \xi, r\eta)$$

such that $b(r, x, y, \tilde{\varrho}, \xi, \tilde{\eta})$ is C^∞ up to $r = 0$ (cf. also 9.3.3 Exercise 7). This can be done for every coordinate neighbourhood of $\mathbb{W}$. By applying the method of solving Exercise 9 we get an associated $B \in Y^{-m}(\mathbb{W}, \underline{h})$. It is then clear that

$$BA - I, \quad AB - I \in L^{-\infty}(\text{int } \mathbb{W}).$$

However, $BA - I \in Y^0(\mathbb{W}, \underline{g}_l)$, $AB - I \in Y^0(\mathbb{W}, \underline{g}_r)$, and $Y^0(\mathbb{W}, \underline{g}_l) \cap L^{-\infty}(\text{int } \mathbb{W}) = Y^0_{M+G}(\mathbb{W}, \underline{g}_l)$, $Y^0(\mathbb{W}, \underline{g}_r) \cap L^{-\infty}(\text{int } \mathbb{W}) = Y^0_{M+G}(\mathbb{W}, \underline{g}_r)$. This proves our theorem.
$\square$

Note that the conditions of Theorem 22 do not imply that

$$A : \mathcal{W}^{s,\gamma}(\mathbb{W}) \to \mathcal{W}^{s-m,\gamma-m}(\mathbb{W})$$

is a Fredholm operator, though the remainders $AB - I$ and $BA - I$ are operators with C^∞-kernels in int $\mathbb{W}$.

9.4 Applications, examples and remarks

The following sections will contain some concluding remarks to the material on edge pseudo-differential operators.

9.4.1 Boundary value problems as particular edge problems

A particular case of a manifold with edges in our sense is a manifold with boundary. The boundary then plays the role of the edge and the local model of the manifold near the boundary is the (closed) half space $\overline{\mathbb{R}}_+ \times \mathbb{R}^q$ that corresponds to the above wedge with the model cone $\overline{\mathbb{R}}_+$. In other words, the base X of the cone disappears in this case (i.e. dim $X = 0$) and the model cone degenerates to the inner normal. The theory of pseudo-differential boundary value problems in the spirit of edge problems may be found in the book [Sc5]. The case of boundary value problems is of independent interest for several reasons. It is a comparatively simple model for the edge theory that contains the basic elements. Moreover, the algebra of boundary value problems on a manifold with boundary can be employed instead of the algebra of pseudo-differential operators on a closed compact manifold to perform a cone theory along the lines of Chapter 8. Then, of course, it is necessary to say more about the concrete algebra of boundary value problems which is by no means canonically defined. A reasonable choice may be Boutet de Monvel's algebra of boundary value problems for pseudo-differential operators with the transmission property. The corresponding cone calculus is contained in the paper [Sr] and in the subsequent Part *II*. Clearly the next step would be the edge theory based on this algebra with the transmission property as a generalization of the above material to a corresponding algebra of boundary value problems on manifolds with edges. This is yet to be done. The question of an adequate choice of a pseudo-differential algebra of boundary value problems is important for starting a procedure for repeatedly forming cones and wedges (to reach higher singularities in this way) and to perform the pseudo-differential calculus based on that algebra. It also leads to a discussion of interesting subalgebras. The boundary problems with the transmission property in the sense of Boutet de Monvel's algebra, cf. [BdM2], [RSc], may be regarded (up to certain modifications) as a subalgebra of edge problems when the model cone is $\overline{\mathbb{R}}_+$. The edge problems with $\mathbb{R}_+$ as model cone allow arbitrary pseudo-differential interior symbols $p(r, x, \varrho, \eta) \in S^m(\mathbb{R}_+ \times \Omega \times \mathbb{R}^{1+q})$ that are C^∞ up to the boundary, i.e. $p \in S^m(\overline{\mathbb{R}}_+ \times \Omega \times \mathbb{R}^{1+q})$, cf. [Sc2]. They will in general not have the transmission property in Boutet de Monvel's sense. Note that the edge problems in this special situation refer even to more general symbols, namely, to the edge degenerate ones that are of the form

$$r^{-m} p(r, y, r\varrho, r\eta)$$

where $p(r, y, \tilde{\varrho}, \tilde{\eta})$ is a symbol of the class $S^m(\overline{\mathbb{R}}_+ \times \Omega \times \mathbb{R}^{1+q}_{\tilde{\varrho}, \tilde{\eta}})$.

Observe also that the transmission property $p(r, y, \varrho, \eta)$ may only be violated for symbols that are not polynomials in (ϱ, η). Let us understand here the transmission property as a condition under which solutions of elliptic boundary value problems are C^∞ up to the boundary once the right-hand sides are in C^∞ up to the boundary and the boundary data in C^∞ on the boundary. Then we might look for an analogous condition on a manifold with edges when the model cone has a non-trivial base. However the smoothness of solutions of elliptic differential edge problems with C^∞ data will in general not be C^∞ up to the edge. As we know, the solutions are of some more general asymptotic behaviour than the Taylor asymptotics. In other words, an analogue of the transmission property in the general edge theory (that may actually be defined) has less importance as in boundary value problems.

If we regard boundary value problems as particular edge problems then we immediately obtain an algebra much larger than Boutet de Monvel's algebra. Remember that [VE1], [VE2], [VE3] and [Es] have investigated pseudo-differential boundary value problems with general interior symbols though not under the algebra aspect.

Another reason why it is interesting to study boundary value problems for pseudo-differential operators without the transmission property is that elliptic mixed boundary problems give rise to such symbols. The Zaremba problem for the Laplace operator in a domain Ω that consists of a boundary value problem with Dirichlet data on one part $\partial\Omega_+$, and Neumann data on the complementary part $\partial\Omega_-$ of the boundary, is an example of an elliptic mixed problem. We get other examples by imposing oblique derivative conditions that jump over a C^∞ submanifold Y of the boundary of codimension 1. The reduction of such problems to the boundary leads typically to pseudo-differential operators on $\partial\Omega_+$ and $\partial\Omega_-$ with violated transmission property with respect to Y. In other words, to solve such classical problems the corresponding pseudo-differential calculus may be employed. On the other hand, mixed problems may also be regarded as edge problems where the (open stretched) wedge in this case is $(\alpha, \beta) \times \mathbb{R}_+ \times Y$ where the base of the model cone is an interval (α, β), $\alpha < \beta$. Although the above wedge calculus has not covered this case, since the base X was assumed to be without boundary, the case with boundary is in principle analogous and we may expect, in particular, similar asymptotics of solutions of mixed problems in a neighbourhood of the edge Y.

9.4.2   The nature of asymptotics in singular configurations

This section discusses some intuitive aspects of the asymptotics of solutions of elliptic equations near geometric singularities. The asymptotics occur even in simplest cases on the half axis $\mathbb{R}_+ \ni r$ for $r \to 0$ with the origin being regarded as a conical singularity. Recall that the asymptotics have the form

$$u(r) \sim \sum_{j=0}^{\infty} \sum_{k=0}^{m_j} c_{jk} r^{-p_j} \ln^k r, \tag{9.197}$$

with $p_j \in \mathbb{C}$, Re $p_j \to -\infty$ as $j \to \infty$. The precise meaning for $u \in \mathcal{K}^{s,\gamma}(\mathbb{R}_+)$ is that for every $\beta \in \mathbb{R}$ there is an $N = N(\beta)$ with

$$\omega(r) \left\{ u(r) - \sum_{j=0}^{N} \sum_{k=0}^{m_j} c_{jk} r^{-p_j} \ln^k r \right\} \in r^{\beta} \mathcal{H}^s(\mathbb{R}_+),$$

for any cut-off function ω. Here, as above, $\mathcal{H}^s(\mathbb{R}_+)$ is the Mellin Sobolev space on $\mathbb{R}_+$ of smoothness $s \in \mathbb{R}$, cf. (7.50). Note that the Taylor expansion (Taylor asymptotics) for $r \to 0$ belongs to the particular cases where $p_j = -j$, $m_j = 0$ for all j. We want to describe once again the way of obtaining asymptotics of solutions in a very simple case. Consider the operator

$$A = r^{-m} \sum_{j=0}^{m} a_j \left(-r \frac{\partial}{\partial r}\right)^j$$

with constant coefficients a_j. Let us solve the equation

$$Au = f \in \mathcal{S}(\overline{\mathbb{R}}_+) \ \ (= \mathcal{S}(\mathbb{R})|_{\mathbb{R}_+})$$

under natural symbol conditions. We have

$$\sum_{j=0}^{m} a_j \left(-r \frac{\partial}{\partial r}\right)^j u = r^m f \in \mathcal{S}(\overline{\mathbb{R}}_+),$$

and hence, by applying the Mellin transform on both sides, cf. (7.43),

$$h(z)\mathcal{M}u(z) = (\mathcal{M}f)(z + m),$$

with $h(z) = \sum_{j=0}^{m} a_j z^j$. We now assume $h(z) \neq 0$ for all $z \in \Gamma_{\frac{1}{2}} = \{\mathrm{Re}\ z = \frac{1}{2}\}$. Then $h^{-1}(z)(\mathcal{M}f)(z+m) \in L_2(\Gamma_{\frac{1}{2}})$, and hence

$$u(r) = \mathcal{M}^{-1}h^{-1}(z)(\mathcal{M}f)(z+m) \in L_2(\mathbb{R}_+)$$

is a solution. Now $(\mathcal{M}f)(z)$ is meromorphic with simple poles at $z = 0, -1, -2, \ldots$. Hence $\mathcal{M}f(z+m)$ is meromorphic with simple poles at $z = -m, -m-1, \ldots$. This implies that $h^{-1}(z)(\mathcal{M}f)(z+m)$ is meromorphic with a resulting pattern of poles including multiplicities. Denote by p_j the poles in $\mathrm{Re}\ z < \frac{1}{2}$ and by $m_j + 1$ the multiplicities, $j = 0, 1, 2, \ldots$. Then there are open neighbourhoods U_j of p_j with

$$\mathcal{M}u(z) = \sum_{k=0}^{m_j} d_{jk}(z - p_j)^{-(k+1)} + g_j(z)$$

with $g_j(z)$ holomorphic in U_j and unique coefficients $d_{jk} = d_{jk}(u)$. On the other hand, if $\omega(r)$ is any cut-off function, then

$$\mathcal{M}(\omega r^{-p} \ln^k r)(z) - (-1)^k k!\ (z - p)^{-(k+1)}$$

is an entire function for every $k \in \mathbb{Z}_+$. Moreover $(\mathcal{M}(1 - \omega)u)(z)$ is holomorphic in $\mathrm{Re}\ z < \frac{1}{2}$. Thus

$$\left(\mathcal{M}\omega\{u - \sum_{j=0}^{N}\sum_{k=0}^{m_j} c_{jk}r^{-p_j}\ln^k r\}\right)(z)$$

for $c_{jk} = (-1)^k(k!)^{-1}d_{jk}$ is holomorphic in $\mathrm{Re}\ z > \frac{1}{2} - \beta$ for any given $\beta > 0$ once N is so large that the strip $\frac{1}{2} - \beta < \mathrm{Re}\ z < \frac{1}{2}$ contains all poles p_j of $\mathcal{M}u(z)$. Furthermore, let $\chi(z)$ be a function in $C^\infty(\mathbb{C})$ with $\chi(z) = 0$ for dist $(z, -p_j) < \varepsilon$, and $\chi(z) = 0$ for dist $(z, -p_j) > 2\varepsilon$ with some $\varepsilon > 0$. Then we also know that

$$\chi(z)\mathcal{M}\omega\{u - \sum_{j=0}^{N}\sum_{k=0}^{m_j} c_{jk}r^{-p_j}\ln^k r\}(z)|_{\Gamma_{\frac{1}{2}-\delta}}$$

belongs to $\mathcal{S}(\Gamma_{\frac{1}{2}-\delta})$ for all $\delta > \beta$. This shows

$$\omega(r)\{u(r) - \sum_{j=0}^{N}\sum_{k=0}^{m_j} c_{jk}r^{-p_j}\ln^k r\} \in r^\delta \mathcal{H}^\infty(\mathbb{R}_+)$$

for all those δ. Since β is arbitrary, we obtain the asymptotics of u in the above sense, here for $s = \infty$. An analogous consideration is possible for arbitrary weights γ. However, we know this anyway from Chapter 8, where we have allowed arbitrary bases X, under the bijectivity of

$$h(z) : H^{s-m}(X) \to H^{s-m}(X)$$

for all $\mathrm{Re}\ z = \frac{n+1}{2} - \gamma$, $n = \dim X$, where $h(z)$ is the conormal symbol of the operator A, cf. (7.62), (7.63).

Let us now return to the above case with dim $X = 0$, and look at the wedge $\overline{\mathbb{R}}_+ \times \mathbb{R}^q \ni (r, y)$ with edge $\mathbb{R}^q$. Then we can start the consideration of generating edge asymptotics with a typical edge degenerate differential operator

$$A = r^{-m} \sum_{j+|\alpha|\leq m} a_{j\alpha}(y)(-r\frac{\partial}{\partial r})^j (rD_y)^\alpha$$

with coefficients $a_{j\alpha}(y) \in \mathcal{S}(\mathbb{R}^q)$. For simplicity we have assumed that the $a_{j\alpha}$ do not depend on r. The construction of solutions to $Au = f \in \mathcal{S}(\overline{\mathbb{R}}_+ \times \mathbb{R}^q)$ is not as immediate as for $q = 0$. However, under the ellipticity in the interior, i.e. when

$$\sum_{j+|\alpha|=m} a_{j\alpha}(y)(-i\tilde{\varrho})^j \tilde{\eta}^\alpha \neq 0 \ \text{ for all } y \text{ and all } (\tilde{\varrho}, \tilde{\eta}) \neq 0,$$

we can construct parametrices B with $BA - I$, $AB - I$ belonging to the corresponding Y^0_{M+G}-classes, cf. 9.3.4 Theorem 22. To be able to conclude elliptic regularity of solutions in the sense of (9.154) we need an extension of A to a matrix $\mathcal{A}$ with additional trace and potential conditions with respect to the boundary $\mathbb{R}^q$. Let us assume that $\mathcal{A}$ is given and let us see how the asymptotics of solutions are then generated. The main contribution comes from the principal conormal symbol of $\sigma^m_\wedge(A)$, cf. (7.95), namely

$$\sigma^m_M \sigma^m_\wedge(A)(y, z) = \sum_{k=0}^m a_{k0}(y)z^k. \tag{9.198}$$

This is a scalar function in contrast to (7.97), since $\dim X = 0$. Analogously to the above consideration the zeros of $h(y, z) = \sum_{k=0}^m a_{k0}(y)z^k$ in the complex z plane are responsible for the exponents of the asymptotics. The condition (9.145) in the above edge calculus was made to have those zeros independent of y. Even in this simple case the asymptotics are much more complicated than in the cone situation. We do not merely have

$$u(r, y) \sim \sum_{j=0}^\infty \sum_{k=0}^{m_j} c_{jk}(y)r^{-p_j} \ln^k \ r \tag{9.199}$$

as might be thought at first. The asymptotics of solutions $u(r, y) \in \mathcal{W}^{s,0}(\mathbb{R}_+ \times \mathbb{R}^q)$ near $r = 0$ have the form

$$u(r, y) \sim \mathcal{F}_{y\to\eta}^{-1} \left\{ \sum_{j-0}^\infty \sum_{k-0}^{m_j} \omega(r[\eta])[\eta]^{\frac{1}{2}} (r[\eta])^{-p_j} \ln^k (r[\eta])\tilde{v}_{jk}(\eta) \right\} \tag{9.200}$$

with $\tilde{v}_{jk}(\eta) = \mathcal{F}_{y\to\eta}v_{jk}$, $v_{jk} \in H^s(\mathbb{R}^q)$. The interpretation of the asymptotics was given above in Section 9.3.1. In our case, $\{(p_j, m_j)\}_{j\in\mathbb{Z}_+}$ are generated from the (translated) poles of the Mellin transform of the right-hand side of the equation in Re $z < \frac{1}{2}$ together with the poles of $h^{-1}(y,z)$. In this sense the picture is very similar to the cone theory of Chapter 8. Observe how the smoothness of the coefficients of the edge asymptotics (9.200) with respect to y depend on Re p_j. This concerns the case of finite s. For $s = -\infty$ the situation is actually like (9.199) with coefficients $c_{jk} \in C^\infty(\mathbb{R}^q)$. Note that the asymptotics (9.200) remind us very much of the finite Taylor expansions of Sobolev distributions $u(r,y) \in H^s(\mathbb{R}^{1+q}_+)$ for $r \to 0$, $s > \frac{1}{2}$, $s - \frac{1}{2} \notin \mathbb{Z}$. If $H^s_0(\overline{\mathbb{R}}^{1+q}_+)$ is the subspace defined as the closure of $C^\infty_0(\mathbb{R}^{1+q}_+)$ in $H^s(\mathbb{R}^{1+q}_+)$, then $H^s_0(\overline{\mathbb{R}}^{1+q}_+)$ has the following complement in $H^s(\mathbb{R}^{1+q}_+)$:

$$V^s(\mathbb{R}^{1+q}_+) = \left\{ \mathcal{F}^{-1}_{\eta \to y}\Big[\sum_{j=0}^{[s-\frac{1}{2}]} \omega(r[\eta])[\eta]^{\frac{1}{2}}(r[\eta])^j \tilde{v}_j(\eta)\Big] : v_j \in H^{s-j-\frac{1}{2}}(\mathbb{R}^q), j = 0,\dots,[s-\frac{1}{2}] \right\}.$$

Here $[s]$ is the largest integer $\le s$ and $\omega(r)$ is an arbitrary fixed cut-off function. The latter assertion follows from

$$H^s(\mathbb{R}^{1+q}_+) = \mathcal{W}^s(\mathbb{R}^q, H^s(\mathbb{R}_+)),$$
$$H^s_0(\overline{\mathbb{R}}^{1+q}_+) = \mathcal{W}^s(\mathbb{R}^q, H^s_0(\overline{\mathbb{R}}_+)),$$

with $(\kappa_\lambda u)(r) = \lambda^{\frac{1}{2}} u(\lambda r)$, cf. Sections 9.1.3, 9.3.1, and from 9.3.1 Exercise 4, together with

$$H^s(\mathbb{R}_+) = H^s_0(\overline{\mathbb{R}_+}) + V^s(\mathbb{R}_+)$$

where

$$V^s(\mathbb{R}_+) = \{\omega(r) \sum_{j=0}^{[s-\frac{1}{2}]} c_j r^j : c_j \in \mathbb{C}, j = 0,\dots,[s-\frac{1}{2}]\}.$$

From (9.200) it is clear that the investigation of asymptotics becomes much more complicated when we drop the condition (9.145). That means that the poles of the y-dependent meromorphic function $h^{-1}(y,z)$ may vary under varying $y \in \mathbb{R}^q$ and that the multiplicities are not in general constant. This happens, for instance, for the operator

$$r^{-2}\{(a(y) + r\frac{\partial}{\partial r})(b(y) + r\frac{\partial}{\partial r}) + \sum_{\substack{j+|\alpha|\le 2 \\ |\alpha|\ge 1}} a_{j\alpha}(y)(-r\frac{\partial}{\partial_l})^j(rD_y)^\alpha\}$$

which is edge-degenerate. In this case, $h(y,z) = (a(y) - z)(b(y) - z)$, and $a(y) \neq b(y)$ will cause simple poles, $a(y) = b(y)$ double poles of $h(y,z)^{-1}$. We observe varying Re $p_j(y)$. They give rise to varying smoothness of coefficients in the asymptotics but also to jumping exponents of r at the points y where $a(y) = b(y)$. It is by no means evident how to formulate the corresponding generalization of the asymptotics. The general discussion of the asymptotics of solutions to elliptic equations with variable and jumping multiplicities of exponents requires a systematic calculus. It requires a generalization of the notion of asymptotics in terms of analytic functionals in the complex plane, the so-called continuous asymptotics. The idea is to replace the meromorphic functions in the complex plane by functions which are holomorphic outside larger sets than discrete ones. If $K \subset \mathbb{C}$ is a compact set and $\mathcal{A}(\mathbb{C} \setminus K)$ the space of all holomorphic functions outside K, then for every holomorphic function $v(z)$ and every $f(z) \in \mathcal{A}(\mathbb{C} \setminus K)$ we can form the analytic functional

$$\langle \zeta_f, v \rangle = \int_C f(z) v(z) dz$$

with a (say C^∞) curve C surrounding the set K clockwise. Assume for simplicity that K is connected. An example is $K = \{p\}$ for some $p \in \mathbb{C}$ and $f(z) = (z-p)^{-(k+1)}$, $k \in \mathbb{Z}_+$. Then

$$\langle \zeta_f, r^{-z} \rangle = c\, r^{-p} \ln^k r$$

with a constant c. In other words, the asymptotics (9.197) can be written as

$$u(r) \sim \sum_{j=0}^{\infty} \langle \zeta_j, r^{-z} \rangle \tag{9.201}$$

with a sequence ζ_j of analytic functionals carried by $K_j = \{p_j\}$ (here being of finite orders, like finite derivatives of the Dirac measures concentrated at p_j). Now by allowing larger compact sets K_j and general analytic functionals ζ_j carried by K_j, (9.201) is the corresponding notion of continuous asymptotics (in contrast to (9.197), that would be called discrete asymptotics in this context). We do not elaborate on the details here. A thorough investigation of the continuous asymptotics may be found in [Sc1], [Sc5]. In [Sc2][parts XII, XIII] the theory of boundary value problems for edge-degenerate symbols was formulated, covering all symbols without the transmission property, in the sense of an algebra with variable (in general branching) asymptotics. This was done in terms of a subalgebra of a corresponding algebra with continuous asymptotics. The edge case with non-trivial base X of the model cone is very similar.

9.4.3 Remarks on the role of the edge trace and potential conditions

Let us now discuss further aspects of the edge conditions of trace and potential type. To illustrate the situation we look mainly at the case of boundary value problems in the half space $\overline{\mathbb{R}}_+ \times \mathbb{R}^q \ni (r, y)$ for edge degenerate differential operators

$$A = r^{-m} \sum_{j+|\alpha| \le m} a_{j\alpha}(r, y) \left(-r \frac{\partial}{\partial r}\right)^j (r D_y)^\alpha,$$

$a_{j\alpha} \in C^\infty(\overline{\mathbb{R}}_+ \times \mathbb{R}^q)$. The interior ellipticity

$$\sum_{j+|\alpha|=m} a_{j\alpha}(r,y)(-i\tilde{\varrho})^j \tilde{\eta}^\alpha \neq 0$$

for all $(r,y) \in \overline{\mathbb{R}}_+ \times \mathbb{R}^q$ and $(\tilde{\varrho}, \tilde{\eta}) \neq 0$ implies that for every $y \in \mathbb{R}^q$ there exists a countable set $D(y) \subset \mathbb{C}$ with finite $D(y) \cap K$ for every compact $K \subset \mathbb{C}$ such that

$$\sigma_M^m(A)(y,\eta) = r^{-m} \sum_{j+|\alpha|\leq m} a_{j\alpha}(0,y)(-r\frac{\partial}{\partial r})^j (r\eta)^\alpha$$

induces Fredholm operators

$$\sigma_\wedge^m(A)(y,\eta) : \mathcal{K}^{s,\gamma}(\mathbb{R}_+) \to \mathcal{K}^{s-m,\gamma-m}(\mathbb{R}_+) \tag{9.202}$$

for all $\gamma \in \mathbb{R}$ with $D(y) \cap \Gamma_{\frac{1}{2}-\gamma} = \emptyset$. This is then true for all $\eta \in \mathbb{R}^q \setminus \{0\}$ and all $s \in \mathbb{R}$. The index of (9.202) is independent of s. The operator family (9.202) was interpreted as the (homogeneous principal) edge symbol. It can be called the boundary symbol of A in the present particular case. From the homogeneity

$$\sigma_\wedge^m(A)(y,\lambda\eta) = \lambda^m \kappa_\lambda \sigma_\wedge^m(A)(y,\eta)\kappa_\lambda^{-1}$$

for all $\lambda > 0$ and all $y \in \mathbb{R}^q$, $\eta \in \mathbb{R}^q \setminus \{0\}$, it follows that the index of (9.202) only depends on $\frac{\eta}{|\eta|} \in S^{q-1}$. It is now possible to choose a map

$$k(y,\eta) : \mathbb{C}^{N_-} \to \mathcal{K}^{s-m,\gamma-m}(\mathbb{R}_+)$$

such that the operator

$$(\sigma_\wedge^m(A)(y,\eta), k(y,\eta)) : \begin{array}{c} \mathcal{K}^{s,\gamma}(\mathbb{R}_+) \\ \oplus \\ \mathbb{C}^{N_-} \end{array} \to \mathcal{K}^{s-m,\gamma-m}(\mathbb{R}_+) \tag{9.203}$$

is surjective. If we assume that $D(y)$ is independent of y, then for every bounded open set $\Omega \subset \mathbb{R}^q$ there is a

$$k(y,\eta) \in C^\infty(\Omega \times S^{q-1}, C_0^\infty(\mathbb{R}_+)) \otimes \mathbb{C}^{N_-} \tag{9.204}$$

such that (9.203) is surjective for all $(y,\eta) \in \Omega \times S^{q-1}$. For convenience let us fix such an Ω. Then, from the general theory of families of Fredholm operators, there follows the existence of a complex C^∞ vector bundle J^+ on $\Omega \times S^{q-1}$ and of a map

$$(t(y,\eta), b(y,\eta)) : \begin{array}{c} \mathcal{K}^{s,\gamma}(\mathbb{R}_+) \\ \oplus \\ \mathbb{C}^{N_-} \end{array} \to J^+_{(y,\eta)}$$

(where $J^+_{(y,\eta)}$ is the fibre of J^+ over (y,η)), which is C^∞ in (y,η), such that

$$\begin{pmatrix} \sigma^m_\wedge(A) & k \\ t & b \end{pmatrix}(y,\eta): \begin{matrix} \mathcal{K}^{s,\gamma}(\mathbb{R}_+) \\ \oplus \\ \mathbb{C}^{N_-} \end{matrix} \to \begin{matrix} \mathcal{K}^{s-m,\gamma-m}(\mathbb{R}_+) \\ \oplus \\ J^+_{(y,\eta)} \end{matrix} \tag{9.205}$$

is an isomorphism. Note that $(t(y,\eta),b(y,\eta))$ can be chosen in such a way that the action has the form

$$(t(y,\eta),b(y,\eta))\begin{pmatrix} u \\ v \end{pmatrix} = \left\{ \int_0^\infty t_j(y,\eta)(r)\overline{u}(r)dr + \sum_{n=1}^{N_-} b_{jn}(y,\eta)v_n \right\}_{j=1,\dots,N_+}$$

with $N_+ = \dim J^+_{(y,\eta)}$, $v = (v_1,\dots,v_{N_-}) \in \mathbb{C}^{N_-}$, and functions $t_j(y,\eta)(r)$ belonging to $C_0^\infty(\mathbb{R}_+)$ with respect to r. The latter observation as well as (9.204) is a consequence of the density of $C_0^\infty(\mathbb{R}_+)$ in $\mathcal{K}^{s,\gamma}(\mathbb{R}_+)$ for every $s,\gamma \in \mathbb{R}$. This is useful for obtaining s,γ-independent k,t,b. However, in applications more specific functions of r occur. Next we pass in the usual manner to a family of isomorphisms (9.205) for all $(y,\eta) \in \Omega \times (\mathbb{R}^q \setminus \{0\})$ by extending the matrix by homogeneity of order m. In other words, if $\underline{a}(y,\eta)$ denotes the left-hand side of (9.205), given for $|\eta| = 1$, we obtain by

$$\underline{a}(y,\lambda\eta) = \lambda^m \begin{pmatrix} \kappa_\lambda & 0 \\ 0 & 1 \end{pmatrix} \underline{a}(y,\eta) \begin{pmatrix} \kappa_\lambda & 0 \\ 0 & 1 \end{pmatrix}^{-1}$$

for $\lambda > 0$ the values of the extension for all $\eta \neq 0$. In the above discussion of edge problems we have assumed for simplicity that J^+ is the trivial bundle $\cong (\Omega \times S^{q-1}) \times \mathbb{C}^{N_+}$. This is not necessary. However, to construct an operator acting between distributional sections of bundles it is necessary to have those bundles over Ω. Thus, a condition is that for sufficiently large N_- the bundle J^+ is the pull-back of some bundle J_0^+ over Ω with respect to the canonical projection $\Omega \times S^{q-1} \to \Omega$. It is known from the general theory of boundary value problems that this condition may be violated for certain differential operators. An example is the Cauchy-Riemann operator in the plane. It can be proved, cf. [Sc2], that the condition is independent of the weight γ. In the following discussion we shall assume that the condition is satisfied, and to simplify notation we write J^+ instead of J_0^+. Now let $\chi(\eta)$ be an excision function. Then

$$\begin{pmatrix} 0 & \chi(\eta)k(y,\eta) \\ \chi(\eta)t(y,\eta) & \chi(\eta)b(y,\eta) \end{pmatrix} \in \mathcal{R}_G^m(\Omega \times \mathbb{R}^q, g; J^-, J^+),$$

here with $J^- = \Omega \times \mathbb{C}^{N_-}$. This allows us to pass to pseudo-differential operators in the y-coordinates. They give together with A in the upper left corner a block matrix

$$\mathcal{A} = \begin{pmatrix} A & K \\ T & B \end{pmatrix} \tag{9.206}$$

that represents a boundary value problem for A. The operator can always be interpreted as a continuous map

$$\mathcal{A}: \quad \begin{matrix} r^{\gamma} C_0^{\infty}(\overline{\mathbb{R}}_+ \times \Omega) \\ \oplus \\ C_0^{\infty}(\Omega, J^-) \end{matrix} \quad \rightarrow \quad \begin{matrix} r^{\gamma-m} C_0^{\infty}(\overline{\mathbb{R}}_+ \times \Omega) \\ \oplus \\ C^{\infty}(\Omega, J^+) \end{matrix}$$

that has extensions by continuity to the corresponding weighted wedge Sobolev spaces. We see in this way that the role of trace and potential conditions on the level of symbols is to complete the boundary symbol $\sigma_{\wedge}^m(A)$ to a family of isomorphisms. Here

$$\operatorname{ind} \sigma_{\wedge}^m(A)(y, \eta) = N_+ - N_- \tag{9.207}$$

with the fibre dimensions $N_{\pm} = N_{\pm}(\gamma)$ of the bundles $J^{\pm}$. The index may be changed under varying γ. The difference $\{N_+(\gamma) - N_-(\gamma)\} - \{N_+(\overline{\gamma}) - N_-(\overline{\gamma})\}$ for different possible $\gamma, \overline{\gamma} \in \mathbb{R}$ can actually be expressed in terms of the pattern of poles and zeros of $\sigma_M^m \sigma_{\wedge}^m(A)(y, z)$ in the strip between $\Gamma_{\frac{1}{2}-\gamma}$ and $\Gamma_{\frac{1}{2}-\overline{\gamma}}$. The weights γ at the spaces where we consider an elliptic operator A, are rather essential with respect to the number of additional trace and potential conditions. In concrete cases the explicit values of the exceptional weights where (9.202) is not a Fredholm operator are not always evident. The task of determining them is analogous to the evaluation of the exponents $-p_j$ of the asymptotics (including the multiplicities and the Laurent coefficients in the Mellin image). This is not the subject of the this exposition. It may also be a separate effort for a concrete operator. The additional trace and potential conditions (for admissible weights) play an important role in the general structure of singular problems, not only for boundaries or edges in the sense of this chapter. Analogous considerations are to be expected when the model cone of the wedge is more complicated. The cone itself, once the base has singularities, will already have edges, where the calculus induces additional trace and potential operators. The edge of the wedge then also leads to extra traces and potentials, so that the skeletons of various dimensions of the polyhedral singularities give rise to their own additional trace and potential conditions. Then the operators will be block matrices with more rows and columns than in (9.206), according to the dimensions of the lower dimensional skeletons. The precise pseudo-differential analysis in this sense for the higher singularities has not yet been elaborated in the literature; it is a program for the future. The general idea is to perform an iterative procedure, that should be arranged as far as possible in an axiomatic way. The present approach to the cone and edge theory may be regarded as the first steps.

Bibliography

[ADN] Agmon S., Douglis A., Nirenberg L. *Estimates near the boundary for solutions of elliptic partial differential equations satisfying general boundary conditions I.* Comm. Pure Appl. Math., **12** (1959), 623–727.

[Ag1] Agranovich M.S. *Elliptic singular integro-differential operators.* Uspehi Mat. Nauk, **20**, 5 (1965), 3–120. (Russian Math. Surveys, **20** (1965), 1–122).

[Ag2] Agranovich M.S. *Boundary value problems for systems with a parameter.* Mat. Sb., **84** (1971), 27–65. (Math. USSR-Sb., **13** (1971), 25–64).

[AV] Agranovich M.S., Vishik M.I. *Elliptic problems with a parameter and parabolic problems of general form.* Uspehi Mat. Nauk, **19**, 3 (1964), 53–161. (Russian Math. Surveys, **19**, 3 (1964), 53–157).

[At] Atiyah M.F. *K-Theory.* Cambridge, Mass., Harvard University, (1965).

[AtBo] Atiyah M.F., Bott R. *A Lefschetz fixed point formula for elliptic complexes I.* Ann. of Math., **86** (1967), 374–407.

[AtS] Atiyah M.F., Singer I.M. *The index of elliptic operators I.* Ann. of Math., **87** (1968), 484–530.

[Be] Beals R. *A general calculus of pseudo-differential operators.* Duke Math. J., **42** (1975), 1–42.

[BeF] Beals R., Fefferman C. *On local solvability of linear partial differential equations.* Ann. of Math., **97** (1973), 482–498.

[BdM1] Boutet de Monvel L. *Comportement d'un opérateur pseudo-différentiel sur une variété à bord.* J. Analyse Math., **27** (1966), 241–304.

[BdM2] Boutet de Monvel L. *Boundary problems for pseudo-differential operators.* Acta Math., **126** (1971), 11–51.

[Ca] Calderón A.P. *Boundary value problems for elliptic equations.* Outlines of the joint Soviet-American symposium on partial differential equations, 303–304, Novosibirsk, (1963).

[Co]   Cordes H.O.   *A global parametrix for pseudo-differential operators over* $\mathbf{R}^n$ *with applications.* SFB 72, Preprints, Univ. Bonn, (1976).

[DH]   Duistermaat J.J., Hörmander L.   *Fourier intergral operators II.* Acta Math., **128** (1972), 183–269.

[Eg1]   Egorov Yu.V.   *Canonical transformations and pseudo-differential operators.* Trudy Mosk. Math., **24** (1971), 3–28. (Trans. Moscow Math. Soc., **24** (1971), 1–28).

[Eg2]   Egorov Yu.V.   *Linear differential equations of principal type.* Nauka, Moscow, (1984). (Plenum Press, New York 1985).

[Eg3]   Egorov Yu.V.   *Lectures on partial differential equations: supplementary chapters.* Moscow University Press, (1985).

[Eg4]   Egorov Yu.V.   *Micro-local analysis.* EMS, **33**, (1990). (Springer Verlag, Berlin Heidelberg, (1991)).

[EgK1]   Egorov Yu.V., Kondratiev V.A.   *On the oblique derivative problem.* Mat. Sb., **78** (1969), 148–176. (Math. USSR-Sb., **7** (1969), 139–169).

[EgK2]   Egorov Yu.V., Kondratiev V.A.   *The negative spectrum of an elliptic operator.* Mat. Sb., **181**, 2 (1990), 147–166. (Math USSR-Sb., **69**, 1 (1991), 155–177).

[EgSh]   Egorov Yu.V., Shubin M.A.   *Partial differential equations I, II, III.* Springer-Verlag Berlin-Heidelberg-New York, (1992).

[Es]   Eskin G.I.   *Boundary value problems for elliptic pseudo-differential equations.* Nauka, Moscow, (1973). (Engl. transl.: Transl. Math. Monogr., **52**, Providence, 1981).

[Fed]   Fedosov B.V.   *Analytic formulas for the index of elliptic operators.* Trudy Mosk. Mat. Obšč., **30** (1974), 159–241. (Transl. Moscow Math. Soc., **30** (1974), 159–240).

[Fef]   Fefferman C.   *The uncertainty principle.* Bull. Amer. Math. Soc., **9** (1986), 129–206.

[G1]   Gohberg I., Goldberg S., Kaashoek M.A.   *Classes of linear Operators, Vol. 1–2.* Birkhäuser Verlag, Basel-Boston-Berlin, (1990).

[G2]   Gohberg I., Krein M.   *Theory and Applications of Volterra Operators in Hilbert space.* Transl. Math. Monographs, **24** (1970), Amer. Math. Soc., Providena, R. I.

[G3] Gohberg I., Krupnik N.Ya. *Einführung in die Theorie der eindimensionalen Singulären Integraloperatoren.* Mathematische Reihe, **63** (1979), Birkhäuser Verlag, Basel.

[Gi] Gilkey P.B. *Invariance theory. The heat equation and Atiyah-Singer index theorem.* Publish or Perish, Inc. Wilmington. Delawaire, (1984).

[Hi] Hirschmann T. *Pseudo-differential operators and asymptotics on manifolds with corners V, Va.* Reports of the Karl-Weierstraß-Institute for Mathematics 07/90, 04/91 Berlin, (1990–1991).

[H1] Hörmander L. *Linear partial differential operators.* Springer-Verlag, Berlin-Heidelberg-New York, (1963).

[H2] Hörmander L. *Pseudo-differential operators.* Comm. Pure Appl. Math., **18** (1965), 501–517.

[H3] Hörmander L. *Pseudo-differential operators and non-elliptic boundary problems.* Ann. of Math., **83** (1965), 129–209.

[H4] Hörmander L. *Pseudo-differential operators and hypoelliptic equations.* Amer. Math. Soc. Symp. on Singular Integrals, (1966, pp. 138–183).

[H5] Hörmander L. *Fourier integral operators I.* Acta Math., **127** (1971), 79–183.

[H6] Hörmander L. *The analysis of linear partial differential operators I, II, III, IV.* Springer-Verlag, Berlin-Heidelberg-New York, (1983–1985).

[Ka] Kannai Y. *An unsolvable hypoelliptic differential operator.* Israel J. Math, **9** (1971), 306–315.

[KA] Kantorovich L.V., Akilov G.P. *Functional analysis in normed spaces (Russian).* Fismatgis Moscow, (1959), (German Transl. Akademie-Verlag Berlin (1964)).

[KN] Kohn J.J., Nirenberg L. *On the algebra of pseudo-differential operators.* Comm. Pure Appl. Math., **18** (1965), 269–305.

[Kon] Kondratiev V.A. *Boundary value problems for elliptic equations in domains with conical or singular points.* Trudy Mosk. Mat. Obšč, **16** (1967), 209–292. (Trans Moscow Math. Soc., **16** (1967), 277–313).

[Koz] Kozhevnikov A.N. *Spectral problems for pseudo-differential systems elliptic in the Douglis-Nirenberg sense and their applications.* Mat. Sb., **92** (1973), 60–88. (Math. USSR-Sb., **21** (1973), 63–90).

[Ku] Kumano-go H. *Pseudo-differential operators.* The MIT Press, Cambridge, Mass. and London, England, (1981).

[Le]   Lewy H. *An example of a smooth linear partial differential equation without solution.* Ann. of Math., **66** (1957), 155–158.

[LM]   Lions J.L., Magenes E. *Problèmes aux limites non-homogènes et applications I.* Dunod, Paris, (1968).

[M1]   Melrose R. Transformation of boundary problems. Acta Math. **147** (1981), 149–236.

[M2]   Melrose R. The Atiyah-Patodi-Singer index theorem. AK Peters, Wellesley, MA 1993.

[M2]   Melrose R., Mendoza G. Elliptic operators of totally characteristic type. Preprint MSRI 047–83, Berkely 1983.

[NT]   Nirenberg L., Treves F. *On local solvability of linear partial differential equations I. Necessary conditions II. Sufficient conditions.* Comm. Pure Appl Math., **23** (1970), 1–38; 459–509.

[P]   Palais R.S. *Seminar on the Atiyah-Singer index theorem.* Princeton University Press, (1965).

[Pl1]   Plamenevskij B.A. On the boundedness of singular integrals in spaces with weight. Mat. Sbornik **76**, 4 (1968), 573–592.

[Pl1]   Plamenevskij B.A. Algebras of pseudo-differential operators. Nauka, Moscow 1986.

[RSc]   Rempel S., Schulze B.-W. *Index theory of elliptic boundary problems.* Akademie-Verlag, Berlin, (1982). (North Oxford Acad. Publ. Comp., Oxford 1985).

[S1]   Schrohe, E. *Spaces of weighted symbols and weighted Sobolev spaces on manifolds.* Springer Lecture Notes in Math., **1256** (1987), 360–377.

[S2]   Schrohe, E. *Spectral invariance, ellipticity, and the Fredholm property for pseudodifferential operators on weighted Sobolev spaces.* Am. Glob. Anal and Geom., **10** (1992), 237–257.

[SKK]   Sato M., Kawai T., Kashiwara *Hyperfunctions and pseudo-differential equations.* Springer Lecture Notes in Math., **287** (1973), 265–529.

[Sf]   Schäfer H.H. *Topological vector spaces.* Macmillan, New York, (1966).

[Sr]   Schrohe E., Schulze B.-W. *Boundary value problems in Boutet de Monvel's algebra for manifolds with conical singularities I, II.* Advances in Part. Diff. Equ. "Pseudo-Differential Calculus and Mathematical Physics" and "Boundary Value Problems, Schrödinger Operators, Deformation Quantization". Akademie Verlag, Berlin (1994), 97–209 and (1995), 70–205.

[Sc1]   Schulze B.-W. *Pseudo-differential operators on manifolds with singularities.* North Holland, Amsterdam, (1991).

[Sc2]   Schulze B.-W. *Pseudo-differential operators and asymptotics on manifolds with corners I–IV, VI–IX.* Reports of the Karl-Weierstrass-Institute for Mathematics 07/89, 02/90, 04/90, 06/90, 08/90, 01/91, 02/91. Berlin (1989–1990). Parts XII, XIII. Reprints of the SFB 256 "Nichtlineare partielle Differentialgleichungen" 224, 230. Bonn, (1992). XII, XIII appeared as:*The variable discrete asymptotics in pseudo-differential boundary value problems I, II.* Advances in Part. Diff. Equ. "Pseudo-Differential Calculus and Mathematical Physics" and "Boundary Value Problems, Schrödinger, Deformation Quantization". Akademie Verlag Berlin, (1994), 9–96 and (1995), 9–69.

[Sc3]   Schulze B.-W. *The Mellin pseudo-differential calculus on manifolds with corners.* Symp. "Analysis on Manifolds with Singularities", Breitenbrunn, 1990. Teubener Texte zur Mathematik. Stuttgart, Leipzig, (1990), 208–289.

[Sc4]   Schulze B.-W. *Mellin representations of pseudo-differential operators on manifolds with corners.* Ann. Glob. Anal. and Geom., **8**, 3 (1990), 261–297.

[Sc5]   Schulze B.-W. *Pseudo-differential boundary value problems, conical singularities, and asymptotics.* Akademie Verlag, Berlin, (1994).

[Se1]   Seeley R.T. *Singular integrals and boundary problems.* Amer. J. Math., **88** (1966), 781–809.

[Se2]   Seeley R.T. *Complex powers of an elliptic operator.* Proc. Symp in Pure Math., **10** (1967), 288–307.

[Sh]   Shubin M.A. *Pseudo-differential operators and spectral theory.* Nauka, Moscow, (1978). (Springer-Verlag, Berlin-Heidelberg, 1987)

[Ta]   Taylor M. *Pseudo-differential operators.* Princeton Univ. Press, (1981).

[Tr]   Treves F. *Introduction to pseudo-differential and Fourier integral operators. Volume 1: Pseudo-differential operators. Volume 2: Fourier integral operators.* Plenum Press, New York and London, (1980).

[V]   Vishik M.I. *On general boundary problems for elliptic differential equations.* Trudy Mosk. Mat. Obšč, **1** (1952), 187–246. (Amer. Math Soc. Transl., **24** (1963), 107–172).

[VE1]   Vishik M.I., Eskin G.I. *Convolution equations in a bounded region.* Uspehi Mat. Nauk, **20**, 3 (1965), 89–152. (Russian Math. Surveys, **20**, 3 (1965), 86–151).

[VE2] Vishik M.I., Eskin G.I. *Convolution equations in a bounded region in spaces with weighted norms.* Mat. Sb., **69**, 1 (1966), 65–110. (American Math. Soc. Transl., **67** (1968), 33–82).

[VE3] Vishik M.I., Eskin G.I. *Elliptic convolution equations in a bounded region and their applications.* Uspehi Mat. Nauk, **22**, 1 (1967), 15–76. (Russian Math. Surveys, **22**, 12 (1967), 13–75).

[W] Weyl H. *Das asymptotische Verteilungsgesetz der Eigenwerte linearer partieller Differentialgleichungen.* Math. Ann., **71** (1912), 441–479.

Index

Titles previously published in the series

OPERATOR THEORY: ADVANCES AND APPLICATIONS
BIRKHÄUSER VERLAG

Edited by **I. Gohberg,**
School of Mathematical Sciences, Tel-Aviv University, Ramat Aviv, Israel

80. **I. Gohberg, H. Langer** (Eds): Operator Theory and Boundary Eigenvalue Problems International Workshop in Vienna, July 27-30, 1993, 1995 (ISBN 3-7643-5259-0)

81. **H. Upmeier**: Toeplitz Operators and Index Theory in Several Complex Variables, 1995 (ISBN 3-7643-5280-5)

82. **T. Constantinescu**: Schur Parameters, Factorization and Dilation Problems, 1996 (ISBN 3-7643-5285-X)

83. **A.B. Antonevich**: Linear Functional Equations. Operator Approach, 1995 (ISBN 3-7643-2931-9)

84. **L.A. Sakhnovich**: Integral Equations with Difference Kernels on Finite Intervals, 1996 (ISBN 3-7643-5267-1)

85. **Y.M. Berezansky, G.F. Us, Z.G. Sheftel**: Functional Analysis, **Vol. I**, 1996 (ISBN 3-7643-5344-9)

86. **Y.M. Berezansky, G.F. Us, Z.G. Sheftel**: Functional Analysis, **Vol. II**, 1996 (ISBN 4-7643-5345-7)

87. **I. Gohberg / P. Lancaster / P.N. Shivakumar** (Eds): Recent Developments in Operator Theory and Its Applications, International Conference in Winnipeg, October 2–6, 1994, 1996 (ISBN 3-7643-5413-5)

88. **J. van Neerven** (Ed.): The Asymptotic Behaviour of Semigroups of Linear Operators 1996 (ISBN 3-7643-5455-0)

89. **Y. Egorov / V. Kondratiev**: On Spectral Theory of Elliptic Operators, 1996 (ISBN 3-7643-5390-2)

90. **A. Böttcher / I. Gohberg** (Eds): Singular Integral Operators and Related Topics. Joint German-Israeli Workshop, Tel Aviv, March 1-10, 1995, 1996 (ISBN 3-7643-5466-6)

91. **A.L. Skubachevskii**: Elliptic Functional Differential Equations and Applications, 1997 (ISBN 3-7643-5404-6)

92. **A. Ya. Shklyar**: Complete Second Order Linear Differential Equations in Hilbert Spaces, 1997 (ISBN 3-7643-5377-5)

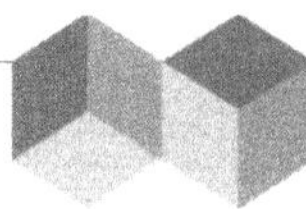

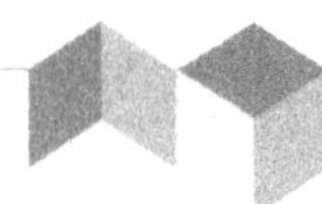

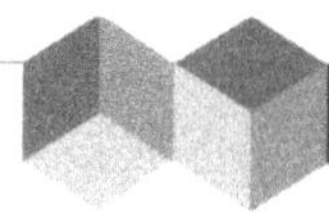